ANALYTICAL MECHANICS
Solutions to Problems in Classical Physics

IOAN MERCHES • DANIEL RADU

CISP

CRC Press
Taylor & Francis Group
Boca Raton London New York

CRC Press is an imprint of the
Taylor & Francis Group, an **informa** business

CRC Press
Taylor & Francis Group
6000 Broken Sound Parkway NW, Suite 300
Boca Raton, FL 33487-2742

First issued in hardback 2019

© 2015 by Taylor & Francis Group, LLC
CRC Press is an imprint of Taylor & Francis Group, an Informa business

No claim to original U.S. Government works

ISBN-13: 978-1-4822-3939-3 (hbk)

This book contains information obtained from authentic and highly regarded sources. Reasonable efforts have been made to publish reliable data and information, but the author and publisher cannot assume responsibility for the validity of all materials or the consequences of their use. The authors and publishers have attempted to trace the copyright holders of all material reproduced in this publication and apologize to copyright holders if permission to publish in this form has not been obtained. If any copyright material has not been acknowledged please write and let us know so we may rectify in any future reprint.

Except as permitted under U.S. Copyright Law, no part of this book may be reprinted, reproduced, transmitted, or utilized in any form by any electronic, mechanical, or other means, now known or hereafter invented, including photocopying, microfilming, and recording, or in any information storage or retrieval system, without written permission from the publishers.

For permission to photocopy or use material electronically from this work, please access www.copyright.com (http://www.copyright.com/) or contact the Copyright Clearance Center, Inc. (CCC), 222 Rosewood Drive, Danvers, MA 01923, 978-750-8400. CCC is a not-for-profit organization that provides licenses and registration for a variety of users. For organizations that have been granted a photocopy license by the CCC, a separate system of payment has been arranged.

Trademark Notice: Product or corporate names may be trademarks or registered trademarks, and are used only for identification and explanation without intent to infringe.

Visit the Taylor & Francis Web site at
http://www.taylorandfrancis.com

and the CRC Press Web site at
http://www.crcpress.com

We can't solve problems by using the same kind of thinking we used when we created them.

Albert Einstein

PREFACE

As the story goes, not everything new is also useful, and not everything old is also obsolete. In theoretical physics, one of the most convincing examples in this respect is offered by Analytical Mechanics. Created by Jean Bernoulli (1654-1705), Pierre Louis Moreau de Maupertuis (1698-1759), Leonhard Euler (1707-1783), Jean le Rond D'Alembert (1717-1783), Joseph Louis Lagrange (1736-1813), Karl Gustav Jacobi (1804-1851), William Rowan Hamilton (1805-1865), Jules Henri Poincaré (1854-1912), and other prominent minds, Analytical Mechanics proved to be a very useful tool of investigation not only in Newtonian Mechanics, but also in almost all classical and modern branches of physics: Electrodynamics, Quantum Field Theory, Theory of Relativity, etc. It can be stated, without exaggeration, that Analytical Mechanics is essential in understanding Theoretical Physics, being a *sine qua non* condition of a profound training of a physicist. Due to its large field of applications, we dare to say that the term "Analytical Mechanics" is somewhat overtaken (out of date).

One of the essential properties of Analytical Mechanics is that it uses abstract, mathematical techniques as methods of investigation. By its object, Analytical Mechanics is a physical discipline, while its methods belong to various branches of mathematics: Analytical and Differential Geometry, Analysis, Differential Equations, Tensor Calculus, Calculus of Variations, Algebra, etc. That is why a physicist who studies analytical formalism must have an appropriate mathematical training.

It is widely spread the idea that it is more important to learn and understand the practical applications of physics, than the theories. In our opinion, connection between Analytical Mechanics and the important chapter devoted to its applications is similar to that between physical discoveries and engineering: if the discovery of electricity, electromagnetic waves, nuclear power, etc., haven't been put into practice, they would have remained within the laboratory frame.

The purpose of this collection of solved problems is intended to give the students possibility of applying the theory (Lagrangian and

Hamiltonian formalisms for both discrete and continuous systems, Hamilton-Jacobi method, variational calculus, theory of stability, etc.) to problems concerning several chapters of Classical Physics. Some problems are difficult to solve, while others are easy. One chapter (the third), as a whole, is dedicated to the gravitational plane pendulum, the problem being solved by all possible analytical formalisms, including, obviously, the Newtonian approach. This way, one can easily observe similarities and differences between various analytical approaches, and their specific efficiency as well.

When needed, some theoretical subjects are developed up to some extent, in order to offer the student possibility to follow solutions to the problems without appealing to other reference sources. This has been done for both discrete and continuous physical systems, or, in analytical terms, systems with finite and infinite degrees of freedom.

A special attention is paid to basics of vector algebra and vector analysis, in Appendix B. Notions like: gradient, divergence, curl, tensor, together with their physical applications, are thoroughly developed and discussed.

This collection of solved problems is a result of many years of teaching Analytical Mechanics, as the first course of theoretical physics, to the students of the Faculty of Physics. There are many excellent textbooks dedicated to applied Analytical Mechanics for both students and their instructors, but we modestly pretend to offer an original view on distribution of the subjects, the thorough analysis of solutions to the problems, and an appropriate choice of applications in various branches of Physics: Mechanics of discrete and continuous systems, Electrodynamics, Classical Field Theory, Equilibrium and small oscillations, etc.

IASI, February 2014

The authors

CONTENTS

CHAPTER I. FUNDAMENTALS OF ANALYTICAL MECHANICS 1
I.1. Constraints 1
    I.1.1. Classification criteria for constraints 2
    I.1.2. The fundamental dynamical problem for a constrained particle 7
    I.1.3. System of particles subject to constraints 9
    I.1.4. Lagrange equations of the first kind 11
I.2. Elementary displacements 12
    I.2.1. Generalities 12
    I.2.2. Real, possible and virtual displacements 13
I.3. Virtual work and connected principles 19
    I.3.1. Principle of virtual work 19
    I.3.2. Principle of virtual velocities 22
    I.3.3. Torricelli's principle 23

CHAPTER II. PRINCIPLES OF ANALYTICAL MECHANICS 26
II.1. D'Alembert's principle 26
    II.1.1. Configuration space 28
    II.1.2. Generalized forces 29
II.2. Hamilton's principle 35

CHAPTER III. THE SIMPLE PENDULUM PROBLEM 47
III.1. Classical (Newtonian) formalism 47
III.2. Lagrange equations of the first kind approach 68
III.3. Lagrange equations of the second kind approach 72
III.4. Hamilton's canonical equations approach 78
III.5. Hamilton-Jacobi method 80
III.6. Action-angle variables formalism 85

CHAPTER IV. PROBLEMS SOLVED BY MEANS OF THE PRINCIPLE OF VIRTUAL WORK 92

CHAPTER V. PROBLEMS OF VARIATIONAL CALCULUS ... 110
V.1. Elements of variational calculus ... 110
    V.1.1. Functionals. Functional derivative ... 110
    V.1.2. Extrema of functionals ... 120
V.2. Problems whose solutions demand elements of variational calculus ... 125
    1. Brachistochrone problem ... 125
    2. Catenary problem ... 129
    3. Isoperimetric problem ... 132
    4. Surface of revolution of minimum area ... 136
    5. Geodesics of a Riemannian manifold ... 139

CHAPTER VI. PROBLEMS SOLVED BY MEANS OF THE LAGRANGIAN FORMALISM ... 156
1. Atwood machine ... 156
2. Double Atwood machine ... 158
3. Pendulum with horizontally oscillating point of suspension ... 163
4. Problem of two identical coupled pendulums ... 172
5. Problem of two different coupled pendulums ... 178
6. Problem of three identical coupled pendulums ... 203
7. Problem of double gravitational pendulum ... 210

CHAPTER VII. PROBLEMS OF EQUILIBRIUM AND SMALL OSCILLATIONS ... 227

CHAPTER VIII. PROBLEMS SOLVED BY MEANS OF THE HAMILTONIAN FORMALISM ... 265

CHAPTER IX. PROBLEMS OF CONTINUOUS SYSTEMS ... 313
A. Problems of Classical Electrodynamics ... 313
B. Problems of Fluid Mechanics ... 332
C. Problems of Magnetofluid Dynamics and Quantum Mechanics ... 360

APPENDICES ... 377

REFERENCES ... 439

CHAPTER I

FUNDAMENTALS OF ANALYTICAL MECHANICS

I.1. Constraints

The concept of *constraint* plays an essential role in analytical mechanics. Both Newtonian and analytical mechanics work with notions like: material point (particle), velocity, acceleration, mass, force, kinetic energy, mechanical work, etc., but the notion of *constraint* is specific to analytical mechanics only. The difference comes from the acceptance of the concept of "freedom". In Newtonian mechanics, a body is *free* if no force acts on it. For instance, consider a body under the influence of gravitational force, $\vec{G} = m\vec{g}$. From the Newtonian mechanics point of view, the gravity acts permanently on the body, meaning that the body cannot be considered as being *free*. But, in view of the analytical mechanics formalism, this body is considered *free*, in the sense that no restriction limits its motion. In other words, within the analytical approach, a body that can move freely along *any* direction in space, and rotate about *any* axis, is said to be *free*.

Definition. A *constraint* is a geometric or kinematic condition that restricts the motion of a body. The constraints are usually given as equalities and/or inequalities, either in explicit, implicit, or parametric forms. As an example, here is a constraint under the implicit form

$$f(\vec{r}, \vec{v}, t) = 0, \qquad (1.1)$$

where f is a function of class C^1 on its domain of definition (as required by the Lagrange equations of the first kind formalism).

Here are a few examples of geometric constraints for a particle (material point): the body is constrained to move on a certain curve (a teleferic moving along its cable of suspension), on a certain surface (a car climbing an inclined street), or inside a certain volume (a little piece of stone traveling inside a soccer ball).

A *constraint* can be assimilated with a *constraint force*. The forces of constraint determine the body to move on a certain curve, a certain surface, or in a certain volume. Unlike Newtonian mechanics, the analytical formalism makes a clear distinction between *applied* and *constraint* forces. While Newtonian mechanics demands for knowledge of *all* kinds of forces, so as to let us to write the fundamental equation of dynamics, the analytical mechanics replaces the *a priori* knowledge of constraint forces by the cognition of the analytical expressions of the constraints (the case of Lagrange equations of the first kind), or, even, such a replacement is not necessary (Lagrange equations of the second kind, Hamilton's canonical equations, and the Hamilton-Jacobi formalism). According to the basic concepts of analytical mechanics, the constraint forces are determined *a posteriori*, that is *after* the law of motion of the mechanical system has been determined. If fact, the analytical formalism allows one to solve even most complicated problems, where the constraint forces are not known from the beginning.

I.1.1. Classification criteria for constraints

There are at least three criteria for classification of the constraints, in terms of the following reasons:

1. Constraints can be expressed either by *equalities*, or by *inequalities*.

2. The *time t* explicitly interferes in the expression of the constraint, or it doesn't.

3. The constraint explicitly depends on the velocity $\vec{v}$, or it doesn't.

1. Constraints expressed by *equalities* are called *bilateral*, while those given by *inequalities* are named *unilateral*.

Here are a few examples of bilateral constraints:

i) A heavy particle of mass m moving on the fixed sphere of radius R (spherical pendulum - see Fig.I.1):

$$x_1^2 + x_2^2 + x_3^2 = R^2,$$

or

$$f(x_1, x_2, x_3) = x_1^2 + x_2^2 + x_3^2 - R^2 = 0. \tag{1.2}$$

ii) A particle moving on the fixed circular cone, of radius R and apex angle 2θ (see Fig.I.2)

$$x_1^2 + x_2^2 = (R - x_3 \tan \theta)^2,$$

or

$$f(x_1, x_2, x_3) = x_1^2 + x_2^2 - x_3^2 \tan^2 \theta + 2R x_3 \tan \theta - R^2 = 0. \tag{1.3}$$

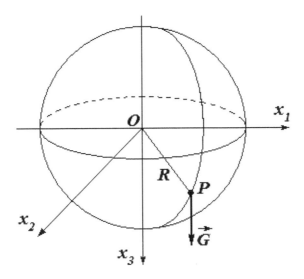

Fig.I.1

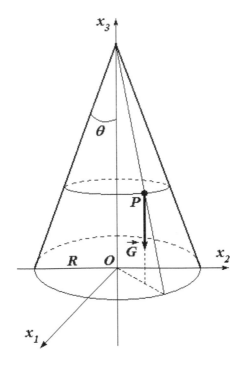

Fig.I.2

iii) A particle moving on a sphere of fixed radius R whose centre moves uniformly in a straight line with velocity $\vec{v}_C = (\vec{a}, \vec{b}, \vec{c})$

$$(x - at)^2 + (y - bt)^2 + (z - ct)^2 = R^2,$$

or
$$f(x_1, x_2, x_3, t) = f(\vec{r}, t) = (x-at)^2 + (y-bt)^2 + (z-ct)^2 - R^2 = 0. \quad (1.4)$$

iv) A body (conceived as a heavy particle of mass m) moving in the gravitational field, suspended at a fixed point O by an inextensible rod of length l (rod pendulum - see Fig.I.3)

$$x_1^2 + x_2^2 = l^2,$$

or

$$f(x_1, x_2) = x_1^2 + x_2^2 - l^2 = 0.$$

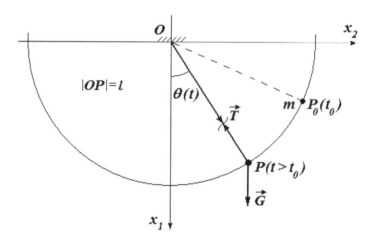

Fig.I.3

Next, we shall give some examples of unilateral constraints:
i) A heavy particle of mass m suspended by means of an inextensible but flexible wire of length l (wire pendulum - see Fig.I.4)

$$x_1^2 + x_2^2 \leq l^2,$$

or

$$f(x_1, x_2) = x_1^2 + x_2^2 - l^2 \leq 0. \quad (1.5)$$

ii) A heavy particle of mass m moving inside a rugby ball at rest. If the ball has the shape of an ellipsoid of rotation of semiaxes a, b, c, the constraint writes

$$\frac{x_1^2}{a^2} + \frac{x_2^2}{b^2} + \frac{x_3^2}{c^2} \leq 1,$$

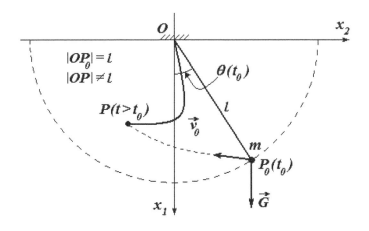

Fig.I.4

or
$$f(x_1, x_2, x_3) = \frac{x_1^2}{a^2} + \frac{x_2^2}{b^2} + \frac{x_3^2}{c^2} - 1 \leq 0. \tag{1.6}$$

iii) A particle of mass m moving inside a soccer ball, conceived as a sphere of radius R, whose center moves uniformly in a straight line with the velocity $\vec{v}_C = (\vec{a}, \vec{b}, \vec{c})$

$$(x - at)^2 + (y - bt)^2 + (z - ct)^2 \leq R^2,$$

or

$$f(x_1, x_2, x_3, t) \equiv f(\vec{r}, t) = (x-at)^2 + (y-bt)^2 + (z-ct)^2 - R^2 \leq 0. \tag{1.7}$$

2. If the constraint equation does not contain the time variable t explicitly [like (1.2), (1.3), (1.5), and (1.6)], it is called *scleronomous* or *stationary*, while a constraint which is time-dependent [like (1.1), (1.4), and (1.7)] is named *rheonomous* or *non-stationary*.

3. A constraint is called *geometric* or *finite* if the components of the velocity do not appear in the constraint equation [*e.g.* (1.2)-(1.7)]. If, on the contrary, the constraint is velocity-dependent, like (1.1), it is named *kinematic* or *differential*. As an example of kinematic constraint, consider a body whose velocity must be permanently tangent to the curve

$$\frac{(x_1 - \varphi)^2}{a^2} - \frac{x_2^2}{b^2} = 1,$$

which is a hyperbola of semi-axes a and b, and focal length φ. The time derivative then yields

$$f(x_1, x_2, \dot{x}_1, \dot{x}_2) = b^2(x_1 - \varphi)\dot{x}_1 - a^2 x_2 \dot{x}_2 = 0. \tag{1.8}$$

In this context, let us take the time derivative of the geometric constraint $f(x_1, x_2, x_2, t) = 0$. The result can be written as

$$\frac{\partial f}{\partial x_i}\dot{x}_i + \frac{\partial f}{\partial t} = 0 \quad (i = \overline{1,3}),$$

showing that *any geometric constraint can be written as a linear kinematic constraint*. (The reciprocal is not true!).

Those differential constraints which can be written in a finite form are called *integrable*. The geometric and integrable constraints are called *holonomic*, while the unilateral and non-integrable constraints are called *non-holonomic*. It is worthwhile to mention that there are no general methods to solve problems implying non-holonomic constrains. Each case is investigated separately, by specific methods, and solution depends on how skillful the researcher is. If such a problem does not admit an analytical solution, then one must appeal to numerical methods.

Usually, a constraint is investigated from all points of view at the same time. In our previous examples, the constraint (1.3) is bilateral, scleronomous and finite, (1.7) is unilateral, rheonomous and geometric, while (1.8) is bilateral, scleronomous and differential.

There exists a very strong connection between constraints and the number of degree of freedom of a mechanical system. By definition, the number of real, independent parameters that uniquely determine spatial position of a body, is called *the number of degree of freedom* of that body. This notion can be generalized to any system of particles or rigid bodies. (N.B. A continuous, deformable medium is considered to have *an infinite number of degrees of freedom*).

Here are some examples. A free particle (material point) in the Euclidean apace E_3 has three degrees of freedom; if the particle is forced to move on a surface or a curve, its number of degrees of freedom reduces by one - or by two, respectively. In its turn, a free rigid body has six degrees of freedom: three associated with translation along some axis, and three corresponding to rotation about the axis. This number diminishes if the body is submitted to one (or more) constraints. In general, *each geometric bilateral constraint reduces by one the number of degrees of freedom of a mechanical system*.

The number of constraints must be smaller than the number of degrees of freedom; otherwise, the mechanical problem would become senseless.

I.1.2. The fundamental problem of Dynamics for a constrained particle

Task. Given the mass m of a particle moving on a fixed curve (Γ), the resultant $\vec{F}$ of applied forces, and the initial conditions compatible with the constraints, $\vec{r}_0 = \vec{r}(t_0)$, $\dot{\vec{r}}_0 = \dot{\vec{r}}(t_0)$, determine the law of motion of the particle $\vec{r} = \vec{r}(t)$, as well as the resultant $\vec{L}$ of the constraint forces.

Solution. Let us consider the following two possibilities:

i) The curve is given parametrically by $x_i = x_i(q)$ $(i = \overline{1,3})$, where q is a real, time-dependent parameter. Most generally, the resultant $\vec{F}$ of the active forces is given as $\vec{F} = \vec{F}(\vec{r}, \dot{\vec{r}}, t)$, so that we have the following parametric dependence:

$$\vec{F}(\vec{r}, \dot{\vec{r}}, t) \xrightarrow{x_i = x_i(q)} \vec{F}(q, \dot{q}, t).$$

To solve the problem, we appeal to the fundamental equation of dynamics

$$m\ddot{\vec{r}} = \vec{F} + \vec{L}, \qquad (1.9)$$

where $\ddot{\vec{r}}$ and $\vec{L}$ are unknown quantities. Projecting (1.9) on the axes of a three-orthogonal reference frame $Oxyz$, we also have

$$m\ddot{x}_i = F_i + L_i \quad (i = \overline{1,3}). \qquad (1.10)$$

This is a system of three ordinary differential equations with four unknowns: L_x, L_y, L_z, and q [because $x_i = x_i(q)$]. Therefore, we need one extra equation. To this end, one decomposes the vector $\vec{L}$ in two mutually orthogonal vector components $\vec{L}_t$ and $\vec{L}_n$, $\vec{L} = \vec{L}_t + \vec{L}_n$, where $\vec{L}_t$ is tangent to the curve (Γ) at the current point P, while $\vec{L}_n$ is situated in a plane normal to the curve at the same point (see Fig.1.5). The component $\vec{L}_t$ is called *force of friction*, and the component $\vec{L}_n$ - *normal reaction*. If $\vec{L}_t = 0$, the motion is named *frictionless* and the curve is *ideal* or *perfectly smooth*. If $\vec{L}_n = 0$ the curve (Γ) is *perfectly rough*, and the force $\vec{L}$ is tangent to the curve.

We shall further suppose a heavy particle of mass m moving on the ideal curve (Γ) $(\vec{L}_t = 0)$. Recalling that the instantaneous velocity $\vec{v} = \dot{\vec{r}}$ is always tangent to the trajectory, which in our case is the curve (Γ), we can write

$$\vec{L} \cdot \vec{v} = \dot{x}L_x + \dot{y}L_y + \dot{z}L_z = 0. \qquad (1.11)$$

This way, the problem is virtually solved: we are left with a system of four differential equations

$$\begin{cases} m\ddot{x}_i = F_i + L_i, \\ \dot{x}_i L_i = 0, \end{cases} \quad (i = \overline{1,3}) \tag{1.12}$$

with four unknowns, q and L_i $(i = \overline{1,3})$.

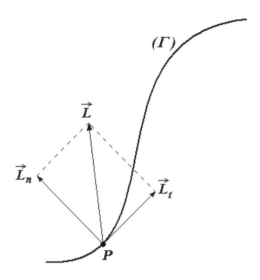

Fig.I.5

ii) The curve is given by the intersection of two implicit surfaces

$$\begin{cases} f_1(x,y,z) = 0, \\ f_2(x,y,z) = 0. \end{cases}$$

There are two ways to obtain the solution in this case:

a) Solve a system of six differential equations

$$\begin{cases} m\ddot{x}_i = F_i + L_i; \\ \dot{x}_i L_i = 0; \\ f_1(x,y,z) = 0; \\ f_2(x,y,z) = 0, \end{cases} \quad (i = \overline{1,3}) \tag{1.13}$$

with six unknowns x, y, x, L_x, L_y, L_z.

b) Decompose the constraint force $\vec{L}$ along two directions, given by ∇f_1 and ∇f_2, which are normal to the two spatial surfaces whose intersection gives rise to curve (Γ) (see Fig.1.6). Supposing, again, that the curve is ideal, we can write

$$\vec{L} = \vec{L}_1 + \vec{L}_2 = \mu \nabla f_1 + \nu \nabla f_2,$$

where μ and ν are two scalar multipliers. The scalar components of the differential equation of motion then write

$$m\ddot{x}_i = F_i + \mu \frac{\partial f_1}{\partial x_i} + \nu \frac{\partial f_2}{\partial x_i} \quad (i = \overline{1,3}). \qquad (1.14)$$

Therefore, we are left with five equations

$$\begin{cases} m\ddot{x}_i = F_i + \mu \dfrac{\partial f_1}{\partial x_i} + \nu \dfrac{\partial f_2}{\partial x_i}, \\ f_1(x,y,z) = 0, \\ f_2(x,y,z) = 0, \end{cases} \quad (i = \overline{1,3}) \qquad (1.15)$$

with five unknown quantities x, y, z, μ, and ν. This procedure allows one to determine both the law of motion $\vec{r} = \vec{r}(t)$, and the constraint force $\vec{L}$.

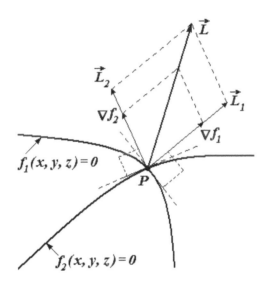

Fig.I.6

Observation. We have restricted our investigation to ideal constraints, but in reality the force of friction cannot be neglected. Nevertheless, our formalism remains valid if the tangential component $\vec{F}_f (\equiv \vec{L}_t)$ of the constraint force is known. So, instead of $\vec{F}$, as resultant of the applied forces is considered $\vec{F}' = \vec{F} + \vec{F}_f$, and the rest of procedure remains unchanged.

I.1.3. System of particles subject to constraints

Consider a system of $N \geq 2$ material points (particles). If at any moment t of the motion the position radius-vectors $\vec{r}_i$ ($i = \overline{1,N}$) of the particles and their velocities $\dot{\vec{r}}_i$ ($i = \overline{1,N}$) can take arbitrary values, we say that the system is *free*. If not, the system is *subject to constraints*.

The most general expression of a (bilateral) constraint for a system of N particles writes

$$f(\vec{r}_1, \vec{r}_2, ..., \vec{r}_n, \dot{\vec{r}}_1, \dot{\vec{r}}_2, ..., \dot{\vec{r}}_n, t) = 0, \tag{1.16}$$

and involves both geometric and kinematic conditions obeyed by the positions and velocities of the particles.

The classification criteria for a system of particles is similar to that for a single material point. For example, relations

$$f_j(\vec{r}_1, \vec{r}_2, ..., \vec{r}_n, \dot{\vec{r}}_1, \dot{\vec{r}}_2, ..., \dot{\vec{r}}_n, t) = 0 \quad (j = \overline{1,l},\ l \leq 3N) \tag{1.17}$$

constitute l bilateral, rheonomous, and differential constraints, while

$$f_j(\vec{r}_1, \vec{r}_2, ..., \vec{r}_n, t) = 0 \quad (j = \overline{1,l},\ l \leq 3N) \tag{1.18}$$

stand for l bilateral, rheonomous, and geometric constraints. The number l of constraints which restricts the possibilities of motion of a system of N particles cannot be bigger that $3N$. If $l = 3N$, the status of mechanical system would be completely determined by the constraints, so that integration of the differential equations of motion would become senseless.

As in the case of one particle, any geometric constraint can be written in a differential form by taking the total derivative with respect to time. For example, the constraints (1.17) can also be written as

$$\sum_{i=1}^{N} (\mathrm{grad}_i\, f_j) \cdot \dot{\vec{r}}_i + \frac{\partial f_j}{\partial t} = 0 \quad (j = \overline{1,l}),$$

which are l kinematic constraints linear in velocities, of the type

$$\sum_{i=1}^{N} \vec{g}_i^{\,j}(\vec{r}_1, \vec{r}_2, ..., \vec{r}_N, t) \cdot \dot{\vec{r}}_i + g_0^{j}(\vec{r}_1, \vec{r}_2, ..., \vec{r}_N, t) = 0 \quad (j = \overline{1,l}).$$

The kinematic constraints can be integrable (*holonomic*) or non-integrable (*non-holonomic*, or *Pfaffian*). All definitions met in the case of one-particle system remain valid.

I.1.4. Lagrange equations of the first kind

As we have seen, the fundamental equation of dynamics for the ith-particle of mass m_i, which belongs to a system of N particles, writes

$$m_i \ddot{\vec{r}}_i = \vec{F}_i + \vec{L}_i \quad (i = \overline{1,N}), \tag{1.19}$$

where $\vec{F}_i$ and $\vec{L}_i$ are the resultants of the applied and constraint forces, respectively. We also recall that, if a particle moves on the ideal surface

$$f(x,y,z) = 0, \tag{1.20}$$

the constraint force writes

$$\vec{L} \equiv \vec{L}_n = \mu \nabla f,$$

while in case of the motion on an ideal curve given by

$$\begin{cases} f_1(x,y,z) = 0, \\ f_2(x,y,z) = 0, \end{cases} \tag{1.21}$$

the constraint force is

$$\vec{L} = \mu \nabla f_1 + \nu \nabla f_2.$$

These results can be generalized for a system of N particles, subject to $l \leq 3N$ ideal constraints. Thus, the constraint force acting on the ith-particle writes

$$\vec{L}_i = \sum_{j=1}^{l} \mu_j \left(\text{grad}_i f_j \right), \tag{1.22}$$

and (1.19) become

$$m_i \ddot{\vec{r}}_i = \vec{F}_i + \sum_{j=1}^{l} \mu_j \left(\text{grad}_i f_j \right) \quad (i = \overline{1,N}). \tag{1.23}$$

Equations (1.23) are called the *Lagrange equations of the first kind*.

The differential equations (1.23), together with equations (1.18) of the constraints, namely

$$\begin{cases} m_i \ddot{\vec{r}}_i = \vec{F}_i + \sum_{j=1}^{l} \mu_j \left(\text{grad}_i f_j \right); \\ f_j(\vec{r}_1, \vec{r}_2, ..., \vec{r}_n, t) = 0; \end{cases} \quad (i = \overline{1,N}, \ j = \overline{1,l}), \tag{1.24}$$

form a system of $3N + l$ scalar equations with $3N + l$ unknowns: x_i, y_i, z_i $(i = \overline{1,N})$ and μ_j $(j = \overline{1,l})$.

This formalism is one of the main procedures used by analytical mechanics, and we shall apply it in the study of gravitational pendulum (see Chap.III).

I.2. Elementary displacements

I.2.1. Generalities

Beside the concept of constraint, analytical mechanics deals with the notion of *elementary displacement*. By means of this new concept, we are able to eliminate even the necessity of knowledge of the analytical expressions of the constraints. This way, the "power" of the analytical formalism sensibly increases and the problem can be solved even in those cases when identification of the constraints is difficult. Schematically, the situation can be presented as follows:

Classical (Newtonian) formalism

Requirement of knowledge of all forces from the beginning

(or, at least, their direction and sense)

↓

Lagrange equations of the first kind formalism

Knowledge of the constraint forces is not necessary, but

the analytical expressions of the constraints is demanded

↓

Lagrange equations of the second kind approach

Knowledge of both the constraint forces, and the analytical

expressions of the constraints is not necessary

Requirement of identification of the number of constraints only

The concept of *elementary displacement* has a pragmatic reason. Obviously, solving a statics problem is much easier than investigating an application on kinematics or dynamics. Indeed, in the first case the problem reduces to a system of algebraic equations, while in the second solution is obtained by solving a system of differential equations. In other words, the main idea is to transform a dynamical problem into a problem of statics: determine the *equilibrium* of a mechanical system, instead of integrating a set of differential equations to find its *law of motion*.

To this end, it is necessary to introduce the concepts of *real, possible*, and *virtual* elementary displacements.

I.2.2. Real, possible and virtual displacements

Consider a system of $N \geq 2$ particles, subject to $l < 3N$ holonomic constraints. During the time interval dt, under the action of applied force $\vec{F}_i$, the ith-particle of the system suffers the elementary displacement $d\vec{r}_i$, subject to both the constraints and the initial conditions compatible with the constraints. This displacement is *unique* and takes place *effectively*. It is called *real elementary displacement*.

The difference between possible and virtual displacements can only be clearly understood if the essential role of the time variable is considered. So, let us suppose that our system is subject to l holonomic, rheonomous constraints of the form

$$f_j(\vec{r}_1, \vec{r}_2, ..., \vec{r}_N, t) = 0 \quad (j = \overline{1,l}). \tag{1.25}$$

If we only set the position of the ith-particle at time t, there can exist an infinite number of velocities of the particle consistent with the constraints. The displacements performed under these conditions are called *possible*. Among all possible displacements only one is real, namely that satisfying both the equation of motion and the initial conditions.

Consider, finally, a class of elementary displacements only consistent with the constraints. These displacements are not real, but *virtual*. They are purely geometric and are *synchronic* (do not depend on the time t). Therefore, if the elementary possible displacements are considered as vectors, then all vectors representing these displacements (whose number is infinite) have the same origin, situated in some point of the variety which represents the constraint at the moment t, and their terminal points lying at any point of this variety, but considered at the moment $t + dt$. In contrast, the virtual displacements are always tangent to the variety representing the constraint and, being elementary, they belong in fact to this variety. In other words, the essential difference between possible and virtual displacements lies in the fact that the first "emerge" from the variety, while the last belong to the variety. This difference can accurately be understood only if one considers the most general case of rheonomous constraints. While the possible elementary displacements perform in time, the virtual ones are "instantaneous", and can be intuitively considered as being taken on a "frozen" constraint (or, similarly, at a "frozen" moment of time).

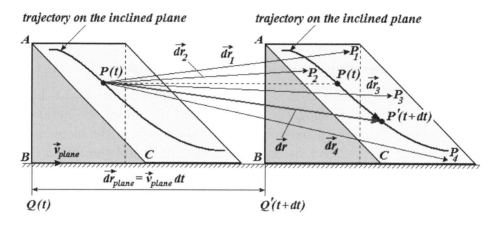

Fig.I.7

To illustrate these concepts, consider a material point (particle) constrained to remain on an inclined plane which moves along a straight line with a constant velocity $\vec{v}_{plane}$ (see Fig.I.7). Here are given two successive positions of the plane, separated by the infinitesimal distance $d\vec{r}_{plane}$, corresponding to the infinitesimal time interval dt. At the moments t and $t + dt$ the side AB of the plane is situated at the points $Q(t)$ and $Q'(t+dt)$, while $P(t)$ and $P'(t+dt)$ denote the positions of the particle on the plane at time t and $t+dt$, respectively. According to our definition, the infinitesimal vectors $d\vec{r}$, $d\vec{r}_1$, $d\vec{r}_2$, $d\vec{r}_3$, etc., with their origin at $P(t)$ and arrow-heads at P', P_1, P_2, etc. represent *possible elementary displacements* of the particle P. Among all these displacements, only one is real, namely $d\vec{r} = \overrightarrow{PP'}$. It satisfies not only the constraint, but also the equations of motion and the initial conditions.

Let us now "freeze" the plane at the moment $t' = t + dt$, when the side AB is at the point $Q'(t') = Q'(t+dt)$. This can be imagined as taking a picture of the plane at the moment t'. Since we are now "out of time", the particle can "move" only on the plane, which becomes - for an instant - a *scleronomous* constraint. The elementary displacements of the particle satisfy, in this case, only the equation of the constraint. They are called *virtual displacements* and represent a very useful tool in analytical mechanics. Some examples of virtual displacements are shown in Fig.I.8. Here the real displacement is denoted by $d\vec{r}$, the possible displacements by $d\vec{r}_1, d\vec{r}_2$, etc., while the virtual displacements are marked by the Greek letter δ: $\delta\vec{r}_1$, $\delta\vec{r}_2$, etc., or $\delta\vec{r}_{12}$, $\delta\vec{r}_{34}$, etc.

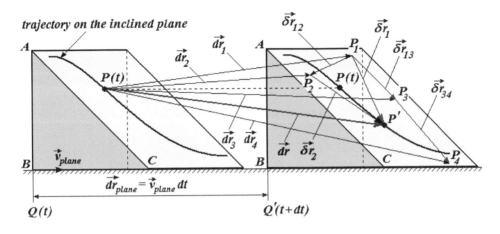

Fig.I.8

Let us show that any elementary virtual displacement can be written as the difference of two possible displacements. To this end, consider a system of particles (material points) subject to l holonomic, rheonomous constraints of the type (1.25):

$$f_j(\vec{r}_1, \vec{r}_2, ..., \vec{r}_N, t) = 0 \quad (j = \overline{1, l}),$$

where $\vec{r}_i = \vec{r}_i(t)$ ($i = \overline{1, N}$). Differentiating these relations, we obtain the conditions that must be satisfied by the possible displacements $d\vec{r}_i$ ($i = \overline{1, N}$)

$$\sum_{i=1}^{N} (\text{grad}_i f_j) \cdot d\vec{r}_i + \frac{\partial f}{\partial t} dt = 0 \quad (j = \overline{1, l}). \tag{1.26}$$

Recalling that $dt \equiv \delta t = 0$ expresses the condition for the elementary displacements $\delta\vec{r}_i$ to be *virtual* (or, equivalently, *atemporal*), equation (1.26) leads to

$$\sum_{i=1}^{N} (\text{grad}_i f_j) \cdot \delta\vec{r}_i = 0 \quad (j = \overline{1, l}). \tag{1.27}$$

Since the vectors $\text{grad}_i f_j$ are orthogonal to the constraint surfaces $f_j(\vec{r}_1, \vec{r}_2, ..., \vec{r}_N, t) = 0$ ($j = \overline{1, l}$), relations (1.27) show that the virtual displacements are always *tangent to the constraints*. Obviously, this statement is not true for the real and possible displacements. For example, given l scleronomous constraints $f_j(\vec{r}_1, \vec{r}_2, ..., \vec{r}_N) = 0$, the possible elementary displacements satisfy the relations

$$\sum_{i=1}^{N} \left(\operatorname{grad}_i f_j\right) \cdot d\vec{r}_i = 0 \quad (j = \overline{1,l}), \tag{1.28}$$

while in case of rheonomous constraints we have

$$\sum_{i=1}^{N} \left(\operatorname{grad}_i f_j\right) \cdot d\vec{r}_i = -\frac{\partial f_j}{\partial t} dt \neq 0 \quad (j = \overline{1,l}).$$

Let us now write the relations (1.26) for two distinct sets of possible elementary displacements, $d\vec{r}_i{}'$ and $d\vec{r}_i{}''$:

$$\sum_{i=1}^{N} \left(\operatorname{grad}_i f_j\right) \cdot d\vec{r}_i{}' + \frac{\partial f_j}{\partial t} dt = 0 \quad (j = \overline{1,l});$$

$$\sum_{i=1}^{N} \left(\operatorname{grad}_i f_j\right) \cdot d\vec{r}_i{}'' + \frac{\partial f_j}{\partial t} dt = 0 \quad (j = \overline{1,l}),$$

and subtract them from each other. The result is

$$\sum_{i=1}^{N} \left(\operatorname{grad}_i f_j\right) \cdot \left(d\vec{r}_i{}' - d\vec{r}_i{}''\right) = 0 \quad (j = \overline{1,l}).$$

Denoting
$$\delta \vec{r}_i = d\vec{r}_i{}' - d\vec{r}_i{}'' \quad (i = \overline{1,N}), \tag{1.29}$$

we still have
$$\sum_{i=1}^{N} \left(\operatorname{grad}_i f_j\right) \cdot \delta \vec{r}_i = 0 \quad (j = \overline{1,l}),$$

which are precisely relations (1.27) satisfied by the virtual displacements $\delta \vec{r}_i$. Consequently, the above statement is proved: *any virtual displacement can be conceived as a difference between two possible displacements.*

This also explains our choice to denote the virtual displacements by two indices in Fig.I.8, such as

$$\delta \vec{r}_{12} = d\vec{r}_2 - d\vec{r}_1, \quad \text{etc.},$$

or, in general
$$\delta \vec{r}_{mn} = d\vec{r}_n - d\vec{r}_m \quad (\forall\, m, n = \overline{1, N}).$$

If one of the two possible displacements is the real one, we have
$$\delta \vec{r}_i = \pm d\vec{r} \mp d\vec{r}_i \quad (i = \overline{1, N}).$$

For example, according to Fig.I.8, the elementary displacement $\delta \vec{r}_1$ writes
$$\delta \vec{r}_1 = -d\vec{r} + d\vec{r}_1,$$
while $\delta \vec{r}_2$ is given by
$$\delta \vec{r}_2 = d\vec{r} - d\vec{r}_2.$$

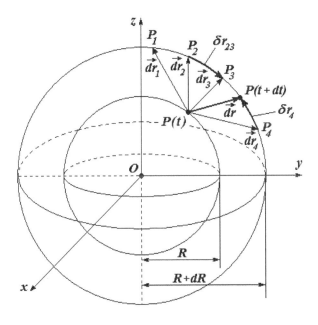

Fig.I.9

As another example of a rheonomous constraint, consider a spherical balloon whose centre is fixed, being in process of inflation, as illustrated in Fig.I.9. Suppose that at two successive moments of time t and $t + dt$ the radii of the balloon are R and $R + dR$, respectively. Let us presume that an ant, considered as a particle (material point) P, moves on the surface of the balloon. If at the moment $t_0 = 0$ the radius of the balloon is zero, and the radius is increasing by a constant radial velocity v_R, then the radius is $R = R(t) = v_R t$, and the analytic expression of the constraint writes
$$f(x, y, z, t) = x^2 + y^2 + z^2 - v_R^2 t^2 = 0.$$

If at the moment t the ant is located at the point P, then at the next moment $t' = t + dt$ the ant can be situated at any point $(P_1, P_2, P_3,$ etc.) on the sphere of radius $R' = R(t') = v_R t' = R + dR$. In this case, the *possible elementary displacements* of the ant are represented by the vectors $d\vec{r}_1, d\vec{r}_2, d\vec{r}_3$, etc., whose number is practically infinite. Among all these possible displacements, there is one which satisfies not only the equation of the constraint, but also the differential equation of motion and the initial conditions as well: *the elementary real displacement $d\vec{r}$*. If, at the moment $t' = t + dt$, the inflation of the balloon would be stopped (or "frozen" - to use a suggestive image), the ant could move only on the sphere of radius $R' = R + dR$. These imaginary, time-independent displacements are purely geometric, and have nothing to do with the real motion of the ant. They are called *virtual elementary displacements*. Two such displacements are indicated in Fig.I.9:

$$\delta \vec{r}_{23} = d\vec{r}_3 - d\vec{r}_2,$$

and

$$\delta \vec{r}_4 = d\vec{r} - d\vec{r}_4.$$

As a last example, let us suppose we are watching a movie. The motion picture is composed by a succession of images, whose frequency is settled at about 16-20 images per second, in accordance with the biological possibility of the human eye to distinguish two successive pictures. If, by any chance, the movie suddenly stops, only one static picture remains on the screen. This picture is very similar to the previous and following ones, but nevertheless distinct, being separated by infinitesimal time intervals. Any imaginary displacement, "performed" at an instant $t\,(\delta t = 0)$ on the static picture, can be considered as a *virtual displacement*.

Therefore, if we denote
— M_{RED} the set of real elementary displacements ;
— M_{PED} the set of possible elementary displacements ;
— M_{VED} the set of virtual elementary displacements ,
in case of the rheonomous constraints we then have

$$M_{RED} \subset M_{PED}, \quad M_{RED} \cap M_{VED} = \emptyset, \quad M_{PED} \cap M_{VED} = \emptyset,$$

while the scleronomous constraints obey the rule

$$M_{RED} \subset M_{PED} \equiv M_{VED}.$$

We nevertheless note that, generally speaking, in the last case the concept of possible elementary displacement cannot be defined.

The table given below displays a synthetic image on the three types of elementary displacements, in case of rheonomous constraints.

	Real elementary displacements	Possible elementary displacements	Virtual elementary displacements
They satisfy the equations of motion	👍	👎	👎
They satisfy the initial conditions	👍	👎	👎
They satisfy the constraints	👍	👍	👍
They take place in time	👍	👍	👎
They are instantaneous	👎	👎	👍
They are pure geometrical	👎	👎	👍
They are tangent to the constraint	👎	👎	👍

Elementary displacements table

I.3. Virtual work and connected principles

I.3.1. Principle of virtual work

As we have seen in the previous paragraph, conversion of a dynamic problem into a static one brings the remarkable advantage of replacing the difficulties of solving a system of differential equations by the easier task of finding solution to an algebraic system of equations. In its turn, any problem of statics implies determination of the equilibrium conditions/positions of the system of particles under discussion. This can be done by means of the *principle of virtual work*.

By definition, the elementary mechanical work of a force $\vec{F}$ by an infinitesimal displacement $d\vec{r}$ of its point of application is given by the dot product $dW = \vec{F} \cdot d\vec{r}$. By analogy, *the elementary virtual work* corresponding to a *virtual* displacement $\delta\vec{r}$ can be defined as $\delta W = \vec{F} \cdot \delta\vec{r}$.

To understand how this notion can be applied, consider a system of $N \geq 2$ particles (material points) and let $\delta\vec{r}_i$ be a virtual elementary displacement of the particle P_i acted upon by the force $\vec{F}_i$. Then, following the above definition, the infinitesimal virtual work "performed" by $\vec{F}_i$ is

$$\delta W = \vec{F}_i \cdot \delta\vec{r}_i. \tag{1.30}$$

Since the material point is at static equilibrium, $\ddot{\vec{r}}_i = 0$, and the fundamental equation of dynamics written for this point

$$m_i \ddot{\vec{r}}_i = \vec{F}_i + \vec{L}_i$$

yields
$$\vec{F}_i + \vec{L}_i = 0.$$

Performing the scalar product of this equation with $\delta \vec{r}_i$ and summing over index i, we obtain

$$\sum_{i=1}^{N} \vec{F}_i \cdot \delta \vec{r}_i + \sum_{i=1}^{N} \vec{L}_i \cdot \delta \vec{r}_i = 0. \tag{1.31}$$

If the constraints are ideal, meaning that the forces of constraint have only normal components, $\vec{L}_i \equiv (\vec{L}_i)_{normal} = (\vec{L}_i)_\perp$, while the virtual elementary displacements are always tangent to the constraint, $\delta \vec{r}_i \equiv (\delta \vec{r}_i)_\parallel$, it then follows that the virtual mechanical work of the constraint forces is null:

$$\sum_{i=1}^{N} \vec{L}_i \cdot \delta \vec{r}_i = \sum_{i=1}^{N} (\vec{L}_i)_\perp \cdot (\delta \vec{r}_i)_\parallel = 0,$$

in which case (1.31) becomes

$$\sum_{i=1}^{N} \vec{F}_i \cdot \delta \vec{r}_i = 0. \tag{1.32}$$

Relation (1.32) expresses the *principle of virtual work: The necessary and sufficient condition for static equilibrium of a scleronomous system subject to ideal constraints is that the virtual work of the applied forces, for virtual displacements consistent with the constraints, be null.*

Let our mechanical system of N particles be subject to l ideal constraints of the form

$$f_j(\vec{r}_1, \vec{r}_2, ..., \vec{r}_N, t) = 0 \quad (j = \overline{1, l}).$$

As we have seen, the virtual elementary displacements satisfy the relations

$$\sum_{i=1}^{N} (\text{grad}_i f_j) \cdot \delta \vec{r}_i = 0 \quad (j = \overline{1, l}), \tag{1.33}$$

or
$$\sum_{i=1}^{N}\left(\frac{\partial f_j}{\partial x_i}\delta x_i + \frac{\partial f_j}{\partial y_i}\delta y_i + \frac{\partial f_j}{\partial z_i}\delta z_i\right) = 0 \quad (j = \overline{1,l}). \tag{1.34}$$

On the other hand, the principle of virtual work (1.32) can be written as
$$\sum_{i=1}^{N}\left(X_i \delta x_i + Y_i \delta y_i + Z_i \delta z_i\right) = 0, \tag{1.35}$$

where $\vec{F}_i = (X_i, Y_i, Z_i)$ $(i = \overline{1,N})$. If the elementary displacements $\delta x_i, \delta y_i, \delta z_i$ would be linearly independent, then (1.35) would lead to the following system of algebraic equations
$$\begin{cases} X_i = 0, \\ Y_i = 0, \quad (i = \overline{1,N}) \\ Z_i = 0, \end{cases} \tag{1.36}$$

whose solution would furnish the equilibrium positions x_i, y_i, z_i $(i = \overline{1,N})$ for the mechanical system. In fact, the virtual displacements $\delta x_i, \delta y_i, \delta z_i$ are not independent, but have to satisfy the set of l equations (1.34).

Under these circumstances, to determine the position of equilibrium of the system we shall use the method of *Lagrangian multipliers*. Multiplying (1.34) by the non-zero scalars λ_j $(j = \overline{1,l})$, then performing summation over j and adding to (1.35), one obtains

$$\sum_{i=1}^{N}\left[\left(X_i + \sum_{j=1}^{l}\lambda_j \frac{\partial f_j}{\partial x_i}\right)\delta x_i + \left(Y_i + \sum_{j=1}^{l}\lambda_j \frac{\partial f_j}{\partial y_i}\right)\delta y_i \right.$$
$$\left. + \left(Z_i + \sum_{j=1}^{l}\lambda_j \frac{\partial f_j}{\partial z_i}\right)\delta z_i\right] = 0. \tag{1.37}$$

This relation has to be satisfied by any $\delta x_i, \delta y_i, \delta z_i$ $(i = \overline{1,N})$. Since the virtual displacements $\delta x_i, \delta y_i, \delta z_i$ are submitted to the l constraints (1.34), only $3N - l$ virtual displacements are linearly independent. To solve the problem, one equalizes to zero l round parentheses out of the total of $3N$ interfering in (1.37), which produces l algebraic equations for the unknowns λ_j $(j = \overline{1,l})$. This way, in (1.37) remain only $3N - l$ terms (containing the already determined λ_j), corresponding to the $3N - l$ linearly independent virtual displacements. The coefficients of these virtual displacements, namely the $3N - l$ round parentheses remaining in (1.37) have to cancel, because (1.37) now stands for a linear combination of linearly independent "vectors". Consequently, all the $3N$ round parentheses in (1.37) must cancel, which leads to a system of $3N$ algebraic equations with $3N+l$ unknowns [3N equilibrium coordinates x_i, y_i, z_i $(i = \overline{1,N})$ and l Lagrangian multipliers λ_j $(j = \overline{1,l})$]. To have a unique solution, the system of $3N$ algebraic equations has to be completed with l analytical relations of the constraints $f_j(\vec{r}_1, \vec{r}_2, ..., \vec{r}_N, t) = 0$ $(j = \overline{1,l})$, that is

$$\begin{cases} X_i + \sum_{j=1}^{l} \lambda_j \frac{\partial f_j}{\partial x_i} = 0, \\ Y_i + \sum_{j=1}^{l} \lambda_j \frac{\partial f_j}{\partial y_i} = 0, \quad (i = \overline{1,N} \ ; \ j = \overline{1,l}) \\ Z_i + \sum_{j=1}^{l} \lambda_j \frac{\partial f_j}{\partial z_i} = 0, \\ f_j(\vec{r}_1, \vec{r}_2, ..., \vec{r}_N, t) = 0. \end{cases} \quad (1.38)$$

This is a system of $3N+l$ equations with $3N+l$ unknown quantities. Therefore, the problem is solved.

I.3.2. Principle of virtual velocities

A variant of the principle of virtual work is the so-called *principle of virtual velocities*. By analogy with the usual definition

$$\vec{v} = \frac{d\vec{r}}{dt} = \dot{\vec{r}},$$

one introduces the notion of *virtual velocity*

$$\vec{v} = \frac{\delta \vec{r}}{\delta t}, \quad (1.39)$$

where δt would *formally* represent the time interval corresponding to the virtual elementary displacement $\delta \vec{r}$. (In fact, a virtual displacement is a purely geometric concept, associated with the time interval $\delta t = 0$). Keeping in mind this observation, the principle of virtual velocities proves to be very useful in determination of the conditions[1] of kinematic equilibrium[2] for a system of material points.

---

[1] In the case of static equilibrium we are talking about "equilibrium positions", while the kinematics operates with "equilibrium conditions".

[2] A system of particles is at *kinematic equilibrium* if it performs a uniform motion (the linear acceleration for rectilinear motions is zero, and so is the angular acceleration for circular motions).

All previous considerations and definitions remain valid, since in both cases of static and kinematic equilibria we have $\ddot{\vec{r}}_i = 0$ in case of rectilinear motions, and $\varepsilon_i = \frac{d\omega_i}{dt} = \ddot{\alpha}_i = 0$ for circular motions. Indeed, the condition $\ddot{\vec{r}}_i = 0$ (or/and $\varepsilon_i = 0$) stands not only for static equilibrium, but also for kinematic equilibrium, expressed by

$$\vec{v}_i = \dot{\vec{r}}_i = \overrightarrow{const.} \quad (\omega_i = const.). \tag{1.40}$$

In other words, the principle of virtual velocities can be considered as an extension of the principle of virtual work for the case when the constant appearing in (1.40) is not necessarily zero.

The mathematical expression for the principle of virtual velocities is easily obtained by using the principle of virtual work

$$\delta W = \sum_{i=1}^{N} \vec{F}_i \cdot \delta \vec{r}_i = 0$$

and dividing it by the infinitesimal "virtual" time interval (formally, $\delta t \neq 0$)

$$\frac{\delta W}{\delta t} = \sum_{i=1}^{N} \vec{F}_i \cdot \frac{\delta \vec{r}_i}{\delta t} = \sum_{i=1}^{N} \vec{F}_i \cdot \vec{v}_i = 0. \tag{1.41}$$

Applying the same procedure for the circular motion, we have

$$\frac{\delta W}{\delta t} = \frac{\sum_{i=1}^{N} M_i \delta \alpha_i}{\delta t} = \sum_{i=1}^{N} M_i \frac{\delta \alpha_i}{\delta t} = \sum_{i=1}^{N} M_i \omega_i = 0. \tag{1.42}$$

Since in (1.41) and (1.42) the quantity $\frac{\delta W}{\delta t}$ has units of power, this principle is also called *principle of virtual power*.

I.3.3. Torricelli's principle

If, in particular, only the gravity acts on the system of material points, the principle of virtual work can also be expressed in a form given by Torricelli: *The necessary and sufficient condition for a system of particles subject to ideal constraints and acted only by their own forces of gravity to be in equilibrium, is that the virtual variation of the quota of the center of gravity of the system be zero: $\delta z_G = 0$.*

To solve a problem of static equilibrium by means of this principle, one can use the following algorithm:

1) An orthogonal reference frame, preferably Cartesian $Oxyz$, with z-axis oriented vertically, is attached to the system of N particles.

2) One identifies the number of constraints, l, settle the number of degrees of freedom of the system, $n = 3N - l$, and associate with each degree of freedom an independent variable quantity characteristic for the system, $\xi_1, \xi_2, ..., \xi_n$.

3) One determines the quota of the centre of gravity[1] of the system by means of one of the following definitions:

$$i)\ z_G = \frac{\sum_{i=1}^{N} m_i z_i}{\sum_{i=1}^{N} m_i} = \frac{1}{M} \sum_{i=1}^{N} m_i z_i, \text{ if the system is discrete, or}$$

$$ii)\ z_G = \frac{\int_{(D)} z\rho(x,y,z)\,dx\,dy\,dz}{\int_{(D)} \rho(x,y,z)\,dx\,dy\,dz} = \frac{1}{M} \int_{(D)} z\rho(x,y,z)\,dx\,dy\,dz, \text{ if}$$

the system is continuously distributed in the spatial domain (D).

4) The determined quota of the centre of gravity is expressed in terms of the n linearly independent parameters $\xi_1, \xi_2, ..., \xi_n$:

$$z_G = z_G(\xi_1, \xi_2, ..., \xi_n).$$

We then calculate the differential dz_G

$$dz_G = X_1(\xi_1, \xi_2, ..., \xi_n)d\xi_1 + X_2(\xi_1, \xi_2, ..., \xi_n)d\xi_2$$
$$+ ... + X_n(\xi_1, \xi_2, ..., \xi_n)d\xi_n,$$

identify dz_G with δz_G

$$\delta z_G = X_1(\xi_1, \xi_2, ..., \xi_n)\delta\xi_1 + X_2(\xi_1, \xi_2, ..., \xi_n)\delta\xi_2$$
$$+ ... + X_n(\xi_1, \xi_2, ..., \xi_n)\delta\xi_n, \tag{1.43}$$

and finally solve the equation $\delta z_G = 0$. Since the parameters $\xi_1, \xi_2, ...,$

---

[1] If the spatial extent of the system of particles is not very large, the centre of mass of the system can be identified with its centre of gravity. The following relations are written within this approximation.

ξ_n are linearly independent, the virtual variations $\delta\xi_1, \delta\xi_2, ..., \delta\xi_n$ are also linearly independent, and (1.43) yields the following system of n algebraic equations in n variables $\xi_1, \xi_2, ..., \xi_n$:

$$\begin{cases} X_1(\xi_1, \xi_2, ..., \xi_n) = 0, \\ X_2(\xi_1, \xi_2, ..., \xi_n) = 0, \\ \cdot \\ \cdot \\ \cdot \\ X_n(\xi_1, \xi_2, ..., \xi_n) = 0. \end{cases}$$

The real solutions of this system provide the equilibrium position of the system of particles.

CHAPTER II

PRINCIPLES OF ANALYTICAL MECHANICS

II.1. D'Alembert's principle

As shown in the last paragraph, the principle of virtual work allows one to determine either *positions* of static equilibrium or *conditions* of kinematic equilibrium for a system of N material points (particles). So, this principle offers possibility of solving a statics problem emerging from the original dynamical one (at least, principially) by means of another very important principle - the D'Alembert principle. As we shall see, this can be done by the help of *force of inertia* concept. Therefore, a dynamical problem can be replaced by a statics one on the real space of motion only [the space of position vectors $\vec{r}_i$ ($i = \overline{1,N}$)], and working only in a non-inertial frame. Besides, this method proves to be useful only as a first step, since - as we shall see - the result is finally obtained by solving a dynamical problem, but this time in the so-called "configuration space". In this space, solution to the problem is obtained in an easier way. It is to be mentioned that the problem is considerably simplified by the existence of constraints (the number of second order differential equations is diminished by $2l$, where l is the number of constraints).

Recalling the Lagrange equations of the first kind formalism, it is not difficult to realize that the bigger the numbers N of particles and l of constraints, the more difficult the task is. Under these circumstances, the system of $3N + l$ equations could become unsolvable. Observing that the number of degrees of freedom $n = 3N - l$ of the system diminishes with the increasing of the number l of constraints, Lagrange had a brilliant idea of replacing the big number of $\vec{r}_i$ ($i = \overline{1,N}$) and parameters λ_j ($j = \overline{1,l}$) by a smaller number of unknowns q_k ($k = \overline{1,n}$). Namely, to each degree of freedom of the mechanical system Lagrange associates a parameter q. These quantities are called *Lagrange variables*, or *generalized coordinates*, and define an abstract space, named *configuration space*.

The Lagrangian formalism is useful not only in Mechanics, but also in various fields of science.

The choice of the generalized coordinates is not unique, and depends to some extent on the ability of the researcher, but there are certain requirements which have to be obeyed:

1) Compulsory requirements:

  i) to allow the position vectors of the material points to be expressed in terms of q_k (and, eventually, of time):

$$\vec{r}_i = \vec{r}_i(q_1, q_2, ..., q_n, t) \equiv \vec{r}_i(q, t) \quad (i = \overline{1, N}); \qquad (2.1)$$

  ii) to satisfy the equations of constraints, that is

$$f_j(q_1, q_2, ..., q_n, t)$$
$$\equiv f_j(\vec{r}_1(q,t), \vec{r}_2(q,t), ..., \vec{r}_N(q,t), t) = 0 \quad (j = \overline{1, l});$$

  iii) to exist the inverse transformation of (2.1), expressed as

$$q_k = q_k(\vec{r}_1, \vec{r}_2, ..., \vec{r}_N, t) \quad (k = \overline{1, n}). \qquad (2.2)$$

2) Optional requirements:

  i) to take into account the symmetry proprieties of the physical system;

  ii) to lead to the solution in the new variables as easy as possible;

  iii) to allow construction of the Lagrangian function (for natural systems - see further) or the kinetic energy (for the systems which do not admit a simple or generalized potential) as simple as possible.

Like the real coordinates $\vec{r}_i$ $(i = \overline{1, N})$, the generalized coordinates must be continuous functions of time, and at least twice differentiable [to assure the existence of $\ddot{q}_k$ $(k = \overline{1, n})$, which are the analogous to the accelerations $\vec{a}_i = \ddot{\vec{r}}_i$ in the real space]. But unlike the real coordinates, which have the units of distance, as generalized coordinates one can take: arc elements, surface elements, angles, entropy, angular velocity, scalar or vector potential, etc. More than that, the choice of the generalized coordinates is not unique: there is usually possible to define another set of generalized coordinates $q'_1, q'_2, ..., q'_n$, so that

$$q_k \longrightarrow q'_k = q'_k(q_1, q_2, ..., q_n, t) \quad (k = \overline{1, n}).$$

If the system of particles is not subject to constraints, then as generalized coordinates can be taken the Cartesian coordinates x_i, y_i, z_i

($i = \overline{1, N}$), or spherical coordinates r, θ, φ, or cylindrical coordinates ρ, φ, z, etc. For example, the system shown in Fig.I.2 is subject to the constraint

$$f(x_1, x_2, x_3) = x_1^2 + x_2^2 - x_3^2 \tan^2 \theta + 2Rx_3 \tan \theta - R^2 = 0,$$

meaning that the particle has $n = 3 - 1 = 2$ degrees of freedom. To these degrees of freedom one associate either the pair $(q_1, q_2) = (x_1, x_3)$, or the pair $(q_1, q_2) = (\varphi, x_3)$, where φ is the angle between Ox_1 and the projection on x_1Ox_2-plane of the radius vector of the particle, etc. (There exists at least two more possibilities, and it is up to the reader to identify them). In case of the first choice, relations (2.1) write

$$\begin{cases} x_1 = x_1, \\ x_2 = \sqrt{\left[(R/\tan\theta) - x_3\right]^2 - x_1^2}, \\ x_3 = x_3, \end{cases}$$

while the second choice yields

$$\begin{cases} x_1 = (R - x_3 \tan\theta)\cos\varphi, \\ x_2 = (R - x_3 \tan\theta)\sin\varphi, \\ x_3 = x_3. \end{cases}$$

II.1.1. Configuration space

As we have mentioned in the previous paragraph, the generalized coordinates q_k ($k = \overline{1, n}$) define a space called *configuration space*. This is an abstract space, usually denoted by R^n, and has a physical (real) meaning only in some particular cases. A point P in configuration space is defined by the set of generalized coordinates $q_1, q_2, ..., q_n$. The name "configurations" comes from the fact that the points of this space stand for "configurations" of the system of particles in the real, physical space. Indeed, the system of position vectors $\vec{r}_1, \vec{r}_2, ..., \vec{r}_N$, at a certain time t, define the so-called *configuration* of the system of particles, at time t. According to relations (2.1), the knowledge of the set of vectors $\vec{r}_i$ ($i = \overline{1, N}$) is equivalent to the cognition of the set of generalized coordinates q_k ($k = \overline{1, n}$). This way, to know the positions of the N particles in the real space means to know the position of a single point, called *representative point*, in the configuration space R^n. More than that, to know the N real trajectories of all particles of the system, expressed parametrically by $\vec{r}_i = \vec{r}_i(t)$ ($i = \overline{1, N}$), is equivalent to know the trajectory of a single point (the representative point) in the configuration space, describing a single trajectory $q_k = q_k(t)$ ($k = \overline{1,n}$), called *generalized trajectory*.

Observations

1. The generalized trajectory does not represent any of the N real trajectories described by the system of particles, but "integrates" all the real trajectories of the system.

2. The generalized trajectory can be conceived as a succession of representative points, each of them representing the configuration of the system at the considered moment of time.

II.1.2. Generalized forces

By differentiating (2.1) with respect to time, we have

$$d\vec{r}_i = \sum_{k=1}^{n} \frac{\partial \vec{r}_i}{\partial q_k} dq_k + \frac{\partial \vec{r}_i}{\partial t} dt \quad (i = \overline{1,N}), \tag{2.3}$$

which might express:

i) the real elementary displacement of the particle P_i during the time interval dt, if $\vec{r}_i = \vec{r}_i(q,t)$ satisfy the differential equations of motion, the equations of constraints, and the initial conditions compatible with the constraints, or

ii) a possible elementary displacement of the same particle subject to the rheonomous constraints

$$f_j\bigl(\vec{r}_1(q,t), \vec{r}_2(q,t), ..., \vec{r}_N(q,t), t\bigr) = 0 \quad (j = \overline{1,l}), \tag{2.4}$$

if $\vec{r}_i = \vec{r}_i(q,t)$ satisfy only these constraints.

If we set $dt \to \delta t = 0$ in (2.3), then $d\vec{r}_i \to \delta\vec{r}_i$, $dq_k \to \delta q_k$, and we obtain

$$\delta\vec{r}_i = \sum_{k=1}^{n} \frac{\partial \vec{r}_i}{\partial q_k} \delta q_k \quad (i = \overline{1,N}), \tag{2.5}$$

which is an infinitesimal virtual displacement of the particle P_i, subject to the rheonomous constraints (2.4).

The significance of the quantities dq_k and δq_k appearing in (2.3) and (2.5), respectively, is similar to the displacements $d\vec{r}_i$ and $\delta\vec{r}_i$ in the real space. Therefore, dq_k signifies

i) the real elementary displacement of the representative point in configuration space, if $q_k = q_k(t)$ ($k = \overline{1,n}$) satisfy the equations of motion in R^n (Lagrange equations - see further), the equations of constraints (2.4), and the initial conditions compatible with the constraints $q_{k0} = q_k(t_0)$, $\dot{q}_{k0} = \dot{q}_k(t_0)$ ($k = \overline{1,n}$);

ii) a possible elementary displacement of the representative point, if only the rheonomous constraint equations (2.4) are satisfied, while δq_k designates a virtual elementary displacement of the representative point in the configuration space. If $q_1, q_2, ..., q_n$ are linearly independent, then $\delta q_1, \delta q_2, ..., \delta q_n$ have the same property.

Let us now return to the concept of virtual work. In the real space, the virtual mechanical work of the forces $\vec{F}_i$ ($i = \overline{1, N}$) applied to $N \geq 2$ particles writes

$$\delta W = \sum_{i=1}^{N} \vec{F}_i \cdot \delta \vec{r}_i,$$

or, by means of (2.5),

$$\delta W = \sum_{i=1}^{N} \vec{F}_i \cdot \left(\sum_{k=1}^{n} \frac{\partial \vec{r}_i}{\partial q_k} \delta q_k \right)$$

$$= \sum_{i=1}^{N} \left(\sum_{k=1}^{n} \vec{F}_i \cdot \frac{\partial \vec{r}_i}{\partial q_k} \right) \delta q_k = \sum_{k=1}^{n} Q_k \delta q_k. \quad (2.6)$$

The quantities Q_k defined by

$$Q_k = \sum_{i=1}^{N} \vec{F}_i \cdot \frac{\partial \vec{r}_i}{\partial q_k} \quad (2.7)$$

are called *generalized forces*. Therefore, in the configuration space the principle of virtual work writes

$$\delta W = \sum_{k=1}^{n} Q_k \delta q_k = 0. \quad (2.8)$$

Since the applied forces $\vec{F}_i$ have the functional dependency of the form

$$\vec{F}_i = \vec{F}_i(\vec{r}_1, \vec{r}_2, ..., \vec{r}_N, \dot{\vec{r}}_1, \dot{\vec{r}}_2, ..., \dot{\vec{r}}_N, t),$$

in view of (2.1) we conclude that the generalized forces Q_k ($k = \overline{1, n}$) obey the following functional dependency

$$Q_k = Q_k(q_1, q_2, ..., q_n, \dot{q}_1, \dot{q}_2, ..., \dot{q}_n, t) \quad (k = \overline{1, n}), \quad (2.9)$$

where the quantities

$$\dot{q}_k = \frac{dq_k}{dt} \quad (k = \overline{1, n}) \quad (2.10)$$

are called *generalized velocities*. Dividing (2.3) by dt, we obtain the connection between real and generalized velocities:

$$\dot{\vec{r}}_i = \sum_{k=1}^{n} \frac{\partial \vec{r}_i}{\partial q_k} \dot{q}_k + \frac{\partial \vec{r}_i}{\partial t} \quad (i = \overline{1, N}). \tag{2.11}$$

Observation. As observed, the generalized forces are not forces in the Newtonian sense of the term (except for the case when the generalized coordinates have unit of length), and, consequently, their units are given by the associated generalized coordinates. Whatever the choice of the generalized coordinates, the product between a generalized coordinate and its associated generalized force has units of mechanical work [see (2.8)]. For example, if q stands for an angular variable, then the associated generalized force *must* have the dimension of a force moment $(N \cdot m)$.

This observation is also valid for the principle of virtual velocities. To write this principle in the configuration space R^n, let us divide δW by δt:

$$\frac{\delta W}{\delta t} = \frac{\sum_{i=1}^{N} \vec{F}_i \cdot \delta \vec{r}_i}{\delta t} = \sum_{i=1}^{N} \vec{F}_i \cdot \frac{\delta \vec{r}_i}{\delta t} = \sum_{i=1}^{N} \vec{F}_i \cdot \frac{\sum_{k=1}^{n} \frac{\partial \vec{r}_i}{\partial q_k} \delta q_k}{\delta t}$$

$$= \sum_{i=1}^{N} \vec{F}_i \cdot \sum_{k=1}^{n} \frac{\partial \vec{r}_i}{\partial q_k} \frac{\delta q_k}{\delta t} = \sum_{k=1}^{n} \left(\sum_{i=1}^{N} \vec{F}_i \cdot \frac{\partial \vec{r}_i}{\partial q_k} \right) \frac{\delta q_k}{\delta t} = \sum_{k=1}^{n} Q_k w_k, \tag{2.12}$$

where by $w_k = \frac{\delta q_k}{\delta t}$ $(k = \overline{1, n})$ we denoted the *virtual generalized velocities*. Resuming the above observation, it is useful to realize that if w_k is an "ordinary" velocity, then the associated generalized force Q_k is a Newtonian force, while if w_k is an angular velocity, then Q_k stands for a force moment. Consequently, in the configuration space the principle of virtual velocities writes

$$\frac{\delta W}{\delta t} = \sum_{k=1}^{n} Q_k w_k = 0. \tag{2.13}$$

Let us now consider the D'Alembert's principle. First of all, we have to mention that it belongs to the *differential principles* of analytical mechanics, being expressed in terms of the elementary variations of the generalized coordinates and velocities. It is the dynamic analogue to the principle of virtual work for applied forces in a static system.

Suppose, as usual, a system of $N \geq 2$ particles $P_1, P_2, ..., P_N$, subject to the applied forces $\vec{F}_i$ ($i = \overline{1,N}$) and to constrained forces $\vec{L}_i$ ($i = \overline{1,N}$). (The constraints are supposed to be holonomic). The fundamental equation of Newtonian mechanics for the particle P_i (*lex secunda*), in an inertial reference frame, then writes

$$m_i \ddot{\vec{r}}_i = \vec{F}_i + \vec{L}_i. \qquad (2.14)$$

By definition, the quantity $(-m_i \ddot{\vec{r}}_i)$ (no summation) is denoted by

$$\vec{J}_i = -m_i \ddot{\vec{r}}_i \qquad (2.15)$$

and is called *force of inertia*. According to (2.15), this force is oriented along the acceleration vector $\ddot{\vec{r}}_i$, but its sense is opposite. By means of this notation, the fundamental equation of dynamics (2.14) becomes

$$\vec{F}_i + \vec{L}_i + \vec{J}_i = 0 \quad (i = \overline{1,N}) \qquad (2.16)$$

and expresses *D'Alembert's principle: At any moment, there is an equilibrium between the applied, the constraint, and the inertia forces acting on a particle.*

This principle is valid in a non-inertial frame only. Indeed, equation (2.16) expresses the equilibrium condition for each particle of the system, which can also be written as

$$m_i \vec{a}_i = \vec{F}_i + \vec{L}_i + \vec{J}_i = 0 \quad (i = \overline{1,N}). \qquad (2.17)$$

This way, the dynamical problem has been converted to a kinematic one (if $\dot{\vec{r}}_i = \vec{v}_i = \overline{(const.)}_i$), or even a static one (if $\dot{\vec{r}}_i = \vec{v}_i = 0$), because in this frame the acceleration vector $\vec{a}_i$ ($i = \overline{1,N}$) of each particle is zero.

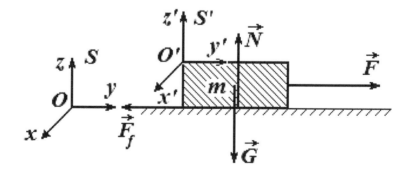

Fig.II.1

To illustrate this situation, here is a simple example. Consider a parallelepipedic body of mass m performing a uniformly accelerated linear motion ($\vec{a} = \overrightarrow{const.} > 0$) on a horizontal plane, subject to the active force $\vec{F} > \vec{F}_f$ (see Fig.II.1).

Since $N = G = mg$, and $|\vec{F}_f| = \mu|\vec{N}|$, it follows that the acceleration of the body is

$$a = |\vec{a}| = \frac{|\vec{F}| - \mu mg}{m} = \frac{F}{m} - \mu g > 0. \qquad (2.18)$$

In the inertial frame (the so-called "laboratory reference frame") the body of mass m moves to the right with a uniform acceleration given by (2.18) (see Fig.II.2).

Consider now a non-inertial frame S', invariably connected to the body. This means that the body is at rest with respect to S' (in fact, the body itself is the reference frame S'). An observer attached to the body (that is to S') moves to the right with body's acceleration $a > 0$, with respect to the laboratory frame (see Fig.II.3).

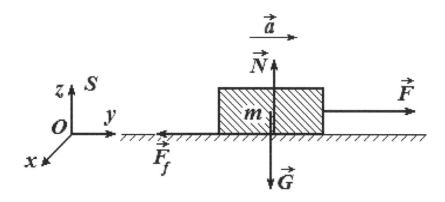

Fig.II.2

Forces $\vec{F}, \vec{F}_f, \vec{G}$, and $\vec{N}$ acting on the body in the frame S will also act in the frame S', but the fact that the body is at rest in S' can only be explained if we accept that in S' also acts the force[1]

$$\vec{F}_{in} = -(\vec{F} + \vec{F}_f) = -m\vec{a}. \qquad (2.19)$$

---

[1] Here we have to take into account that $\vec{G} + \vec{N} = 0$

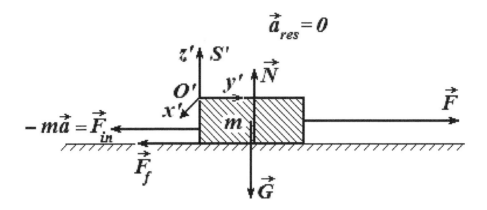

Fig.II.3

This is *the force of inertia*. It is a real, applied force in S', and by means of $\vec{F}_{in}$ a problem of dynamics has been turned into a statics problem. Indeed, in the frame S' we have

$$m\vec{a}_{res} = \vec{F} + \vec{F}_f + \vec{G} + \vec{N} + \vec{F}_{in} = 0, \qquad (2.20)$$

therefore the resultant acceleration is $\vec{a}_{res} = \ddot{\vec{r}} = 0$. Unlike the forces $\vec{F}, \vec{F}_f, \vec{G}, \vec{N}$, acting in both frames S and S', the force $\vec{F}_{in}$ acts only in the non-inertial frame S'. In other words, in the inertial frame S the force of inertia *does not exist*.

This conversion of a dynamical problem into a statics one does not entirely solve the problem. Indeed, in order to write the principle of virtual work we should know *all* forces acting upon the body in the frame S'. But, unfortunately, the force of inertia[1] $\vec{F}_{in} = -m\ddot{\vec{r}}$ is not known (otherwise, the dynamic problem would be solved). It seems to be a vicious circle: to solve the problem, one must know the solution! Consequently, the answer has to be searched elsewhere. The one who solved the problem was *Joseph-Louis Lagrange*.

In this respect, let us perform the dot product of relation (2.16) expressing D'Alembert's principle by the virtual displacement $\delta \vec{r}_i$ ($i = \overline{1,N}$). The result is

$$\sum_{i=1}^{N} \vec{F}_i \cdot \delta \vec{r}_i + \vec{J}_i \cdot \delta \vec{r}_i = \sum_{i=1}^{N} (\vec{F}_i + \vec{J}_i) \cdot \delta \vec{r}_i = 0, \qquad (2.21)$$

where we took into account the fact that, in case of ideal constraints,

$$\sum_{i=1}^{N} \vec{L}_i \cdot \delta \vec{r}_i = \sum_{i=1}^{N} \left(\vec{L}_i\right)_{normal} \cdot \left(\delta \vec{r}_i\right)_{tangential} = 0.$$

---

[1] Here index "*in*" comes from "inertia".

Relation (2.21) can also be written as

$$\sum_{i=1}^{N} \left(\vec{F}_i - m_i \ddot{\vec{r}}_i \right) \cdot \delta \vec{r}_i = 0 \tag{2.22}$$

and expresses D'Alembert's principle in the form given by Lagrange:

The sum of the elementary virtual works of applied and inertial forces acting on a system subject to ideal constraints is zero.

This form of D'Alembert's principle is more useful, because it does not contain the constraint forces. It can be successfully used to obtain the second-order differential equations of motion of the system in configuration space - the celebrated *Lagrange equations of the second kind*, or simply, *Lagrange equations*. These equations write:

$$\frac{d}{dt}\left(\frac{\partial T}{\partial \dot{q}_k}\right) - \frac{\partial T}{\partial q_k} = Q_k \quad (k = \overline{1, n}) \tag{2.23}$$

for systems which do not admit a potential, and

$$\frac{d}{dt}\left(\frac{\partial L}{\partial \dot{q}_k}\right) - \frac{\partial L}{\partial q_k} = 0 \quad (k = \overline{1, n}) \tag{2.24}$$

for natural systems (*i.e.* systems possessing a Lagrangian). Here $T = T(q, \dot{q}, t)$ is the kinetic energy of the system, Q_k - the generalized forces, and $L(q, \dot{q}, t)$ - the Lagrangian of the problem. We shall derive these equations in the next paragraph, by means of an integral principle.

As one can see, solving the dynamical problem in the real space cannot be totally avoided. Nevertheless, transition through a problem of statics (formulated in a non-inertial frame of the real space) to a problem of dynamics (in the configuration space) is convenient, because a problem of dynamics is easier to be solved in configuration space, than in the real space. Among other reasons, we mention the fact that the number of the second order differential equations is smaller by $2l$, where l stands for the number of constraints.

II.2. Hamilton's principle

This principle is of fundamental importance in many areas of physics, not only in analytical mechanics. It is an integral, variational principle. Unlike the differential principles, the integral principles consider evolution of the physical system in a *finite* time interval. These principles operate with *global* variations, either in real or in configuration spaces. It is worth mentioning the strong connection existing between differential and integral principles of analytical mechanics.

In order to facilitate the presentation, we shall begin our investigation in the real space R^3. Consider a system of $N \geq 2$ particles, subject to l holonomic constraints

$$f_k(\vec{r}_1, \vec{r}_2, ..., \vec{r}_N, t) = 0 \quad (k = \overline{1, l}) \qquad (2.25)$$

and suppose that the law of motion of the system

$$\vec{r}_i = \vec{r}_i(t) \quad (i = \overline{1, N}) \qquad (2.26)$$

is known, within the time interval $t_1 \leq t \leq t_2$.

Consider, also, another law of motion of the system

$$\vec{r}_i^{\,*} = \vec{r}_i^{\,*}(t) \quad (i = \overline{1, N}) \qquad (2.27)$$

which satisfies only[1] the equations of constraints (2.25). Making allowance for the terminology used in case of elementary displacements, we call this motion *a possible motion*. Following the general presentation of the variational principles (see paragraph V.1), and adapting the corresponding quantities, we demand that both trajectories given by (2.26) and (2.27) pass through the same two points P_1 and P_2, corresponding to the time moments t_1 and t_2 (see Fig.II.4), that is

$$\vec{r}_i(t_1) = \vec{r}_i^{\,*}(t_1), \text{ and } \vec{r}_i(t_2) = \vec{r}_i^{\,*}(t_2) \quad (i = \overline{1, N}). \qquad (2.28)$$

Let us define the virtual elementary displacement $\delta \vec{r}_i$ as being the variation of the position vector of the particle P_i from one point of the real trajectory (C_{real}) to an infinitely closed point of the possible trajectory $(C_{possible}) \equiv (C^*)$ as (see Fig.II.4)

$$\delta \vec{r}_i = \vec{r}_i(t) - \vec{r}_i^{\,*}(t) \quad (i = \overline{1, N}). \qquad (2.29)$$

In this case, the conditions (2.28) write

$$\delta \vec{r}_i(t_1) = 0, \quad \delta \vec{r}_i(t_2) = 0 \quad (i = \overline{1, N}). \qquad (2.30)$$

---

[1] It satisfies neither the differential equations of motion, nor the initial conditions compatible with the constraints.

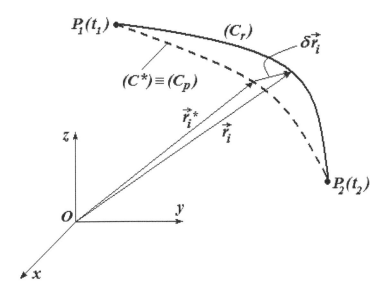

Fig.II.4

To arrive at Hamilton's principle, we shall use D'Alembert's principle in the form given by Lagrange:

$$\sum_{i=1}^{N}(\vec{F}_i - m_i\ddot{\vec{r}}_i) \cdot \delta\vec{r}_i = 0,$$

or

$$\sum_{i=1}^{N} m_i\ddot{\vec{r}}_i \cdot \delta\vec{r}_i = \sum_{i=1}^{N} \vec{F}_i \cdot \delta\vec{r}_i = \delta W.$$

Using some simple mathematical manipulations, we still have:

$$\sum_{i=1}^{N} m_i\ddot{\vec{r}}_i \cdot \delta\vec{r}_i = \frac{d}{dt}\left(\sum_{i=1}^{N} m_i\dot{\vec{r}}_i \cdot \delta\vec{r}_i\right) - \sum_{i=1}^{N} m_i\dot{\vec{r}}_i \cdot \frac{d}{dt}(\delta\vec{r}_i)$$

$$= \frac{d}{dt}\left(\sum_{i=1}^{N} m_i\dot{\vec{r}}_i \cdot \delta\vec{r}_i\right) - \sum_{i=1}^{N} m_i\dot{\vec{r}}_i \cdot \delta\dot{\vec{r}}_i = \sum_{i=1}^{N} \vec{F}_i \cdot \delta\vec{r}_i = \delta W. \quad (2.31)$$

As can be observed, the term $\sum_{i=1}^{N} m_i\dot{\vec{r}}_i \cdot \delta\dot{\vec{r}}_i$ comes from the virtual variation of the kinetic energy $T = \frac{1}{2}\sum_{i=1}^{N} m_i|\dot{\vec{r}}_i|^2$ of the system of particles. Indeed,

$$T^* = \frac{1}{2}\sum_{i=1}^{N} m_i|\dot{\vec{r}}_i^*|^2 = \frac{1}{2}\sum_{i=1}^{N} m_i|\dot{\vec{r}}_i - \delta\dot{\vec{r}}_i|^2$$

$$= \frac{1}{2}\sum_{i=1}^{N} m_i \left(|\dot{\vec{r}}_i|^2 - 2\dot{\vec{r}}_i \cdot \delta\dot{\vec{r}}_i + |\delta\dot{\vec{r}}_i|^2\right) \approx \frac{1}{2}\sum_{i=1}^{N} m_i \left(|\dot{\vec{r}}_i|^2 - 2\dot{\vec{r}}_i \cdot \delta\dot{\vec{r}}_i\right)$$

$$= \frac{1}{2}\sum_{i=1}^{N} m_i |\dot{\vec{r}}_i|^2 - \sum_{i=1}^{N} m_i \dot{\vec{r}}_i \cdot \delta\dot{\vec{r}}_i = T - \sum_{i=1}^{N} m_i \dot{\vec{r}}_i \cdot \delta\dot{\vec{r}}_i,$$

which yields

$$\sum_{i=1}^{N} m_i \dot{\vec{r}}_i \cdot \delta\dot{\vec{r}}_i = T - T^* = \delta T.$$

Introducing this result into (2.31), we can write

$$\frac{d}{dt}\left(\sum_{i=1}^{N} m_i \dot{\vec{r}}_i \cdot \delta\vec{r}_i\right) - \sum_{i=1}^{N} m_i \dot{\vec{r}}_i \cdot \delta\dot{\vec{r}}_i$$

$$= \frac{d}{dt}\left(\sum_{i=1}^{N} m_i \dot{\vec{r}}_i \cdot \delta\vec{r}_i\right) - \delta T = \sum_{i=1}^{N} \vec{F}_i \cdot \delta\vec{r}_i = \delta W,$$

and, therefore

$$\frac{d}{dt}\left(\sum_{i=1}^{N} m_i \dot{\vec{r}}_i \cdot \delta\vec{r}_i\right) = \delta T + \delta W = \delta(T + W). \tag{2.32}$$

Integrating this equation between the time moments t_1 and t_2, and taking into account (2.30), we still have

$$\int_{t_1}^{t_2} \frac{d}{dt}\left(\sum_{i=1}^{N} m_i \dot{\vec{r}}_i \cdot \delta\vec{r}_i\right) dt = \int_{t_1}^{t_2} \delta(T + W)\, dt,$$

or

$$\left(\sum_{i=1}^{N} m_i \dot{\vec{r}}_i \cdot \delta\vec{r}_i\right)\bigg|_{t_1}^{t_2}$$

$$= \sum_{i=1}^{N} m_i \dot{\vec{r}}_i \cdot \delta\vec{r}_i(t_2) - \sum_{i=1}^{N} m_i \dot{\vec{r}}_i \cdot \delta\vec{r}_i(t_1) = 0 = \int_{t_1}^{t_2} \delta(T+W)\, dt,$$

that is

$$\int_{t_1}^{t_2} \delta(T + W)\, dt = 0.$$

But the operators $\int \cdot \, dt$ and $\delta \cdot$ are independent, therefore they commute with each other, so that the previous relation also writes

$$\delta \int_{t_1}^{t_2} (T + W) \, dt = 0, \qquad (2.33)$$

and represents the mathematical expression for the *generalized Hamilton's principle*.

If the applied force is generated by a potential function $V(\vec{r}, t)$, then the potential $V(\vec{r}^*, t)$ corresponding to a possible motion writes

$$V(\vec{r}^*, t) \equiv V(\vec{r}^*_1, \vec{r}^*_2, ..., \vec{r}^*_N, t)$$

$$= V(\vec{r}_1 - \delta \vec{r}_1, \vec{r}_2 - \delta \vec{r}_2, ..., \vec{r}_N - \delta \vec{r}_N, t)$$

$$= V(\vec{r}_1, \vec{r}_2, ..., \vec{r}_N, t) - \sum_{i=1}^{N} (\text{grad}_i V) \cdot \delta \vec{r}_i + \mathcal{O}(|\delta \vec{r}_i|^2)$$

$$\approx V(\vec{r}_1, \vec{r}_2, ..., \vec{r}_N, t) - \sum_{i=1}^{N} (\text{grad}_i V) \cdot \delta \vec{r}_i$$

$$= V(\vec{r}_1, \vec{r}_2, ..., \vec{r}_N, t) + \sum_{i=1}^{N} \vec{F}_i \cdot \delta \vec{r}_i = V(\vec{r}_1, \vec{r}_2, ..., \vec{r}_N, t) + \delta W$$

or

$$V(\vec{r}^*, t) \equiv V^* = V + \delta W,$$

which yields

$$\delta W = -(V - V^*) = -\delta V,$$

and (2.33) leads to

$$\delta \int_{t_1}^{t_2} (T - V) \, dt = \delta \int_{t_1}^{t_2} L \, dt = 0, \qquad (2.34)$$

where

$$L(\vec{r}, \dot{\vec{r}}, t) = T(\vec{r}, \dot{\vec{r}}, t) - V(\vec{r}, t)$$

is the *Lagrangian function* expressed in the real space.

Relation (2.34) expresses *Hamilton's principle* in real space. Since the quantities $\vec{r}_i$ and $\dot{\vec{r}}_i$ are not independent, but must satisfy the same constraints (2.25), we have to resume this reasoning in configuration space R^n, where both generalized coordinates q_j and generalized velocities $\dot{q}_j$ are independent.

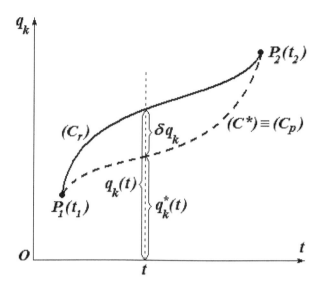

Fig.II.5

Following a similar procedure, but this time in the configuration space, one observes that the analogues of (2.26) and (2.27) are

$$q_k = q_k(t) \quad (k = \overline{1,n}) \qquad (2.26')$$

and

$$q_k^* = q_k^*(t) \quad (k = \overline{1,n}), \qquad (2.27')$$

respectively, conditions (2.30) correspond to

$$\delta q_k(t_1) = 0, \quad \delta q_k(t_2) = 0 \quad (k = \overline{1,n}), \qquad (2.30')$$

while the relations similar to (2.29) are

$$\delta q_k = q_k(t) - q_k^*(t) \quad (k = \overline{1,n}). \qquad (2.29')$$

Since the generalized coordinates q_k are independent, except for conditions (2.29'), the virtual variations δq_k ($k = \overline{1,n}$) are also independent. Furthermore, according to the definition, these variations are instantaneous. A formal graphical representation of the set of q_k coordinates as functions of time is given in Fig.II.5.

Using these observations, relation (2.34) shall be written in configuration space as

$$\delta \int_{t_1}^{t_2} L \, dt = \delta \int_{t_1}^{t_2} L(q, \dot{q}, t) \, dt = 0. \qquad (2.35)$$

Let us denote
$$S \equiv S(q) = \int_{t_1}^{t_2} L(q, \dot{q}, t) \, dt \tag{2.36}$$

and call this integral *action*. With this notation, (2.35) finally writes

$$\delta S = 0. \tag{2.37}$$

This simple relation comprises the essence of analytical mechanics. It expresses Hamilton's principle for natural systems subject to holonomic constraints: *Among all possible generalized trajectories passing through two fixed points in configuration space, corresponding to two moments of time t_1 and t_2, the trajectory associated with the real motion corresponds to a stationary action.* Since in most cases the extremum of the action integral is a minimum, Hamilton's principle is also called *principle of minimum action*. It was published in 1834 by *Willian Rowan Hamilton*.

Let us now prove that all the important results of analytical mechanics can be obtained by means of Hamilton's principle.

1. Lagrange equations of the second kind, written for both cases of simple and generalized potentials, on the one hand, and non-potential forces, on the other.

i) To deduce Lagrange equations of the second kind for natural systems that admit a simple potential function $V = V(q, t)$ we use Hamilton's principle (2.37)

$$\delta S = \delta \int_{t_1}^{t_2} L(q, \dot{q}, t) \, dt = 0,$$

written for virtual and independent[1] variations δq of the generalized coordinates. We therefore have

$$\delta S = \delta \int_{t_1}^{t_2} L(q, \dot{q}, t) \, dt = \int_{t_1}^{t_2} \delta L(q, \dot{q}, t) \, dt$$

$$= \int_{t_1}^{t_2} \left(\frac{\partial L}{\partial q_k} \delta q_k + \frac{\partial L}{\partial \dot{q}_k} \delta \dot{q}_k \right) dt$$

$$= \int_{t_1}^{t_2} \frac{\partial L}{\partial q_k} \delta q_k \, dt + \int_{t_1}^{t_2} \frac{d}{dt} \left(\frac{\partial L}{\partial \dot{q}_k} \delta q_k \right) dt - \int_{t_1}^{t_2} \frac{d}{dt} \left(\frac{\partial L}{\partial \dot{q}_k} \right) \delta q_k \, dt$$

---

[1] except for the moments of time t_1 and t_2, when $\delta q_k(t_1) = 0$ and $\delta q_k(t_2) = 0$.

$$= \int_{t_1}^{t_2} \frac{\partial L}{\partial q_k} \delta q_k \, dt + \left(\frac{\partial L}{\partial \dot{q}_k} \delta q_k \right) \bigg|_{t_1}^{t_2} - \int_{t_1}^{t_2} \frac{d}{dt} \left(\frac{\partial L}{\partial \dot{q}_k} \right) \delta q_k \, dt$$

$$= \int_{t_1}^{t_2} \frac{\partial L}{\partial q_k} \delta q_k \, dt - \int_{t_1}^{t_2} \frac{d}{dt} \left(\frac{\partial L}{\partial \dot{q}_k} \right) \delta q_k \, dt$$

$$= - \int_{t_1}^{t_2} \left[\frac{d}{dt} \left(\frac{\partial L}{\partial \dot{q}_k} \right) - \frac{\partial L}{\partial q_k} \right] \delta q_k \, dt = 0,$$

where conditions (2.30') have been considered. Since the virtual variations δq_k are arbitrary, the last relation yields the Lagrange equations

$$\frac{d}{dt} \left(\frac{\partial L}{\partial \dot{q}_k} \right) - \frac{\partial L}{\partial q_k} \quad (k = \overline{1,n}). \tag{2.38}$$

ii) If our natural system admits a generalized potential $V = V(q, \dot{q}, t)$, we have to show that Hamilton's principle (2.33) is equivalent to (2.37), on condition that the generalized forces Q_k are written as

$$Q_k = \frac{d}{dt} \left(\frac{\partial V}{\partial \dot{q}_k} \right) - \frac{\partial V}{\partial q_k} \quad (k = \overline{1,n}).$$

Indeed, we have:

$$0 = \delta \int_{t_1}^{t_2} (T + W) \, dt = \int_{t_1}^{t_2} (\delta T + \delta W) \, dt = \int_{t_1}^{t_2} \delta T \, dt + \int_{t_1}^{t_2} Q_k \delta q_k \, dt$$

$$= \int_{t_1}^{t_2} \delta T \, dt + \int_{t_1}^{t_2} \left[\frac{d}{dt} \left(\frac{\partial V}{\partial \dot{q}_k} \right) - \frac{\partial V}{\partial q_k} \right] \delta q_k \, dt$$

$$= \int_{t_1}^{t_2} \delta T \, dt + \int_{t_1}^{t_2} \frac{d}{dt} \left(\frac{\partial V}{\partial \dot{q}_k} \delta q_k \right) dt - \int_{t_1}^{t_2} \frac{\partial V}{\partial \dot{q}_k} \delta \dot{q}_k \, dt - \int_{t_1}^{t_2} \frac{\partial V}{\partial q_k} \delta q_k \, dt$$

$$= \int_{t_1}^{t_2} \delta T \, dt + \left(\frac{\partial V}{\partial \dot{q}_k} \delta q_k \right) \bigg|_{t_1}^{t_2} - \int_{t_1}^{t_2} \left(\frac{\partial V}{\partial \dot{q}_k} \delta \dot{q}_k + \frac{\partial V}{\partial q_k} \delta q_k \right) dt$$

$$= \int_{t_1}^{t_2} \delta T \, dt - \int_{t_1}^{t_2} \delta V \, dt = \int_{t_1}^{t_2} \delta(T - V) \, dt = \delta \int_{t_1}^{t_2} L \, dt = \delta S = 0,$$

which completes the proof.

iii) To deduce Lagrange equations of the second kind for mechanical systems acted on by non-potential forces, we start with the form (2.33) of Hamilton's principle:

$$0 = \delta \int_{t_1}^{t_2} (T + W) \, dt = \int_{t_1}^{t_2} \delta T \, dt + \int_{t_1}^{t_2} \delta W \, dt$$

$$= \int_{t_1}^{t_2} \left(\frac{\partial T}{\partial q_k} \delta q_k + \frac{\partial T}{\partial \dot{q}_k} \delta \dot{q}_k \right) dt + \int_{t_1}^{t_2} Q_k \delta q_k \, dt$$

$$= \int_{t_1}^{t_2} \frac{\partial T}{\partial q_k} \delta q_k \, dt + \int_{t_1}^{t_2} \frac{d}{dt} \left(\frac{\partial T}{\partial \dot{q}_k} \delta q_k \right) dt$$

$$- \int_{t_1}^{t_2} \frac{d}{dt} \left(\frac{\partial T}{\partial \dot{q}_k} \right) \delta q_k \, dt + \int_{t_1}^{t_2} Q_k \delta q_k \, dt$$

$$= \int_{t_1}^{t_2} \frac{\partial T}{\partial q_k} \delta q_k \, dt + \left(\frac{\partial T}{\partial \dot{q}_k} \delta q_k \right) \bigg|_{t_1}^{t_2} - \int_{t_1}^{t_2} \frac{d}{dt} \left(\frac{\partial T}{\partial \dot{q}_k} \right) \delta q_k \, dt + \int_{t_1}^{t_2} Q_k \delta q_k \, dt$$

$$= - \int_{t_1}^{t_2} \left[\frac{d}{dt} \left(\frac{\partial T}{\partial \dot{q}_k} \right) - \frac{\partial T}{\partial q_k} - Q_k \right] \delta q_k \, dt \quad (k = \overline{1, n}),$$

where conditions (2.30') have been taken into account. Since the virtual variations δq_k are arbitrary [except for (2.30')], we are left with

$$\frac{d}{dt} \left(\frac{\partial T}{\partial \dot{q}_k} \right) - \frac{\partial T}{\partial q_k} = Q_k \quad (k = \overline{1, n}). \tag{2.39}$$

As one can see, by means of relation (2.33) both Lagrange equations for natural systems subject to generalized-potential and non-potential forces can be obtained. For this reason, (2.33) expresses - as we previously asserted - *the generalized Hamilton's principle*.

2. Let us define the *Hamiltonian function*, usually known as *Hamiltonian*, $H(q, p, t)$, as

$$H(q, p, t) = p_k \dot{q}_k - L(q, \dot{q}, t).$$

Using Hamilton's principle (2.37) and the Hamiltonian, one can obtain the system of *Hamilton's canonical equations*, which are the "heart" of the Hamiltonian formalism. We have:

$$0 = \delta S = \delta \int_{t_1}^{t_2} L \, dt = \delta \int_{t_1}^{t_2} (p_k \dot{q}_k - H) \, dt$$

$$= \int_{t_1}^{t_2} (p_k \delta \dot{q}_k + \dot{q}_k \delta p_k) \, dt - \int_{t_1}^{t_2} \delta H \, dt$$

$$= \int_{t_1}^{t_2} p_k \delta \dot{q}_k \, dt + \int_{t_1}^{t_2} \dot{q}_k \delta p_k \, dt - \int_{t_1}^{t_2} \left(\frac{\partial H}{\partial q_k} \delta q_k + \frac{\partial H}{\partial p_k} \delta p_k \right) dt$$

$$= \int_{t_1}^{t_2} \frac{d}{dt}(p_k \delta q_k) \, dt - \int_{t_1}^{t_2} \dot{p}_k \delta q_k \, dt + \int_{t_1}^{t_2} \dot{q}_k \delta p_k \, dt$$

$$- \int_{t_1}^{t_2} \left(\frac{\partial H}{\partial q_k} \delta q_k + \frac{\partial H}{\partial p_k} \delta p_k \right) dt$$

$$= (p_k \delta q_k) \Big|_{t_1}^{t_2} - \int_{t_1}^{t_2} \dot{p}_k \delta q_k \, dt + \int_{t_1}^{t_2} \dot{q}_k \delta p_k \, dt$$

$$- \int_{t_1}^{t_2} \left(\frac{\partial H}{\partial q_k} \delta q_k + \frac{\partial H}{\partial p_k} \delta p_k \right) dt$$

$$= \int_{t_1}^{t_2} \left[-\left(\dot{p}_k + \frac{\partial H}{\partial q_k} \right) \delta q_k + \left(\dot{q}_k - \frac{\partial H}{\partial p_k} \right) \delta p_k \right] dt.$$

Since the virtual variations δq_k and δp_k are arbitrary [except for (2.30')], the last relation yields *Hamilton's canonical equations*

$$\begin{cases} \dot{q}_k = \dfrac{\partial H}{\partial p_k}; \\ \dot{p}_k = -\dfrac{\partial H}{\partial q_k}. \end{cases} \quad (k = \overline{1,n}) \qquad (2.40)$$

Observations

1. Hamilton's principle is *a variational principle*. Indeed, comparing our investigation with the results displayed in §1, Chap.V, concerning functionals and their extrema, the following correspondence can be observed:

Independent variable x	$\longleftrightarrow$	Independent variable t
Dependent variable (variational parameter) $y = y(x)$	$\longleftrightarrow$	Dependent variable (variational parameter) $q = q(t)$
Function of class C^2 $f = f(x, y, y')$	$\longleftrightarrow$	Lagrange function $L = L(t, q, \dot{q})$
Functional $J = \int_{x_1}^{x_2} f(x, y, y') \, dx$	$\longleftrightarrow$	Action functional $S(q) = \int_{t_1}^{t_2} L(t, q, \dot{q}) \, dt$
Euler-Lagrange equations $\frac{d}{dx}\left(\frac{\partial f}{\partial y_i'}\right) - \frac{\partial f}{\partial y_i} = 0$ $(i = \overline{1,n})$	$\longleftrightarrow$	Lagrange equations of the 2nd kind $\frac{d}{dt}\left(\frac{\partial L}{\partial \dot{q}_k}\right) - \frac{\partial L}{\partial q_k} = 0$ $(k = \overline{1,n})$

As we have seen in the paragraph dedicated to variational calculus, the solutions of the Euler-Lagrange equations are those "points" in which the basic functional has an extremum (minimum or maximum). These points are called *extremals*. In this regard, solutions of the Lagrange equations of the 2nd kind are extremals of the action functional $S = \int_{t_1}^{t_2} L \, dt$. In the light of the above considerations, Hamilton's principle can be enounced as follows: *The real motion of a physical system is described by the extremal of the action integral.* In other words, among all functions $y_i = y_i(x)$ of class C^2 defined on the interval $x \in [x_1, x_2]$, and satisfying conditions $y_i(x_1) = a$, $y_i(x_2) = b$ $(i = 1, 2, ...)$, where a and b are two real constants, the real motion of the system corresponds to that function which extremizes the action integral.

2. Since the quantities used in this paragraph can be defined in any reference frame, it follows that Hamilton's principle does not depend on the choice of coordinates.

3. A single scalar function, the Lagrangian L, contains the whole information about the studied physical system. Once we know the Lagrangian, we can easily determine the differential equations of motion, and the associated conservation laws as well.

4. Hamilton's principle can be also used to describe in a unitary manner some other systems, such as physical fields. This property is due to the fact that the terms contained by the Lagrangian have the units of *energy*. This quantity can be defined in *any* physical system, while not all motions/interactions can be described by means of the concept of force. This way are obtained, for example, the fundamental equations of: electrodynamics (Maxwell's equations), theory of elasticity (Lamé's equations), non-relativistic quantum mechanics (Schrödinger's equation), etc.

5. Hamilton's variational principle of stationary action is widely applied in various fields of science, being one of the most general principles of nature. In order to construct the Lagrangian of a certain physical system, one must consider the following criteria (principles):

- *Superposition principle*. If the system is composed by at least two particles, then the Lagrangian must contain three groups of terms: (i)terms describing each particle of the system, supposed to be alone; (ii)terms describing interaction between any particle and the rest of particles of the system; (iii)terms describing interaction between each particle with the external force fields (if they exist).

- *Invariance principle*. The action has to be invariant with respect to the symmetry group characterizing the mechanical system: the Galilei-Newton group in case of Newtonian mechanics, the Lorentz-

Einstein group in case of relativistic mechanics, etc.

- *Correspondence principle.* All results obtained for a particular domain included in a more general one should be achieved by means of Hamilton's principle applied to the more general domain.

- *Symmetry principle.* The Lagrangian has to be not only as simple as possible, but also constructed in a way leading to differential equations of motion displaying the symmetry properties of the physical system. This can be done by a suitable choice of the generalized coordinates.

CHAPTER III

THE SIMPLE PENDULUM PROBLEM

The aim of this section is to solve the fundamental problem of mechanics for a constrained system, namely the simple pendulum, by means of six different methods connected to: (i) classical (Newtonian) approach; (ii) Lagrange equations of the first kind; (iii) Lagrange equations of the second kind; (iv) Hamilton's canonical equations; (v) the Hamilton-Jacobi formalism, and (vi) the action-angle formalism. This way, we shall put into evidence the resemblances and differences between these formalisms, on the one side, and show the generality and potency of the analytical formalism, as compared to the classical one, on the other.

The physical system we are going to study is represented by a material point (particle), suspended by a massless rod, constrained to move without friction on a circle in a vertical plane, under the influence of a uniform and homogeneous gravitational field. It is our purpose to study the motion of such a system only in case of free, harmonic, and non-amortized oscillations, performing motions of an arbitrary amplitude. The study of the most general case (non-linear damped and - eventually - forced oscillations) implies knowledge of notions like: bifurcation points, strange attractors, Lyapunov coefficients, etc., which overpass our approach. Our discussion stands for an application of the various methods offered by analytical mechanics which can be used for solving a problem.

III.1. Classical (Newtonian) formalism

According to this approach, the known elements are: the mass of the body, the acting forces (including the constraint forces), and the initial conditions compatible with the constraints. The reader is asked to find the equation of motion and the elements/characteristics of the motion: trajectory, period, frequency, etc. As well-known, determination of the solution is based on the second Newtonian low (*lex secunda*), which furnishes the differential equation of motion. As initial conditions are usually given the position and velocity of the body at some initial moment. Solution of the differential equation of motion is *the low of motion*, commonly written as $\vec{r} = \vec{r}(t)$.

One of the most important differences between the classical (Newtonian) and analytical formalisms is connected to the constraints. The classical approach demands knowledge of the constrained forces (at least their number and orientation), while the analytical procedure allows one to determine these forces at the end of calculation.

Let us denote by $\vec{G}$ the force of gravity, *i.e.* the *applied force*, and by $\vec{T}$ the constrained force. It is convenient to choose the reference system with its origin O at the point of suspension, while the Ox and Oy axes are oriented as shown in Fig.III.1.

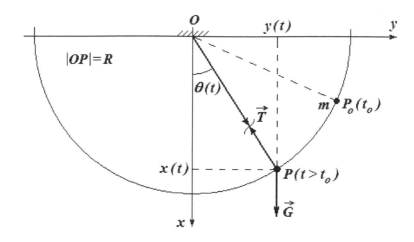

Fig.III.1

Suppose that, at the moment $t = 0$, the body is at the point $P_0(x_0, y_0)$ at rest ($v_{0x} = 0, v_{0y} = 0$). At the moment t, the angle between the rod and Ox is $\theta(t)$. Projecting the equation of motion $m\ddot{\vec{r}} = \vec{F}$ on axes, one obtains

$$\begin{cases} Ox: & m\ddot{x} = mg - T\cos\theta\ ; \\ Oy: & m\ddot{y} = -T\sin\theta, \end{cases}$$

or

$$\begin{cases} \ddot{x} = g - \frac{T}{m}\cos\theta\ ; \\ \ddot{y} = -\frac{T}{m}\sin\theta. \end{cases} \qquad (3.1)$$

This is a system of two differential equations with four unknowns: x, y, θ, T. To solve it, we use the parametric equations of trajectory of the body of mass m : $x = R\cos\theta$; $y = R\sin\theta$, straightforwardly leading to
$$\begin{cases} \ddot{x} = -R\,\ddot{\theta}\sin\theta - R\,\dot{\theta}^2\cos\theta \; ; \\ \ddot{y} = R\,\ddot{\theta}\cos\theta - R\,\dot{\theta}^2\sin\theta, \end{cases}$$
and equations (3.1) become
$$\begin{cases} -R\,\ddot{\theta}\tan\theta - R\,\dot{\theta}^2 = \frac{g}{\cos\theta} - \frac{T}{m}, \\ R\,\ddot{\theta} - R\,\dot{\theta}^2\tan\theta = -\frac{T}{m}\tan\theta. \end{cases} \qquad (3.2)$$

This way, we are left with two equations with two unknown quantities. By eliminating T, we have
$$\ddot{\theta} + \frac{g}{R}\sin\theta = 0, \qquad (3.3)$$
which is the differential equation of motion. If $\theta_0(\theta_{t_0=0}) < 4°$, one can approximate $\sin\theta \approx \theta$ and equation (3.3) goes to the well-known *linear harmonic oscillator* equation
$$\ddot{\theta} + \frac{g}{R}\theta = 0. \qquad (3.4)$$

Our result shows that, in this particular case, the body of mass m performs free, non-amortized, isochronous oscillations. Solution of the equation (3.4) can be written at least in four ways:
$$\theta = A_1 e^{i\omega_0 t} + A_2 e^{-i\omega_0 t} \; ; \qquad (3.5\ a)$$
$$\theta = B_1 \sin\omega_0 t + B_2 \cos\omega_0 t \; ; \qquad (3.5\ b)$$
$$\theta = C_1 \sin(\omega_0 t + C_2) \; ; \qquad (3.5\ c)$$
$$\theta = D_1 \cos(\omega_0 t + D_2) \, , \qquad (3.5\ d)$$
where $\omega_0^2 = g/R$, while the arbitrary constants $A_1, A_2, B_1, B_2, C_1, C_2, D_1, D_2$, are determined by means of initial conditions. For example, solution (3.5 c) leads to: $\theta_0 = \theta_{t=0} = C_1 \sin C_2$, $\dot{\theta}_0 = \omega_0 C_1 \cos C_2 = 0$. Since C_1 cannot vanish, this gives $C_2 = \pi/2$, $C_1 = \theta_0$, and the final solution reads
$$\theta(t) = \theta_0 \sin\left(\omega_0 t + \frac{\pi}{2}\right).$$

The period of the harmonic pendulum is therefore
$$T_0 = \frac{2\pi}{\omega_0} = 2\pi\sqrt{\frac{R}{g}}, \qquad (3.6)$$

and depends only on the length R of the rod. In short, in the case of small oscillations ($\theta_0 < 4°$), the period τ_0 does not depend on the amplitude. Such motions are called *isochronous* or *tautochronous*. (In Greek, *tauto* means *the same*, and *chronos* means *time*).

Let us now consider an arbitrary value of the angle θ_0. At the initial moment $t_0 = 0$, the angle between the rod and the vertical is θ_0, while the point is at rest ($\dot{\theta}_0 = \dot{\theta}(0) = 0$) at the point $P_0(t_0)$ (see Fig.III.1). As a result, from the mathematical point of view, we have to solve the Cauchy's problem for equation (3.4), with the initial conditions

$$\begin{cases} \theta_{t=0} = \theta_0 = const. \\ \dot{\theta}_{t=0} = \dot{\theta}_0 = 0. \end{cases} \quad (3.7)$$

If θ_0 has an arbitrary value $\theta_0 \in (0, \pi/2)$, integration of the differential equation of motion is more complicated. In addition, we have to be sure that the motion remains periodical. To this end, we shall go from the second derivative with respect to time to the first derivative with respect to θ, that is

$$\ddot{\theta} \equiv \frac{d^2\theta}{dt^2} = \frac{d}{dt}\left(\frac{d\theta}{dt}\right) = \frac{d}{d\theta}\left(\frac{d\theta}{dt}\right)\frac{d\theta}{dt} = \frac{d\dot{\theta}}{d\theta}\dot{\theta} = \frac{1}{2}\frac{d\dot{\theta}^2}{d\theta}.$$

Introducing now $\ddot{\theta}$ into (3.3), separating variables and integrating, we can write

$$\int_{\dot{\theta}(0)=0}^{\dot{\theta}(t)} d\dot{\theta}^2 = -\frac{2g}{R}\int_{\theta(0)=\theta_0}^{\theta(t)} \sin\theta \, d\theta,$$

or, by integration

$$\dot{\theta}^2 = \frac{2g}{R}(\cos\theta - \cos\theta_0), \quad (3.8)$$

that is

$$\dot{\theta} = \pm\sqrt{\frac{2g}{R}(\cos\theta - \cos\theta_0)}. \quad (3.9)$$

Since the angular velocity $\dot{\theta}$ has to be real, we must have $(\cos\theta - \cos\theta_0) \geq 0$. which means $|\theta| \leq \theta_0$. Consequently, there are two *turning points*

$$\theta_1 = -\theta_0, \qquad \theta_2 = +\theta_0.$$

As a result, the "plus" sign in (3.9) corresponds to the intervals where the angular velocity $\dot{\theta}$ is positive: $E \to D \to C$ [or, equivalently, $\theta \in (-\theta_0, 0)$], and $C \to B \to A$ [or, $\theta \in (0, \theta_0)$] (see Fig.III.2, Fig.III.3, and Table III.1), while the "minus" sign is associated with the intervals with negative angular velocity: $A \to B \to C$ [or, equivalently, $\theta \in (\theta_0, 0)$], and $C \to D \to E$ [or, $\theta \in (0, -\theta_0)$].

Next, let us analyze the manner of variation of the quantities $\theta = \theta(t)$ and $\dot\theta = \dot\theta(t)$. To this end, we shall consider the vertical axis Ox as being a zero-axis for the values of θ: the angles determined on the right hand side of the circle are positive, while the left-oriented angles are negative. We shall start our analysis from the initial point, characterized by $\theta_i = \theta_0, \dot\theta_i = \dot\theta_0 = 0$. As one can see, it is useful to denote the starting point (position) by A, the point corresponding to $\theta = 0$ by C, and so on (see Fig.III.2).

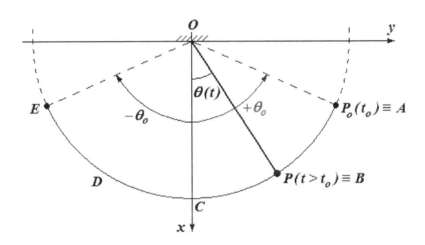

Fig.III.2

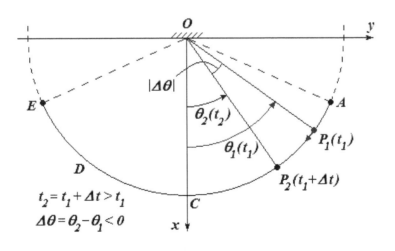

Fig.III.3

Inside the interval $A \to B \to C$ the angle θ decreases by positive values until zero, while the angular velocity $\dot{\theta}$ decreases from zero (at A) to its minimum value $\dot{\theta}_{min}$, by negative values. This behavior can be easily explained by making use of the definition of the derivative at a point, and Fig.III.3 as well. Indeed,

$$\dot{\theta} = \frac{d\theta}{dt} = \lim_{\Delta t \to 0} \frac{\Delta\theta}{\Delta t} = \lim_{t_2 \to t_1} \frac{\theta_2 - \theta_1}{t_2 - t_1} < 0.$$

Within the interval $C \to D \to E$ the angle θ decreases by negative values up to its minimum value $-\theta_0$. The body cannot pass over this point, going to greater negative values of θ, because $\theta = -\theta_0$ is a turning point, while $\dot{\theta}$ increases, also by negative values, up to value zero.

On the interval $E \to D \to C$ angle θ increases by negative values up to value zero, while $\dot{\theta}$ increases, this time by positive values, up to its maximum value, corresponding to point C. Within the next interval $C \to B \to A$ angle θ increases by positive values up to its maximum (initial) value θ_0 and cannot pass over this point because $\theta = \theta_0$ is a turning point. Angular velocity $\dot{\theta}$ decreases in this interval from its maximum value up to zero, by positive values. From this moment on, the motion resumes, meaning that it remains periodical.

Table III.1

Point →	A		C		E		C		A
$\theta(t)$	$+\theta_0$	↘(+)	0	↘(-)	$-\theta_0$	↗(-)	0	↗(+)	$+\theta_0$
$\dot{\theta}(t)$	0	↘(-)	$\dot{\theta}_{min}$	↗(-)	0	↗(+)	$\dot{\theta}_{max}$	↘(+)	0
E_c	0	↗	$E_{c\,max}$	↘	0	↗	$E_{c\,max}$	↘	0
E_p	$E_{p\,max}$	↘	0	↗	$E_{p\,max}$	↘	0	↗	$E_{p\,max}$

This analysis can be completed by the energetic approach. As reference level for the potential gravitational energy can be considered the plane orthogonal to Ox axis, passing through point C. Since the constraint is ideal (there are no energy losses), the maximum potential energy at point A is "spent" on the account of mechanical work performed by the gravitational field. This leads to an increase of the kinetic energy of the body of mass m (which is zero at point A) up to its maximum value in the point C, where the potential energy is zero. This maximum value is numerically equal to the mechanical work performed by the gravitational field, and it is also equal to the initial potential energy at point A. Since the gravitational field is conservative, the total energy conserves. There exists a continuous "transformation" of energy, the kinetic and potential energies successively passing through their maximum and minimum (zero) values, alternatively attained at points A, C, E. A synthesis of this analysis is shown in Table III.1.

Let us now turn back to equation (3.9). Here the variables can easily be separated

$$\dot{\theta} = \frac{d\theta}{dt} = \pm\sqrt{\frac{2g}{R}(\cos\theta - \cos\theta_0)} \Rightarrow dt = \pm\frac{d\theta}{\sqrt{\frac{2g}{R}(\cos\theta - \cos\theta_0)}},$$

which gives

$$\int_{t_0}^{t} dt = \pm \int_{\theta_0}^{\theta} \frac{d\theta}{\sqrt{\frac{2g}{R}(\cos\theta - \cos\theta_0)}}. \qquad (3.10)$$

It is more convenient to replace θ by a new variable ψ, defined as

$$\sin\frac{\theta}{2} = \sin\frac{\theta_0}{2}\sin\psi \equiv k\sin\psi,$$

meaning

$$d\theta = \frac{2k\cos\psi\, d\psi}{\cos(\theta/2)}.$$

The quantity $(\cos\theta - \cos\theta_0)$ appearing in (3.10) then writes

$$(\cos\theta - \cos\theta_0) = -(1-\cos\theta) + (1-\cos\theta_0) = -2\sin^2\frac{\theta}{2} + 2\sin^2\frac{\theta_0}{2}$$

$$= -2k^2\sin^2\psi + 2k^2 = 2k^2\cos^2\psi.$$

Taking as initial moment $t_0 = 0$, we then have:

$$t = -\sqrt{\frac{R}{2g}}\int_{\theta_0}^{\theta} \frac{d\theta}{\sqrt{(\cos\theta - \cos\theta_0)}}$$

$$= -\sqrt{\frac{R}{2g}}\int_{\pi/2}^{\arcsin\left(\frac{1}{k}\sin\frac{\theta}{2}\right)} \frac{2k\cos\psi\, d\psi}{\cos\frac{\theta}{2}\sqrt{2}k\cos\psi}$$

$$= -\sqrt{\frac{R}{g}}\int_{\pi/2}^{\arcsin\left(\frac{1}{k}\sin\frac{\theta}{2}\right)} \frac{d\psi}{\sqrt{1-\sin^2\frac{\theta}{2}}}$$

$$= -\sqrt{\frac{R}{g}} \int_{\pi/2}^{\arcsin\left(\frac{1}{k}\sin\frac{\theta}{2}\right)} \frac{d\psi}{\sqrt{1-k^2\sin^2\psi}}$$

$$= \sqrt{\frac{R}{g}} \left[\int_0^{\pi/2} \frac{d\psi}{\sqrt{1-k^2\sin^2\psi}} - \int_0^{\arcsin\left(\frac{1}{k}\sin\frac{\theta}{2}\right)} \frac{d\psi}{\sqrt{1-k^2\sin^2\psi}} \right]$$

$$= \sqrt{\frac{R}{g}} \left[F\left(\frac{\pi}{2}, k\right) - F\left(\arcsin\left(\frac{1}{k}\sin\frac{\theta}{2}\right), k\right) \right]$$

$$= \sqrt{\frac{R}{g}} \left[K(k) - F\left(\arcsin\left(\frac{1}{k}\sin\frac{\theta}{2}\right), k\right) \right]. \tag{3.11}$$

Here we took into account that, since $k = \sin\frac{\theta_0}{2} > 0$, for $\theta \in (-\theta_0, \theta_0)$, we have $|k\cos\psi| = k\cos\psi$. We also considered that at the initial moment the pendulum departs from A, and chose the minus sign in (3.10). In (3.11) have been used the following notations:

$$F(\psi_0, k) = \int_0^{\psi_0} \frac{d\psi}{\sqrt{1-k^2\sin^2\psi}} \tag{3.12}$$

for the *elliptic integral of first species*, of amplitude ψ_0 and modulus k, and

$$K(k) = F\left(\frac{\pi}{2}, k\right) = \int_0^{\pi/2} \frac{d\psi}{\sqrt{1-k^2\sin^2\psi}} \tag{3.13}$$

for the *complete elliptic integral of the first species*. In general, elliptic integrals cannot be expressed in terms of elementary functions, and are exposed in special tables.

Denote $g/R \equiv \omega_0^2$ and observe that for a given θ_0 we have $K(k) \equiv$ the initial phase $\varphi_0 =$ const. Formula (3.11) then shows that the "phase" of the periodical motion of pendulum is

$$\varphi(t) = \omega_0 t - K(k) = -F\left(\arcsin\left(\frac{1}{k}\sin\frac{\theta}{2}\right), k\right), \tag{3.14}$$

which is the implicit form of the equation of motion of pendulum. This equation can be solved in terms of θ by the help of special functions $JA(u, m)$ and $JSN(u, m)$, called *amplitude of the Jacobi elliptic functions*, and *Jacoby elliptic function "sn"*, respectively, defined as:

(1) Amplitude of the Jacobi elliptic functions is "inversion" of the first kind elliptic integral, that is, if $u = F(\phi, m)$, then $\phi = am(u, m)$.

In other words, the special function "amplitude of the Jacobi elliptic functions" "extracts" the amplitude ϕ of a first kind elliptic integral.

(2) The Jacobi elliptic function "sn" expresses the sinus of the amplitude of a first species elliptic integral: if ϕ is the amplitude, then $sn(u, m) = \sin \phi$.

As known, the rational functions involving square roots of some quadratic forms can be integrated by the help of inverse trigonometric functions. As a result, the trigonometric functions can be defined as being inversions of the functions given by these integrals.

By analogy, the *elliptic functions* are defined as being the inverted functions of those given by the elliptic integrals. An example has already been given above at point (1), by introducing the special function $JA(u, m)$, called *amplitude of the Jacobi elliptic functions*. Here the argument m is often omitted, so that $am(u, m)$ gets a simpler form: $am(u)$.

The Jacobi elliptic functions $JSN(u, m)$ and $JCN(u, m)$ are given by the relations $sn(u) = \sin(\phi)$ and $cn(u) = \cos(\phi)$, respectively, where $\phi = am(u, m)$. In its turn, the function $JDN(u, m)$ is defined as $dn(u) = \sqrt{1 - m \sin^2 \phi} = \Delta(\phi)$.

The total number of existing Jacobi elliptic functions is twelve: $JPQ(u, m)$, where P and Q belong to the set of four letters S, C, D, N. Each Jacobi elliptic function $JPQ(u, m)$ satisfies the relation $pq(u) = \frac{pn(u)}{qn(u)}$, with $nn(u) = 1$.

There are several relations between the Jacobi elliptic functions, similar to those between the ordinary trigonometric functions. As a matter of fact, in the limit, the Jacobi elliptic functions turn into the usual trigonometric functions. Here are some examples: $sn(u, 0)$ = sin(u), $sn(u, 1)$=tanh(u), $cn(u, 0)$ = cos(u), $cn(u, 1)$ = sech(u), $dn(u, 0) = 1$, $dn(u, 1)$= sech(u), etc.

One of the most important properties of the elliptic functions is the fact that they are *double-periodic* in their complex arguments. The ordinary trigonometric functions are simple-periodic, that is they satisfy the relation $f(z+s\omega) = f(z)$, for any integer s. Double periodicity is expressed by $f(z + r\omega + s\omega') = f(z)$, for any integers r and s. The Jacobi elliptic functions $sn(u, m), cn(u, m)$, etc. are double-periodic in the complex plane of variable u. Their periods are determined by $\omega = 4K(m)$ and $\omega' = 4iK(1-m)$, where the complete elliptic integral of the first species K is given by (3.13).

Taking into account these definitions, equation (3.14) can be written as

$$F\left(\arcsin\left(\frac{1}{k}\sin\frac{\theta}{2}\right), k\right) = K(k) - \omega_0 t, \qquad (3.15)$$

which yields

$$JA\Big(K(k) - \omega_0 t, k\Big) = \arcsin\Big(\frac{1}{k}\sin\frac{\theta}{2}\Big)$$

as well as

$$JSN\Big(K(k) - \omega_0 t, k\Big) = \sin\Big[\arcsin\Big(\frac{1}{k}\sin\frac{\theta}{2}\Big)\Big] = \frac{1}{k}\sin\frac{\theta}{2},$$

therefore

$$\sin\frac{\theta}{2} = k\ JSN\Big(K(k) - \omega_0 t, k\Big),$$

and, finally,

$$\theta \equiv \theta(t) = 2\ \arcsin\Big[k\ JSN\Big(K(k) - \omega_0 t, k\Big)\Big], \qquad (3.16)$$

which is the explicit form of the equation of motion of the simple pendulum.

Let us now show that, if we choose the "minus" sign in (3.10), meaning that the starting point is E (see Fig.III.2), the result is the same. Indeed, we can write:

$$t = \sqrt{\frac{R}{2g}} \int_{-\theta_0}^{-\theta} \frac{d\theta}{\sqrt{(\cos\theta - \cos\theta_0)}}$$

$$= \sqrt{\frac{R}{2g}} \int_{-\pi/2}^{-\arcsin\left(\frac{1}{k}\sin\frac{\theta}{2}\right)} \frac{2k\cos\psi\ d\psi}{\cos\frac{\theta}{2}\sqrt{2}k\cos\psi}$$

$$= \sqrt{\frac{R}{g}} \int_{-\pi/2}^{-\arcsin\left(\frac{1}{k}\sin\frac{\theta}{2}\right)} \frac{d\psi}{\sqrt{1 - \sin^2\frac{\theta}{2}}}$$

$$= \sqrt{\frac{R}{g}} \int_{-\pi/2}^{-\arcsin\left(\frac{1}{k}\sin\frac{\theta}{2}\right)} \frac{d\psi}{\sqrt{1 - k^2\sin^2\psi}}$$

$$= \sqrt{\frac{R}{g}}\Big[\int_{-\pi/2}^{0} \frac{d\psi}{\sqrt{1 - k^2\sin^2\psi}} + \int_{0}^{-\arcsin\left(\frac{1}{k}\sin\frac{\theta}{2}\right)} \frac{d\psi}{\sqrt{1 - k^2\sin^2\psi}}\Big]$$

$$= \sqrt{\frac{R}{g}}\Big[-F\Big(-\frac{\pi}{2}, k\Big) + F\Big(-\arcsin\Big(\frac{1}{k}\sin\frac{\theta}{2}\Big), k\Big)\Big]$$

$$= \sqrt{\frac{R}{g}} \left[K(k) - F\left(\arcsin\left(\frac{1}{k}\sin\frac{\theta}{2}\right), k \right) \right],$$

or

$$\omega_0 t = K(k) - F\left(\arcsin\left(\frac{1}{k}\sin\frac{\theta}{2}\right), k \right),$$

and, finally

$$F\left(\arcsin\left(\frac{1}{k}\sin\frac{\theta}{2}\right), k \right) = K(k) - \omega_0 t, \tag{3.17}$$

which completes the proof.

As we have discussed at the beginning of this paragraph, if the angular amplitude θ is small ($\theta \leq \theta_0 = 4°$), one can approximate $\sin\theta \approx \theta$, and equation (3.3) reduces to (3.4). Denoting $\omega_0^2 = g/R$, this equation can also be written as

$$\ddot{\theta} + \omega_0^2 \theta = 0, \tag{3.18}$$

which is the well known *differential equation of the linear harmonic oscillator*. This is an ordinary differential equation, homogeneous, with constant coefficients. The general solution is obtained by solving the attached characteristic equation

$$r^2 + \omega_0^2 = 0,$$

with the roots $r_{1,2} = \pm i\omega_0$. Solution is therefore [see (3.5a)]

$$\theta \equiv \theta(t) = A_1 e^{r_1 t} + A_2 e^{r_2 t} = A_1 e^{i\omega_0 t} + A_2 e^{-i\omega_0 t},$$

where the arbitrary constants A_1 and A_2 are determined by using the initial conditions (3.7). As previously shown, solution can also be written in three equivalent forms (3.5b), (3.5c) and (3.5d). We leave to the reader to verify the equivalence of all these possible solutions.

Taking (3.5c) as the solution, the initial conditions (3.7) write

$$\theta_0 = \theta_{t=0} = C_1 \sin C_2;$$

$$\dot{\theta}_0 = \omega_0 C_1 \cos C_2 = 0,$$

and the solution is

$$\theta(t) = \theta_0 \sin\left(\omega_0 t + \frac{\pi}{2}\right) = \theta_0 \cos(\omega_0 t). \tag{3.19}$$

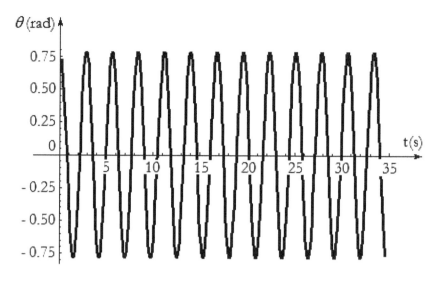

Fig.III.4

Using the *Mathematica* software, specialized in both analytical and numerical calculations, one can give graphical representations for $\theta = \theta(t)$ for arbitrary oscillations [*i.e.* any angular amplitude $\theta_0 \in (0, \pi/2)$], described by (3.16), as well as for small oscillations ($\theta_0 < 4^o$), associated with equation (3.19) (see Figs.III.4, III.5). To make the difference between the two types of oscillations more obvious, their superposition have also been represented graphically (see Fig.III.6).

For θ_0 and ω we chose the following values: $\theta_0 = \pi/4\ rad$, $\omega = 3\pi/4\ rad/s$. As can be seen, at the beginning the two oscillations are

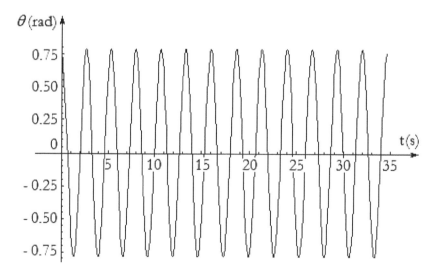

Fig.III.5

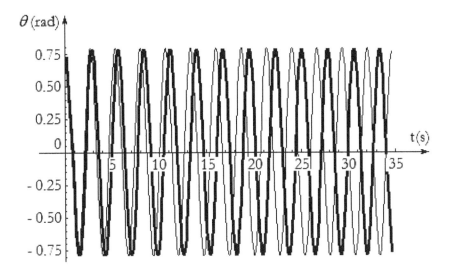

Fig.III.6

similar (they are "in-phase"). After some time, e.g. at $t \simeq 11\pi$ s, the two graphic representations are in antiphase. The explanation is given by the fact that the oscillations of pendulum remain periodical, but they are not harmonic anymore.

As we have discussed, if $\theta_0 \leq 4^o$, the period of motion of pendulum is $\tau_0 = 2\pi/\omega_0 = 2\pi\sqrt{R/g}$. We call these oscillation free, non-amortized, isochronous, with amplitude and initial phase depending on initial conditions. If $\theta > 4^o$, we have to appeal to (3.11).

Since the motion of pendulum is identical on the four intervals $(ABC), (CDE), (EDC), (CBA)$, the period of motion is

$$\tau = 4(\Delta t)_{(ABC)}.$$

Since $(\Delta t)_{(ABC)}$ is precisely the value of t given by (3.11), corresponding to the point $C(\theta = 0)$, we have:

$$(\Delta t)_{(ABC)} = t_{\theta=0} = \sqrt{\frac{R}{g}} \left[K(k) - F\left(\arcsin\left(\frac{1}{k} \sin \frac{\theta}{2} \right), k \right) \right]_{\theta=0}$$

$$= \sqrt{\frac{R}{g}} K(k),$$

and therefore

$$\tau = 4\sqrt{\frac{R}{g}} K(k) = \frac{4}{\omega_0} \int_0^{\pi/2} \frac{d\psi}{\sqrt{1 - k^2 \sin^2 \psi}}. \qquad (3.20)$$

We shall use this formula to express the period τ of the motion of pendulum for arbitrary amplitudes, that is $\theta_0 \in (0, \pi/2)$, in terms of a series of even powers of θ_0. This procedure is very useful in applications, because allows one to determine the period of the unharmonic pendulum with any precision (*Mathematica* can handle approximate real numbers with any number of digits).

To do this, we first observe that, ignoring the singular value $\theta_0 = \pi$, we have $|k| < 1$. A series expansion of the integrand of (3.20) gives:

$$\frac{1}{\sqrt{1 - k^2 \sin^2 \psi}} = 1 + \frac{1}{2} k^2 \sin^2 \psi + \frac{1 \cdot 3}{2 \cdot 4} k^4 \sin^4 \psi + \ldots$$

$$= \sum_{n=1}^{\infty} \frac{1 \cdot 3 \cdot 5 \cdots (2n - 1)}{2 \cdot 4 \cdot 6 \cdots 2n} k^{2n} \sin^{2n} \psi \qquad (3.21)$$

$$= 1 + \sum_{n=1}^{\infty} \frac{(2n-1)!!}{(2n)!!} k^{2n} \sin^{2n} \psi = 1 + \sum_{n=1}^{\infty} \frac{(2n-1)!!}{2^n n!} k^{2n} \sin^{2n} \psi.$$

Since $|k| < 1$, the series (3.21) is uniformly, absolutely convergent on the interval $(0, \pi)$. This fact allows us to integrate it term by term (operations of integration and summation commute, that is the series of integrals equals the integral of series). To this end, we shall first deduce some relation known as *Wallis' formula*.

Let us consider the integrals of the form $I_{2n} = \int_0^{\pi/2} \sin^{2n} \psi \, d\psi$, where n is an integer. We have:

$$I_{2n} = \int_0^{\pi/2} \sin^{2n} \psi \, d\psi = \int_0^{\pi/2} \sin^{2n-2} \psi (1 - \cos^2 \psi) \, d\psi$$

$$= \int_0^{\pi/2} \sin^{2n-2} \psi \, d\psi - \int_0^{\pi/2} \sin^{2n-2} \psi \cos^2 \psi \, d\psi$$

$$= I_{2n-2} - \int_0^{\pi/2} \sin^{2n-2} \psi \cos^2 \psi \, d\psi.$$

Integrating by parts the last term, we still have

$$\int_0^{\pi/2} \sin^{2n-2} \psi \cos^2 \psi \, d\psi = \int_0^{\pi/2} \cos \psi \, d\left(\frac{1}{2n-1} \sin^{2n-1} \psi\right)$$

$$= \frac{1}{2n-1} \left[\sin^{2n-1} \psi \cos \psi \right]_0^{\pi/2}$$

$$-\int_0^{\pi/2} \left(\frac{1}{2n-1} \sin^{2n-1}\psi\right)(-\sin\psi)\, d\psi$$

$$= \frac{1}{2n-1}\int_0^{\pi/2} \sin^{2n}\psi\, d\psi,$$

therefore

$$I_{2n} = I_{2n-2} - \frac{1}{2n-1}\int_0^{\pi/2}\sin^{2n}\psi\, d\psi = I_{2n-2} - \frac{1}{2n-1}I_{2n},$$

which yields the following recurrence relation

$$2n I_{2n} = (2n-1) I_{2n-2}.$$

Taking $n = 1, 2, 3$, etc., we arrive at:

$$\begin{array}{ll}
n = 1 & 2I_2 = I_0 \\
n = 2 & 4I_4 = 3I_2 = \frac{1\cdot 3}{2} I_0 \\
n = 3 & 6I_6 = 5I_4 = \frac{1\cdot 3\cdot 5}{2\cdot 4} I_0 \\
\vdots & \vdots \\
n = n-1 & 2(n-1)I_{2n-2} = (2n-3)I_{2n-4} \\
& \qquad\qquad\quad = \frac{1\cdot 3\cdot 5\cdots (2n-3)}{2\cdot 4\cdot 6\cdots (2n-4)} I_0 \\
n = n & 2n I_{2n} = (2n-1)I_{2n-2} = \frac{1\cdot 3\cdot 5\cdots (2n-1)}{2\cdot 4\cdot 6\cdots (2n-2)} I_0,
\end{array}$$

leading to

$$I_{2n} = \frac{1\cdot 3\cdot 5\cdots (2n-1)}{2\cdot 4\cdot 6\cdots 2n} I_0 = \frac{1\cdot 3\cdot 5\cdots (2n-1)}{2\cdot 4\cdot 6\cdots 2n}\int_0^{\pi/2} d\psi$$

$$= \frac{1\cdot 3\cdot 5\cdots (2n-1)}{2\cdot 4\cdot 6\cdots 2n}\cdot \frac{\pi}{2},$$

which is *Wallis' formula*.

We are now able to calculate $K(k)$. Indeed,

$$K(k) = \int_0^{\pi/2} \frac{d\psi}{\sqrt{1-k^2\sin^2\psi}}$$

$$= \int_0^{\pi/2}\left[1 + \sum_{n=1}^{\infty}\frac{1\cdot 3\cdot 5\cdots (2n-1)}{2\cdot 4\cdot 6\cdots 2n} k^{2n}\sin^{2n}\psi\right] d\psi$$

$$= \frac{\pi}{2} + \sum_{n=1}^{\infty} \left[\frac{1 \cdot 3 \cdot 5 \cdots (2n-1)}{2 \cdot 4 \cdot 6 \cdots 2n} k^{2n} \int_0^{\pi/2} \sin^{2n} \psi \, d\psi \right]$$

$$= \frac{\pi}{2} + \sum_{n=1}^{\infty} \left[\frac{(2n-1)!!}{2n!!} k^{2n} \int_0^{\pi/2} \sin^{2n} \psi \, d\psi \right]$$

$$= \frac{\pi}{2} + \sum_{n=1}^{\infty} \left[\frac{(2n-1)!!}{2^n n!} k^{2n} \int_0^{\pi/2} \sin^{2n} \psi \, d\psi \right]$$

$$= \frac{\pi}{2} + \sum_{n=1}^{\infty} \left[\left(\frac{(2n-1)!!}{2^n n!} \right)^2 k^{2n} \right] \cdot \frac{\pi}{2} = \frac{\pi}{2} \left\{ 1 + \sum_{n=1}^{\infty} \left[\left(\frac{(2n-1)!!}{2^n n!} \right)^2 k^{2n} \right] \right\}.$$

The period of swing of a simple gravity pendulum, for an arbitrary amplitude, is then:

$$\tau = 4\sqrt{\frac{R}{g}} K(k) = 4\sqrt{\frac{R}{g}} \frac{\pi}{2} \left\{ 1 + \sum_{n=1}^{\infty} \left[\left(\frac{(2n-1)!!}{2^n n!} \right)^2 k^{2n} \right] \right\}$$

$$= 2\pi \sqrt{\frac{R}{g}} \left\{ 1 + \sum_{n=1}^{\infty} \left[\left(\frac{(2n-1)!!}{2^n n!} \right)^2 k^{2n} \right] \right\} \qquad (3.22)$$

$$= \tau_0 \left\{ 1 + \sum_{n=1}^{\infty} \left[\left(\frac{(2n-1)!!}{2^n n!} \right)^2 k^{2n} \right] \right\}$$

$$= \tau_0 \left\{ 1 + \sum_{n=1}^{\infty} \left[\left(\frac{(2n-1)!!}{2^n n!} \right)^2 \sin^{2n} \frac{\theta_0}{2} \right] \right\}.$$

This result is *exact*, meaning that we did not perform any approximation so far. If $\theta_0 < \pi/2$, then $\sin(\theta_0/2)$ can also be developed in series according to the well-known formula

$$\sin x = x - \frac{1}{3!} x^3 + \frac{1}{5!} x^5 - \ldots = \sum_{l=0}^{\infty} \frac{(-1)^l}{(2l+1)!} x^{2l+1},$$

therefore

$$\sin \frac{\theta_0}{2} = \sum_{l=0}^{\infty} \frac{(-1)^l}{(2l+1)!} \left(\frac{\theta_0}{2} \right)^{2l+1} = \frac{\theta_0}{2^1 \cdot 1!} - \frac{\theta_0^3}{2^3 \cdot 3!} + \frac{\theta_0^5}{2^5 \cdot 5!} - \ldots$$

Sometimes, in practical applications, only the first 4-5 terms of this series expansion are chosen. Let us write

$$\tau = \tau_0 (c_1 + c_2 \theta_0^2 + c_3 \theta_0^4 + c_4 \theta_0^6 + \ldots).$$

According to (3.22), $c_1 = 1$. To write the rest of the coefficients c_2, c_3, c_4, etc. one must observe that, for example, c_4 involves the terms corresponding to $n = 1, n = 2$, and $n = 3$. We then have

$$\tau = \tau_0 \left(1 + \frac{1}{4}\sin^2\frac{\theta_0}{2} + \frac{9}{64}\sin^4\frac{\theta_0}{2} + \frac{225}{2304}\sin^6\frac{\theta_0}{2} + \ldots \right). \quad (3.23)$$

Since

$$\sin\frac{\theta_0}{2} = \frac{\theta_0}{2^1 \cdot 1!} - \frac{\theta_0^3}{2^3 \cdot 3!} + \frac{\theta_0^5}{2^5 \cdot 5!} - \frac{\theta_0^7}{2^7 \cdot 7!} + \ldots$$

$$= \frac{\theta_0}{2} - \frac{\theta_0^2}{48} + \frac{\theta_0^5}{3840} - \frac{\theta_0^7}{645120} + \ldots, \quad (3.24)$$

we can write

$$\sin^2\frac{\theta_0}{2} = \left(\frac{\theta_0}{2^1 \cdot 1!} - \frac{\theta_0^3}{2^3 \cdot 3!} + \frac{\theta_0^5}{2^5 \cdot 5!} - \frac{\theta_0^7}{2^7 \cdot 7!} + \ldots \right)^2 = \frac{\theta_0^2}{4} + \mathcal{O}(\theta_0^4). \quad (3.25)$$

Therefore, in view of (3.23), the coefficient of θ_0^2 is: $c_2 = \frac{1}{4} \cdot \frac{1}{4} = \frac{1}{16}$.

Next, let us determine c_3. In view of (3.24), we have:

$$\sin^4\frac{\theta_0}{2} = \left(\frac{\theta_0}{2^1 \cdot 1!} - \frac{\theta_0^3}{2^3 \cdot 3!} + \frac{\theta_0^5}{2^5 \cdot 5!} - \frac{\theta_0^7}{2^7 \cdot 7!} + \ldots \right)^4$$

$$= \left(\frac{\theta_0}{2^1 \cdot 1!} - \frac{\theta_0^3}{2^3 \cdot 3!} + \frac{\theta_0^5}{2^5 \cdot 5!} - \frac{\theta_0^7}{2^7 \cdot 7!} + \ldots \right)^2$$

$$\times \left(\frac{\theta_0}{2^1 \cdot 1!} - \frac{\theta_0^3}{2^3 \cdot 3!} + \frac{\theta_0^5}{2^5 \cdot 5!} - \frac{\theta_0^7}{2^7 \cdot 7!} + \ldots \right)^2$$

$$= \left(\frac{\theta_0}{2} - \frac{\theta_0^3}{48} + \frac{\theta_0^5}{3840} + \mathcal{O}(\theta_0^7) \right)^2 \left(\frac{\theta_0}{2} - \frac{\theta_0^3}{48} + \frac{\theta_0^5}{3840} + \mathcal{O}(\theta_0^7) \right)^2$$

$$= \left(\frac{\theta_0^2}{4} + \frac{\theta_0^6}{2304} - 2\frac{\theta_0}{2}\frac{\theta_0^3}{48} + 2\frac{\theta_0}{2} \cdot \frac{\theta_0^5}{3840} + \mathcal{O}(\theta_0^8) \right)$$

$$\times \left(\frac{\theta_0^2}{4} + \frac{\theta_0^6}{2304} - 2\frac{\theta_0}{2}\frac{\theta_0^3}{48} + 2\frac{\theta_0}{2} \cdot \frac{\theta_0^5}{3840} + \mathcal{O}(\theta_0^8) \right)$$

$$= \left(\frac{\theta_0^2}{4} - \frac{\theta_0^4}{48} + \frac{\theta_0^6}{1440} + \mathcal{O}(\theta_0^8) \right) \left(\frac{\theta_0^2}{4} - \frac{\theta_0^4}{48} + \frac{\theta_0^6}{1440} + \mathcal{O}(\theta_0^8) \right)$$

$$= \frac{\theta_0^4}{16} - \frac{\theta_0^6}{96} + \mathcal{O}(\theta_0^8). \quad (3.26)$$

To put into evidence the coefficients of θ_0^4 and θ_0^6, we write (3.25) as

$$\sin^2\frac{\theta_0}{2} = \left(\frac{\theta_0}{2} - \frac{\theta_0^3}{3! \cdot 2^3} + \frac{\theta_0^5}{5! \cdot 2^5} - \frac{\theta_0^7}{7! \cdot 2^7} + \ldots \right)^2$$

$$= \left(\frac{\theta_0}{2} - \frac{\theta_0^3}{48} + \frac{\theta_0^5}{3840} - \mathcal{O}(\theta_0^7)\right)^2$$

$$= \frac{\theta_0^2}{4} + \frac{\theta_0^6}{2304} - 2\frac{\theta_0}{2}\cdot\frac{\theta_0^3}{48} + 2\frac{\theta_0}{2}\cdot\frac{\theta_0^5}{3840} + \mathcal{O}(\theta_0^8)$$

$$= \frac{\theta_0^2}{4} - \frac{\theta_0^4}{48} + \frac{\theta_0^6}{1440} + \mathcal{O}(\theta_0^8). \tag{3.27}$$

Since, according to (3.23), the coefficients of $\sin^2\frac{\theta_0}{2}$ and $\sin^4\frac{\theta_0}{2}$ are $\frac{1}{4}$ and $\frac{9}{64}$, respectively, the coefficient c_3 writes

$$c_3 = \frac{1}{4}\cdot\left(-\frac{1}{48}\right) + \frac{9}{64}\cdot\frac{1}{16} = \frac{11}{3072}.$$

Finally, in order to determine c_4 one must consider the contribution of $\sin^6\frac{\theta_0}{2}$ to the term in θ_0^6, that is:

$$\sin^6\frac{\theta_0}{2} = \left(\frac{\theta_0^2}{4} - \frac{\theta_0^4}{48} + \frac{\theta_0^6}{1440} + \mathcal{O}(\theta_0^8)\right)^3$$

$$= \left(\frac{\theta_0^2}{4} - \frac{\theta_0^4}{48} + \frac{\theta_0^6}{1440} + \mathcal{O}(\theta_0^8)\right)^2$$

$$\times \left(\frac{\theta_0^2}{4} - \frac{\theta_0^4}{48} + \frac{\theta_0^6}{1440} + \mathcal{O}(\theta_0^8)\right) \tag{3.28}$$

$$= \left(\frac{\theta_0^4}{16} - \frac{\theta_0^6}{96} + \mathcal{O}(\theta_0^8)\right)\left(\frac{\theta_0^2}{4} - \frac{\theta_0^4}{48} + \frac{\theta_0^6}{1440} + \mathcal{O}(\theta_0^8)\right)$$

$$= \frac{\theta_0^6}{64} + \mathcal{O}(\theta_0^8).$$

By means of (3.23) and (3.26)-(3.28), we then have

$$c_4 = \frac{1}{4}\cdot\frac{1}{1440} + \frac{9}{64}\cdot\left(-\frac{1}{96}\right) + \frac{225}{2304}\cdot\frac{1}{64} = \frac{173}{737280}.$$

Consequently, for an arbitrary angle $\theta_0 < \frac{\pi}{2}$, the period increases gradually with angular amplitude, according to the formula

$$\tau = 2\pi\sqrt{\frac{R}{g}}\left(1 + \frac{1}{16}\theta_0^2 + \frac{11}{3072}\theta_0^4 + \frac{173}{737280}\theta_0^6 + ...\right)$$

$$= \tau_0\left[1 + \frac{1}{16}\theta_0^2 + \frac{11}{3072}\theta_0^4 + \frac{173}{737280}\theta_0^6 + \mathcal{O}(\theta_0^8)\right]. \tag{3.29}$$

If the angular amplitude of oscillation is small ($\theta_0 < 4°$), the terms involving θ_0^2, θ_0^4, etc. can be neglected as compared to unity, and we fall back on the formula corresponding to isochronous oscillations

$$\tau|_{\theta_0 < 4°} \simeq \tau_0 = 2\pi \sqrt{\frac{R}{g}}. \tag{3.30}$$

This formula is used in the construction of astronomical instruments, with $\theta_0 = 1°30'$.

Observation. We could have arrived at the same result by using a different procedure. Namely, instead of determining the exact solution of the differential equation of motion (3.4) and its period, followed by their approximation in order to find the results associated with the harmonic pendulum [*i.e.* relations (3.5a) and (3.30)], one can first expand in series $\sin\theta$ in equation (3.4) and then, in terms of the chosen approximation, find solutions for the differential equation of motion for various models of non-harmonic pendulums. For example, if we consider only the first two terms between parentheses in equation

$$\ddot{\theta} + \frac{g}{R}\sin\theta = \ddot{\theta} + \frac{g}{R}\left(\theta - \frac{\theta^3}{3!} + \frac{\theta^5}{5!} - \frac{\theta^7}{7!} + ...\right) = 0, \tag{3.31}$$

this corresponds to the first order approximation of the non-linear equation of non-harmonic pendulum, which is

$$\ddot{\theta} + \frac{g}{R}\theta - \frac{g}{6R}\theta^3 = 0. \tag{3.32}$$

An approximate solution to this equation can be written as

$$\theta(t) = \theta_0 \sin\omega t + \varepsilon\theta_0 \sin 3\omega t, \tag{3.33}$$

where ε is a dimensionless constant much smaller than 1. One can find

$$\omega^2 = \omega_0^2\left(1 - \frac{1}{8}\theta_0^2\right), \tag{3.34}$$

so that

$$\omega \simeq \omega_0\left(1 - \frac{1}{16}\theta_0^2\right),$$

and, therefore,

$$\tau\left(=\frac{2\pi}{\omega}\right) \simeq \tau_0\left(1 + \frac{1}{16}\theta_0^2\right),$$

which is precisely the first-order approximation of (3.29).

Presence of the term proportional to $\sin 3\omega t$ is suggested by the trigonometric identity $\sin^3 x = (3/4) \sin x - (1/4) \sin 3x$. This way, the term involving θ^3 in (3.32) will produce a term proportional to $\sin 3\omega t$, which must be compensated by the term $\varepsilon \sin 3\omega t$. The third power of this new term shall generate a term proportional to $\varepsilon^3 \sin 9\omega t$, and so on. If $\varepsilon \ll 1$, a rapid convergence of the series is expected. Solution (3.33) is, therefore, an approximate solution.

A better approximation can be obtained if we also consider the term proportional to θ^5 in (3.31). In this case, using the trigonometric identity $\sin^5 u = (5/8) \sin u - (5/16) \sin 3u + (1/16) \sin 5u$, we search for a solution in the form

$$\theta(t) = \theta_0 \sin \omega t + \varepsilon \theta_0 \sin 3\omega t + \lambda \theta_0 \sin 5\omega t.$$

As a result, the term proportional to θ^5 generates a new series, which is more rapidly convergent than the first one, and so on.

Let us now impose the solution (3.33) to verify equation (3.32). We have:

$$\ddot{\theta}(t) = \frac{d^2}{dt^2}(\theta_0 \sin \omega t + \varepsilon \theta_0 \sin 3\omega t)$$
$$= -\omega^2 \theta_0 \sin \omega t - 9\omega^2 \varepsilon \theta_0 \sin 3\omega t,$$

as well as

$$[\theta(t)]^3 = \theta_0^3 (\sin \omega t + \varepsilon \sin 3\omega t)^3$$
$$= \theta_0^3 \left(\sin^3 \omega t + 3\varepsilon \sin^2 \omega t \sin 3\omega t + \mathcal{O}(\varepsilon^2) \right)$$
$$\simeq \theta_0^3 \left(\frac{3}{4} \sin \omega t - \frac{1}{4} \sin 3\omega t + 3\varepsilon \sin^2 \omega t \sin 3\omega t \right),$$

where the identity $\sin^3 x = (3/4) \sin x - (1/4) \sin 3x$ has been used.

Introducing the last two relations together with (3.33) into (3.32), one obtains

$$-\omega^2 \theta_0 \sin \omega t - 9\omega^2 \varepsilon \theta_0 \sin 3\omega t + \omega_0^2 (\theta_0 \sin \omega t + \varepsilon \theta_0 \sin 3\omega t)$$

$$-\frac{1}{6} \omega_0^2 \theta_0^3 \left(\frac{3}{4} \sin \omega t - \frac{1}{4} \sin 3\omega t + 3\varepsilon \sin^2 \omega t \sin 3\omega t \right) = 0.$$

In order that (3.33) is an approximate solution at any time t, the sums of the coefficients of $\sin \omega t$ and $\sin 3\omega t$ - separately - must be zero, that is

$$-\omega^2 + \omega_0^2 - \frac{3\omega_0^2}{24} \theta_0^2 = 0, \tag{3.35}$$

and
$$-9\omega^2\varepsilon\theta_0 + \omega_0^2\varepsilon\theta_0 + \frac{\omega_0^2}{24}\theta_0^3 - \frac{\omega_0^2}{2}\theta_0^3\varepsilon\sin^2\omega t = 0. \qquad (3.36)$$

As one observes, relation (3.34) follows immediately from (3.35), while (3.36) yields
$$\varepsilon = \frac{(\omega_0^2/24)\theta_0^2}{9\omega^2 - \omega_0^2} \simeq \frac{\theta_0^2}{192},$$

where the last term (which is small as compared to the other three) has been neglected, and the approximation $\omega^2 \simeq \omega_0^2$ has been used. Parameter ε gives the relative contribution of the term containing $\sin 3\omega t$ in a solution "dominated" by $\sin \omega t$. For example, if $\theta_0 = 0.3\,rad$, the relative modification of frequency is $\frac{\Delta\omega}{\omega} = \frac{(\omega-\omega_0)}{\omega} \approx -10^{-2}$, and $\varepsilon \approx 10^{-3}$.

It can be easily verified that using an approximate solution of the form
$$\theta(t) = \theta_0 \sin \omega t + \lambda \theta_0 \sin 2\omega t,$$

and performing the required calculations, one obtains $\lambda = 0$. This means that the pendulum generates *mostly* the third harmonic (that is terms in $\sin 3\omega t$), while the second harmonic (corresponding to terms in $\sin 2\omega t$), as well as any other even harmonic, are absent. This result comes from the fact that the equation of motion contains a term involving θ^3 (but no term in θ^2). Here "mostly" means that some other odd harmonics are also generated.

These results can be interpreted as follows. In case of big amplitudes, does not exist a unique frequency of the periodical motion of pendulum. The most "powerful" term is expressed as a function of $\sin \omega t$, and we call ω *the fundamental frequency* of the pendulum. In the first order of approximation, the fundamental frequency is given by (3.34), while (3.29) offers better approximations. The exact form of ω is given by (3.20) or (3.22). The term involving $\sin 3\omega t$ is called *the third harmonic of the fundamental frequency*. Obviously, the real motion presents an infinite number of harmonics: the higher is their order, the smaller is their contribution (as an example, the amplitude of the third harmonic is about one thousand times smaller than the fundamental one).

As we have seen at the beginning of this paragraph, to solve the problem within this formalism one must know the constraint force $\vec{T}$ (tension in the rod). The geometric properties of $\vec{T}$ (direction and sense) are shown in Fig.III.1, while its magnitude can be determined by means of (3.1), (3.3), and (3.8). Indeed,

$$|\vec{T}| = T = -\frac{m\ddot{y}}{\sin\theta} = -\frac{m(R\ddot{\theta}\cos\theta - R\dot{\theta}^2\sin\theta)}{\sin\theta}$$

$$-\frac{m\left\{R\left(-\frac{g}{R}\sin\theta\right)\cos\theta - R\left[\frac{2g}{R}(\cos\theta - \cos\theta_0)\right]\sin\theta\right\}}{\sin\theta}$$

$$= mg(3\cos\theta - 2\cos\theta_0). \tag{3.37}$$

This way, all about the problem of simple pendulum, studied in terms of Newtonian mechanics, has been said. Next, we shall discuss the same application, by means of the analytical formalism.

III.2. Lagrange equations of the first kind approach

Within this formalism, the problem is presented as follows: given the mass of the body, the applied forces, the analytical equations of constraints and the initial conditions compatible with the constraints, find the low of motion of the body, the constraint forces, and, eventually, the elements of the motion like: trajectory, period of the motion, etc.

As one can see, this approach (as all the other analytical formalisms) does not demand the *a priori* knowledge of the constrained forces; they are determined *a posteriori*. As a matter of fact, this is the main advantage of the analytical formalism. The Lagrangian equations of the first kind method allows to replace the knowledge of constraint forces by the analytical expressions of the constraints (given in either explicit, or implicit forms). In most cases, the constraint forces are not known from the beginning and the Newtonian approach is not able to solve the problem.

Here is the algorithm for solving a problem within the Lagrangian equations of the first kind method:

i) Identification of the applied forces $\vec{F}_i$ ($i = \overline{1, N}$), where N is the number of the bodies (particles) composing the system, and their reference to a conveniently chosen coordinate system.

ii) Fixing of the initial conditions, compatible with the constraints: $\vec{r}_{i0} = \vec{r}_i(t_0)$, and $\dot{\vec{r}}_{i0} = \dot{\vec{r}}_i(t_0)$.

iii) Identification of the constraints, as well as their analytical expressions: $f_k(\vec{r}_1, \vec{r}_2, ..., \vec{r}_N; t) = 0$ ($k = \overline{1, s}$), where s is the number of constraints.

iv) Establishment and finding the solutions of the system of $3N+s$ equations in $3N+s$ unknowns: N Lagrange equations of the first kind

$$m_i \ddot{\vec{r}}_i = \vec{F}_i + \sum_{k=1}^{s} \lambda_k grad_i f_k \quad (i = \overline{1,N}),$$

together with expressions of s constraints, while the unknowns are: $3N$ coordinates $x_i = x_i(t)$, $y_i = y_i(t)$, $z_i = z_i(t)$ and s non-zero constants λ_k ($k = \overline{1,s}$).

v) Determination of the constraint forces $\vec{L}_i$ ($i = \overline{1,N}$), using their definition $\vec{L}_i = \sum_{k=1}^{s} \lambda_k grad_i f_k$.

Before applying this algorithm in our case, we point out that the Lagrange equations of the first kind formalism is valid for ideal constraints only. This is a very restrictive condition, and does not correspond to reality, but, as far as the friction forces are known[1], these can be considered as applied forces and added to them.

Using the same reference frame as in the previous paragraph, let us choose xOy as the plane of motion. Then, the only applied force (force of gravity)

$$\vec{G} = (mg, 0, 0)$$

is accompanied by the initial conditions

$$\vec{r}_0 = (x_0, y_0, 0) = (R\cos\theta_0, R\sin\theta_0, 0),$$

and

$$\dot{\vec{r}}_0 = 0.$$

---

[1] The friction forces are usually determined by empirical methods. For example, in case of the relative motion of two bodies, one determines experimentally that the force of friction is proportional to the normal reaction, $\vec{F}_f = \mu \vec{N}$, where μ is the *coefficient of friction*. If the motion takes place in a viscous fluid, the experiment shows that for small velocities (laminar flow) the force of friction is proportional to the relative velocity between body and fluid, $\vec{F}_f = -k\vec{v}_r$, where the coefficient k depends on viscosity of the fluid and and the form of the transverse section of the body. At high velocities (turbulent flow), the same force is proportional to the squared relative velocity, $\vec{F}_f = -k|\vec{v}_r|^2 \vec{\nu}$, where $\vec{\nu}$ is the unit vector of direction of motion of the body, and k - a "constant" which depends on the medium density, the area of the cross section of the body, and the so-called *form coefficient*, tightly connected to the aerodynamic characteristics of the body.

The body is submitted to two constraints:

$$f_1(x, y) = x^2 + y^2 - R^2 = 0, \qquad (3.38\ a)$$

and

$$f_2(z) = z = 0. \qquad (3.38\ b)$$

Since our body is a particle (point mass), the index i can be dropped. The system of equations leading to the law of motion of the body of mass m then is

$$\begin{cases} m\ddot{\vec{r}} = \vec{G} + \sum_{k=1}^{2} \lambda_k grad\, f_k, \\ f_1(x, y) = x^2 + y^2 - R^2 = 0, \\ f_2(z) = z = 0. \end{cases} \qquad (3.39)$$

Since the constraints are given by

$$grad\, f_1 = 2x\vec{i} + 2y\vec{j}, \qquad (3.40\ a)$$

and

$$grad\, f_2 = 0, \qquad (3.40\ b)$$

the system (3.39) becomes

$$\begin{cases} m\ddot{x} = mg + 2\lambda x, \\ m\ddot{y} = 2\lambda y, \\ x^2 + y^2 - R^2 = 0, \\ z = 0, \end{cases} \qquad (3.41)$$

which is a system of four equations with four unknowns: x, y, z, λ. Here λ stands for λ_1.

Relation (3.41 c) allows one to parametrize the constraint $f_1(x, y)$ as

$$\begin{cases} x = R\cos\theta, \\ y = R\sin\theta, \end{cases}$$

and the first two equations of the system (3.41) write

$$\begin{cases} m(-R\ddot{\theta}\sin\theta - R\dot{\theta}^2\cos\theta) = mg + 2\lambda R\cos\theta, \\ m(R\ddot{\theta}\cos\theta - R\dot{\theta}^2\sin\theta) = 2\lambda R\sin\theta. \end{cases} \qquad (3.42)$$

This way, the problem reduces to two differential equations with two unknowns: $\theta(t)$ and the scalar λ. A rearrangement of terms, followed by division of the equations, yield:

$$\frac{-R\ddot{\theta}\sin\theta - R\dot{\theta}^2\cos\theta - g}{R\ddot{\theta}\cos\theta - R\dot{\theta}^2\sin\theta} = \frac{\cos\theta}{\sin\theta},$$

leading to the differential equation for $\theta(t)$

$$\ddot{\theta} + \frac{g}{R}\sin\theta = 0, \qquad (3.3')$$

which is precisely equation (3.3) obtained within the Newtonian approach. From now on, the problem is solved following classical formalism. The only difference consists in determination of the force of constraint $\vec{L}$, which in Lagrangian method is not previously known (neither its direction, nor its sense or magnitude). This can be found by means of its definition

$$\vec{L} = \sum_{k=1}^{2} \lambda_k \mathrm{grad}\, f_k = \lambda_1 \mathrm{grad}\, f_1 + \lambda_2 \mathrm{grad}\, f_2 = \lambda \mathrm{grad}\, f_1$$

$$= 2\lambda x \vec{i} + 2\lambda y \vec{j} = 2\lambda R(\vec{i}\cos\theta + \vec{j}\sin\theta).$$

The magnitude of $\vec{L}$ is

$$|\vec{L}| = \sqrt{L_x^2 + L_y^2} \equiv L = \sqrt{4\lambda^2 R^2 \cos^2\theta + 4\lambda^2 R^2 \sin^2\theta} = 2\lambda R. \quad (3.43)$$

The parameter λ can be determined by means of (3.42 b). Using (3.3') and (3.7), we have

$$\lambda = \frac{m(R\ddot{\theta}\cos\theta - R\dot{\theta}^2\sin\theta)}{2R\sin\theta}$$

$$= \frac{m\left\{R\left(-\frac{g}{R}\sin\theta\right)\cos\theta - R\left[\frac{2g}{R}(\cos\theta - \cos\theta_0)\right]\sin\theta\right\}}{2R\sin\theta}$$

$$= \frac{mg(2\cos\theta_0 - 3\cos\theta)}{2R}.$$

The magnitude of the constraint force therefore is

$$L = 2R\lambda = mg(2\cos\theta_0 - 3\cos\theta). \qquad (3.44)$$

Except for a sign difference (which shall be explained in the next paragraph), this formula coincides with (3.37) expressing $|\vec{T}| = T$. This fact can also be proved if we write (3.43) as

$$|\vec{L}| \equiv L = \sqrt{L_x^2 + L_y^2} = \sqrt{4\lambda^2 R^2 \cos^2\theta + 4\lambda^2 R^2 \sin^2\theta} = |2\lambda R|,$$

which obviously leads to
$$|L| = mg|2\cos\theta_0 - 3\cos\theta| = |3\cos\theta - 2\cos\theta_0| = |T|.$$

To conclude, the constraint force $\vec{L}$ and tension in the rod $\vec{T}$ have the same magnitude. The difference in sign appears because the two vectors have opposite senses: the gradient is oriented along the external normal to the constraint (trajectory), while the vector $\vec{T}$ is, by definition, oriented along the internal normal (see Fig.III.1).

III.3. Lagrange equations of the second kind approach

The problem to solve is the same as in the previous approach. The difference appears in the procedure of finding the result. Namely, the use of Lagrange equations of the second kind demand neither knowledge of the constraint forces, nor the analytical expressions of the constraints (essential for the first kind of Lagrange equations method), but only *the number* of constraints.

The algorithm for solving a problem within the Lagrangian equations of the second kind method demands the following steps:

i) Identification of applied forces $\vec{F}_i$, $(i = \overline{1,N})$, where N stands for the number of the bodies (particles) of the system.

ii) Establishment of the initial conditions (compatible with the constraints) in real space: $\vec{r}_{i0} = \vec{r}_i(t_0)$, and $\dot{\vec{r}}_{i0} = \dot{\vec{r}}_i(t_0)$.

iii) Identification of *the number* s of constraints, the number of degrees of freedom of the system, $n = 3N - s$, and the convenient choice of generalized coordinates $q_j = q_j(t)$ $(j = \overline{1,n})$.

iv) Determination of the relations connecting the real and configuration spaces $\vec{r}_i = \vec{r}_i(q_1, q_2, ..., q_n, t) \equiv \vec{r}_i(q,t)$ $(i = \overline{1,N})$, together with their inversions $q_j = q_j(\vec{r}_1, \vec{r}_2, ..., \vec{r}_N, t)$ $(j = \overline{1,n})$, as well as the initial conditions in configuration space, $q_{j0} = q_j(t_0)$, $\dot{q}_{j0} = \dot{q}_j(t_0)$.

v) Determination of the kinetic energy $T = T(q, \dot{q}, t)$ and the generalized forces $Q_j = Q_j(q,t)$ for non-potential systems, or the Lagrangian function $L(q, \dot{q}, t) = T(q, \dot{q}, t) - V(q, \dot{q}, t)$ in the case of natural systems. (A system that admits a simple or generalized potential is called *natural*).

vi) Writing and solving the Lagrange equations of the second kind $\frac{d}{dt}\left(\frac{\partial T}{\partial \dot{q}_j}\right) - \frac{\partial T}{\partial q_j} = Q_j$ $(j = \overline{1,n})$, if the applied forces are not potential, and $\frac{d}{dt}\left(\frac{\partial L}{\partial \dot{q}_j}\right) - \frac{\partial L}{\partial q_j} = 0$ $(j = \overline{1,n})$, in case of natural systems. Using the initial conditions, one determines the law of motion in configuration space $q_j = q_j(t)$ $(j = \overline{1,n})$, which also represents the parametric equations of generalized trajectory in this space.

vii) Establishment of the law of motion in the real space $\vec{r}_i = \vec{r}_i(q_j(t); t) \equiv \vec{r}_i(t)$ $(i = \overline{1,N}; j = \overline{1,n})$.

viii) Finding the constraint forces, by means of the fundamental equation of dynamics, written for systems subject to constraints:

$$\vec{L}_i = m_i \ddot{\vec{r}}_i - \vec{F}_i \quad (i = \overline{1,N}).$$

We shall use this algorithm in order to study the problem of mathematical pendulum, performing free, non-amortized oscillations, for an arbitrary value of the angular amplitude. The only applied force is the force of gravity $\vec{G} = (mg, 0, 0)$, its components being expressed with regard to the already chosen reference frame xOy (see Fig.III.7).

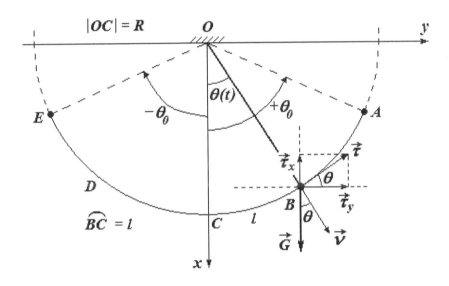

Fig.III.7

The initial conditions in the real space remain, obviously, the same: $\vec{r}_0 = (R\cos\theta_0, R\sin\theta_0, 0)$, and $\dot{\vec{r}}_0 = 0$.

Liberty of the motion is limited by two constraints (already discussed in the previous paragraph), so that the system possesses a single degree of freedom: $n = 3N - s = 3 \cdot 1 - 2 = 1$. The choice of the generalized coordinate associated to this degree of freedom is not unique. In this respect, one can choose the coordinate x, or coordinate y, or an arc length determined relative to an arbitrary origin, or, more convenient, the angle θ between Ox-axis and the rod. Generally speaking, the appropriate choice of the generalized coordinate(s) depends on the experience and skillfulness of the person engaged in solving a problem. In other words, if the initial choice is not accurate (convenient), meaning that solving the Lagrange equations is difficult (or even impossible), there always exists a possibility of choosing another set of generalized coordinates according to the "point" transformation in configuration space:

$$q_j \to q'_j = q'_j(q_1, q_2, ..., q_n; t) \quad (j = \overline{1, n}).$$

So, we choose the angular coordinate θ as generalized coordinate associated to the single degree of freedom of the simple pendulum. The relations connecting x, y, on the one hand, and θ on the other, are

$$x = R\cos\theta, \qquad y = R\sin\theta,$$

while the initial conditions write

$$\theta_{t_0=0} = \theta_0, \qquad \dot{\theta}_{t_0=0} = \dot{\theta}_0 = 0.$$

In general, Lagrange equations have the form

$$\frac{d}{dt}\left(\frac{\partial T}{\partial \dot{q}_j}\right) - \frac{\partial T}{\partial q_j} = Q_j \ (j = \overline{1, n}). \tag{3.45}$$

Here $Q_j(q, \dot{q}, t) = \sum_{i=1}^{N} \vec{F}_i \cdot \frac{\partial \vec{r}_i}{\partial q_j}$ are the generalized forces, and $T(q, \dot{q}, t)$ the kinetic energy. Our system is a natural system, that is one can define a Lagrangian function $L = T - V$, where the potential V can be either simple $V = V(\vec{r}, t)$, or generalized $V = V(\vec{r}, \dot{\vec{r}}, t)$. In our case, V can be deduced as follows:

$$\vec{F} = -\text{grad}V(\vec{r}, t) \overset{(formally)}{=} -\frac{dV}{d\vec{r}}.$$

Then

$$dV = -\vec{F} \cdot d\vec{r} = -mg\,dx,$$

and, by integration

$$V = -mgx + V_0.$$

The arbitrary constant V_0 is determined by a convenient choice of the reference level of the potential energy. Since $\text{grad}(V + V_0) = \text{grad}V$, where V_0 is a constant, the potential energy is not uniquely determined, so it is up to us to choose a useful value for V_0. Let us take

$$V_{x=0} = 0,$$

in which case we have

$$V = V(x) = -mgx, \tag{3.46}$$

or, since $x = R\cos\theta$,

$$V(\theta) = -mgR\cos\theta.$$

The kinetic energy T is

$$T = \frac{1}{2}\sum_{i=1}^{N} m_i|\dot{\vec{r}}_i|^2 = \frac{1}{2}m(\dot{x}^2 + \dot{y}^2) = \frac{1}{2}mR^2\dot{\theta}^2, \qquad (3.47)$$

so that the Lagrangian can be written as

$$L = \frac{1}{2}mR^2\dot{\theta}^2 + mgR\cos\theta. \qquad (3.48)$$

The Lagrange equation of the second kind then writes

$$\frac{d}{dt}\left(\frac{\partial L}{\partial \dot{\theta}}\right) - \frac{\partial L}{\partial \theta} = 0. \qquad (3.49)$$

Since $\frac{\partial L}{\partial \dot{\theta}} = mR^2\dot{\theta}$, $\frac{\partial L}{\partial \theta} = -mgR\sin\theta$, we arrive at the already known equation

$$\ddot{\theta} + \frac{g}{R}\sin\theta = 0. \qquad (3.50)$$

From now on, the problem is solved as shown in the previous paragraphs. The only difference appears in determination of the constraint force, which, in view of the fundamental equation of Mechanics for a system subject to constraints, can be written as

$$\vec{L}_i = m_i\ddot{\vec{r}}_i - \vec{F}_i \quad (i = \overline{1,N}). \qquad (3.51)$$

Before going further, it is useful to mention that $\dot{\theta}$ can be determined in a less complicated way than it was within the Newtonian approach. This can be done by using a first integral suggested by the problem. Since the time t does not explicitly appear in the Lagrangian (3.48), equation (3.49) admits the first integral

$$\dot{\theta}\frac{\partial L}{\partial \dot{\theta}} - L = const.$$

We have:

$$\dot{\theta}\frac{\partial L}{\partial \dot{\theta}} - L = mR^2\dot{\theta}^2 - \frac{1}{2}mR^2\dot{\theta}^2 - mgR\cos\theta$$

$$= \frac{1}{2}mR^2\dot{\theta}^2 - mgR\cos\theta = const. = E,$$

where E is *the total energy* of the system. It can be determined in view of the initial conditions $\theta_{t_0=0} = \theta_0$, $\dot{\theta}_{t_0=0} = \dot{\theta}_0 = 0$. It then follows

$$E = const = -mgR\cos\theta_0$$

and the first integral yields

$$\dot{\theta} = \pm\sqrt{\frac{2g}{R}(\cos\theta - \cos\theta_0)},$$

as expected.

In general, one can also use another type of first integrals of the Lagrange equations of the second kind, namely those associated with cyclic variables (if there are any): if the Lagrangian does not explicitly depend on a certain variable (called *cyclic*), then the generalized momentum associated (or conjugated) with this variable is conserved. Indeed, if q_α is a cyclic generalized coordinate, then $\partial L/\partial q_\alpha = 0$, and, following the Lagrange equation for this coordinate, one obtains

$$\frac{\partial L}{\partial \dot{q}_\alpha} = p_\alpha = const.$$

Let us now turn back to the determination of the constraint force. Since in our case $N = 1$, the index i in (3.51) can be dropped, so that

$$\vec{L} = m\ddot{\vec{r}} - \vec{F} = m\ddot{\vec{r}} - \vec{G}. \tag{3.52}$$

Unlike the Lagrange equations of the first kind formalism, where the projections on axes of the constraint force were first determined, this time we take as a reference frame a system of two orthogonal axes, connected to the body trajectory: one axis tangent to the trajectory and pointing the direction of increasing θ, of versor $\vec{\tau}$, and the other axis normal to the trajectory, of versor $\vec{\nu}$, oriented along the external normal (see Fig.III.7). The components of the constraint force on these axes then are

$$L_\tau = \vec{L} \cdot \vec{\tau}, \tag{3.53}$$

and

$$L_\nu = \vec{L} \cdot \vec{\nu}. \tag{3.54}$$

In order to determine L_τ and L_ν, we shall first write $\vec{\tau}$ and $\vec{\nu}$ in terms of $\vec{i}$ and $\vec{j}$. With the help of Fig.III.7, we have

$$\vec{\tau} = \vec{\tau}_x + \vec{\tau}_y = -\vec{i}\sin\theta + \vec{j}\cos\theta,$$

and
$$\vec{\nu} = \vec{\nu}_x + \vec{\nu}_y = \vec{i}\cos\theta + \vec{j}\sin\theta.$$

According to (3.52), the components of $\vec{L}$ are
$$\vec{L} = (m\ddot{x} - mg,\ m\ddot{y},\ 0),$$

which allows us to write
$$L_\tau = \vec{L}\cdot\vec{\tau} = [(m\ddot{x} - mg)\vec{i} + m\ddot{y}\vec{j}]\cdot(-\vec{i}\sin\theta + \vec{j}\cos\theta)$$
$$= m(g - \ddot{x})\sin\theta + m\ddot{y}\cos\theta,$$

and
$$L_\nu = \vec{L}\cdot\vec{\nu} = [(m\ddot{x} - mg)\vec{i} + m\ddot{y}\vec{j}]\cdot(\vec{i}\cos\theta + \vec{j}\sin\theta)$$
$$= m(\ddot{x} - g)\cos\theta + m\ddot{y}\sin\theta.$$

Since
$$\ddot{x} = -R\ddot{\theta}\sin\theta - R\dot{\theta}^2\cos\theta;$$
$$\ddot{y} = R\ddot{\theta}\cos\theta - R\dot{\theta}^2\sin\theta,$$

where
$$\dot{\theta}^2 = \frac{2g}{R}(\cos\theta - \cos\theta_0); \qquad \ddot{\theta} = -\frac{g}{R}\sin\theta,$$

we have:
$$L_\tau = m(g + R\ddot{\theta}\sin\theta + R\dot{\theta}^2\cos\theta)\sin\theta + mR\ddot{\theta}\cos^2\theta - mR\dot{\theta}^2\sin\theta\cos\theta$$
$$= mg\sin\theta + mR\ddot{\theta} = mR\left(\ddot{\theta} + \frac{g}{R}\sin\theta\right) = 0. \qquad (3.55)$$

This result has been expected, because the constraint is ideal (the force of friction, given by the tangent component of the constraint force, *must* be zero). Also,
$$L_\nu = -m(g + R\ddot{\theta}\sin\theta + R\dot{\theta}^2\cos\theta)\cos\theta + mR\ddot{\theta}\sin\theta\cos\theta - mR\dot{\theta}^2\sin^2\theta$$
$$= -mg\cos\theta - 2mg(\cos\theta - \cos\theta_0) = mg(2\cos\theta_0 - 3\cos\theta). \qquad (3.56)$$

The magnitude of the constraint force is therefore
$$L = |L_\nu| = mg|2\cos\theta_0 - 3\cos\theta|. \qquad (3.57)$$

As one can see, this result is identical to that obtained in the previous paragraph, but differs in sign from that found by the Newtonian approach. The explanation is that the tension $\vec{T}$ in the rod (see

Fig.III.1), and the external normal $\vec{\nu}$ (see Figg.III.7) have opposite senses.

III.4. Hamilton's canonical equations approach

Within this formalism, the enunciation of the problem is the same as in the previous two paragraphs. To solve a problem by method offered by Hamilton's canonical equations, one uses the following algorithm:

i) Identification of applied forces $\vec{F}_i$ ($i = \overline{1,N}$), where N is the number of bodies (particles) of the system.

ii) Setting the initial conditions, compatible with the constraints, in real space: $\vec{r}_{i0} = \vec{r}_i(t_0)$, and $\dot{\vec{r}}_{i0} = \dot{\vec{r}}_i(t_0)$.

iii) Identification of the number of degrees of freedom of the studied system $n = 3N - s$ and a convenient choice of the generalized coordinates $q_j = q_j(t)$ ($j = \overline{1,n}$).

iv) Establishment of connection relations between the real and configuration spaces $\vec{r}_i = \vec{r}_i(q_1, q_2, ..., q_n, t) \equiv \vec{r}_i(q, t)$ ($i = \overline{1, N}$).

v) Determination of the Lagrange function $L = L(q, \dot{q}, t)$ and of the generalized momenta $p_j = \partial L/\partial \dot{q}_j$ ($j = \overline{1,N}$) as well.

vi) Writing the inverse of relation $\vec{r}_i = \vec{r}_i(q, t)$, that is $q_j = q_j(\vec{r}_1, \vec{r}_2, ..., \vec{r}_N, t) \equiv q_j(\vec{r}, t)$, and setting the initial conditions in phase space $q_{j0} = q_j(t_0)$, $p_{j0} = p_j(t_0)$ ($j = \overline{1,n}$).

vii) Writing the Hamiltonian by means of the definition $H = \sum_{j=1}^{n} p_j \dot{q}_j - L = H(p, q, t)$.

viii) Writing and solving Hamilton's canonical equations

$$\dot{q}_j = +\frac{\partial H}{\partial p_j}; \qquad \dot{p}_j = -\frac{\partial H}{\partial q_j}, \qquad (3.58)$$

which furnishes the law of motion of the system in phase space $q_j = q_j(t)$, $p_j = p_j(t)$ ($j = \overline{1,n}$). These relations also represent the parametric equations of the generalized trajectory of the representative point in phase space.

ix) Determination of the law of motion in the real space: $\vec{r}_i = \vec{r}_i(t)$.

x) Determination of the constraint forces: $\vec{L}_i = m_i \ddot{\vec{r}}_i - \vec{F}_i$ ($i = \overline{1,N}$).

In the light of this algorithm, we shall resume the problem of simple gravitational pendulum performing free, harmonic, non-amortized oscillations, for an arbitrary value θ_0 of the oscillation angular amplitude.

The only applied force is the force of gravity $\vec{G} = (mg, 0, 0)$, whose components are expressed with respect to the reference system xOy (see Fig.III.7). The initial conditions are, obviously, the same:

$$\vec{r}_0 = (x_0, y_0, 0) = (R\cos\theta_0, R\sin\theta_0, 0),$$

and

$$\dot{\vec{r}}_0 = 0,$$

As we previously did, since the system possesses a single degree of freedom ($n = 3N - s = 3 \cdot 1 - 2 = 1$), as generalized coordinate one can choose the angular coordinate θ. The Lagrangian being

$$L = \frac{1}{2}mR^2\dot{\theta}^2 + mgR\cos\theta,$$

the associated generalized momentum is

$$p_\theta = \frac{\partial L}{\partial \dot{\theta}} = mR^2\dot{\theta}. \tag{3.59}$$

Consequently, the phase space has two dimensions. In other words, a point in this space has two coordinates (θ, p_θ). The initial conditions write

$$\theta_{t_0=0} = \theta_0;$$

$$p_{\theta(t_0=0)} = (p_\theta)_0 = mR^2\dot{\theta}_{t_0=0} = 0,$$

while the Hamiltonian is

$$H = p_\theta\dot{\theta} - L = p_\theta \frac{p_\theta}{mR^2} - \frac{1}{2}mR^2\frac{p_\theta^2}{m^2R^4} - mgR\cos\theta$$

$$= \frac{1}{2}\frac{p_\theta^2}{mR^2} - mgR\cos\theta = H(\theta, p_\theta). \tag{3.60}$$

By virtue of (3.58), we have

$$\dot{\theta} = \frac{\partial H}{\partial p_\theta}; \quad \dot{p}_\theta = -\frac{\partial H}{\partial \theta}, \tag{3.61}$$

or, in view of (3.60),

$$\dot{\theta} = \frac{p_\theta}{mR^2}; \quad \dot{p}_\theta = -mgR\sin\theta. \tag{3.62}$$

This is a coupled systems of two linear differential equations for variables θ and p_θ. To obtain equation for θ, one takes the derivative with respect to time of the first equation of (3.62), then introduces the result into the second equation. The result is

$$\ddot{\theta} + \frac{g}{R}\sin\theta = 0, \tag{3.63}$$

which is precisely the equation we were looking for.

From now on, the procedure follows the way already shown in the previous paragraph (Lagrange equations of the second kind), concerning determination of both the elements of motion[1], and the constraint force $\vec{L} = m\ddot{\vec{r}} - \vec{G}$.

III.5. Hamilton-Jacobi method

To solve a problem by means of the Hamilton-Jacobi formalism, one must follow the steps of this algorithm:

i) Identification of applied forces $\vec{F}_i$ ($i = \overline{1,N}$), where N is the number of bodies (particles) composing the system.

ii) Establishment of the initial conditions (compatible with the constraints) in real space $\vec{r}_{i0} = \vec{r}_i(t_0)$; $\dot{\vec{r}}_{i0} = \dot{\vec{r}}_i(t_0)$.

iii) Identification of the degrees of freedom of the physical system $n = 3N - s$ and a convenient choice of the generalized coordinates $q_j = q_j(t)$ ($j = \overline{1,n}$).

iv) Establishment of relations between the real and configuration spaces $\vec{r}_i = \vec{r}_i(q_1, q_2, ..., q_n, t) \equiv \vec{r}_i(q,t)$ ($i = \overline{1,N}$), their inverses $q_j = q_j(\vec{r}_1, \vec{r}_2, ..., \vec{r}_N, t) \equiv q_j(\vec{r}, t)$, as well as the initial conditions in configuration space $q_{j0} = q_j(t_0)$, $\dot{q}_{j0} = \dot{q}_j(t_0)$ ($j = \overline{1,n}$).

v) Writing the Lagrangian of the system $L(q, \dot{q}, t)$, the generalized momenta associated with the generalized coordinates $p_j = \frac{\partial L}{\partial \dot{q}_j}$ ($j = \overline{1,n}$), and establishment of the $2n$ initial conditions in configuration space $q_{j0} = q_j(t_0)$, $p_{j0} = p_j(t_0)$ ($j = \overline{1,n}$).

vi) Determination of the Hamiltonian of the system

$$H = \sum_{j=1}^{n} p_j \dot{q}_j - L = H(q, p, t).$$

---

[1] We note that even if Hamilton's canonical equations and second kind Lagrange equations are equivalent, in this paragraph we cannot take advantage of the two first integrals established within Lagrangian approach. As a result, $\dot{\theta}$ is determined by means of the Newtonian formalism.

vii) Writing and solving the Hamilton-Jacobi equation

$$\frac{\partial S}{\partial t} + H\left(q_1, q_2, ..., q_n, \frac{\partial S}{\partial q_1}, \frac{\partial S}{\partial q_2}, ..., \frac{\partial S}{\partial q_n}, t\right) = 0, \quad (3.64)$$

where $S(q,t)$ is *Hamilton's principal function*.

viii) Determination of the law of motion in configuration space $q_j = q_j(t)$ $(j = \overline{1,n})$, by means of Jacobi's theorem.

ix) Setting the law of motion in real space $\vec{r}_i = \vec{r}_i(t)$ $(i = \overline{1,N})$, in view of relations established at the point (iv) of this algorithm.

x) Finding the constraint forces, by means of the fundamental equation of Mechanics for systems subject to constraints

$$\vec{L}_i = m_i \ddot{\vec{r}}_i - \vec{F}_i \quad (i = \overline{1,N}).$$

We shall use this algorithm to solve the problem of plane (simple) pendulum, performing free, non-amortized oscillations, of an arbitrary amplitude θ_0. First of all, we observe that the only applied (active) force is the force of gravity $\vec{G} = (mg, 0, 0)$, whose components are expressed in the xOy coordinate system (see Fig.III.7).

The initial conditions are, obviously, the same as in the preceding approaches:

$$\vec{r}_0 = (x_0, y_0, 0) = (R\cos\theta_0, R\sin\theta_0, 0);$$

$$\dot{\vec{r}}_0 = 0.$$

Since the motion is subject to two constraints (see the previous methods), there is a single degree of freedom: $n = 3N - s = 3 \cdot 1 - 2 = 1$. As generalized coordinate, associated with this degree of freedom, one chooses the angle θ. The coordinates of the real space x, y and the generalized coordinate θ are related by

$$x = R\cos\theta; \quad y = R\sin\theta,$$

while the initial conditions in configuration space write

$$\theta_{t=0} = \theta_0, \quad \dot{\theta}_{t=0} = \dot{\theta}_0 = 0.$$

The Lagrangian and the Hamiltonian have been already deduced in paragraphs III.3 [see (3.48)] and III.4 [see (3.60)]:

$$L = \frac{1}{2}mR^2\dot{\theta}^2 + mgR\cos\theta,$$

$$H = \frac{1}{2}\frac{p_\theta^2}{mR^2} - mgR\cos\theta. \tag{3.65}$$

According to Jacobi's theorem (see further), one can write

$$p_j = \frac{\partial S}{\partial q_j} \quad (j = \overline{1,n}),$$

where $S(q,t)$ is *Hamilton's principal function*. In our case, since our system has only one degree of freedom, the generalized momentum is $p_\theta = \frac{\partial S}{\partial \theta}$, and the Hamiltonian becomes

$$H = \frac{1}{2mR^2}\left(\frac{\partial S}{\partial \theta}\right)^2 - mgR\cos\theta.$$

The Hamilton-Jacobi equation then writes

$$\frac{\partial S}{\partial t} + \frac{1}{2mR^2}\left(\frac{\partial S}{\partial \theta}\right)^2 - mgR\cos\theta = 0. \tag{3.66}$$

There are no general methods for solving Hamilton-Jacobi equation (3.64). Nevertheless, in some particular cases one can use the method of separation of variables. In this respect, a great help comes from Jacobi's theorem. We shall not prove it, but apply it in our case. To do this, one must recall several important definitions.

A *complete integral* (or *complete solution*) of a first-order partial differential equation is a solution of this equation which contains as many arbitrary, independent constants as there are independent variables in the equation. In Hamilton-Jacobi equation interfere $(n+1)$ independent variables: $q_1, q_2, ..., q_n, t$, but, since S appears only as partial derivatives, one of the arbitrary constants is purely additive. Therefore, a complete integral of the Hamilton-Jacobi equation writes

$$S(q_1, q_2, ..., q_n, a_1, a_2, ..., a_n, a_{n+1}, t)$$
$$= S(q_1, q_2, ..., q_n, a_1, a_2, ..., a_n, t) + const.,$$

where as an additive constant has been taken, for example, a_{n+1}. Because in Hamilton-Jacobi equation appear only partial derivatives of S, this constant can be dropped. To conclude, a complete integral of the Hamilton-Jacobi equation writes $S(q_1, q_2, ..., q_n, a_1, a_2, ..., a_n, t)$, the constants $a_1, ..., a_n$ being called *essential*.

According to *Jacobi's theorem*, if $S(q_1, q_2, ..., q_n, a_1, a_2, ..., a_n, t)$ is a complete integral (complete solution) of the Hamilton-Jacobi equation (3.64), then the general solution of Hamilton's canonical equations (3.58) is given by

$$p_j = \frac{\partial S}{\partial q_j}; \quad b_j = \frac{\partial S}{\partial a_j} \quad (j = \overline{1,n}), \tag{3.67}$$

where b_j ($j = \overline{1,n}$) are new constants. It is not our purpose to prove this theorem, but we shall use it. Introducing $S(q,a,t)$ into (3.67b), we obtain the law of motion in the configuration space:

$$q_j = q_j(a_1, a_2, ..., a_n, b_1, b_2, ..., b_n, t) \ (j = \overline{1,n}).$$

The constants $a_1, ..., a_n, b_1, ..., b_n$ are determined by means of the initial conditions. This way, the law of motion in real space $\vec{r}_i = \vec{r}_i(q(t), t) = \vec{r}_i(t)$ ($i = \overline{1,N}$) can be found.

In our particular case, we have to determine a complete integral $S = S(\theta, E, t)$ of equation (3.66). It depends on a single essential constant E, which has been chosen to be the total energy of the system. Since there are two independent variables θ and t, and the Hamiltonian does not explicitly depend on time, this suggests a solution of the form

$$S(\theta, a, t) = W(\theta, a) + S_1(a, t), \tag{3.68}$$

where a is an essential arbitrary constant, while W and S_1 are two functions, unknown for the moment. Introducing (3.68) into the Hamilton-Jacobi equation

$$\frac{\partial S}{\partial t} + H\left(\theta, \frac{\partial S}{\partial \theta}\right) = 0,$$

we arrive at

$$\frac{\partial S_1(a,t)}{\partial t} + H\left(\theta, \frac{\partial W(\theta, a)}{\partial \theta}\right) = 0. \tag{3.69}$$

Since the first term depends only on t, and the second only on θ, equation (3.69) is satisfied only if each of the two terms equals the same constant, say E. Consequently, we have:

$$H\left(\theta, \frac{\partial W}{\partial \theta}\right) = E, \tag{3.70}$$

called *restricted* or *abbreviated* Hamilton-Jacobi equation, and

$$\frac{\partial S_1}{\partial t} = -E. \tag{3.71}$$

Since a complete solution of (3.69) must depend on one essential constant only, this means that a must be precisely E. The constraint is scleronomous, and the Hamiltonian does not explicitly depend on time. This means that the total energy E of the system is conserved.

It is now clear why we denoted by E the constant in Eqs. (3.70) and (3.71).

Equation (3.71) yields

$$S_1(a,t) \equiv S_1(E,t) = -Et,$$

so that (3.68) becomes

$$S(\theta, a, t) \equiv S(\theta, E, t) = -Et + W(\theta, E). \tag{3.72}$$

To determine $W(\theta, E)$ one must solve the abbreviated Hamilton-Jacobi equation (3.70). As we shall see, in our case is not necessary to know the explicit form of this function. Introducing

$$p_\theta = \frac{\partial S}{\partial \theta} = \frac{\partial W}{\partial \theta}$$

into Hamilton's function, and the result into (3.70), we have

$$\frac{1}{2mR^2}\left(\frac{\partial W}{\partial \theta}\right)^2 - mgR\cos\theta = E,$$

and, by integration

$$W(\theta, E) = \pm \int \sqrt{2mR^2(E + mgR\cos\theta)}\, d\theta.$$

According to Hamilton-Jacobi theorem, $\left(\frac{\partial S}{\partial E}\right) = b$ [see (3.67-2)], and we have

$$-t + \frac{\partial W}{\partial E} = b. \tag{3.73}$$

Since b must have units of time, we denote it by t_1, and (3.73) becomes

$$\frac{\partial W}{\partial E} = t + t_1,$$

that is

$$\pm \frac{\partial}{\partial E}\left[\int \sqrt{2mR^2(E + mgR\cos\theta)}\,d\theta\right] = t + t_1,$$

or, recalling that E is an essential constant, and θ an independent variable,

$$\pm \int \frac{\partial}{\partial E}\left[\sqrt{2mR^2(E + mgR\cos\theta)}\right]d\theta = t + t_1.$$

Performing the derivative, we still have

$$\pm \int \frac{d\theta}{\sqrt{\frac{2E}{mR^2} + \frac{2g}{R}\cos\theta}} = t + t_1.$$

The last equation says that $t + t_1$ is the primitive of

$$\pm \left[\frac{2E}{mR^2} + \frac{2g}{R}\cos\theta\right]^{-1/2},$$

which means

$$\frac{d}{d\theta}(t + t_1) = \pm \frac{1}{\sqrt{\frac{2E}{mR^2} + \frac{2g}{R}\cos\theta}},$$

or

$$\dot\theta = \frac{d\theta}{dt} = \pm\sqrt{\frac{2E}{mR^2} + \frac{2g}{R}\cos\theta}.$$

Let us introduce a new constant

$$const. = \cos\theta_0 = -\frac{E}{mgR}. \tag{3.74}$$

This finally leads to

$$\dot\theta = \pm\sqrt{\frac{2g}{R}(\cos\theta - \cos\theta_0)}. \tag{3.75}$$

This equation has been already obtained within the frame of the Newtonian and Lagrange equations of the second kind approaches. From now on, the problem can be solved as in paragraph III.3.

Observation. The way we have introduced the constant $\cos\theta_0$ is in agreement with the expression of the total energy of the system, $E = -mgR\cos\theta_0$, deduced by the Lagrange equations of the second kind formalism.

III.6. Action-angle variables formalism

This approach is successfully used in case of the systems performing periodic motions. It emerges from the Hamilton-Jacobi method and consists in defining a new set of variables: J_j - *action variables*, and w_j - *angle variables*, instead of the canonical parameters q_j, p_j $(j = \overline{1,n})$, where n is the number of degrees of freedom of the system.

Let us briefly explain the essence of this formalism. Consider a conservative system with one degree of freedom. Denoting by E the total energy of the system, we can write

$$H(p, q) = E. \tag{3.76}$$

This is the implicit form of a curve in the phase plane (p, q) and represents the generalized trajectory of the conservative system characterized by the constant energy E. There are two possible periodical motions:

1. If the generalized trajectory is a closed curve, the motion is called *vibration*. In this case, the generalized coordinate q oscillates between two constant values, both p and q being functions of time with the same period (see Fig.III.8).

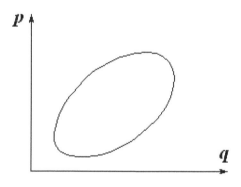

Fig.III.8

2. If the solution $p = p(q, E)$ of equation (3.76) can be written as a periodical function of q with period q_0, that is $p(q + kq_0, E) = p(q, E)$ ($k \in \mathbb{Z}$), the periodical motion is called *rotation*, or *revolution*. In this case, the coordinate q can take any value (see Fig.III.9).

There are systems capable to perform both vibration and rotation motions. As we shall see, such a system is precisely our case of the simple pendulum.

Let us define *the action variable* as

$$J = \oint p \, dq, \tag{3.77}$$

where the integral is taken over a complete cycle of variation of q. It stands for both the closed area shown in Fig.III.8, and the shaded area drawn in Fig.III.9, corresponding to a period of motion of rotation.

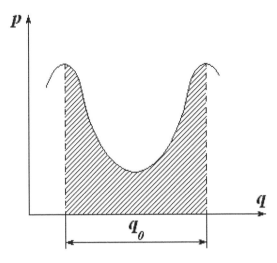

Fig.III.9

According to the definition, the quantity J and the action $S = \int_{t_1}^{t_2} L\, dt$ have the same dimension. Taking into account the relations $p = p(q, E)$ and (3.77), we can write $J = J(E)$ or, conversely, $E = E(J)$. The complete integral corresponding to Hamiltonian (3.76) is $S = S(q, E)$, with a single essential constant E. Since $E = E(J)$, this yields

$$S = S(q, J). \tag{3.78}$$

The canonical variable w associated with J is defined by

$$w = \frac{\partial S}{\partial J}, \tag{3.79}$$

being called *angle variable*. The name comes from the fact that w is associated with an angular momentum, so that its dimension is that of an angle. In view of (3.76) and $E = E(J)$, the new Hamiltonian is

$$H = \mathcal{H}(J). \tag{3.80}$$

With this choice, canonical equations lead to

$$\begin{cases} \dot{J} = -\frac{\partial \mathcal{H}}{\partial w} = 0, \\ \dot{w} = \frac{\partial \mathcal{H}}{\partial J} = \nu(J). \end{cases} \tag{3.81}$$

It follows from $(3.81)_1$ that $J = const.$ (which we already know). As a result, ν is also a constant, depending on J. Integrating $(3.81)_2$, we obtain

$$w = \nu t + \alpha, \tag{3.82}$$

where α is an arbitrary constant of integration.

To find the significance of ν, we shall determine the way of variation of w for a complete cycle of variation of q, for both vibration and rotation motions. Thus,

$$\Delta w = \oint \frac{\partial w}{\partial q}\, dq = \oint \frac{\partial^2 S}{\partial q \partial J}\, dq = \oint \frac{\partial p}{\partial J}\, dq = \frac{d}{dJ}\oint p\, dq = 1.$$

Denoting by τ the period corresponding to a complete cycle, the last two relations yield $\Delta w = \nu \Delta t = \nu \tau = 1$, which means

$$\nu = \frac{1}{\tau}. \tag{3.83}$$

This result shows that ν is the frequency of the periodical variation of q. This displays the fact that the period of any motion can be determined without solving the equation of motion, if one knows the dependence on time of the Hamiltonian. Inverting (3.78) and using (3.82), one can also determine the time dependence of coordinate q.

Let us now apply these considerations to the "vibration" motion of a simple pendulum. As seen in paragraph III.3, the Lagrangian of the system is

$$L = \frac{1}{2}mR^2\dot{\theta}^2 + mgR\cos\theta.$$

In our case $q = \theta$, while the associated generalized momentum is $p = \frac{\partial L}{\partial \dot{\theta}} = mR^2\dot{\theta}$. The Hamiltonian therefore is

$$H = p\dot{\theta} - L = \frac{1}{2}mR^2\dot{\theta}^2 - mgR\cos\theta = \frac{p^2}{2mR^2} - mgR\cos\theta = E.$$

This allows us to write the equation of generalized trajectory in the phase plane (θ, p) as

$$p = \pm\sqrt{2mR^2(E + mgR\cos\theta)}, \tag{3.84}$$

as well as the action variable J

$$J = -2\int_{\theta_0}^{-\theta_0}[2mR^2(E + mgR\cos\theta)]^{1/2}\, d\theta$$

$$= 4\int_0^{\theta_0}[2mR^2(E + mgR\cos\theta)]^{1/2}\, d\theta.$$

Equation (3.81)$_2$ can be written as

$$\nu = \nu(J) = \frac{\partial \mathcal{H}}{\partial J} = \frac{1}{\frac{\partial J}{\partial \mathcal{H}}} = \left(\frac{\partial J}{\partial E}\right)^{-1}.$$

According to (3.83), we then have

$$\tau = \frac{1}{\nu} = \frac{\partial J}{\partial E} = 4\frac{d}{dE}\left(\int_0^{\theta_0} \sqrt{2mR^2(E + mgR\cos\theta)}\, d\theta\right)$$

$$= 4\sqrt{\frac{R}{2g}} \int_0^{\theta_0} \frac{d\theta}{\sqrt{\cos\theta - \cos\theta_0}}. \qquad (3.85)$$

This result has been already obtained by means of Newtonian approach, and the remaining part of investigation can be found within this formalism. There is, nevertheless, one problem: the angle-action method is not able to determine the constraint forces (in our case, tension in the rod). Due to this inconvenience, the angle-action procedure is used only for systems performing periodical motions, in which case the main problem is determination of the period of motion.

In the case of small oscillations ($\theta_0 < 4°$), one can approximate

$$\cos\theta = 1 - \frac{\theta^2}{2} + \mathcal{O}(\theta^4) \simeq 1 - \frac{\theta^2}{2},$$

and the Hamiltonian writes

$$H = \frac{p^2}{2mR^2} - mgR\cos\theta \simeq \frac{p^2}{2mR^2} + mgR\left(\frac{\theta^2}{2}\right) - mgR,$$

or, by omitting the constant term $(-mgR)$,

$$H = \frac{p^2}{2mR^2} + \frac{mgR\theta^2}{2} = E,$$

so that

$$p = \pm\sqrt{2mR^2\left(E - \frac{mgR\theta^2}{2}\right)},$$

leading to

$$J = -2\int_{\theta_0}^{-\theta_0} \left[2mR^2\left(E - \frac{mgR\theta^2}{2}\right)\right]^{1/2} d\theta$$

$$= \sqrt{2} \int_{-\theta_0}^{\theta_0} \left(4mER^2 - 2m^2gR^3\theta^2\right)^{1/2} d\theta = 4E\sqrt{\frac{R}{g}} \int_{-\xi_1}^{\xi_1} (1-\xi^2)^{1/2} d\xi$$

$$= 4E\sqrt{\frac{R}{g}} \int_{\beta_1}^{\beta_1+\pi} \cos^2\beta \, d\beta = 4E\sqrt{\frac{R}{g}} \int_{\beta_1}^{\beta_1+\pi} \frac{1+\cos 2\beta}{2} d\beta$$

$$= 4E\sqrt{\frac{R}{g}} \left[\frac{\pi}{2} + \left(\frac{\sin 2\beta}{4}\right)_{\beta_1}^{\beta_1+\pi}\right] = 2\pi E\sqrt{\frac{R}{g}},$$

where the following changes of variable have been used

$$\theta\sqrt{\frac{mgR}{2E}} = \xi = \sin\beta, \quad \theta_0\sqrt{\frac{mgR}{2E}} = \xi_1, \quad \beta_1 = \arcsin(-\xi_1).$$

Therefore,

$$J = 2\pi E\sqrt{\frac{R}{g}}, \qquad (3.86)$$

leading to the well-known formula of small, isochronous oscillations of a simple pendulum

$$\tau = \frac{dJ}{dE} = 2\pi\sqrt{\frac{R}{g}}. \qquad (3.87)$$

Let us now turn back to the generalized trajectory given by (3.84). The value $p = 0$ corresponds to

$$\cos\theta_0 = -\frac{E}{mgR}. \qquad (3.88)$$

If $E < mgR$, the pendulum performs a periodical motion called *vibration*, with angle θ varying between $-\theta_0$ and θ_0, with θ_0 given by (3.88). If $E > mgR$, there are no restrictions on θ, that is the motion is the same for any $\theta + 2k\pi$ ($k \in \mathbb{Z}$). A special situation is given by the limit case $E = mgR$, corresponding to $\theta = \pm\pi$ [or, in general, $\theta = (2k+1)\pi$]. These values of θ, together with $p = 0$, designate the position of unstable equilibrium of the system.

* * *

To conclude this chapter, let us calculate the deviation from the real time of an *astronomical pendulum* ($\theta_0 = 1°30'$) in one year, assuming that the oscillations are isochronous, with period

$$T_{iso} = 2\pi\sqrt{\frac{R}{g}}.$$

The "exact" period, which takes into account the amplitude θ_0 of the oscillations, is given by (3.9)

$$T_{non-iso} = 4\sqrt{\frac{R}{g}} \int_0^{\pi/2} \frac{d\psi}{\sqrt{1 - k^2 \sin^2 \psi}}$$

with $k = \sin \frac{\theta_0}{2}$. The time difference between the two values, during one period, is

$$\Delta T = T_{non-iso} - T_{iso} = 2\sqrt{\frac{R}{g}} \left[2 \int_0^{\pi/2} \frac{d\psi}{\sqrt{1 - k^2 \sin^2 \psi}} - \pi \right] > 0.$$

In other words, the "ideal" clock (which performs isochronous oscillations) advances by ΔT seconds for each period. This clock performs

$$\nu_{non-iso} = \frac{1}{T_{non-iso}}$$

complete oscillations per unit time. During one year, the number of complete oscillations is

$$N_{exact} = 365 \times 24 \times 3600 \times \nu_{non-iso} = \frac{365 \times 24 \times 3600}{T_{non-iso}}.$$

The time difference between the two clocks, in one year's time, is

$$\Delta t_{1\,year} = N_{exact} \Delta T = 365 \times 24 \times 3600 \times \frac{T_{non-iso} - T_{iso}}{T_{non-iso}}$$

$$= 365 \times 24 \times 3600 \times \left(1 - \frac{\pi}{2K(k)}\right)$$

$$= 365 \times 24 \times 3600 \times \left[1 - \pi \left(2 \int_0^{\pi/2} \frac{d\psi}{\sqrt{1 - \sin^2 \frac{\theta_0}{2} \sin^2 \psi}}\right)^{-1}\right].$$

Performing numerical calculations, one obtains

$$\Delta t_{1\,year} = 1350.897\,s = 22.515\,min.$$

The time difference in one day is therefore

$$\Delta t_{1\,day} = 3.701\,s.$$

The difference is too big to consider "good" a clock performing isochronous oscillations. In case of a *second pendulum* (a pendulum which beats once every second, *i.e.* for which $T = 2s$), characterized by $R = \frac{g}{\pi^2}$, the difference $T_{non-iso} - T_{iso}$ is

$$\Delta T_{(2s)} = T_{non-iso} - T_{iso} = T_{non-iso} - 2$$

$$= 2.000157465 - 2 = 157.465\,\mu s.$$

CHAPTER IV

PROBLEMS SOLVED BY MEANS OF THE PRINCIPLE OF VIRTUAL WORK

Problem 1
Using the virtual work principle, determine the equilibrium position of the system shown in Fig.IV.1. We are given the following data: masses m_1 and m_2 of the two bodies, the angle α of the inclined plane, and the distance between the top of the inclined plane and the pulley axis. The friction phenomenon and the pulley radius are neglected, while the wire is supposed to be inextensible. It is presumed that during the motion the body of mass m_1 remains on the inclined plane, and the body of mass m_2 only moves vertically.

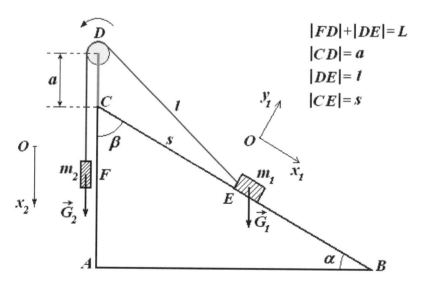

Fig.IV.1

Solution
Let us first identify the constraints affecting the degrees of freedom of the system. To this end, we attach to each body a reference

frame, as shown in Fig.IV.1. The equations of constraints then are:

i) $f_1(z_1) = z_1 = C_1$;

ii) $f_2(y_1) = y_1 = 0$ (the body of mass m_1 does not leave the inclined plane);

iii) $f_3(y_2) = y_2 = C_2$;

iv) $f_4(z_2) = z_2 = C_3$;

v) $f_5(x_1, x_2) = \sqrt{x_1^2 + 2ax_1 \sin\alpha + a^2} + x_2 + a - L = 0$.

The constants C_1, C_2, C_3, without loss of generality, can be considered equal to zero. We denoted by L the constant length of the wire connecting the two bodies, and f_5 was written by means of generalized Pythagoras theorem in triangle CED. We have also supposed that the two reference frames have the same origin at C, the highest point of the inclined plane. The system has, therefore, $3 \cdot 2 - 5 = 1$ degree of freedom.

According to the principle of virtual work

$$\sum_{i=1}^{N} \vec{F}_i \cdot \delta \vec{r}_i = 0, \qquad (4.1)$$

where N is number of bodies of the system, and $\vec{F}_i$ ($i = \overline{1, N}$) the applied forces. In our case, the applied forces are $\vec{G}_1$ and $\vec{G}_2$, so that (4.1) writes

$$\vec{G}_1 \cdot \delta \vec{r}_1 + \vec{G}_2 \cdot \delta \vec{r}_2 = 0. \qquad (4.2)$$

The vector quantities appearing in (4.2) have the following components:

$$\vec{G}_1 = (m_1 g \sin\alpha, \ -m_1 g \cos\alpha, \ 0),$$

$$G_2 = (m_2 g, \ 0, \ 0),$$

$$\delta \vec{r}_1 = (-\delta x_1, \ 0, \ 0),$$

$$\delta \vec{r}_2 = (\delta x_2, \ 0, \ 0),$$

where we took advantage of the choice of reference frames as regarded the motion of the bodies (the body of mass m_1 climbs up the inclined plane, moving in the negative sense of Ox_1 axis). Using these observations, (4.2) becomes:

$$-m_1 g \sin\alpha \, \delta x_1 + m_2 g \, \delta x_2 = 0. \qquad (4.3)$$

This is a null linear combination of the quantities δx_1 and δx_2, with constant coefficients $m_1 g \sin\alpha$ and $m_2 g$. Obviously, δx_1 and δx_2 cannot be linearly independent: otherwise we would have $m_1 g \sin\alpha =$

0 and $m_2 g = 0$, which would be absurd. This is also shown by the fact that system has only one degree of freedom. Therefore, since the two virtual variations δx_1 and δx_2 are linearly dependent, one of them must be eliminated.

To this end, we observe (see Fig.IV.1) that an infinitesimal variation of the coordinate x_1 of the body with mass m_1 identifies with an elementary variation of $s = |CE|$, that is

$$\delta x_1 \equiv \delta s. \tag{4.4}$$

We also observe that a small variation of the coordinate x_2 of the body with mass m_2 equals a small variation of $l = |DE|$ (the wire is inextensible), so that

$$\delta x_2 \equiv \delta l. \tag{4.5}$$

Consequently, a relation between δx_1 and δx_2, on the one hand, and a formula connecting δs and δl, on the other, are equivalent. Using the generalized Pythagoras theorem in triangle CED, we have

$$l^2 = a^2 + s^2 - 2as\cos(\pi - \beta) = a^2 + s^2 + 2as\cos\beta$$

$$= a^2 + s^2 + 2as\cos\left(\frac{\pi}{2} - \alpha\right) = a^2 + s^2 + 2as\sin\alpha,$$

and, by differentiation,

$$l\, dl = (s + a\sin\alpha)\, ds.$$

This means that we also have

$$l\, \delta l = (s + a\sin\alpha)\, \delta s,$$

or, in view of (4.4) and (4.5),

$$l\, \delta x_2 = (s + a\sin\alpha)\, \delta x_1.$$

Eliminating δx_2 between this relation and (4.3), we are left with

$$[-m_1 g l \sin\alpha + m_2 g(s + a\sin\alpha)]\, \delta x_1 = 0. \tag{4.6}$$

Since this relation is valid for *any* virtual infinitesimal variation δx_1, the square bracket vanishes

$$-m_1 g l \sin\alpha + m_2 g(s + a\sin\alpha) = 0,$$

which is the equilibrium condition. Dividing this relation by $g \neq 0$, we still have

$$m_1\sqrt{a^2 + s^2 + 2as\sin\alpha}\;\sin\alpha = m_2(s + a\sin\alpha).$$

Squaring this relation and rearranging the terms, we are left with the following quadratic equation in s:

$$s^2(m_2^2 - m_1^2\sin^2\alpha) - 2sa\sin\alpha(m_1^2\sin^2\alpha - m_2^2) + a^2\sin^2\alpha(m_2^2 - m_1^2) = 0,$$

with the solutions

$$s_{1,2} = \frac{a\sin\alpha(m_1^2\sin^2\alpha - m_2^2)}{m_2^2 - m_1^2\sin^2\alpha} \pm$$

$$\frac{\sqrt{a^2\sin^2\alpha(m_1^2\sin^2\alpha - m_2^2)^2 - a^2\sin^2\alpha(m_2^2 - m_1^2)(m_2^2 - m_1^2\sin^2\alpha)}}{m_2^2 - m_1^2\sin^2\alpha},$$

or

$$s_{1,2} = a\sin\alpha$$

$$\times\left[-1 \pm \frac{\sqrt{(m_1^2\sin^2\alpha - m_2^2)^2 - (m_2^2 - m_1^2)(m_2^2 - m_1^2\sin^2\alpha)}}{m_2^2 - m_1^2\sin^2\alpha}\right].$$

The only physically acceptable solution is

$$s \equiv s_1 = a\sin\alpha$$

$$\times\left[\frac{\sqrt{(m_1^2\sin^2\alpha - m_2^2)^2 - (m_2^2 - m_1^2)(m_2^2 - m_1^2\sin^2\alpha)}}{m_2^2 - m_1^2\sin^2\alpha} - 1\right],$$

which can also be written as

$$s \equiv s_1 = a\sin\alpha\left[\frac{m_1\cos\alpha\sqrt{m_2^2 - m_1^2\sin^2\alpha}}{m_2^2 - m_1^2\sin^2\alpha} - 1\right]$$

$$= a\sin\alpha\left[\frac{m_1\cos\alpha}{\sqrt{m_2^2 - m_1^2\sin^2\alpha}} - 1\right]. \quad (4.7)$$

As observed, in order that the system has an equilibrium position, the following two conditions

$$m_2^2 - m_1^2 \sin^2 \alpha > 0;$$

$$\frac{m_1 \cos \alpha}{\sqrt{m_2^2 - m_1^2 \sin^2 \alpha}} \geq 1$$

have simultaneously to be fulfilled. In other words, the system

$$\begin{cases} m_2^2 - m_1^2 \sin^2 \alpha > 0; \\ m_1 \cos \alpha \geq \sqrt{m_2^2 - m_1^2 \sin^2 \alpha} \end{cases} \tag{4.8}$$

must be compatible.

The system (4.8) can also be written as

$$\begin{cases} |\sin \alpha| < \dfrac{m_2}{m_1}, \\ m_1 \geq m_2. \end{cases}$$

We easily realize that the system is compatible. Since $\alpha \in \left(0, \frac{\pi}{2}\right)$, it also follows a condition regarding angle α:

$$\alpha < \arcsin \frac{m_2}{m_1}, \quad \text{with } m_1 > m_2.$$

To conclude, the system is at equilibrium for

$$s = a \sin \alpha \left[\frac{m_1 \cos \alpha}{\sqrt{m_2^2 - m_1^2 \sin^2 \alpha}} - 1 \right]. \tag{4.9}$$

Problem 2

Solve Problem 1 by means of Torricelli's principle.

Solution

Since the only force acting on the mechanical system is the force of gravity, our problem can also be solved by using Torricelli's principle. In fact, Torricelli's principle is a particular form of the principle of virtual work, suitable to be applied to our problem. According to this principle, we have

$$\delta z_G = 0, \tag{4.10}$$

where z_G is the vertical height of the centre of gravity of the system. Following the already known algorithm, it is convenient to attach a unique three-orthogonal reference frame $Oxyz$ to the system of particles (bodies, in our case), with z-axis oriented vertically (see Fig.IV.2).

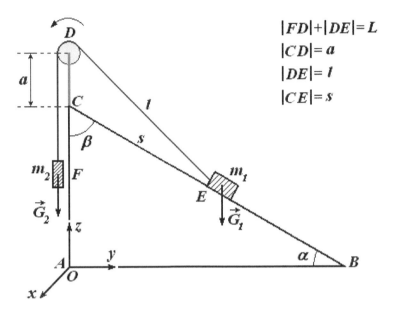

Fig.IV.2

If the xOy-plane is chosen as reference level for the potential energy of the system, then the height z_G of the centre of gravity is[1]

$$z_G = \frac{\sum_{i=1}^{N} z_i m_i}{\sum_{i=1}^{N} m_i} = \frac{1}{M} \sum_{i=1}^{N} z_i m_i, \qquad (4.11)$$

where N is the number of the material points of the system, and

$$M = \sum_{i=1}^{N} m_i$$

is the mass of the system. In our case $N = 2$, so that

$$z_G = \frac{m_1 z_1 + m_2 z_2}{m_1 + m_2}. \qquad (4.12)$$

---

[1] It is supposed that the force of gravity does not vary within the domain occupied by the system.

This way, our problem turns into an application of plane geometry. To ease the reasoning we shall simplify the figure, keeping only the elements necessary to determine z_1 and z_2 (see Fig.IV.3).

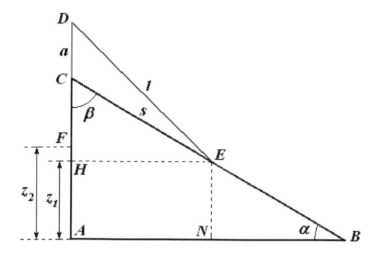

Fig.IV.3

We then have:

$$a = |CD|, \quad l = |DE|, \quad s = |CE|, \quad z_1 = |EN| = |AH|, \quad z_2 = |AF|,$$

also

$$|FC| + a + l = L,$$
$$|AC| - z_1 = s \sin\alpha,$$
$$|AC| = z_2 + |FC|.$$

Therefore,

$$|FC| = L - a - l,$$
$$z_1 = |AC| - s \sin\alpha,$$
$$z_2 = |AC| - |FC| = |AC| - L + a + l,$$

and formula (4.12) becomes

$$z_G = \frac{m_1 z_1 + m_2 z_2}{m_1 + m_2}$$

$$= \frac{1}{m_1 + m_2}\left[m_1\left(|AC| - s\sin\alpha\right) + m_2\left(|AC| - L + a + l\right)\right]$$

$$= \frac{|AC|(m_1 + m_2) - m_2(L - a)}{m_1 + m_2} + \frac{m_2 l - m_1 s \sin\alpha}{m_1 + m_2}.$$

Since $|AC| = const.$, we can write

$$z_G = C_0 + \frac{m_2 l - m_1 s \sin\alpha}{m_1 + m_2}, \qquad (4.13)$$

where by C_0 we denoted the constant quantity

$$C_0 = \frac{|AC|(m_1 + m_2) - m_2(L - a)}{m_1 + m_2}.$$

Let us now differentiate (4.13), then replace differentials by virtual infinitesimal variations. The result is

$$\delta z_G = \frac{m_2 \delta l - m_1 \delta s \sin\alpha}{m_1 + m_2}.$$

Using the relation between δl and δs achieved in the previous problem

$$l\,\delta l = (s + a\sin\alpha)\,\delta s,$$

we still have

$$\delta z_G = \frac{1}{l(m_1 + m_2)}\bigl[m_2(s + a\sin\alpha) - m_1 l \sin\alpha\bigr]\delta s.$$

Since δs is arbitrary, the last relation yields

$$m_2(s + a\sin\alpha) - m_1 l \sin\alpha = 0,$$

or

$$m_2(s + a\sin\alpha) = m_1\sqrt{s^2 + a^2 + 2as\sin\alpha}\,\sin\alpha.$$

Squaring this equation and rearranging the terms, we finally have

$$s^2(m_2^2 - m_1^2 \sin^2\alpha) - 2sa\sin\alpha(m_1^2 \sin^2\alpha - m_2^2)$$
$$+ a^2 \sin^2\alpha(m_2^2 - m_1^2) = 0,$$

which has been already obtained in the preceding problem. From now on, the way to solution is known.

Problem 3

Using Torricelli's principle, determine the equilibrium position of the homogeneous, rigid rod shown in Fig.IV.4. The length of the rod is $2l$, the radius of the fixed semi-cylinder is R, while friction is neglected.

Solution

As observed, the equilibrium position of the rod is univoquely determined by the angle $\alpha \in \left(0, \frac{\pi}{2}\right)$ only. One can intuitively anticipate that the rod can be at equilibrium only if its centre of gravity falls inside the semi-cylinder. Otherwise, due to the force of gravity, the rod would fall down outside the semi-cylinder.

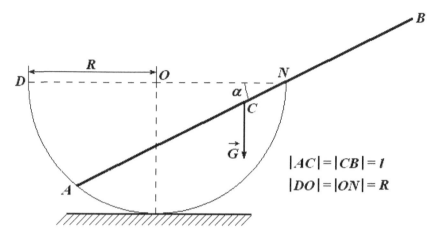

Fig.IV.4

Since the only force applied to the rod is its own gravity $\vec{G} = m\vec{g}$, where m is the mass of the homogeneous rod, the problem can be solved by means of Torricelli's principle. Choosing a Cartesian reference frame with axes oriented as shown in Fig.IV.5, and the xOy-plane passing through DN as the reference level for the potential energy, our task is to determine the segment $|MC| = z_G$. This way, up to a certain point, our physical problem becomes a problem of plane geometry.

We have:

$$|MC| = |NC|\sin\alpha = (|AN| - |AC|)\sin\alpha$$

$$= (2|AP| - l)\sin\alpha = (2R\cos\alpha - l)\sin\alpha.$$

The vertical height of the centre of gravity of the rod then is

$$z_G = (2R\cos\alpha - l)\sin\alpha. \tag{4.14}$$

According to Torricelli's principle, it is required to have $\delta z_G = 0$. Differentiating (4.14), then passing to virtual elementary variation, we easily obtain

$$\delta z_G = \left[-2R\sin^2\alpha + (2R\cos\alpha - l)\cos\alpha\right]\delta\alpha = 0.$$

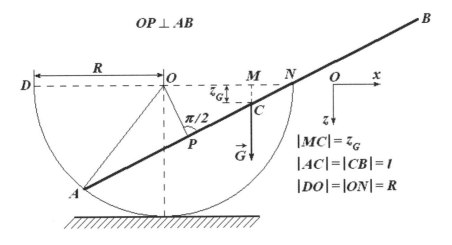

Fig.IV.5

Since $\delta\alpha$ is arbitrary, the last relation implies

$$-2R\sin^2\alpha + (2R\cos\alpha - l)\cos\alpha = 4R\cos^2\alpha - l\cos\alpha - 2R = 0,$$

which is an algebraic equation of the second degree in $\cos\alpha$, with the roots

$$(\cos\alpha)_{1,2} = \frac{l \pm \sqrt{l^2 + 32R^2}}{8R}.$$

Due to the fact that $\alpha \in \left(0, \frac{\pi}{2}\right)$, the only acceptable solution is

$$\cos\alpha = \frac{l + \sqrt{l^2 + 32R^2}}{8R}. \qquad (4.15)$$

Therefore, the equilibrium position of the rod, inside the semi-cylinder, is given by the angle

$$\alpha = \arccos\left(\frac{l + \sqrt{l^2 + 32R^2}}{8R}\right). \qquad (4.16)$$

Since $\forall \alpha$, $-1 \leq \cos\alpha \leq 1$, we must have

$$\cos\alpha = \frac{l + \sqrt{l^2 + 32R^2}}{8R} \leq 1,$$

which is equivalent to

$$l^2 + 32R^2 \leq 64R^2 + l^2 - 16Rl,$$

or
$$l \leq 2R, \qquad (4.17)$$

in agreement with our intuitive presumption.

Problem 4

Two material points (particles) P_1 and P_2, of masses m_1 and m_2, are connected by a massless, inextensible, perfectly malleable wire of length l. Neglecting friction and supposing that the two material points can move on the semicircle $x^2 + y^2 = R^2$, $z = 0$, $y > 0$ (see Fig.IV.6), determine the equilibrium position of the system by means of the principle of virtual work, if $2l > \pi R$.

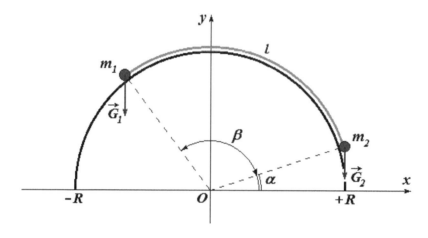

Fig.IV.6

Solution

The system is subject to the following constraints (see Fig.IV.7):
i) $z_1 = 0$;
ii) $z_2 = 0$;
iii) $x_1^2 + y_1^2 - R^2 = 0$; \qquad (4.18)
iv) $x_2^2 + y_2^2 - R^2 = 0$;
v) $(R^2 - x_1^2)(R^2 - x_2^2) - \left(R^2 \cos \frac{l}{R} - x_1 x_2\right)^2 = 0$.

According to analytical formalism, the system possesses $3N - s = 3 \cdot 2 - 5 = 1$ degree of freedom. An intuitive analysis shows that angle α cannot be bigger than $\pi/2$; otherwise, the system could not be in equilibrium whatever α is (both points would be situated on the same side of the semicircle, with respect to y-axis). In addition, even if $0 < \alpha < \frac{\pi}{2}$, to avoid the above mentioned situation, we must have $\frac{\pi}{2} < \alpha + \beta < \pi$. In other words, to be in equilibrium, the two points must be on both sides with respect to y-axis. Let us prove that our intuitive observations are consistent with the results of calculation.

The equilibrium position of the system can be indicated by any one of the following parameters: angle α, x_1, x_2, y_1, y_2, s, etc., where s is the arc BC of the circle (see Fig.IV.7).

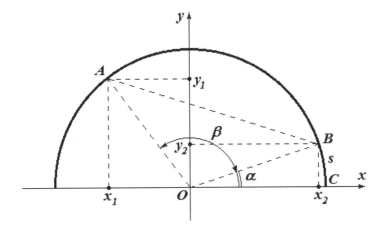

Fig.IV.7

The applied forces are:

$$\vec{G}_1 = (0, -m_1 g, 0), \quad \vec{G}_2 = (0, -m_2 g, 0), \qquad (4.19)$$

while the virtual displacements of the two points are

$$\delta \vec{r}_1 = (\delta x_1, \delta y_1, 0), \quad \delta \vec{r}_2 = (\delta x_2, \delta y_2, 0). \qquad (4.20)$$

The mathematical expression of the principle of virtual work

$$\sum_{i=1}^{N} \vec{F}_i \cdot \delta \vec{r}_i = 0 \qquad (4.21)$$

then yields

$$-m_1 g \delta y_1 - m_2 g \delta y_2 = 0. \qquad (4.22)$$

The virtual displacements δy_1 and δy_2 are not linearly independent (there is one degree of freedom). To determine the relationship between δy_1 and δy_2, we observe that

$$\begin{cases} y_1 = R \sin[\pi - (\alpha + \beta)] = R \sin(\alpha + \beta), \\ y_2 = R \sin \alpha, \end{cases} \qquad (4.23)$$

where β is given (fixed) due to the length l of the wire. Differentiating (4.23) and replacing $d\alpha$ by $\delta\alpha$, we have

$$\begin{cases} \delta y_1 = R\cos(\alpha + \beta)\,\delta\alpha, \\ \delta y_2 = R\cos\alpha\,\delta\alpha. \end{cases} \quad (4.24)$$

Replacing now δy_1, δy_2 in (4.22) with their values given by (4.24), we have

$$[m_1 \cos(\alpha + \beta) + m_2 \cos\alpha]\delta\alpha = 0. \quad (4.25)$$

Since $\delta\alpha$ is arbitrary, we have

$$m_1 \cos(\alpha + \beta) + m_2 \cos\alpha = 0,$$

or, if $\cos(\alpha + \beta)$ is developed,

$$\tan\alpha = \frac{m_1 \cos\beta + m_2}{m_1 \sin\beta}. \quad (4.26)$$

Therefore, the angle α corresponding to the equilibrium of the system is

$$\alpha = \arctan\left[\frac{m_1 \cos\beta + m_2}{m_1 \sin\beta}\right], \quad (4.27)$$

or, in terms of R and l,

$$\alpha = \arctan\left[\frac{m_1 \cos(l/R) + m_2}{m_1 \sin(l/R)}\right]. \quad (4.28)$$

Let us now go back to equation (4.25). It can be written as

$$\frac{m_1}{m_2} = \frac{1}{\tan\alpha \sin\beta - \cos\beta}.$$

Since the ratio $\frac{m_1}{m_2}$ must be positive and different from zero, we must have

$$\tan\alpha \sin\beta - \cos\beta > 0,$$

or, equivalently,

$$-\cos(\alpha + \beta) > \cos\alpha. \quad (4.29)$$

Taking into account the initial "constraint" $0 < \alpha < \frac{\pi}{2}$, the inequality (4.29) can be satisfied only if $\frac{\pi}{2} < \alpha + \beta < \pi$, which is in full agreement with our initial intuitive consideration. To conclude, the system can be at equilibrium only if the two bodies (conceived as material points) are on the one side and the other with respect to vertical Oy, that is, if $\alpha + \beta > \frac{\pi}{2}$.

Problem 5

Solve Problem 4 by means of Torricelli's principle.

Solution

Since the only applied forces to the bodies are the forces of their own gravity, the problem can also be solved by means of Torricelli's principle. As we know, this principle requires that the virtual variation of the height of the centre of gravity of the system is zero:

$$\delta z_G = 0. \tag{4.30}$$

To this end, let us first rename the coordinate axes, as shown in Fig.IV.8. With these new notations, the centre of mass of the system is given by

$$z_G = \frac{m_1 z_1 + m_2 z_2}{m_1 + m_2}.$$

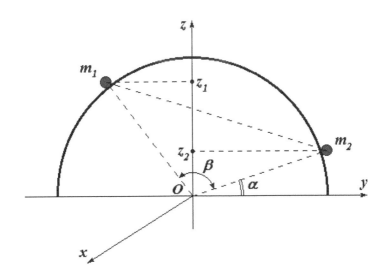

Fig.IV.8

As easily seen,

$$\begin{cases} z_1 = R\sin[\pi - (\alpha + \beta)] = R\sin(\alpha + \beta), \\ z_2 = R\sin\alpha, \end{cases}$$

so that

$$z_G = \frac{m_1 R \sin(\alpha + \beta) + m_2 R \sin\alpha}{m_1 + m_2}.$$

Differentiating this relation, and then replacing the differential symbol with the symbol of first variation, we have

$$\delta z_G = \frac{R}{m_1 + m_2}[m_1 \cos(\alpha + \beta) + m_2 \cos \alpha]\delta\alpha.$$

The Torricelli's principle requires

$$\frac{R}{m_1 + m_2}[m_1 \cos(\alpha + \beta) + m_2 \cos \alpha]\delta\alpha = 0, \qquad (4.31)$$

which is equivalent to (4.25). From now on, the calculation follows the same way as in problem 4.

Problem 6

Two material points (particles) of masses m_1 and m_2, connected by a spring of length l_0 (at rest) and elastic constant k, can move without friction on the walls of a gutter with the opening angle α (see Fig.IV.9). Supposing that the motion of the particles can be performed only in xy-plane, being permanently in contact with the walls of the gutter, determine the equilibrium position of the system. It is assumed that the spring is oriented along a straight line.

Solution

The analytical expressions of the constraints are (see Fig.IV.10):
i) $y_1 = -x_1 \cot \frac{\alpha}{2}$;
ii) $y_2 = x_2 \cot \frac{\alpha}{2}$;
iii) $z_1 = 0$;
iv) $z_2 = 0$.

The system possesses $3N - s = 3 \cdot 2 - 4 = 2$ degrees of freedom. The principle of virtual work writes

$$\sum_{i=1}^{2} \vec{F_i} \cdot \delta \vec{r_i} = 0, \qquad (4.32)$$

where

$$\begin{cases} \vec{F_1} = \vec{G_1} + \vec{F}_{e1}, \\ \vec{F_2} = \vec{G_2} + \vec{F}_{e2}, \\ \vec{G_1} = (0, -m_1 g, 0), \\ \vec{G_2} = (0, -m_2 g, 0), \\ \vec{F}_{e1} = (-F_e \cos \beta, F_e \sin \beta, 0), \\ \vec{F}_{e2} = -\vec{F}_{e1} = (F_e \cos \beta, -F_e \sin \beta, 0), \\ \delta \vec{r_1} = (\delta x_1, \delta y_1, 0), \\ \delta \vec{r_2} = (\delta x_2, \delta y_2, 0), \end{cases} \qquad (4.33)$$

with
$$|\vec{F}_e| = |\vec{F}_{e_1}| = |\vec{F}_{e_2}| = k(l_o - l). \quad (4.34)$$

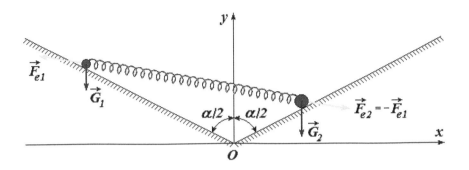

Fig.IV.9

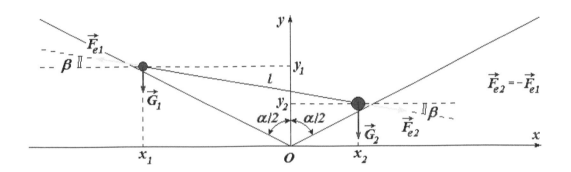

Fig.IV.10

Since $x_1 < 0$ and $x_2 > 0$, the equilibrium length of the deformed spring is
$$l = \frac{x_2 - x_1}{\cos \beta}, \quad (4.35)$$

so that the modulus of the elastic force acting on the two particles is
$$F_e = |\vec{F}_e| = k\left(l_0 - \frac{x_2 - x_1}{\cos \beta}\right), \quad (4.36)$$

and the principle of virtual work (4.32) writes

$$-m_1 g \delta y_1 - F_e \cos\beta \delta x_1 + F_e \sin\beta \delta y_1$$
$$-m_2 g \delta y_2 + F_e \cos\beta \delta x_2 - F_e \sin\beta \delta y_2 = 0. \quad (4.37)$$

Since the system possesses two degrees of freedom, only two variables out of five x_1, x_2, y_1, y_2, and β which interfere in (4.37) are linearly independent. It is convenient to choose as linearly independent parameters the Cartesian coordinates x_1 and x_2. Observing that

$$\begin{cases} y_1 = -x_1 \cot \dfrac{\alpha}{2} \quad \Rightarrow \quad \delta y_1 = -\delta x_1 \cot \dfrac{\alpha}{2}, \\ y_2 = x_2 \cot \dfrac{\alpha}{2} \quad \Rightarrow \quad \delta y_2 = \delta x_2 \cot \dfrac{\alpha}{2}, \\ \sin \beta = -\dfrac{(x_1 + x_2) \cot \frac{\alpha}{2}}{\sqrt{(x_2 - x_1)^2 + (x_1 + x_2)^2 \cot^2 \frac{\alpha}{2}}}, \\ \cos \beta = \dfrac{x_2 - x_1}{\sqrt{(x_2 - x_1)^2 + (x_1 + x_2)^2 \cot^2 \frac{\alpha}{2}}}, \\ \tan \beta = -\dfrac{x_1 + x_2}{x_2 - x_1} \cot \dfrac{\alpha}{2}, \end{cases} \quad (4.38)$$

and using (4.36), equation (4.37) writes:

$$\delta x_1 \left[m_1 g \cot \dfrac{\alpha}{2} - \dfrac{kl_0(x_2 - x_1)}{\sqrt{(x_2 - x_1)^2 + (x_1 + x_2)^2 \cot^2 \frac{\alpha}{2}}} \right.$$

$$\left. + k(x_2 - x_1) + \dfrac{kl_0(x_1 + x_2) \cot^2 \frac{\alpha}{2}}{\sqrt{(x_2 - x_1)^2 + (x_1 + x_2)^2 \cot^2 \frac{\alpha}{2}}} - k(x_1 + x_2) \cot^2 \dfrac{\alpha}{2} \right]$$

$$+ \delta x_2 \left[-m_2 g \cot \dfrac{\alpha}{2} + \dfrac{kl_0(x_2 - x_1)}{\sqrt{(x_2 - x_1)^2 + (x_1 + x_2)^2 \cot^2 \frac{\alpha}{2}}} - k(x_2 - x_1) \right.$$

$$\left. + \dfrac{kl_0(x_1 + x_2) \cot^2 \frac{\alpha}{2}}{\sqrt{(x_2 - x_1)^2 + (x_1 + x_2)^2 \cot^2 \frac{\alpha}{2}}} - k(x_1 + x_2) \cot^2 \dfrac{\alpha}{2} \right] = 0. \quad (4.39)$$

Since the virtual displacements δx_1 and δx_2 are linearly independent, relation (4.39) lead to the following system of two algebraic equations for the unknowns x_1 and x_2:

$$\begin{cases} m_1 g \cot \dfrac{\alpha}{2} - \dfrac{kl_0(x_2 - x_1)}{R} + k(x_2 - x_1) \\ + \dfrac{kl_0(x_1 + x_2) \cot^2 \frac{\alpha}{2}}{R} - k(x_1 + x_2) \cot^2 \dfrac{\alpha}{2} = 0; \\ -m_2 g \cot \dfrac{\alpha}{2} + \dfrac{kl_0(x_2 - x_1)}{R} - k(x_2 - x_1) \\ + \dfrac{kl_0(x_1 + x_2) \cot^2 \frac{\alpha}{2}}{R} - k(x_1 + x_2) \cot^2 \dfrac{\alpha}{2} = 0, \end{cases} \quad (4.40)$$

where we denoted

$$R = \sqrt{(x_2 - x_1)^2 + (x_1 + x_2)^2 \cot^2 \frac{\alpha}{2}}.$$

Adding and subtracting the two equations (4.40), they can be written in a simpler form, as

$$\begin{cases} (m_1 + m_2)g \cot \frac{\alpha}{2} + 2k(x_2 - x_1) = \frac{2kl_0}{R}(x_2 - x_1); \\ (m_1 - m_2)g \cot \frac{\alpha}{2} + \frac{2kl_0}{R}(x_2 + x_1) \cot^2 \frac{\alpha}{2} = 2k(x_1 + x_2) \cot^2 \frac{\alpha}{2}. \end{cases} \quad (4.41)$$

As a result of some not difficult but long calculations, the solution which is acceptable from the physical point of view (out of two possible solutions) is found to be

$$\begin{cases} x_1 = \frac{m_1 + m_2 \cos \alpha}{2k \sin \alpha} \left(g - \frac{2kl_o \sin \frac{\alpha}{2}}{\sqrt{m_1^2 + m_2^2 + 2m_1 m_2 \cos \alpha}} \right), \\ x_1 = \frac{m_2 + m_1 \cos \alpha}{2k \sin \alpha} \left(-g + \frac{2kl_o \sin \frac{\alpha}{2}}{\sqrt{m_1^2 + m_2^2 + 2m_1 m_2 \cos \alpha}} \right). \end{cases} \quad (4.42)$$

These relations give the equilibrium position of the system. We leave up to the reader to analyze relations between the characteristic quantities m_1, m_2, α, k, and l_0, so that $x_2 > 0$ and $x_1 < 0$. If, in particular, $m_1 = m_2 = m$ and $\alpha = \frac{\pi}{2}$, the solution given by (4.42) becomes

$$\begin{cases} x_1 = \frac{mg}{2k} - \frac{l_0}{2}, \\ x_2 = \frac{l_0}{2} - \frac{mg}{2k} = -x_1, \end{cases} \quad (4.43)$$

which is obvious. In this particular case the equilibrium conditions $x_1 < 0$ and $x_2 > 0$ coincide and write

$$kl_0 > mg. \quad (4.44)$$

The last relation shows that the equilibrium position of the system (different from the trivial one, when the bodies "fall" in the edge of the gutter) is attained only if the elastic constant k of the spring is big enough ($k > \frac{G}{l_0}$) to exceed the weight of the two bodies.

CHAPTER V

PROBLEMS OF VARIATIONAL CALCULUS

V.1. Elements of variational calculus

V.1.1. Functionals. Functional derivative

Definition. A *functional* is an application defined on a linear space X, or on a part of X, with values in the field of scalars on which the linear space X is defined. (N.B. The French term for a field is *corps*; in English, a field is a *corps commutatif*). In other words, a functional can be considered as a "function" whose domain consists of a set of functions, and whose codomain is a set of scalars (numbers, figures, etc.). This definition shows that, in fact, a functional is an operator.

Let $J(y)$ be a functional defined on the domain $\mathcal{D}$ of a Banach space $\mathcal{B}$, with values on the real straight line $\mathbb{R}$.

Definition. A *Banach space* is a normed linear space that is a complete metric space with respect to the metric derived from its norm.

Definition. A *complete space* (or *Cauchy space*) is a linear space in which any Cauchy sequence of points converges to an element of the space.

Definition. A *Cauchy sequence* in the metric space (X, d) is a sequence $\{X_n\}_{n \in N}$ of elements of X that satisfy the property: $\forall \, \varepsilon > 0$, $\exists \, n_\varepsilon \in N$, so that $d(x_n, x_m) < \varepsilon$, $\forall n, m \geq n_\varepsilon$.

Definition. A *distance* on a set X is any application d defined on $X \times X$ with positive real values, $d : X \times X \to \mathbb{R}_+$, that satisfies the following properties:

$D_1 : d(x, y) = 0$ if and only if $x = y$, $(x, y \in X)$;
$D_2 : d(x, y) = d(y, x)$, $\forall x, y \in X$;
$D_3 : d(x, y) \leq d(x, z) + d(z, y)$, $\forall x, y, z \in X$.

Definition. A *metric space* is any pair (X, d), where X is a set, and d a distance defined on X.

Definition. A *normed space* is a linear space possessing a norm, p.

Definition. A *norm* is any seminorm that satisfies the property $p(x) = 0 \Rightarrow x = 0$.

Definition. A *seminorm* is a function with real values, $p : X \to \mathbb{R}$, that satisfies the following properties:

$SN_1 : p(x+y) \leq p(x) + p(y), \ \forall x, y \in X$;
$SN_2 : p(\lambda x) = |\lambda| p(x), \ \forall x, y \in X$, and λ a non-zero scalar.

A norm is usually denoted by: $p(x) \equiv \|x\|$.

Let $J(y)$ be a functional defined as previously shown. Let also $y \in \mathcal{D}$, $h \in \mathcal{B}$, so that $y + h \in \mathcal{D}$

Definition (differentiability 1): The functional $J(y)$ is *differentiable* at the "point" y if the difference $J(y+h) - J(y)$ can be written as

$$J(y+h) - J(y) = \delta(y, h, J) + r(y, h, J),$$

where $\delta(y, h, J)$ is a functional linear in h

$$\delta(y, \alpha h_1 + \beta h_2, J) = \alpha\, \delta(y, h_1, J) + \beta\, \delta(y, h_2, J),$$

where α and β are two scalars, while $r(y, h, J)$ is a functional that satisfies the condition

$$\lim_{\|h\| \to 0} \frac{|r(y, h, J)|}{\|h\|} = 0.$$

(The "point" y is, in general, a function).

Definition (differentiability 2): The functional $J(y)$ is *differentiable* at the "point" y if there exists a functional linear in h, $\delta(y, h, J)$, so that

$$\lim_{\|h\| \to 0} [J(y+h) - J(y) - \delta(y, h, J)] = 0.$$

Definition. If the conditions of the differentiability of a functional at a "point" are fulfilled, then the functional $\delta(y, h, J)$ is called *differential* of the functional $J(y)$.

Theorem: If $\delta(y, h, J)$ exists, then it is unique. (The demonstration is based on *reductio ad absurdum*. We skip this proof).

Definition: If the functional $r(y, h, J)$ can be written as

$$r(y, h, J) = \frac{1}{2} \delta_2(y, h, J) + r_2(y, h, J),$$

where

$$\delta_2(y, th, J) = t^2 \delta_2(y, h, J)$$

and
$$\lim_{\|h\|\to 0} \frac{|r_2(y,h,J)|}{\|h\|^2} = 0,$$

then the functional $\delta_2(y, h, J)$ is called *the second-order differential* of functional $J(y)$.

For a better understanding of these definitions, next we shall present the analogy between functionals and the usual functions. Let $f : I \subset \mathbb{R} \to \mathbb{R}$ be a function defined on interval I of the real axis, with values in the set of real numbers, differentiable at the point $x_0 \in I$. Then, for any $x \neq x_0$, we can write

$$f(x) - f(x_0) = f'(x_0)(x - x_0) + \alpha(x)(x - x_0). \tag{$*$}$$

Since $f(x)$ is differentiable at the point $x_0 \in I$, we have

$$\lim_{x \to x_0} \frac{f(x) - f(x_0)}{x - x_0} = f'(x_0) + \lim_{x \to x_0} \alpha(x) = f'(x_0),$$

which necessarily yields

$$\lim_{x \to x_0} \alpha(x) = 0.$$

Denoting $x - x_0 = h$, then $x = x_0 + h$ and we have

$$f(x_0 + h) - f(x_0) = hf'(x_0) + h\alpha(x_0 + h) \stackrel{\lim_{h \to 0} \alpha(x_0+h)=0}{=} hf'(x_0).$$

The linear function $g = hf'(x_0)$ is called *the differential* of $f(x)$ at the point x_0, and it is usually denoted by $g = df(x_0)$. The linearity in h then writes:

$$(\alpha h_1 + \beta h_2)f'(x_0) \equiv df(x_0, \alpha h_1 + \beta h_2)$$

$$= \alpha h_1 f'(x_0) + \beta h_2 f'(x_0) = \alpha df(x_0, h_1) + \beta df(x_0, h_2)$$

$$\equiv \alpha h_1 f'(x_0) + \beta h_2 f'(x_0).$$

We then have the following obvious analogy:

$$\begin{array}{ccc} J & \to & f \\ y & \to & x_0 \\ h & \to & h \end{array}$$

$$\delta(y, h, J) \to d(x_0, h, f)$$
$$r(y, h, J) \to r(x_0, h, f)$$

where the following notations have been used:

$$f'(x_0)(x - x_0) = hf'(x_0) = df(x_0, h) \equiv d(x_0, h, f);$$

$$\alpha(x)(x - x_0) = h\alpha(x_0 + h) \equiv r(x_0, h, f).$$

To the limit condition obeyed by the functional $r(y, h, J)$,

$$\lim_{\|h\| \to 0} \frac{|r(y, h, J)|}{\|h\|} = 0,$$

shall therefore correspond the following condition on $\alpha(x)$

$$\lim_{h \to 0} \frac{h\,\alpha(x_0 + h)}{h} = \lim_{h \to 0} \alpha(x) \equiv \lim_{x \to x_0} \alpha(x) = 0.$$

One usually prefers the notation $h = dx$, so that the differential of f at the point x_0 writes

$$df(x_0) = f'(x_0)\,dx.$$

To realize the analogy regarding the second differential of a functional, let us expand function $f(x)$ in Taylor series about the point x_0:

$$f(x) = f(x_0) + f'(x_0)(x - x_0) + \frac{1}{2}f''(x_0)(x - x_0)^2 + \mathcal{O}(x - x_0)^3$$

$$= f(x_0) + f'(x_0)(x - x_0) + \frac{1}{2}f''(x_0)(x - x_0)^2 + \beta(x)(x - x_0)^2, \quad (**)$$

where we have used the notation

$$\beta(x)(x - x_0)^2 = \frac{1}{3!}f'''(x_0)(x - x_0)^3 + \frac{1}{4!}f^{(iv)}(x_0)(x - x_0)^4 + \ldots$$

with obvious condition
$$\lim_{x \to x_0} \beta(x) = 0.$$

In view of (*) and (**), we still have

$$\alpha(x)(x - x_0) = \frac{1}{2}f''(x_0)(x - x_0)^2 + \beta(x)(x - x_0)^2.$$

Using notation $h = x - x_0$, we also have

$$\alpha(x)(x - x_0) = h\alpha(x_0 + h) \equiv r(x_0, h, f) = \frac{1}{2}h^2 f''(x_0) + h^2\beta(x_0 + h),$$

or, if we define the second order variation (also called "the second order differential") of $f(x)$ at the point x_0 as

$$\delta_2 f = f''(x_0)(x - x_0)^2 = h^2 f''(x_0) \equiv \delta_2(x_0, h, f),$$

we can write

$$r(x_0, h, f) = \frac{1}{2}\delta_2(x_0, h, f) + r_2(x_0, h, f),$$

where we denoted

$$r_2(x_0, h, f) = h^2 \beta(x_0 + h).$$

Obviously,

$$\delta_2(x_0, th, f) = (th)^2 f''(x_0) = t^2 h^2 f''(x_0) = t^2 \delta_2(x_0, h, f),$$

as well as

$$\lim_{h \to 0} \frac{r_2(x_0, h, f)}{h^2} = \lim_{h \to 0} \frac{h^2 \beta(x_0 + h)}{h^2}$$
$$= \lim_{h \to 0}[\beta(x_0 + h)] = \lim_{x \to x_0} \beta(x) = 0.$$

As one can see, the analogy completes if on the table of correspondences we add the following two relations

$$\delta_2(y, h, J) \quad \to \quad \delta_2(x_0, h, f),$$

$$r_2(y, h, J) \quad \to \quad r_2(x_0, h, f).$$

A functional commonly used in Analytical Mechanics is $J(y) = \int_a^b f(x, y, y')\,dx$, where function $f(x)$ has to satisfy certain conditions of continuity. Let then $f(x, y, z)$ be a function of class C^2 on the domain $\mathcal{D}: a \le x \le b,\ -\infty < y, z < +\infty$, with uniformly continuous partial derivatives with respect to y, z.

Definition (continuity at a point in the language of $(\varepsilon, \eta(\varepsilon))$): let $f : I \subset \mathbb{R} \to \mathbb{R}$ be a function, and $x_0 \in I$ a point within its domain of definition. We then say that $f(x)$ is *continuous at the point* x_0 if $\forall\,\varepsilon > 0,\ \exists \eta(\varepsilon) > 0$, so that $|f(x) - f(x_0)| < \varepsilon,\ \forall x \in I$, as soon as $|x - x_0| < \eta(\varepsilon)$.

Observation. η depends only on ε if we consider continuity at a single point. But, if we refer to the continuity on the whole interval $I \subset \mathbb{R}$, η depends on x_0 as well.

Definition (continuity at a point in the language of vicinities). Consider $f : I \subset \mathbb{R} \to \mathbb{R}$, and $x_0 \in I$. Function $f(x)$ is said to be continuous at the point x_0 if $\forall U(f(x_0))$, $\exists V(x_0)$, so that $\forall\, x \in I \cap V(x_0)$, $f(x) \in U(f(x_0))$.

Definition (continuity at a point in the language of sequences): Consider $f : I \subset \mathbb{R} \to \mathbb{R}$, and $x_0 \in I$. Function $f(x)$ is said to be *continuous at the point* x_0 if $\forall \{x_n\}_{n \in N}$ a sequence convergent to x_0, the sequence $f(x_n)_{n \in N}$ is convergent to $f(x_0)$.

Definition. Consider $f : I \subset \mathbb{R} \to \mathbb{R}$. Function $f(x)$ is said to be *uniformly continuous* on $I \subset \mathbb{R}$ if $\forall \varepsilon > 0, \exists \eta(\varepsilon) > 0$, so that $|f(x'') - f(x')| < \varepsilon, \forall\, x, x' \in I$ as soon as $|x - x'| < \eta(\varepsilon)$.

Observation: η depends only on ε, and is independent of x' and x'' which satisfy relation $|x' - x''| < \eta(\varepsilon)$. Let $f(x)$ be only continuous on $I \subset \mathbb{R}$ and $x' \in I$. If x' is maintained fix, to a given ε corresponds an η that changes together with ε, but depends on x' as well, $\eta = \eta(\varepsilon, x')$. If x' covers the multitude I, ε being fixed, then the multitude of values of $\eta(\varepsilon, x')$ has an inferior margin,

$$\eta_0(\varepsilon) = \inf_{\substack{\forall x' \in I \\ \varepsilon\ fixed}} \{\eta(\varepsilon, x')\}.$$

If $\eta_0(\varepsilon) > 0$, then function $f(x)$ is said to be *uniformly continuous* on I.

Theorem. A continuous function on a compact (i.e. closed and bounded) interval is *uniformly continuous* on that interval. Example: $f(x) = x^3$, $x \in [1, 3]$.

Theorem of finite increments (Lagrange): Consider the function $f : I \subset \mathbb{R} \to \mathbb{R}$, and $a, b \in I$ any two points in I. If
L_1: function f is continuous on $[a, b]$, and
L_2: function f is differentiable on (a, b),
then[1] $\exists^*\ c \in (a, b)$, so that

$$f(b) - f(a) = (b - a)f'(c).$$

Observation. There are some equivalent formulations for the relation $f(b) - f(a) = (b - a)f'(c)$:
1. $f(x) - f(a) = (x - a)f'(\xi),\ a < \xi < x$;

---

[1] The symbol $\exists^*$ stands for "exists at least"

2. $f(a+h) - f(a) = hf'(a+\theta h)$, $0 < \theta < 1$.

Definition. It is said that the *function $f : I \subset \mathbb{R} \to \mathbb{R}$ is derivable at the point $x_0 \in I$*, if the ratio $\frac{f(x)-f(x_0)}{x-x_0}$ has a finite limit at the point x_0. This limit is called *the derivative* of $f(x)$ at x_0 being denoted as $f'(x_0)$:

$$\lim_{x \to x_0} \frac{f(x) - f(x_0)}{x - x_0} = f'(x_0).$$

Consider now the functional

$$J(y) = \int_a^b f(x, y, y') \, dx$$

defined on the set of functions $y = y(x) \in C^2[a, b]$, which satisfies the conditions $y(a) = 0$, $y(b) = 0$. If we define the quantity

$$\|y\| = \max \left\{ \sup_{x \in [a,b]} |y(x)|, \ \sup_{x \in [a,b]} |y'(x)| \right\},$$

as *norm* on this space, then the set of these functions form a Banach space[1] $\mathcal{B}$.

Let $h(x)$ be a function of the Banach space $\mathcal{B}$. Obviously, $y(x) + h(x) \in \mathcal{B}$ and the functional

$$J(y + h) = \int_a^b f(x, y + h, y' + h') \, dx$$

is well-defined, and we can calculate the difference

$$J(y+h) - J(y) = \int_a^b f(x, y+h, y'+h') \, dx - \int_a^b f(x, y, y') \, dx$$

$$= \int_a^b \left[f(x, y+h, y'+h') - f(x, y, y') \right] dx.$$

Applying the theorem of finite increments, we still have

$$J(y+h) - J(y) = \int_a^b \left[h \frac{\partial f}{\partial y}(x, y+\theta h, y'+\theta h') \right.$$

---

[1] We have considered the definition of a functional on a Banach space, because on this mathematical structure all the quantities useful in analytical mechanics (convergence, continuity, differentiability and integrability) are defined.

$$+h'\frac{\partial f}{\partial y'}(x,y+\theta h,y'+\theta h')\Big]\,dx, \qquad (5.1)$$

with $0<\theta<1$. Denoting

$$\left(\frac{\partial f}{\partial y}\right)^* = \frac{\partial f}{\partial y}(x,y+\theta h,y'+\theta h'), \quad 0<\theta<1 \qquad (5.2)$$

and

$$\left(\frac{\partial f}{\partial y'}\right)^* = \frac{\partial f}{\partial y'}(x,y+\theta h,y'+\theta h'), \quad 0<\theta<1, \qquad (5.3)$$

we still have

$$J(y+h)-J(y) = \int_a^b \left[h\left(\frac{\partial f}{\partial y}\right)^* + h'\left(\frac{\partial f}{\partial y'}\right)^*\right] dx, \quad 0<\theta<1.$$

Introducing the new notations

$$\frac{\partial f}{\partial y}(x,y+\theta h,y'+\theta h') - \frac{\partial f}{\partial y}(x,y,y') = \left(\frac{\partial f}{\partial y}\right)^* - \frac{\partial f}{\partial y} = f_0$$

and

$$\frac{\partial f}{\partial y'}(x,y+\theta h,y'+\theta h') - \frac{\partial f}{\partial y'}(x,y,y') = \left(\frac{\partial f}{\partial y'}\right)^* - \frac{\partial f}{\partial y'} = f_1,$$

and taking into account the uniform continuity of $\frac{\partial f}{\partial y}$ and $\frac{\partial f}{\partial y'}$, it then follows that $\forall \varepsilon>0$, $\exists\,\eta(\varepsilon)>0$, so that

$$\left|\left(\frac{\partial f}{\partial y}\right)^* - \frac{\partial f}{\partial y}\right| = |f_0| < \varepsilon$$

and

$$\left|\left(\frac{\partial f}{\partial y'}\right)^* - \frac{\partial f}{\partial y'}\right| = |f_1| < \varepsilon,$$

as soon as

$$|(y+\theta h)-y| = |\theta h| < |h| < \eta(\varepsilon)$$

and, respectively

$$|(y'+\theta h')-y'| = |\theta h'| < |h'| < \eta(\varepsilon)$$

or - if one takes into account the definition of norm - as soon as $\|h\|<\eta(\varepsilon)$. By means of (5.2) and (5.3), we still have

$$J(y+h)-J(y) = \int_a^b \left(h\frac{\partial f}{\partial y} + h'\frac{\partial f}{\partial y'}\right) dx + \int_a^b (hf_0 + h'f_1)\,dx. \qquad (5.4)$$

Using the properties of integrals, one can easily verify that the integral
$$L(h) = \int_a^b \left(h\frac{\partial f}{\partial y} + h'\frac{\partial f}{\partial y'} \right) dx$$
is linear in h. Indeed,
$$L(\alpha h_1 + \beta h_2) = \int_a^b \left[(\alpha h_1 + \beta h_2)\frac{\partial f}{\partial y} + (\alpha h_1' + \beta h_2')\frac{\partial f}{\partial y'} \right] dx$$
$$= \alpha \int_a^b \left(h_1\frac{\partial f}{\partial y} + h_1'\frac{\partial f}{\partial y'} \right) dx + \beta \int_a^b \left(h_2\frac{\partial f}{\partial y} + h_2'\frac{\partial f}{\partial y'} \right) dx$$
$$= \alpha L(h_1) + \beta L(h_2).$$

We also have
$$\left| \int_a^b (h f_0 + h' f_1)\, dx \right| \leq \max\left(|h|, |h'|\right) \int_a^b \left(|f_0| + |f_1|\right) dx,$$
that is
$$\frac{1}{\|h\|} \left| \int_a^b (h f_0 + h' f_1)\, dx \right| \leq \int_a^b \left(|f_0| + |f_1|\right) dx \leq 2\varepsilon(b-a). \quad (5.5)$$

Therefore, (5.4) gives a decomposition of $J(y+h) - J(y)$ in a functional linear in terms of $h(x)$, and another one that satisfies the condition (5.5). This means that the differentiability conditions are satisfied, and the differential of $J(y)$ is
$$\delta(y, h, J) = \int_a^b \left(h\frac{\partial f}{\partial y} + h'\frac{\partial f}{\partial y'} \right) dx. \quad (5.6)$$

Since the function $f(x, y, y')$ is of class C^2 on its domain of definition, $f \in C^2(\mathcal{D})$, where $\mathcal{D}: a \leq x \leq b$, $\infty < y, y' < +\infty$, we can integrate by parts the second term in (5.6) and obtain
$$\int_a^b h' \frac{\partial f}{\partial y'} dx = \left(h\frac{\partial f}{\partial y'} \right)_a^b - \int_a^b h\frac{d}{dx}\left(\frac{\partial f}{\partial y'} \right) dx.$$

But
$$\left(h\frac{\partial f}{\partial y'} \right)_a^b = h(b)\frac{\partial f}{\partial y'}(b, y(b), y'(b)) - h(a)\frac{\partial f}{\partial y'}(a, y(a), y'(a)),$$

and, since $h(x) \in \mathcal{B}$, we have $h(a) = 0$ and $h(b) = 0$, which yields

$$\delta(y, h, J) = \int_a^b h(x) \left[\frac{\partial f}{\partial y} - \frac{d}{dx}\left(\frac{\partial f}{\partial y'}\right) \right] dx. \qquad (5.7)$$

This result can be generalized to functionals of the type

$$\int_a^b f\left(x, y, y', y'', ..., y^{(n)}\right) dx,$$

where $f\left(x, y, y', y'', ..., y^{(n)}\right)$ is a function of $(n+2)$ variables of class C^n on its domain of definition, $\mathcal{D}: a \leq x \leq b, -\infty < y, y', ..., y^{(n)} < +\infty$. A similar reasoning leads to the differential of this functional, which is

$$\delta(y, h, J) = \int_a^b h(x) \left[\frac{\partial f}{\partial y} - \frac{d}{dx}\left(\frac{\partial f}{\partial y'}\right) + \frac{d^2}{dx^2}\left(\frac{\partial f}{\partial y''}\right) \right.$$

$$\left. +... + (-1)^n \frac{d^n}{dx^n}\left(\frac{\partial f}{\partial y^{(n)}}\right) \right] dx. \qquad (5.8)$$

Here are two other examples of most frequently met functionals, together with their differentials:

i) $J(x) = \int_{t_1}^{t_2} f\left(t, x(t), \dot{x}(t)\right) dt$, where $x = x(t)$ is a vector function with n components, $x(t) = (x_1(t), x_2(t), ..., x_n(t))$, and $\dot{x}(t) = \frac{dx}{dt}$. Following a similar reasoning, the differential of the functional $J(y)$ is found as

$$\delta(x, h, J) = \int_{t_1}^{t_2} \sum_{i=1}^n h_i(t) \left[\frac{\partial f}{\partial x_i} - \frac{d}{dt}\left(\frac{\partial f}{\partial \dot{x}_i}\right) \right] dt. \qquad (5.9)$$

ii) $J(z) = \int_\Omega f\left(x, y, z(x,y), \frac{\partial z}{\partial x}(x,y), \frac{\partial z}{\partial y}(x,y)\right) dx\, dy$. In this case one obtains

$$\delta(z, h, J) = \int_\Omega h \left[\frac{\partial f}{\partial z} - \frac{\partial}{\partial x}\left(\frac{\partial f}{\partial p}\right) - \frac{\partial}{\partial y}\left(\frac{\partial f}{\partial q}\right) \right] dx\, dy, \qquad (5.10)$$

where $p = \frac{\partial z}{\partial x}$, $q = \frac{\partial z}{\partial y}$.

V.1.2. Extrema of functionals

A) Necessary conditions for extremum

Let $J(y)$ be a differentiable functional, defined on a Banach space $\mathcal{B}$.

Definition. A value y_0 is called *"point" of relative maximum* of $J(y)$ if $\exists\, V(y_0)$ a vicinity of y_0, so that $\forall\, y \in (y_0)$,

$$J(y) \leq J(y_0).$$

Definition. A value y_0 is called *"point" of relative minimum* of $J(y)$ if $\exists\, V(y_0)$ a vicinity of y_0, so that $\forall\, y \in (y_0)$,

$$J(y) \geq J(y_0).$$

Theorem: If $J(y)$ is a differentiable functional, and y_0 a "point" of extremum for $J(y)$, then, $\forall h \in \mathcal{B}$,

$$\delta(y_o, h, J) = 0. \tag{5.11}$$

Therefore, a necessary condition for y_0 to be a point of extremum for $J(y)$ is that $\delta(y, h, J)$ equals zero at that point.

Definition. The "points" y_0 where condition (5.11) is satisfied are called *extremals*.

In other words, the extremals of a functional are those functions $y(x_0)$ which cancel the first differential of the functional.

B) Sufficient conditions for extremum

Theorem: If the functional $J(y)$ admits the first differential $\delta(y, h, J)$ and the second differential $\delta_2(y, h, J)$, then the necessary and sufficient condition for $J(y)$ to achieve its maximum at the "point" y_0, $\forall h \in \mathcal{B}$, is to have

$$\delta(y_0, h, J) = 0, \quad r(y_0, h, J) \leq 0,$$

while the condition for $J(y)$ to have a minimum at the "point" y_0 writes

$$\delta(y_0, h, J) = 0, \quad r(y_0, h, J) \geq 0.$$

Indeed, since $J(y)$ admits the first differential, we have

$$J(y_0 + h) - J(y_0) = \delta(y_0, h, J) + r(y_0, h, J).$$

But $\delta(y_0, h, J) = 0$ is the necessary condition for y_0 to be a point of extremum for $J(y)$, so that we are left with

$$J(y_0 + h) - J(y_0) = r(y_0, h, J).$$

Consequently, in order for y_0 to be
- a point of relative maximum, i.e. $J(y_0 + h) - J(y_0) \leq 0$, it is necessary and sufficient that $r(y_0, h, J) \leq 0$;
- a point of relative minimum, i.e. $J(y_0 + h) - J(y_0) \geq 0$, it is necessary and sufficient that $r(y_0, h, J) \geq 0$.

It can be shown that these conditions are fulfilled as follows:
1) For $r(y_0, h, J) \leq 0$,
$$\begin{cases} - \text{the necessary condition is}: \delta_2(y_0, h, J) \leq 0, \\ - \text{the sufficient condition is}: \delta_2(y_0, h, J) \leq -C\|h\|^2; \end{cases}$$

2) For $r(y_0, h, J) \geq 0$,
$$\begin{cases} - \text{the necessary condition is}: \delta_2(y_0, h, J) \geq 0, \\ - \text{the sufficient condition is}: \delta_2(y_0, h, J) \geq C\|h\|^2, \end{cases}$$

where $C > 0$ is a constant.

Theorem. Let $\Omega \subset \mathbb{R}^n$ be a domain of the abstract n-dimensional space $\mathbb{R}^n$, $x = (x_1, x_2, ..., x_n) \in \Omega$ a point in this domain, and $\mathcal{M}$ the set of functions $h(x) = h(x_1, x_2, ..., x_n)$, which are continuous on Ω and cancel on the frontier $\partial \Omega$ of Ω. Under these conditions, if $A(x)$ is a continuous function on Ω and

$$\int_\Omega h(x) A(x) \, dx = 0, \quad \forall h(x) \in \mathcal{M},$$

then $A(x) = 0, \forall x \in \Omega$.

Consider the functional $J(y) = \int_a^b f(x, y, y') \, dx$. The necessary condition for an extremum of this functional is that its differential, given by (5.7), must be zero

$$\delta(y, h, J) = \int_a^b h(x) \left[\frac{\partial f}{\partial y} - \frac{d}{dx}\left(\frac{\partial f}{\partial y'}\right) \right] dx \equiv G(y) = 0, \quad \forall h(x).$$

Since
- $h \in C^2[a, b]$;
- $h(a) = 0$, $h(b) = 0$;
- $f \in C^2(a \leq x \leq b, -\infty < y, y' < \infty)$, that is $\frac{\partial f}{\partial y} - \frac{d}{dx}\left(\frac{\partial f}{\partial y'}\right)$ is continuous, the conditions imposed by the last theorem are fulfilled. Consequently, in order that

$$\int_a^b h(x) \left[\frac{\partial f}{\partial y} - \frac{d}{dx}\left(\frac{\partial f}{\partial y'}\right) \right] dx = 0, \quad \forall h(x),$$

it is necessary and sufficient to have

$$\frac{\partial f}{\partial y} - \frac{d}{dx}\left(\frac{\partial f}{\partial y'}\right) = 0. \tag{5.12}$$

This is the *Euler* (sometimes called *Euler-Lagrange*) equation of extremals. This equation can also be written as

$$\frac{\partial^2 f}{\partial y'^2}y'' + \frac{\partial^2 f}{\partial y \partial y'}y' + \frac{\partial^2 f}{\partial x \partial y'} - \frac{\partial f}{\partial y} = 0. \tag{5.13}$$

Equations of extremals for the functional

$$J(x) = \int_{t_1}^{t_2} f(t, x(t), \dot{x}(t))\, dt,$$

where $x(t) \equiv (x_1(t), x_2(t), ..., x_n(t))$, are written as

$$\frac{\partial f}{\partial x_i} - \frac{d}{dt}\left(\frac{\partial f}{\partial \dot{x}_i}\right) = 0, \quad (i = \overline{1, n}).$$

These equations were first obtained by Euler in 1744, and then generalized by Lagrange in Mechanics. They are known as the *Euler-Lagrange equations*.

The necessary condition for an extremum of the functional

$$\int_a^b f(x, y, y', y'', ..., y^{(n)})\, dx$$

is given by Euler's equation

$$\frac{\partial f}{\partial y} - \frac{d}{dx}\left(\frac{\partial f}{\partial y'}\right) + \frac{d^2}{dx^2}\left(\frac{\partial f}{\partial y''}\right) + ...$$

$$+ (-1)^n \frac{d^n}{dx^n}\left(\frac{\partial f}{\partial y^{(n)}}\right) = 0.$$

Let

$$F(x, y, y', ..., y^{(n)}) = 0$$

be the implicit form of an n-th order differential equation.

Definition. A differential equation of order $n - k$ that contains $k \geq 1$ arbitrary constants

$$\psi(x, y, y', ..., y^{(n-k)}, C_1, C_2, ..., C_k) = 0,$$

being verified by the general integral

$$\varphi(x, y, C_1, ..., C_n) = 0$$

of the given differential equation, is called *intermediate integral of order k* or, simply, *integral of order k* of this equation. For $k = 1$, the differential equation of order $n - 1$

$$\chi(x, y, y', ..., y^{(n-1)}, C) = 0,$$

that depends on a single constant of integration C, is called a *first integral* of the differential equation $F(x, y, y', ..., y^{(n)}) = 0$.

First integrals of the Euler-Lagrange equations

$$\frac{\partial f}{\partial y_i} - \frac{d}{dx}\left(\frac{\partial f}{\partial \dot{y}_i}\right) = 0, \quad (i = \overline{1, n}).$$

1. Suppose that the function f does not explicitly depend on y_i. Then $\frac{\partial f}{\partial y_i} = 0$, and the Euler-Lagrange equations yield

$$\frac{\partial f}{\partial \dot{y}_i} = const. \tag{5.14}$$

2. Suppose that the function f does not explicitly depend on the independent variable x. In this case,

$$\dot{y}_i \frac{\partial f}{\partial \dot{y}_i} - f = const. \tag{5.15}$$

is a first integral of the Euler-Lagrange equations. Indeed, taking the first derivative of (5.15) with respect to x, we have

$$\frac{d}{dx}\left(\dot{y}_i \frac{\partial f}{\partial \dot{y}_i}\right) - \frac{d}{dx} f = 0,$$

or

$$\ddot{y}_i \frac{\partial f}{\partial \dot{y}_i} + \dot{y}_i \frac{d}{dx}\left(\frac{\partial f}{\partial \dot{y}_i}\right) - \frac{\partial f}{\partial y_i}\dot{y}_i - \frac{\partial f}{\partial \dot{y}_i}\ddot{y}_i = 0,$$

that is

$$-\dot{y}_i \left[\frac{\partial f}{\partial y_i} - \frac{d}{dx}\left(\frac{\partial f}{\partial \dot{y}_i}\right)\right] = 0,$$

which completes the proof.

Observation. To determine the differential of the functional

$$J(y) = \int_a^b f(x, y, y')\, dx,$$

defined on the set of functions $y = y(x) \in C^2[a, b]$, we have considered $y(a) = 0$, $y(b) = 0$. Our theory remains also valid for less restrictive conditions, such as $y(a) = a_1 \neq 0$, and $y(b) = b_1 \neq 0$. It is essential, nevertheless, to have $h(a) = 0$, $h(b) = 0$ or, in other words, if the function $y_0 = \varphi(x)$ passes through the points $P(a, a_1)$ and $Q(b, b_1)$, then the curve $y(x) = \varphi(x) + h(x) = y_0(x) + h(x)$ must also pass through the same points (see Fig.V.1).

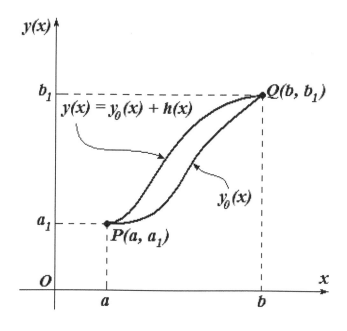

Fig.V.1

Conditional extrema of functionals

Consider the functional $J(y) = \int_a^b f(x, y, y')\, dx$, where the function $y = y(x)$ is subject to a restrictive condition (*constraint*) defined as

$$G(y) = \int_a^b g(x, y, y')\, dx = const. \equiv C.$$

As can be seen, $G(y)$ is a functional of the same type.

To determine the conditional extrema of $J(y)$ we shall appeal to the following

Theorem. In order that y_0 is a "point" of extremum of the functional $J(y)$, where $y(x)$ satisfies the supplementary condition $G(y) = C$, it is necessary that the differential of the functional $W(y) = J(y) - \lambda G(y)$ cancels for $y = y_0$. Here λ is some non-zero scalar.

Therefore, to determine the conditional extrema of the functional $J(y)$, one search for the ordinary extrema of a new functional $W(y)$ defined as

$$W(y) = J(y) - \lambda G(y) = \int_a^b \left[f(x, y, y') - \lambda g(x, y, y') \right] dx$$

$$= \int_a^b w(x, y, y') \, dx,$$

where $w = f - \lambda g$. Solution to such a problem implies determination of both the conditional extrema of $J(x)$ and the scalar $\lambda \neq 0$. To this end, one uses the above theorem, which furnishes the following system of "equations"

$$\begin{cases} \delta(y, h, W) = 0; \\ G(y) = C, \end{cases}$$

with two unknowns: y_0 and λ.

V.2. Problems whose solutions demand elements of variational calculus

1. Brachistochrone problem

Chronologically, this is the first problem of variational calculus. The word *brachistochrone* comes from Greek: *brachistos* (the shortest) and *chronos* (time). It was formulated by Jean Bernoulli in 1696 as follows: among all curves situated in a vertical plane and passing through two fixed points, determine the curve on which a heavy particle moves without friction in shortest time.

Solution

Choosing a reference frame as shown in Fig.V.2, where $P_0(0,0)$ and $P_1(x_1, y_1)$ are any two arbitrary points, we can write

$$ds = v \, dt, \qquad (5.16)$$

where v is the magnitude of the velocity of particle along the curve, and dt the infinitesimal time interval for the particle to cover the infinitesimal distance ds.

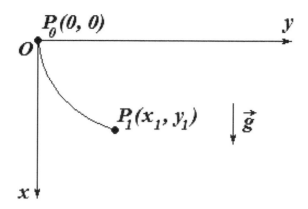

Fig.V.2

According to our choice, P_0 is higher than P_1. The time interval required for the particle to move from P_0 to P_1 then is

$$t = \int_0^t dt = \int_{P_0}^{P_1} \frac{ds}{v}. \tag{5.17}$$

Velocity v can be obtained by means of either the energy conservation law, or the kinetic energy theorem. Since the particle starts from rest, both ways yield

$$\frac{1}{2}mv^2 = mgx, \tag{5.18}$$

that is

$$v = \sqrt{2gx}. \tag{5.19}$$

On the other hand,

$$(ds)^2 = (dx)^2 + (dy)^2 = (dx)^2(1 + y'^2),$$

where $y' = dy/dx$. Therefore,

$$ds = dx\sqrt{1 + y'^2}. \tag{5.20}$$

Using (5.17), (5.19), and (5.20), we have

$$t = \frac{1}{\sqrt{2g}} \int_0^{x_1} \sqrt{\frac{1 + y'^2}{x}}\, dx = \int_0^{x_1} f(x, y')\, dx. \tag{5.21}$$

As one can see, the time t has been obtained as a functional whose "characteristic" function $f(x, y')$ does not depend on $y = y(x)$. This means that the Euler equation

$$\frac{d}{dx}\left(\frac{\partial f}{\partial y'}\right) - \frac{\partial f}{\partial y} = 0$$

admits the first integral
$$\frac{\partial f}{\partial y'} = const, \qquad (5.22)$$

with
$$f(x, y') = \sqrt{\frac{1 + y'^2}{x}}. \qquad (5.23)$$

Take now the value $\frac{1}{\sqrt{2a}}$ for the constant appearing in (5.22), where a is a new constant. In view of (5.23), we then have

$$\frac{y'}{\sqrt{x(1 + y'^2)}} = \frac{1}{\sqrt{2a}},$$

and, by separation of variables,

$$y = \int_0^x \sqrt{\frac{x}{2a - x}}\, dx. \qquad (5.24)$$

To solve this integral, we shall make the following change of variables

$$\sqrt{\frac{x}{2a - x}} = u. \qquad (5.25)$$

We then have

$$\frac{\frac{1}{2\sqrt{x}}\sqrt{2a - x} + \sqrt{x}\frac{1}{\sqrt{2a-x}}}{2a - x}\, dx = du,$$

which yields

$$dx = \frac{1}{a}(2a - x)^{3/2}\sqrt{x}\, du. \qquad (5.26)$$

Using (5.25), we have
$$x = \frac{2au^2}{1 + u^2}, \qquad (5.27)$$

so that (5.26) leads to
$$dx = \frac{4au}{(1 + u^2)^2}\, du. \qquad (5.28)$$

Introducing (5.25) and (5.28) into (5.24), we can write

$$y = \int_0^u \frac{4au^2}{(1 + u^2)^2}\, du = -2a \int_0^u u\, d\left(\frac{1}{1 + u^2}\right)$$

$$= -2a\left[\left(\frac{u}{1+u^2}\right)\Big|_0^u - \int_0^u \frac{du}{1+u^2}\right]$$

$$= -2a\left[\frac{u}{1+u^2} - (\arctan u)\Big|_0^u\right] = -2a\left(\frac{u}{1+u^2} - \arctan u\right).$$

A new change of variables

$$u = \tan\frac{\theta}{2}$$

yields

$$y = -2a\left(\frac{\tan\frac{\theta}{2}}{1+\tan^2\frac{\theta}{2}} - \frac{\theta}{2}\right) = -2a\left(\sin\frac{\theta}{2}\cos\frac{\theta}{2} - \frac{\theta}{2}\right)$$

$$= -2a\left(\frac{\sin\theta}{2} - \frac{\theta}{2}\right) = a(\theta - \sin\theta),$$

so that

$$x = \frac{2au^2}{1+u^2} = 2a\frac{\tan^2\frac{\theta}{2}}{1+\tan^2\frac{\theta}{2}} = 2a\sin^2\frac{\theta}{2} = a(1-\cos\theta).$$

Summarizing our results, we have obtained

$$\begin{cases} x = a(1-\cos\theta), \\ y = a(\theta - \sin\theta), \end{cases} \tag{5.29}$$

which are the parametric equations of a cycloid, with y-axis as basis and concavity oriented upwards (see Fig.V.3).

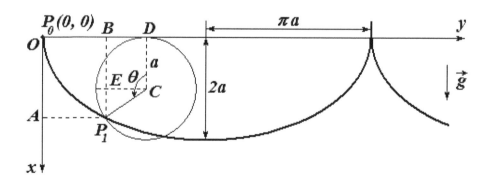

Fig.V.3

The constant a is not arbitrary anymore, but it represents the radius of the circle that generates the cycloid, by rolling without friction on y-axis. In our picture, the cycloid is generated by the point P_1, which is fixed with respect to the circle of radius a. According to Fig.V.3, we can write

$$x = \|OA\| = \|DC\| + \|EP_1\| = a + \|CP_1\|\sin\left(\theta - \frac{\pi}{2}\right) = a(1 - \cos\theta),$$

and

$$y = \|OB\| = \|OD\| - \|BD\| = \widehat{DP_1} - \|BD\| = \widehat{DP_1} - \|EC\|$$

$$= a\theta - a\cos\left(\theta - \frac{\pi}{2}\right) = a(\theta - \sin\theta).$$

To conclude, the brachistochrone curve is a cycloid (arc of a cycloid). If the body is given an initial velocity at P_0, or if friction is taken into account, then the curve that minimizes time will differ from the one described above.

2. Catenary problem

Determine the curve formed by a rope or chain of uniform density and perfect flexibility, hanging freely between two points of suspension, not in the same vertical line. The rope has a fixed length (it is not extensible).

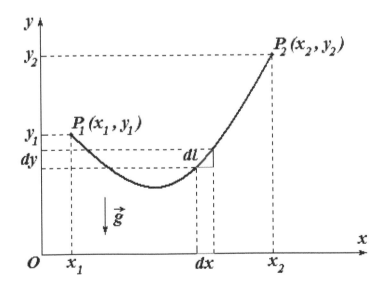

Fig.V.4

Solution.
Let l be the length of the chain, m its mass, and $P_1(x_1, y_1)$, $P_2(x_2, y_2)$ the points of suspension (see Fig.V.4). The equilibrium condition of the chain demands an extremum for the potential energy (minimum, in our case). To evaluate the potential energy, we split the chain in infinitesimal elements so that one may neglect the variation of the vertical coordinate y.

If the plane xOz is taken as the reference level for the potential energy, then gravitational potential energy of the length element dl represented in Fig.V.4 is

$$dE_p = dm\,g\,y = dl\,\lambda\,g\,y, \tag{5.30}$$

where $\lambda = l/m$ is the mass per unit length of the chain. Since $(dl)^2 = (dx)^2 + (dy)^2$, we then have

$$dl = \sqrt{1+y'^2}\,dx; \quad y' = \frac{dy}{dx}, \tag{5.31}$$

where only the positive solution has been considered. The potential energy of the whole chain then is

$$E_p = \int_{P_1}^{P_2} dE_p = \int_{P_1}^{P_2} \lambda g y\,dl = \lambda g \int_{x_1}^{x_2} y\sqrt{1+y'^2}\,dx. \tag{5.32}$$

Leaving aside the multiplicative factor λg, which in no way interferes in the extrema of the potential energy functional, one can consider the functional

$$E(y) = \int_{x_1}^{x_2} y\sqrt{1+y'^2}\,dx. \tag{5.33}$$

Following the variational procedure, we have to demand that the functional (5.33) has an extremum. Since the chain has a constant length

$$l = \int_{P_1}^{P_2} dl = \int_{x_1}^{x_2} \sqrt{1+y'^2}\,dx, \tag{5.34}$$

this condition tells us that we have to look for a *conditional extremum*. As easily seen, our constraint can be expressed through the functional

$$L(y) = \int_{x_1}^{x_2} \sqrt{1+y'^2}\,dx = const. \tag{5.35}$$

Therefore, to solve the problem means to find the extrema of the new functional

$$F(y) = E(y) - \mu L(y), \tag{5.36}$$

where μ is an arbitrary (but non-zero) scalar. Using (5.33) and (5.35), we can write

$$F(y) = \int_{x_1}^{x_2} \left[(y-\mu)\sqrt{1+y'^2}\right] dx = \int_{x_1}^{x_2} f(y,y')\,dx,$$

where
$$f(y,y') = (y-\mu)\sqrt{1+y'^2}. \tag{5.36'}$$

The extremals of the functional (5.36) are found by solving the Euler-Lagrange equation

$$\frac{d}{dx}\left(\frac{\partial f}{\partial y'}\right) - \frac{\partial f}{\partial y} = 0, \tag{5.37}$$

where $f(y,y')$ is given by (5.36'). Since f does not explicitly depend on the independent variable x, we have the first integral

$$y'\frac{\partial f}{\partial y'} - f = const. \equiv C, \tag{5.38}$$

which yields
$$y - \mu = -C\sqrt{1+y'^2},$$

or, by squaring and choosing the positive solution

$$y' = \frac{dy}{dx} = \frac{1}{C}\sqrt{(y-\mu)^2 - C^2}.$$

Separation of variables leads to the following differential equation

$$\frac{dy}{\sqrt{(y-\mu)^2 - C^2}} = \frac{1}{C}\,dx.$$

To integrate this equation, it is convenient to make the change of variable
$$y - \mu = C\cosh u, \tag{5.39}$$

and the solution is
$$u = \frac{1}{C}x + C_1,$$

so that solution (5.39) finally writes

$$y = \mu + C\cosh\left(\frac{x}{C} + C_1\right). \tag{5.40}$$

If fact, we deal with an infinite number of functions, obtained as solutions of (5.40), but only one curve passes through the points P_1 and P_2 and, in addition, obeys the constraint (5.34). Imposing these conditions, we are left with the following system of three algebraic equations for the unknown constants μ, C, and C_1:

$$\begin{cases} y_1 = \mu + C \cosh\left(\dfrac{x_1}{C} + C_1\right), \\ y_2 = \mu + C \cosh\left(\dfrac{x_2}{C} + C_1\right), \\ l = \displaystyle\int_{x_1}^{x_2} \sqrt{1 + \sinh^2\left(\dfrac{x}{C} + C_1\right)}\, dx, \end{cases}$$

or

$$\begin{cases} y_1 = \mu + C \cosh\left(\dfrac{x_1}{C} + C_1\right), \\ y_2 = \mu + C \cosh\left(\dfrac{x_2}{C} + C_1\right), \\ l = C \sinh\left(\dfrac{x_2}{C} + C_1\right) - C \sinh\left(\dfrac{x_1}{C} + C_1\right). \end{cases}$$

We leave the task of solving this system up to the reader. The curve given by equation (5.40) is called *catenary* (from the Latin *catena*, meaning *chain*).

3. Isoperimetric problem

Determine the form of a plane curve of a given length l which encloses a surface of maximum area.

Solution.

According to Fig.V.5, the area $\mathcal{A}$ of the surface S bounded by the closed curve (Γ) is given by

$$\mathcal{A} = \int_a^b \varphi_2(x)\, dx - \int_a^b \varphi_1(x)\, dx = -\int_b^a \varphi_2(x)\, dx - \int_a^b \varphi_1(x)\, dx$$

$$= -\left[\int_a^b \varphi_1(x)\, dx + \int_b^a \varphi_2(x)\, dx\right]$$

$$= -\left[\int_{\widehat{ANB}} \varphi_1(x)\, dx + \int_{\widehat{BMA}} \varphi_2(x)\, dx\right] = -\oint_{(\Gamma)} y\, dx,$$

where $\mathcal{A}$ has been written as a difference of two areas: one bounded by the graphic of $y = \varphi_2(x)$ between points A and B, the x-axis, and the straight lines $x = a$, $x = b$ (the line-hatched area), and the other one bounded by the graphic of $y = \varphi_1(x)$, between the same points A and B, and the straight lines $x = a$, $x = b$ (the cross-hatched area).

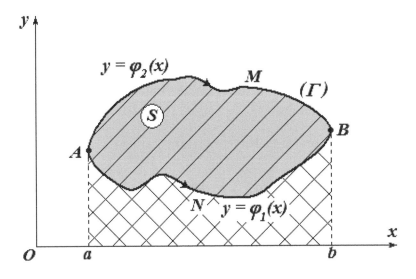

Fig.V.5

To express in a more symmetric form the above relation, we shall add a "zero" written as

$$0 = \frac{1}{2} \oint_{(\Gamma)} d(xy) = \frac{1}{2} \oint_{(\Gamma)} x \, dy + \frac{1}{2} \oint_{(\Gamma)} y \, dx.$$

The result is

$$\mathcal{A} = -\oint_{(\Gamma)} y \, dx + \frac{1}{2} \oint_{(\Gamma)} x \, dy + \frac{1}{2} \oint_{(\Gamma)} y \, dx$$

$$= \frac{1}{2} \oint_{(\Gamma)} (x \, dy - y \, dx). \tag{5.41}$$

Suppose that the curve is defined by the parametric equations

$$\begin{cases} x = x(t); \\ y = y(t), \quad t \in [a, b], \end{cases}$$

with $a = x(t_1)$ and $b = x(t_2)$. Then area $\mathcal{A}$ writes

$$\mathcal{A} = \frac{1}{2} \oint_{(\Gamma)} (x\dot{y} - y\dot{x}) \, dt, \tag{5.42}$$

with $\dot{x} = \frac{dx}{dt}$, $\dot{y} = \frac{dy}{dt}$. The problem reduces, therefore, to the determination of extremals of the functional

$$F(x,y) = \frac{1}{2} \oint_{(\Gamma)} (x\dot{y} - y\dot{x})\, dt, \qquad (5.43)$$

subject to the condition that the length l of (Γ) is given (and fixed):

$$l = \oint_{(\Gamma)} dl = \oint_{(\Gamma)} \sqrt{(dx)^2 + (dy)^2} = \oint_{(\Gamma)} \sqrt{\dot{x}^2 + \dot{y}^2}\, dt.$$

The functional expressing the constraint can be written as

$$L(x,y) = \oint_{(\Gamma)} \sqrt{\dot{x}^2 + \dot{y}^2}\, dt = const. \equiv l. \qquad (5.44)$$

Toward the general theory, we construct a new functional

$$G(x,y) = F(x,y) - \lambda L(x,y) = \oint_{(\Gamma)} \left[\frac{1}{2}(x\dot{y} - y\dot{x}) - \lambda\sqrt{\dot{x}^2 + \dot{y}^2}\right] dt$$

$$= \oint_{(\Gamma)} g(x(t), y(t), \dot{x}(t), \dot{y}(t))\, dt,$$

where λ is an arbitrary (but non-zero) scalar. To determine the extremals of the functional $G(x,y)$, we have to solve the corresponding Euler-Lagrange equations

$$\begin{cases} \dfrac{d}{dt}\left(\dfrac{\partial g}{\partial \dot{x}}\right) - \dfrac{\partial g}{\partial x} = 0; \\ \dfrac{d}{dt}\left(\dfrac{\partial g}{\partial \dot{y}}\right) - \dfrac{\partial g}{\partial y} = 0, \end{cases} \qquad (5.45)$$

with

$$g(x, y, \dot{x}, \dot{y}) = \frac{1}{2}(x\dot{y} - y\dot{x}) - \lambda\sqrt{\dot{x}^2 + \dot{y}^2}.$$

Performing the derivatives in (5.45), we are left with

$$\begin{cases} \dfrac{d}{dt}\left(-\dfrac{1}{2}y - \lambda\dfrac{\dot{x}}{\sqrt{\dot{x}^2 + \dot{y}^2}}\right) - \dfrac{1}{2}\dot{y} = 0; \\ \dfrac{d}{dt}\left(+\dfrac{1}{2}x - \lambda\dfrac{\dot{y}}{\sqrt{\dot{x}^2 + \dot{y}^2}}\right) + \dfrac{1}{2}\dot{x} = 0, \end{cases}$$

or

$$\begin{cases} -\dot{y} - \lambda \dfrac{\ddot{x}\sqrt{\dot{x}^2+\dot{y}^2} - \dot{x}\frac{\dot{x}\ddot{x}+\dot{y}\ddot{y}}{\sqrt{\dot{x}^2+\dot{y}^2}}}{\dot{x}^2+\dot{y}^2} = 0; \\ +\dot{x} - \lambda \dfrac{\ddot{y}\sqrt{\dot{x}^2+\dot{y}^2} - \dot{y}\frac{\dot{x}\ddot{x}+\dot{y}\ddot{y}}{\sqrt{\dot{x}^2+\dot{y}^2}}}{\dot{x}^2+\dot{y}^2} = 0, \end{cases}$$

and, still

$$\begin{cases} \dot{y}\left[-1 - \lambda\dfrac{\ddot{x}\dot{y} - \ddot{y}\dot{x}}{(\dot{x}^2+\dot{y}^2)^{3/2}}\right] = 0; \\ \dot{x}\left[+1 + \lambda\dfrac{\ddot{x}\dot{y} - \ddot{y}\dot{x}}{(\dot{x}^2+\dot{y}^2)^{3/2}}\right] = 0. \end{cases} \quad (5.46)$$

Since the solutions $\dot{x} = 0$, $\dot{y} = 0$ are not acceptable, the only remaining possibility is cancellation of the square brackets, that is

$$\frac{\ddot{x}\dot{y} - \ddot{y}\dot{x}}{(\dot{x}^2+\dot{y}^2)^{3/2}} = -\frac{1}{\lambda}. \quad (5.47)$$

The nature of parameter t has not been specified so far. Let us assume that t is precisely the arc length l of the curve (Γ), determined with respect to a fixed point on (Γ), so that

$$\dot{x}^2 + \dot{y}^2 = \left(\frac{dx}{dt}\right)^2 + \left(\frac{dy}{dt}\right)^2 = \frac{(dx)^2 + (dy)^2}{(dl)^2} = 1. \quad (5.48)$$

With this choice, equation (5.47) yields

$$\dot{y}\ddot{x} - \dot{x}\ddot{y} = -\frac{1}{\lambda}.$$

To solve the problem, we need one more equation in $x(t)$ and $y(t)$. This is done by taking the derivative of (5.48) with respect to t:

$$\dot{x}\ddot{x} + \dot{y}\ddot{y} = 0.$$

Thus, we have obtained the following system of two second-order differential equations in two variables $x(t)$ and $y(t)$:

$$\begin{cases} \dot{y}\ddot{x} - \dot{x}\ddot{y} = -\frac{1}{\lambda}; \\ \dot{x}\ddot{x} + \dot{y}\ddot{y} = 0. \end{cases} \quad (5.49)$$

To facilitate the integration of (5.49), let us extract $\ddot{x}$ from the second equation, and introduce the result into the first one. We then have

$$\ddot{y} = \frac{1}{\lambda}\dot{x}. \tag{5.50}$$

Repeating the procedure for $\ddot{y}$, one obtains

$$\ddot{x} = -\frac{1}{\lambda}\dot{y}. \tag{5.51}$$

This way, we are left with a simpler system of second order differential equations, namely

$$\begin{cases} \ddot{x} = -\lambda^{-1}\dot{y}, \\ \ddot{y} = \lambda^{-1}\dot{x}. \end{cases} \tag{5.52}$$

Integrating once gives

$$\begin{cases} \dot{x} = -\lambda^{-1}y + C_1, \\ \dot{y} = \lambda^{-1}x + C_2. \end{cases}$$

With these values of $\dot{x}$ and $\dot{y}$, (5.48) becomes

$$1 = \dot{x}^2 + \dot{y}^2 = \left(C_1 - \frac{y}{\lambda}\right)^2 + \left(C_2 + \frac{x}{\lambda}\right)^2,$$

or

$$(x + \lambda C_2)^2 + (y - \lambda C_1)^2 = \lambda^2. \tag{5.53}$$

This is the equation of a circle of radius $R = \lambda$, located in the xy-plane, with centre at the point $x_C = -\lambda C_2$, $y_C = \lambda C_1$. So, our problem is solved: for *any* constants λ, C_1, C_2 (which remain undetermined), *the plane curve of a given length enclosing a maximum area is a circle.*

4. Surface of revolution of minimum area

Let $P_1(x_1, y_1)$ and $P_2(x_2, y_2)$ be two given (and fixed) points situated in the xy-plane of a three-orthogonal trieder $Oxyz$. Determine the curve $y = y(x)$ passing through P_1 and P_2 and generating by revolution about an axis (say, x), a surface of minimum area.

Solution.

The surface produced as a result of revolution about Ox is formed by three parts: areas of the two circles of the "truncated cone", of radii

y_1 and y_2, and the lateral area of the "truncated cone" (see Fig.V.6). Since y_1 and y_2 are fixed, our problem is to determine the plane curve $y = y(x)$ which, by its revolution about the x-axis, generates a lateral surface of minimum area.

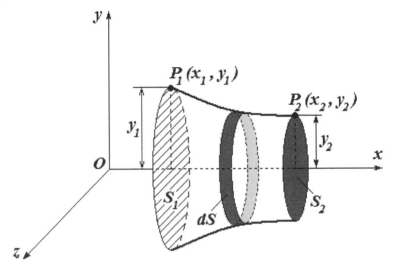

Fig.V.6

As shown in Fig.V.6, the areas of the two circular "bases" are $S_1 = \pi y_1^2 = const.$ and $S_2 = \pi y_2^2 = const.$, while the area of the elementary surface of width[1] ds, and length $2\pi y$ (see Fig.V.7) is

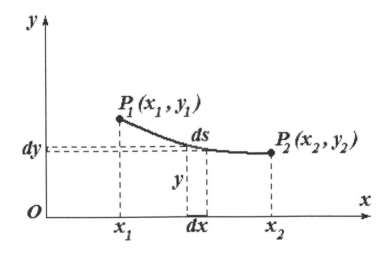

Fig.V.7

---

[1] This width is small enough so that one can suppose that all points on ds have the same coordinate y

$$dS = 2\pi y \, ds, \tag{5.54}$$

where
$$ds = \sqrt{(dx)^2 + (dy)^2} = \sqrt{1 + y'^2} \, dx,$$

with $y' = dy/dx$. Then the lateral surface area of the "truncated cone" is

$$S = 2\pi \int_{x_1}^{x_2} y\sqrt{1 + y'^2} \, dx. \tag{5.55}$$

Ignoring the multiplicative constant 2π, which in no way affects our result, the area (5.55) is expressed as a functional, namely

$$S(y) = \int_{x_1}^{x_2} y\sqrt{1 + y'^2} \, dx = \int_{x_1}^{x_2} f(y, y') \, dx. \tag{5.56}$$

To determine extrema of the functional (5.56), we have to solve Euler's equation

$$\frac{d}{dx}\left(\frac{\partial f}{\partial y'}\right) - \frac{\partial f}{\partial y} = 0. \tag{5.57}$$

Since $f(y, y')$ does not explicitly depend on x, there exists the first integral of (5.57)

$$y'\frac{\partial f}{\partial y'} - f = const. \equiv C. \tag{5.58}$$

We have:
$$\frac{\partial f}{\partial y'} = \frac{yy'}{\sqrt{1 + y'^2}},$$

so that
$$\frac{yy'}{\sqrt{1 + y'^2}} - y\sqrt{1 + y'^2} = C,$$

or, after simple algebraic manipulations

$$y' = \frac{dy}{dx} = \frac{1}{C}\sqrt{y^2 - C^2},$$

Considering only positive solution and integrating, we still have

$$\int \frac{dy}{\sqrt{y^2 - C^2}} = \frac{x}{C} + C_1. \tag{5.59}$$

The integral on the l.h.s. can be easily solved by substitution

$$y = C \cosh u. \tag{5.60}$$

Indeed,
$$\int \frac{dy}{\sqrt{y^2 - C^2}} = \int \frac{C \sinh u}{\sqrt{C^2(\cosh^2 u - 1)}} du = u.$$

In view of (5.59), we can write
$$u = \frac{x}{C} + C_1,$$

and (5.60) finally leads to
$$y = C \cosh\left(\frac{x}{C} + C_1\right). \tag{5.60'}$$

To give the answer to our problem, one must determine the constants $C \neq 0$ and C_1. This is done by imposing condition that the curve (5.60') passes through $P_1(x_1, y_1)$ and $P_2(x_2, y_2)$, which means
$$\begin{cases} y_1 = C \cosh\left(\frac{x_1}{C} + C_1\right); \\ y_2 = C \cosh\left(\frac{x_2}{C} + C_1\right). \end{cases}$$

To conclude, the curve passing through two given points, that generates by its revolution about x-axis a surface of minimum area, is a *catenary*.

5. Geodesics of a Riemannian manifold
Determine the geodesics of a n-dimensional Riemannian manifold.

Solution.
This problem concerns the general theory of relativity, but we have considered it here because it is an important application of the variational calculus. Before going further, let us define the terms *geodesic* and *Riemannian manifold* (or *space*). In geometric approach, a *geodesic* is a generalization of the notion of a "straight line" to "curved spaces", being (locally) the shortest path between two points in the space. From the dynamical point of view, a geodesic is the trajectory described by a free particle (material point) in space.

Obviously, the shape of the geodesic depends on the structure of the space on which it is defined. For instance, in an Euclidean three-dimensional space E_3, the shortest distance between two points is determined by a straight line passing through the two points. From the dynamical point of view, a free particle moves uniformly in a straight line, relative to an inertial frame. All these definitions can be obtained by a variational procedure as follows.

In the Euclidean space E_3, the infinitesimal distance between two points is given by

$$ds = |d\vec{r}| = \sqrt{(dx)^2 + (dy)^2 + (dz)^2} = \sqrt{dx_i\, dx_i}, \quad (i = \overline{1.3}), \quad (5.61)$$

where $x_1 = x$, $x_2 = y$, $x_3 = z$, and Einstein's summation convention has been used. The squared ds, that is

$$ds^2 = |d\vec{r}|^2 = d\vec{r} \cdot d\vec{r} = (dx)^2 + (dy)^2 + (dz)^2 = dx_i dx_i \quad (i = \overline{1.3}) \quad (5.62)$$

is called *metric* of the three-dimensional Euclidean space E_3.

Let us parametrize some curve in E_3 by

$$x_i = x_i(t) \quad (i = \overline{1,3}),$$

where t is the parameter. Then the distance between any two points $P'(x', y', z')$ and $P''(x'', y'', z'')$ on this curve is given by

$$\Delta s = \int_{P'}^{P''} ds = \int_{P'}^{P''} \sqrt{dx_i(t)dx_i(t)} = \int_{t_1}^{t_2} \sqrt{\dot{x}_i \dot{x}_i}\, dt, \quad (5.63)$$

where

$$\dot{x}_i = \frac{dx_i}{dt}, \quad x'_i = x_i(t_1), \quad x''_i = x_i(t_2).$$

As one observes, the distance Δs is written as a functional of the type

$$J(x) = \int_{t_1}^{t_2} f(t, x, \dot{x})\, dt,$$

where

$$f(\dot{x}) = \sqrt{\dot{x}_i \dot{x}_i}.$$

Since f does not explicitly depend on $x(t)$, the Euler equation of extremals

$$\frac{d}{dt}\left(\frac{\partial f}{\partial \dot{x}_j}\right) - \frac{\partial f}{\partial x_j} = 0 \quad (j = \overline{1,3})$$

admits the first integral

$$\frac{\partial f}{\partial \dot{x}_j} = (const.)_j \equiv C_j \quad (j = \overline{1,3}) \quad (5.64)$$

or
$$\frac{\partial}{\partial \dot{x}_j}\left(\sqrt{\dot{x}_i\dot{x}_i}\right) = \frac{\dot{x}_j}{\sqrt{\dot{x}_i\dot{x}_i}} = C_j \quad (i,j = \overline{1,3}). \tag{5.65}$$

Let us consider the following two cases:

(a) If the arc length s is taken as the parameter t, then
$$\sqrt{\dot{x}_i\dot{x}_i} = \sqrt{\frac{dx_i}{ds}\frac{dx_i}{ds}} = \frac{\sqrt{dx_i dx_i}}{\sqrt{(ds)^2}} = 1.$$

With this result, (5.65) leads to
$$\dot{x}_j = C_j \quad (j = \overline{1,3}), \tag{5.66}$$

or
$$x_j = C_j s + C'_j, \tag{5.67}$$

where C'_j ($j = \overline{1,3}$) are three new constants of integration. To determine the six constants of integration C_j and C'_j, one imposes the condition that P' and P'' are on the curve.

Relations (5.67) can be written explicitly as
$$x = C_1 s + C'_1,$$
$$y = C_2 s + C'_2,$$
$$z = C_3 s + C'_3.$$

Eliminating the parameter s, we obtain
$$y = C_2 \frac{x - C'_1}{C_1} + C'_2 = \frac{C_2}{C_1}x + \left(-\frac{C_2 C'_1}{C_1} + C'_2\right) = k_1 x + k_2; \tag{5.68}$$

$$z = C_3 \frac{y - C'_2}{C_2} + C'_3 = \frac{C_3}{C_2}x + \left(-\frac{C_3 C'_2}{C_2} + C'_3\right) = k_3 x + k_4, \tag{5.69}$$

where the new constants k_i ($i = \overline{1,4}$) are expressed in terms of C_j and C'_j. The relations
$$\begin{cases} y = k_1 x + k_2; \\ z = k_3 x + k_4, \end{cases} \tag{5.70}$$

represent a straight line in E_3. In other words, *the geodesics of the Euclidean three-dimensional space E_3 are straight lines.*

(b) Let us now consider parameter t as being the time. Then (5.65) writes
$$\frac{v_j}{\sqrt{|\vec{v}|^2}} = C_j \quad (j = \overline{1,3}),$$

or, by multiplying both numerator and denominator of the fraction by the mass $m \neq 0$ of the body

$$\frac{p_j}{\sqrt{|\vec{p}|^2}} = C_j \quad (j = \overline{1,3}). \tag{5.71}$$

If the resultant force acting on the body is zero (the body moves freely), then according to the fundamental principle of dynamics

$$\vec{F} = \frac{d\vec{p}}{dt} = 0,$$

meaning that $\vec{p} = const.$, and (5.71) yields

$$p_j = K_j \quad (j = \overline{1,3}),$$

where K_j $(j = \overline{1,3})$ are constants. Therefore,

$$m\dot{x}_j = K_j \quad (j = \overline{1,3}). \tag{5.72}$$

On the other hand, the kinetic energy theorem says that the kinetic energy of a free body is a constant. In this case, since $m|\vec{v}|^2 = C'$ and $|\vec{p}|^2 = m^2|\vec{v}|^2 = C''$, we have $m = C''/C' = const.$, where C' and C'' are also constants. Then (5.72) can be written as

$$\dot{x}_j = K'_j = (const.)_j,$$

or

$$x_j = K'_j t + x_{0j} \quad (j = \overline{1,3}), \tag{5.73}$$

where the arbitrary constants of integration x_{0j} $(j = \overline{1,3})$ can be determined by the initial conditions. In vector notation, equation (5.73) writes

$$\vec{r} = \vec{K}'t + \vec{r}_0. \tag{5.73'}$$

Therefore, in the Euclidean three-dimensional space E_3, a free body moves uniformly, in a straight line.

* * *

To define a *Riemannian manifold* one must introduce the notions of metric of a space, metric tensor, as well as *covariance* and *contravariance* of tensors. For a better understanding of the theory, we shall first consider a three-dimensional space. Here, in addition to the simplest coordinate system (the Cartesian frame), there are two more (but not most!) general reference frames: (i) orthogonal curvilinear coordinates (a system of curvilinear coordinates in which each family of surfaces intersects the others at right angles); (ii) non-orthogonal rectilinear coordinates (the coordinate axes are straight lines, but the angles of intersection are different from 90°). As the most general case one can consider a combination of i) and ii): the coordinate axes are curvilinear, while the angles of intersection are different from 90° (see Fig.V.8).

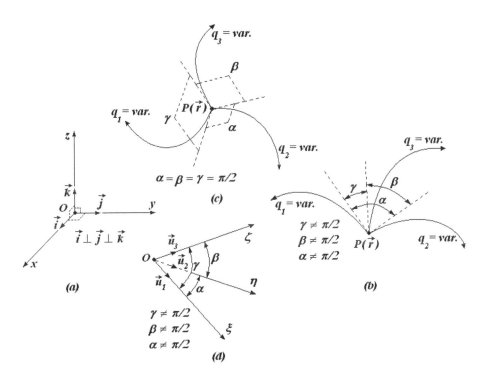

Fig.V.8

Since the orthogonal system of coordinates is thoroughly discussed in textbooks, we shall analyze the case (ii), showing how non-orthogonality of axes requires introduction of the concept of *variance*. To facilitate the graphic representations, let us first take into account a two-dimensional space (generalization to three dimensions is trivial). Consider a vector $\vec{a}$ and write it in component forms as reported to two different reference systems: one orthogonal, and the other non-orthogonal (see Fig.V.9).

In the first case (see Fig.V.9a) we can write
$$\begin{cases} a_1 = \vec{a} \cdot \vec{u}_1, \\ a_2 = \vec{a} \cdot \vec{u}_2. \end{cases} \tag{5.74}$$

Using the notations $\vec{a}_1 = a_1\vec{u}_1$ and $\vec{a}_2 = a_2\vec{u}_2$, we also have

$$\vec{a} = \vec{a}_1 + \vec{a}_2, \qquad (5.75)$$

where $\vec{u}_1$ and $\vec{u}_2$ are the unit vectors of the two axes, and a_1, a_2 - the components of the vector $\vec{a}$ (the orthogonal projections of $\vec{a}$ on the coordinate axes).

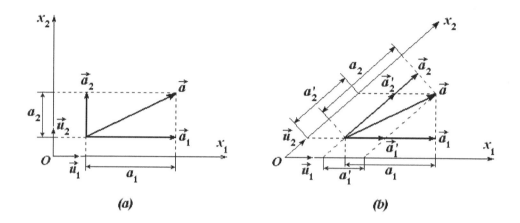

Fig.V.9

In the second case (see Fig.V.9b), there are two possibilities to define the components of the vector $\vec{a}$: either by tracing straight lines perpendicular to the axes (components a_1 and a_2), or by drawing straight lines parallel to the axes (components a'_1 and a'_2). The first possibility allows us to write

$$\begin{cases} a_1 = \vec{a} \cdot \vec{u}_1, \\ a_2 = \vec{a} \cdot \vec{u}_2, \end{cases} \qquad (5.76)$$

but, if we denote $\vec{a}_1 = a_1\vec{u}_1$ and $\vec{a}_2 = a_2\vec{u}_2$, the relation (5.75) is not valid anymore, since now

$$\vec{a} \neq \vec{a}_1 + \vec{a}_2. \qquad (5.77)$$

Using the other procedure, if we denote

$$\begin{cases} \vec{a}'_1 = a'_1\vec{u}_1, \\ \vec{a}'_2 = a'_2\vec{u}_2, \end{cases} \qquad (5.78)$$

then relation (5.75) remains valid, that is

$$\vec{a} = \vec{a}'_1 + \vec{a}'_2, \qquad (5.79)$$

but this time
$$\begin{cases} a'_1 \neq \vec{a} \cdot \vec{u}_1, \\ a'_2 \neq \vec{a} \cdot \vec{u}_2, \end{cases} \quad (5.80)$$

so that (5.74) loses its validity. In other words, the components a'_1 and a'_2 are not obtained as scalar products of the vector $\vec{a}$, on the one hand, and the unit vectors $\vec{u}_1$ and $\vec{u}_2$, on the other.

To "conciliate" these two possible choices for both cases (*i.e.* for the two possible choices/possibilities), one introduces the so-called *dual basis*. Consider, in this respect, a system of three non-coplanar, linearly independent vectors $\vec{u}$, $\vec{v}$, and $\vec{w}$, which form a *basis* in the Euclidean three-dimensional space E_3. Then any vector $\vec{a} \in E_3$ can be written as
$$\vec{a} = \lambda \vec{u} + \mu \vec{v} + \nu \vec{w}, \quad (5.81)$$

where the three scalars (not all zero) are called *the components* of the vector $\vec{a}$ in basis $\{\vec{u}, \vec{v}, \vec{w}\}$.

Let us now introduce another set of three vectors $\vec{u}^*, \vec{v}^*$, and $\vec{w}^*$, defined as
$$\begin{cases} \vec{u}^* \cdot \vec{u} = 1, \\ \vec{u}^* \cdot \vec{v} = 0, \\ \vec{u}^* \cdot \vec{w} = 0, \end{cases} \begin{cases} \vec{v}^* \cdot \vec{u} = 0, \\ \vec{v}^* \cdot \vec{v} = 1, \\ \vec{v}^* \cdot \vec{w} = 0, \end{cases} \begin{cases} \vec{w}^* \cdot \vec{u} = 0, \\ \vec{w}^* \cdot \vec{v} = 0, \\ \vec{w}^* \cdot \vec{w} = 1. \end{cases} \quad (5.82)$$

It can be easily verified that the triplet of vectors $\vec{u}^*$, $\vec{v}^*$, and $\vec{w}^*$
$$\vec{u}^* = \frac{\vec{v} \times \vec{w}}{(\vec{u}, \vec{v}, \vec{w})}; \quad \vec{v}^* = \frac{\vec{w} \times \vec{u}}{(\vec{u}, \vec{v}, \vec{w})}; \quad \vec{w}^* = \frac{\vec{u} \times \vec{v}}{(\vec{u}, \vec{v}, \vec{w})}, \quad (5.83)$$

where $(\vec{u}, \vec{v}, \vec{w}) = \vec{u} \cdot (\vec{v} \times \vec{w})$ is the mixed product of the three vectors, satisfying the definition conditions (5.82).

According to their definition, the vectors $\vec{u}^*$, $\vec{v}^*$, and $\vec{w}^*$ are, in their turn, linearly independent, which means that they also form a basis in R_3, called *dual basis* of the direct basis $\{\vec{u}, \vec{v}, \vec{w}\}$. It can also be shown that the dual of dual of a vector yields the original vector, that is
$$\left(\vec{u}^*\right)^* = \vec{u}, \quad \left(\vec{v}^*\right)^* = \vec{v}, \quad \left(\vec{w}^*\right)^* = \vec{w}. \quad (5.84)$$

The components of any vector $\vec{a} \in E_3$ with respect to the direct basis (*e.g.* the scalars λ, μ and ν that appear in (5.81)), can be expressed as scalar products of $\vec{a}$ and the corresponding vectors of dual basis. Indeed, by means of (5.81) and (5.82), it is easy to show that
$$\lambda = \vec{a} \cdot \vec{u}^*, \quad \mu = \vec{a} \cdot \vec{v}^*, \quad \nu = \vec{a} \cdot \vec{w}^*. \quad (5.85)$$

Coming now back to our problem, in case of the frames with rectilinear, non-orthogonal coordinates we can write both a relation of type (5.75) for the unprimed components

$$\vec{a} = \vec{a}_1 + \vec{a}_2,$$

but where

$$\vec{a}_1 = a_1 \vec{u}_1^*, \quad \vec{a}_2 = a_2 \vec{u}_2^*, \tag{5.86}$$

and a relation of type (5.74) for the primed components $a_1' = \vec{a} \cdot \vec{u}_1$, $a_2' = \vec{a} \cdot \vec{u}_2$, but where $\vec{u}_1$ and $\vec{u}_2$ must be replaced by $\vec{u}_1^*$ and $\vec{u}_2^*$, that is

$$a_1' = \vec{a} \cdot \vec{u}_1^*, \quad a_2' = \vec{a} \cdot \vec{u}_2^*. \tag{5.87}$$

Indeed, in the first case we have $a_1 = \vec{a} \cdot \vec{u}_1$, $a_2 = \vec{a} \cdot \vec{u}_2$, and $\vec{a} \neq \vec{a}_1 + \vec{a}_2$. But, if $\vec{a}_1$ and $\vec{a}_2$ are given by (5.86), even now we can write a relation of type (5.75)

$$\vec{a} = a_1 \vec{u}_1^* + a_2 \vec{u}_2^* = \vec{a}_1 + \vec{a}_2,$$

because, in view of (5.85) and (5.84) we have

$$\begin{cases} a_1 = \vec{a} \cdot \left(\vec{u}_1^*\right)^* = \vec{a} \cdot \vec{u}_1, \\ a_2 = \vec{a} \cdot \left(\vec{u}_2^*\right)^* = \vec{a} \cdot \vec{u}_2, \end{cases}$$

which are precisely relations (5.76). This way, the two types of relations are in "agreement" with each other, in the sense that they are simultaneously valid.

In the second case, we have $\vec{a} = \vec{a}_1' + \vec{a}_2'$, but $\vec{a}_1' \neq \vec{a} \cdot \vec{u}_1$, and $\vec{a}_2' \neq \vec{a} \cdot \vec{u}_2$. But, if we replace $\vec{u}_1$ and $\vec{u}_2$ by $\vec{u}_1^*$ and $\vec{u}_2^*$, then the two relations become equalities ($a_1' = \vec{a} \cdot \vec{u}_1^*$, and $a_2' = \vec{a} \cdot \vec{u}_2^*$). This means that, in this second case, a relation of type (5.76) is also valid. Indeed, according to (5.81) and (5.85), we have

$$\vec{a} = a_1' \vec{u}_1 + a_2' \vec{u}_2 = \vec{a}_1' + \vec{a}_2',$$

in agreement with (5.79) and (5.79). Therefore, the problem is solved.

In practice, one usually uses a different set of notations, namely:

$$\begin{cases} a_1' = a^1, \\ a_2' = a^2, \end{cases} \text{ and } \begin{cases} \vec{u}_1^* = \vec{u}^1, \\ \vec{u}_2^* = \vec{u}^2. \end{cases} \tag{5.88}$$

We conclude that, in a non-orthogonal frame, any vector $\vec{a}$ has two sets of components: one with *lower* indices

$$\begin{cases} a_1 = \vec{a} \cdot \vec{u}_1, \\ a_2 = \vec{a} \cdot \vec{u}_2, \end{cases} \tag{5.89}$$

and the other one with *upper* indices

$$\begin{cases} a^1 = \vec{a} \cdot \vec{u}^1, \\ a^2 = \vec{a} \cdot \vec{u}^2. \end{cases} \quad (5.90)$$

Components with lower indices are called *covariant*, while those with upper indices are called *contravariant*. The direct basis is formed by covariant vectors, and the dual basis by contravariant vectors[1]. In addition, as shown in Fig.V.9b, if the angle between axes becomes 90°, the two types of components are coincident (the projections orthogonal and parallel to axes become identical). In other words, in case of orthogonal coordinates, there is no difference between covariant and contravariant components of a vector.

To summarize, in a non-orthogonal vector space any vector has two sets of components: covariant, if the vector is expressed in the dual (contravariant) basis $\{\vec{u}^1, \vec{u}^2, \vec{u}^3\}$, and contravariant, if the vector is expressed in the direct (covariant) basis $\{\vec{u}_1, \vec{u}_2, \vec{u}_3\}$. For example, the radius-vector $\vec{r} = x\vec{i} + y\vec{j} + z\vec{k}$, in a non-orthogonal space, can be written either in terms of contravariant (dual) basis

$$\vec{r} = x_1 \vec{u}^1 + x_2 \vec{u}^2 + x_3 \vec{u}^3,$$

or in terms of covariant (direct) basis

$$\vec{r} = x^1 \vec{u}_1 + x^2 \vec{u}_2 + x^3 \vec{u}_3.$$

In light of the above definitions, assuming that the unit vectors of both bases are constant, while the axes of the (non-orthogonal) frame are rectilinear (so that the unit vectors of the axes are the same at any point of the space), the differential $d\vec{r}$ of a radius-vector $\vec{r}$ can also be written in two ways:

$$d\vec{r} = dx_1 \vec{u}^1 + dx_2 \vec{u}^2 + dx_3 \vec{u}^3,$$

and

$$d\vec{r} = dx^1 \vec{u}_1 + dx^2 \vec{u}_2 + dx^3 \vec{u}_3,$$

in which case the metric of the space is

$$ds^2 = |d\vec{r}|^2 = d\vec{r} \cdot d\vec{r}$$

---

[1] By virtue of (5.84), the terms "direct" and "dual" used to designate the bases are relative. In fact, the two bases $\{\vec{u}_1, \vec{u}_2\}$ and $\{\vec{u}^1, \vec{u}^2\}$ are dual to each other.

$$\left(dx_1\vec{u}^1 + dx_2\vec{u}^2 + dx_3\vec{u}^3\right) \cdot \left(dx_1\vec{u}^1 + dx_2\vec{u}^2 + dx_3\vec{u}^3\right),$$

$$= \nearrow \atop \rightarrow \atop \searrow \quad \left(dx^1\vec{u}_1 + dx^2\vec{u}_2 + dx^3\vec{u}_3\right) \cdot \left(dx^1\vec{u}_1 + dx^2\vec{u}_2 + dx^3\vec{u}_3\right),$$

$$\left(dx_1\vec{u}^1 + dx_2\vec{u}^2 + dx_3\vec{u}^3\right) \cdot \left(dx^1\vec{u}_1 + dx^2\vec{u}_2 + dx^3\vec{u}_3\right).$$

Using the new notations, we can write (5.82) in the condensed form

$$\vec{u}_i \cdot \vec{u}^j = \begin{cases} 1, & i = j, \\ 0, & i \neq j, \end{cases} \tag{5.91}$$

so that

$$ds^2 = |d\vec{r}|^2$$

$$= \nearrow \atop \rightarrow \atop \searrow \quad \begin{aligned} \left(dx_i\vec{u}^i\right) \cdot \left(dx_j\vec{u}^j\right) &= \left(\vec{u}^i \cdot \vec{u}^j\right)dx_i dx_j, \\ \left(dx^i\vec{u}_i\right) \cdot \left(dx^j\vec{u}_j\right) &= \left(\vec{u}_i \cdot \vec{u}_j\right)dx^i dx^j, \\ \left(dx_i\vec{u}^i\right) \cdot \left(dx^j\vec{u}_j\right) &= \left(\vec{u}^i \cdot \vec{u}_j\right)dx_i dx^j. \end{aligned} \tag{5.92}$$

Let us denote

$$\begin{cases} \vec{u}^i \cdot \vec{u}^j = g^{ij}, \\ \vec{u}_i \cdot \vec{u}_j = g_{ij}. \end{cases} \tag{5.93}$$

In view of (5.91), the metric (5.92) can be expressed in one of the following three forms:

$$ds^2 = g^{ij}dx_i dx_j = g_{ij}dx^i dx^j = dx_i dx^i. \tag{5.94}$$

The quantities g^{ij} and g_{ij} are components of a contravariant and a covariant tensor, respectively, called *contravariant (covariant) metric tensor*. As seen, relations (5.93) show that the contravariant (covariant) components of the metric tensor are given by the scalar products of the vectors of dual (direct) bases.

Relation (5.94) yields, on the one hand

$$dx_i = g_{ij}dx^j, \tag{5.95}$$

and

$$dx^i = g^{ij}dx_j, \tag{5.96}$$

on the other. This shows that, by means of a suitable choice of the metric tensor, an index can be lowered or raised, respectively.

Let us now consider an Euclidean m-dimensional space E_m, and let $y_1, y_2, ..., y_m$ be the Cartesian coordinates of some point in this space. The metric ds^2 in E_m is

$$ds^2 = dy_j dy_j \quad (j = \overline{1,m}). \tag{5.97}$$

Consider now in E_m a n-dimensional variety (subspace) $\mathcal{R}_n$ $(n < m)$, and let $x^1, x^2, ..., x^n$ be the coordinates of a point in $\mathcal{R}_n$. Since $y_j = y_j(x^1, x^2, ..., x^n)$ $(j = \overline{1,m})$, we can write

$$ds^2 = \frac{\partial y_j}{\partial x^i} \frac{\partial y_j}{\partial x^k} dx^i dx^k = g_{ik} dx^i dx^k \quad (j = \overline{1,m}; \; i, k = \overline{1,n}), \tag{5.98}$$

where by

$$g_{ik}(x^1, x^2, ..., x^n) = g_{ki} = \frac{\partial y_j}{\partial x^i} \frac{\partial y_j}{\partial x^k} \quad (j = \overline{1,m}; \; i, k = \overline{1,n})$$

have been denoted the covariant components of the *metric tensor*. This is a symmetric, second rank tensor. If $g_{ik} = \delta_{ik}$, that is, if the manifold $\mathcal{R}_n$ is Euclidean, we get back to the metric (5.97).

If the metric (5.98) is invariant with respect to the general coordinate transformation

$$x'^i = x'^i(x^1, x^2, ..., x^n) \quad (i = \overline{1,n}),$$

then the manifold $\mathcal{R}_n$ is called *Riemannian*.

It is now our purpose to determine the differential equations of geodesics of a Riemannian manifold $\mathcal{R}_n$. Let x^i $(i = \overline{1,n})$ be the coordinates of a particle moving in $\mathcal{R}_n$, and

$$x^i = x^i(s) \quad (i = \overline{1,n}) \tag{5.99}$$

the parametric equations of a curve passing through two given points P_1 and P_2. The arc length of the curve between the two points is

$$L(x) = \int_{P_1}^{P_2} ds = \int_{P_1}^{P_2} \sqrt{g_{ik} dx^i dx^k} = \int_{P_1}^{P_2} \sqrt{g_{ik} \dot{x}^i \dot{x}^k} \, ds, \tag{5.100}$$

where $\dot{x}^i = dx^i/ds$. The curve given by (5.99) is a geodesic, if the functional (5.100) has an extremum (which usually is a minimum). To this end, the function

$$f(x, \dot{x}, s) = \sqrt{g_{ik} \dot{x}^i \dot{x}^k} \quad (i, k = \overline{1,n}) \tag{5.101}$$

must satisfy the Euler-Lagrange equations

$$\frac{d}{ds}\left(\frac{\partial f}{\partial \dot{x}^j}\right) - \frac{\partial f}{\partial x^j} = 0 \quad (j = \overline{1,n}). \tag{5.102}$$

Since

$$\sqrt{g_{ik}\dot{x}^i\dot{x}^k} = \sqrt{g_{ik}\frac{dx^i}{ds}\frac{dx^k}{ds}} = \frac{1}{ds}\sqrt{g_{ik}dx^idx^k} = \frac{ds}{ds} = 1, \tag{5.103}$$

we have:

$$\frac{\partial f}{\partial \dot{x}^j} = \frac{1}{\sqrt{g_{ik}\dot{x}^i\dot{x}^k}} g_{ik}\delta^i_j \dot{x}^k = \frac{1}{\sqrt{g_{ik}\dot{x}^i\dot{x}^k}} g_{jk}\dot{x}^k = g_{jk}\dot{x}^k;$$

$$\frac{d}{ds}\left(\frac{\partial f}{\partial \dot{x}^j}\right) = \frac{d}{ds}\left(g_{jk}\dot{x}^k\right)$$

$$= \frac{\partial g_{jk}}{\partial x^i}\frac{dx^i}{ds}\dot{x}^k + g_{jk}\ddot{x}^k = \frac{\partial g_{jk}}{\partial x^i}\dot{x}^i\dot{x}^k + g_{jk}\ddot{x}^k;$$

$$\frac{\partial f}{\partial x^j} = \frac{\partial}{\partial x^j}\left(\sqrt{g_{ik}\dot{x}^i\dot{x}^k}\right) = \frac{1}{2\sqrt{g_{ik}\dot{x}^i\dot{x}^k}}\frac{\partial}{\partial x^j}\left(g_{ik}\dot{x}^i\dot{x}^k\right)$$

$$= \frac{1}{2\sqrt{g_{ik}\dot{x}^i\dot{x}^k}}\frac{\partial g_{ik}}{\partial x^j}\dot{x}^i\dot{x}^k = \frac{1}{2}\frac{\partial g_{ik}}{\partial x^j}\dot{x}^i\dot{x}^k.$$

With these results, equations (5.102) become

$$g_{jk}\ddot{x}^k + \frac{\partial g_{jk}}{\partial x^i}\dot{x}^i\dot{x}^k - \frac{1}{2}\frac{\partial g_{ik}}{\partial x^j}\dot{x}^i\dot{x}^k = 0,$$

or

$$g_{jk}\ddot{x}^k + \frac{1}{2}\left(\frac{\partial g_{jk}}{\partial x^i}\dot{x}^i\dot{x}^k + \frac{\partial g_{jk}}{\partial x^i}\dot{x}^i\dot{x}^k\right) - \frac{1}{2}\frac{\partial g_{ik}}{\partial x^j}\dot{x}^i\dot{x}^k$$

$$= g_{jk}\ddot{x}^k + \frac{1}{2}\left(\frac{\partial g_{ji}}{\partial x^k}\dot{x}^k\dot{x}^i + \frac{\partial g_{jk}}{\partial x^i}\dot{x}^i\dot{x}^k\right) - \frac{1}{2}\frac{\partial g_{ik}}{\partial x^j}\dot{x}^i\dot{x}^k$$

$$= g_{jk}\ddot{x}^k + \frac{1}{2}\left(\frac{\partial g_{jk}}{\partial x^i} + \frac{\partial g_{ij}}{\partial x^k} - \frac{\partial g_{ik}}{\partial x^j}\right)\dot{x}^i\dot{x}^k = 0.$$

If we denote

$$\Gamma_{ik,j} = \frac{1}{2}\left(\frac{\partial g_{jk}}{\partial x^i} + \frac{\partial g_{ij}}{\partial x^k} - \frac{\partial g_{ik}}{\partial x^j}\right), \tag{5.104}$$

the last equation writes

$$g_{jk}\ddot{x}^k + \Gamma_{ik,j}\dot{x}^i\dot{x}^k = 0. \tag{5.105}$$

The quantities (5.104) are called *Christoffel symbols of the first kind*. Multiplying (5.104) by g^{jl} and performing summation over j, we finally obtain the differential equation of geodesic lines in $\mathcal{R}_n$

$$\ddot{x}^l + \Gamma^l_{ik}\dot{x}^i\dot{x}^k = 0, \tag{5.106}$$

where

$$g^{jl}\Gamma_{ik,j} = \Gamma^l_{ik} \tag{5.107}$$

are the *Christoffel symbols of the second kind*. It can be shown that, except for the linear transformations of coordinates, the Christoffel symbols of both kinds *are not tensors*.

The differential equations of geodesics of a n-dimensional Riemannian manifold $\mathcal{R}_n$ are, at the same time, the equations of motion of a free particle in the gravitational field. In fact, the quantities $a^l = \ddot{x}^l + \Gamma^l_{ik}\dot{x}^i\dot{x}^k$ stand for the components of the n-dimensional acceleration vector in $\mathcal{R}_n$. It can be easily proved that the quantities $\ddot{x}^l$ ($l = \overline{1,n}$) *are not* vectors.

Observation. If we denote by

$$\Phi = \frac{1}{2}g_{ik}\dot{x}^i\dot{x}^k,$$

then Euler-Lagrange equations

$$\frac{d}{ds}\left(\frac{\partial\Phi}{\partial\dot{x}^j}\right) - \frac{\partial\Phi}{\partial x^j} = 0 \quad (j = \overline{1,n})$$

lead to the same result. Consequently, the variational principles

$$\delta\int_{P_1}^{P_2}\sqrt{g_{ik}\dot{x}^i\dot{x}^k}\,ds = 0,$$

and

$$\delta\int_{P_1}^{P_2} g_{ik}\dot{x}^i\dot{x}^k\,ds = 0$$

are equivalent.

Application. Determine the geodesics of a sphere of radius one.

In spherical coordinates, the element of the arc length writes

$$ds^2 = dr^2 + r^2 d\theta^2 + r^2 \sin^2\theta \, d\varphi^2, \qquad (5.108)$$

and, if $r = R = 1$ (R is the radius of the sphere)

$$ds^2 = d\theta^2 + \sin^2\theta \, d\varphi^2. \qquad (5.109)$$

The variational principle then writes

$$\delta \int ds = \delta \int \frac{ds^2}{ds^2} ds = \delta \int \frac{d\theta^2 + \sin^2\theta \, d\varphi^2}{ds^2} ds$$

$$= \delta \int (\dot\theta^2 + \dot\varphi^2 \sin^2\theta) \, ds = 0,$$

with $\dot\theta = \frac{d\theta}{ds}$, $\dot\varphi = \frac{d\varphi}{ds}$. Therefore, we have to look for the extremals of the functional

$$F(\theta, \varphi) = \int_{P_1}^{P_2} f(\theta, \dot\theta, \dot\varphi) \, ds, \qquad (5.110)$$

where P_1 and P_2 are two given points on the surface of the sphere, and

$$f(\theta, \dot\theta, \dot\varphi) = \dot\theta^2 + \dot\varphi^2 \sin^2\theta = 1. \qquad (5.111)$$

The Euler-Lagrange equation for the variable φ is

$$\frac{d}{ds}\left(\frac{\partial f}{\partial \dot\varphi}\right) - \frac{\partial f}{\partial \varphi} = 0. \qquad (5.112)$$

Calculating the derivatives

$$\frac{\partial f}{\partial \varphi} = 0; \quad \frac{\partial f}{\partial \dot\varphi} = 2\dot\varphi \sin^2\theta;$$

$$\frac{d}{ds}\left(\frac{\partial f}{\partial \dot\varphi}\right) = 4\dot\theta \, \dot\varphi \sin\theta \cos\theta + 2\ddot\varphi \sin^2\theta$$

we are left with

$$\ddot\varphi + 2\dot\theta \, \dot\varphi \cot\theta = 0. \qquad (5.113)$$

To find the explicit equation of the geodesic, $\varphi = \varphi(\theta)$, we have to eliminate the parameter s from equations (5.111) and (5.113). To this end, we observe that (5.113) can also be written as

$$d\dot\varphi + 2\dot\varphi \cot\theta \, d\theta = 0,$$

or, if the variables are separated

$$\frac{d\dot\varphi}{\dot\varphi} = -2\cot\theta\, d\theta = -2\frac{\cos\theta\, d\theta}{\sin\theta} = -2\frac{d(\sin\theta)}{\sin\theta},$$

and the integration is carried out

$$\ln\dot\varphi = -2\ln(\sin\theta) + \ln C = \ln\left(\frac{C}{\sin^2\theta}\right),$$

which yields

$$\dot\varphi = \frac{C}{\sin^2\theta},$$

where C is an arbitrary constant of integration. Observing that

$$\dot\theta = \frac{d\theta}{ds} = \frac{d\theta}{d\varphi}\frac{d\varphi}{ds} = \frac{d\theta}{d\varphi}\dot\varphi,$$

equation (5.111) yields

$$\dot\theta^2 + \dot\varphi^2\sin^2\theta = \frac{C^2}{\sin^4\theta}\left[\left(\frac{d\theta}{d\varphi}\right)^2 + \sin^2\theta\right] = 1,$$

or, if the variables are separated

$$d\varphi = \frac{C}{\sin\theta}\frac{d\theta}{\sqrt{\sin^2\theta - C^2}}, \tag{5.114}$$

so that

$$C\int\frac{d\theta}{\sin\theta\sqrt{\sin^2\theta - C^2}} = \varphi - \varphi_0, \tag{5.115}$$

where φ_0 is an arbitrary constant of integration. To integrate this equation, it is convenient to make the following change of variable

$$\sin^2\theta - C^2 = \zeta^2.$$

We then have

$$C\int\frac{d\zeta}{(\zeta^2 + C^2)\sqrt{1 - C^2 - \zeta^2}} = \varphi - \varphi_0. \tag{5.116}$$

Here we have an integral of the type

$$\int\frac{dx}{(x^2 + a^2)\sqrt{b^2 - x^2}},$$

with $a = C$ and $b = \sqrt{1-C^2}$. The result can be found in the tables of integrals

$$\int \frac{dx}{(x^2+a^2)\sqrt{b^2-x^2}} = \frac{1}{a\sqrt{a^2+b^2}} \arctan\left(\frac{x}{a}\sqrt{\frac{b^2+a^2}{b^2-x^2}}\right),$$

and (5.116) writes

$$C\int \frac{d\zeta}{(\zeta^2+C^2)\sqrt{1-C^2-\zeta^2}} = \arctan\left(\frac{\zeta}{C\sqrt{1-C^2-\zeta^2}}\right)$$

$$= \arctan\left(\frac{\sqrt{\sin^2\theta - C^2}}{C\cos\theta}\right) = \varphi - \varphi_0$$

so that

$$\frac{\sqrt{\sin^2\theta - C^2}}{C\cos\theta} = \tan(\varphi - \varphi_0),$$

and, finally,

$$\cos(\varphi - \varphi_0) = \frac{1}{\sqrt{1+\tan^2(\varphi-\varphi_0)}} = \frac{C}{\sqrt{1-C^2}} \cot\theta. \qquad (5.117)$$

This is the equation of a plane passing through the origin of the coordinate system, placed at the centre of the sphere. Therefore, the geodesics we are looking for are *great circles of the sphere*, obtained as a result of intersection between the plane (5.117) and the sphere (see Fig.V.10).

To put into evidence the fact that (5.117) represents, indeed, the equation of a plane passing through the origin of the coordinate system, let us rewrite this equation in Cartesian coordinates. To this end, we shall use the well-known transformation relations

$$\begin{cases} x = \sin\theta \cos\varphi, \\ y = \sin\theta \sin\varphi, \\ z = \cos\theta, \end{cases}$$

where we have considered $r = 1$. We have:

$$\cos(\varphi - \varphi_0) = \cos\varphi \cos\varphi_0 + \sin\varphi \sin\varphi_0 = \frac{x}{\sin\theta}\cos\varphi_0 + \frac{y}{\sin\theta}\sin\varphi_0$$

$$= \frac{C}{\sqrt{1-C^2}}\cot\theta = \frac{C}{\sqrt{1-C^2}}\frac{\cos\theta}{\sin\theta} = \frac{C}{\sqrt{1-C^2}}\frac{z}{\sin\theta}.$$

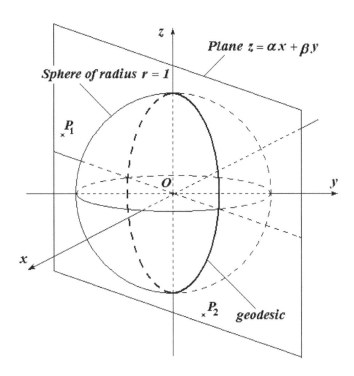

Fig.V.10

Denoting
$$\alpha = \frac{\sqrt{1-C^2}\cos\varphi_0}{C},$$
and
$$\beta = \frac{\sqrt{1-C^2}\sin\varphi_0}{C},$$
we finally obtain
$$z = \alpha x + \beta y, \qquad (5.118)$$
which is the equation of a *plane* passing through tho origin of the coordinate system. Here α and β are two constants depending on C and φ_0, that can be determined by imposing the condition that the plane passes through the two fixed points P_1 and P_2 (and, obviously, through the origin O).

CHAPTER VI

PROBLEMS SOLVED BY MEANS OF THE LAGRANGIAN FORMALISM

1. Atwood machine

The Atwood's machine was invented in 1784 by Rev. George Atwood in order to verify the mechanical laws of motion with constant acceleration. It essentially consists of two bodies of masses m_1 and m_2, connected by an inextensible massless string of length l over an ideal massless pulley of radius r and moment of inertia I (see Fig.VI.1). Neglecting friction between the string and the pulley, find the differential equation of motion of the system.

Solution

Since the only applied force is the force of gravitation, which is conservative, our system is a *natural system*. Supposing the pulley as

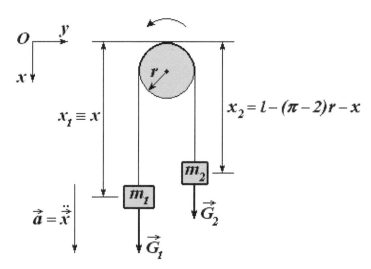

Fig.VI.1

a third body of the system, and taking $z = 0$ as the plane of motion, the system is submitted to the following constraints:

 i) $f_1(z_1) = z_1 = 0$;
 ii) $f_2(z_2) = z_2 = 0$;
 iii) $f_3(z_3) = z_3 = 0$ (the pulley of mass m_3 does not have a motion of translation along z-axis);
 iv) $f_4(y_1) = y_1 = C_1$ (*const.*) (the body of mass m_1 moves only along the x-axis);
 v) $f_5(y_2) = y_2 = C_2$ (*const.*) (the body of mass m_2 moves only along the x-axis);
 vi) $f_6(y_3) = y_3 = C_3$ (*const.*) (the pulley of mass m_3 does not have a motion of translation along y-axis);
 vii) $f_7(x_1, x_2) = x_1 + x_2 + (\pi - 2)r - l = 0$ (the string is inextensible);
 viii) $f_8(x_3) = x_3 = 0$ (the pulley of mass m_3 does not have a motion of translation along x-axis.

Here indices $1, 2$, and 3 are attached to the bodies of masses m_1, m_2, and m_3, respectively.

As one can see, the system has $3 \cdot 3 - 8 = 1$ degree of freedom. Let x be the associated generalized coordinate. To write the Lagrangian of the system it is necessary to determine its kinetic and potential energies.

Since the dimensions of the pulley are finite, its moment of inertia cannot be neglected. In this respect, the kinetic energy of rotation of the pulley about its own axis has also to be considered. Therefore, we have:

$$T = T_1 + T_2 + T_3 = \frac{1}{2}m_1|\vec{v}_1|^2 + \frac{1}{2}m_2|\vec{v}_2|^2 + \frac{1}{2}I\omega^2,$$

where ω is angular frequency of rotation of the pulley about its own axis. Since the string does not slide on the pulley, the linear velocity of an arbitrary point of the discus periphery, and the translation velocity of the bodies of masses m_1 and m_2 are the same: $|\vec{v}_1| = |\vec{v}_2| = \dot{x}$. Then

$$T = \frac{1}{2}m_1\dot{x}_1^2 + \frac{1}{2}m_2\dot{x}_2^2 + \frac{1}{2}I\frac{\dot{x}^2}{r^2}.$$

The potential energy of our conservative system is $V = V_b + V_p$, where V_b is the potential energy of the two bodies, and $V_p = const.$ the potential energy of the pulley. We can write

$$dV_b = -dA = -\vec{G}_1 \cdot d\vec{r}_1 - \vec{G}_2 \cdot d\vec{r}_2 = -g(m_1 dx_1 + m_2 dx_2),$$

so that
$$V_b = -m_1 g x_1 - m_2 g x_2 + V_{b0}.$$

A convenient choice of the reference system for the potential energy makes it possible to consider the integration constant V_{b0} as being zero. Indeed, taking $V_b(x_1 = 0, x_2 = 0) = 0$, one obtains $V_{b0} = 0$. Therefore

$$V = V_b + V_p = -m_1 g x_1 - m_2 g x_2 + V_p = -(m_1 - m_2)gx + V_c,$$

where $V_c = -m_2 g l + m_2 g(\pi - 2)r + V_p = const.$ can be dropped according to the definition/property of equivalent Lagrangians.

The Lagrangian of the system then writes

$$L = T - V = \frac{1}{2}\left(m_1 + m_2 + \frac{I}{r^2}\right)\dot{x}^2 + (m_1 - m_2)gx.$$

The Lagrange equation of the second kind for the generalized coordinate x is

$$\frac{d}{dt}\left(\frac{\partial L}{\partial \dot{x}}\right) - \frac{\partial L}{\partial x} = 0. \tag{6.1}$$

Performing the derivatives and introducing the results into (6.1), one obtains the differential equation of motion of the system

$$\left(m_1 + m_2 + \frac{I}{r^2}\right)\ddot{x} - (m_1 - m_2)g = 0, \tag{6.2}$$

which yields the constant acceleration of the system

$$a = \ddot{x} = \frac{(m_1 - m_2)g}{m_1 + m_2 + \frac{I}{r^2}} = const. \tag{6.3}$$

2. Double Atwood machine

The system schematically presented in Fig.VI.2 is called *double Atwood machine*. Like in the previous paragraph, the masses of the pulleys is negligible, and the friction is neglected. Using the Lagrangian formalism, write the differential equations of motion of the system, and determine the accelerations of the three bodies.

Solution

Here we have, again, a natural system (the only active force is the force of gravity). Neglecting the radii of pulleys as compared to the lengths of the strings, the equations of constraints are:

i) $f_1(z_1) = z_1 = 0$ (without loss of generality, one can assume that the motion takes place in the plane $z = 0$);

ii) $f_2(z_2) = z_2 = 0$ (idem);

iii) $f_3(z_3) = z_3 = 0$ (idem);

iv) $f_4(y_1) = y_1 = C_1(const.)$ (the body of mass m_1 moves only along the x-axis);

v) $f_5(y_2) = y_2 = C_2(const.)$ (the body of mass m_2 moves only along the x-axis);

vi) $f_6(y_3) = y_3 = C_3(const.)$ the body of mass m_3 moves only along the x-axis);

vii) $f_7(x_1, x_2, x_3) = 2x_1 + x_2 + x_3 - 2l_1 - l_2 = 0$ (the wires are inextensible[1]).

This shows that the system has $3 \cdot 3 - 7 = 2$ degrees of freedom. We should mention that, unlike the "simple" Atwood machine, when the pulley was considered one of the three material points of the system, this time the degrees of freedom involved by the existence of the two pulleys is not considered (otherwise, we would have had 13 equations to define the constraints). Let the associated generalized coordinates be $\xi_1(= x_1)$ and ξ_2 (see Fig.VI.2). To write the Lagrangian, we have to know the kinetic and potential energies. These quantities are going to be determined by means of some simplifying conditions: the radii of the pulleys are negligible as compared to lengths of the wires, while their moments of inertia are ignored.

The kinetic energy of the system is then given by the sum of the kinetic energy of translation of the three bodies along the x-axis, that is

$$T = T_1 + T_2 + T_3 = \frac{1}{2}m_1|\vec{v}_1|^2 + \frac{1}{2}m_2|\vec{v}_2|^2 + \frac{1}{2}m_3|\vec{v}_3|^2$$

$$= \frac{1}{2}m_1\dot{x}_1^2 + \frac{1}{2}m_2\dot{x}_2^2 + \frac{1}{2}m_3\dot{x}_3^2 = \frac{1}{2}m_1\dot{\xi}_1^2 + \frac{1}{2}m_2\dot{x}_2^2 + \frac{1}{2}m_3\dot{x}_3^2.$$

---

[1] When writing the equation of this constraint, we neglected the radii of the two pulleys. As in case of the simple Atwood machine, this approximation acts only on the potential energy of the system, which differs from the real one by a constant quantity that can be ignored in the Lagrangian. If, nevertheless, the finite radii of the pulleys are considered, then the constraint f_7 writes: $f_7(x_1, x_2, x_3) = 2x_1 + x_2 + x_3 - 2l_1 - l_2 + 2(\pi - 2)r_1 + \pi r_2 = 0$.

Here we have to express $\dot{x}_2^2$ and $\dot{x}_3^2$ in terms of $\dot{\xi}_1$ and $\dot{\xi}_2$. Neglecting the radii of the pulleys[1], one can write

$$x_2 = l_1 + \xi_2 - x_1 = l_1 + \xi_2 - \xi_1,$$

so that

$$\dot{x}_2^2 = (\dot{\xi}_2 - \dot{\xi}_1)^2.$$

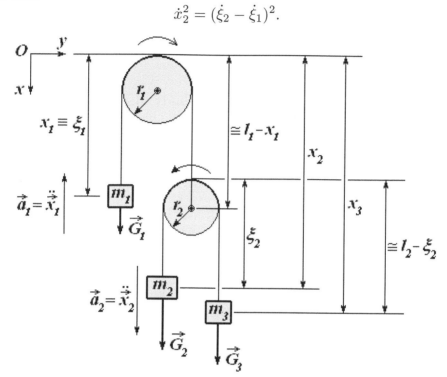

Fig.VI.2

Within the same approximation[2], Fig.VI.2 shows that

$$x_3 = l_2 + x_2 - 2\xi_2,$$

which yields

$$\dot{x}_3^2 = (\dot{x}_2 - 2\dot{\xi}_2)^2 = (\dot{\xi}_2 - \dot{\xi}_1 - 2\dot{\xi}_2)^2 = (-\dot{\xi}_1 - \dot{\xi}_2)^2 = (\dot{\xi}_1 + \dot{\xi}_2)^2,$$

---

[1] The same expression for the total kinetic energy of the system is also obtained if one considers the exact expression for x_2, that is: $x_2 = l_1 + \xi_2 - \xi_1 - (\pi - 2)r_1 - r_2$, because the constant quantities disappear when taking the derivatives.

[2] Writing the exact expression for x_3, that is: $x_3 = x_2 + l_2 - 2\xi_2 - (\pi - 2)r_2$, leads to the same total kinetic energy, for the reason specified in the previous footnote.

and the kinetic energy writes

$$T = \frac{1}{2}m_1\dot{\xi}_1^2 + \frac{1}{2}m_2(\dot{\xi}_2 - \dot{\xi}_1)^2 + \frac{1}{2}m_3(\dot{\xi}_2 + \dot{\xi}_1)^2. \tag{6.4}$$

To determine the potential energy of the system, we make allowance for the usual procedure

$$dV = -dA = -\vec{G}_1 \cdot d\vec{r}_1 - \vec{G}_2 \cdot d\vec{r}_2 - \vec{G}_3 \cdot d\vec{r}_3$$

$$= -m_1 g dx_1 - m_2 g dx_2 - m_3 g dx_3,$$

so that

$$V = -m_1 g x_1 - m_2 g x_2 - m_3 g x_3 + V_0.$$

A convenient choice of the reference level for V [$V(x_1 = 0, x_2 = 0, x_3 = 0) = 0$] yields $V_0 = 0$, and, using the same approximation (the radii of pulleys are negligible as compared to the lengths of the two inextensible wires)[1], the potential energy writes

$$V = -m_1 g \xi_1 - m_2 g(l_1 + \xi_2 - \xi_1) - m_3 g(l_2 + x_2 - 2\xi_2)$$

$$= -m_1 g \xi_1 - m_2 g(l_1 + \xi_2 - \xi_1) - m_3 g(l_2 + l_1 + \xi_2 - \xi_1 - 2\xi_2) \tag{6.5}$$

$$= -g(m_1 - m_2 - m_3)\xi_1 - g(m_2 - m_3)\xi_2 - gl_1(m_2 + m_3) - m_3 g l_2.$$

The Lagrangian of the system therefore is

$$L = T - V = \frac{1}{2}m_1\dot{\xi}_1^2 + \frac{1}{2}m_2(\dot{\xi}_2 - \dot{\xi}_1)^2 + \frac{1}{2}m_3(\dot{\xi}_1 + \dot{\xi}_2)^2$$

$$+ g(m_1 - m_2 - m_3)\xi_1 + g(m_2 - m_3)\xi_2, \tag{6.6}$$

where the constant term $[gl_1(m_2 + m_3) + m_3 g l_2]$ has been dropped.

There are two Lagrange equations associated with the two generalized coordinates:

$$\frac{d}{dt}\left(\frac{\partial L}{\partial \dot{\xi}_1}\right) - \frac{\partial L}{\partial \xi_1} = 0, \tag{6.7}$$

and

$$\frac{d}{dt}\left(\frac{\partial L}{\partial \dot{\xi}_2}\right) - \frac{\partial L}{\partial \xi_2} = 0. \tag{6.8}$$

---

[1] An exact calculation would demand to add the term $g(\pi-2)(m_2 r_1 + m_3 r_2) + gm_2 r_2$ to the already written expression for the potential energy, appearing with changed sign in the Lagrangian. According to definition of equivalent Lagrangians, this constant term can be omitted.

Performing the derivatives, one easily obtains

$$\ddot{\xi}_1(m_1 + m_2 + m_3) + \ddot{\xi}_2(m_3 - m_2) - g(m_1 - m_2 - m_3) = 0, \quad (6.9)$$

$$\ddot{\xi}_1(m_3 - m_2) + \ddot{\xi}_2(m_2 + m_3) - g(m_2 - m_3) = 0. \quad (6.10)$$

We are left with an algebraic system of two equations in two unknowns $\ddot{\xi}_1$ and $\ddot{\xi}_2$. Solving the second equation for $\ddot{\xi}_2$ in terms of $\ddot{\xi}_1$, and substituting the obtained value into the first equation, one gets

$$\ddot{\xi}_1\left(m_1+m_2+m_3-\frac{(m_2-m_3)^2}{m_2+m_3}\right)-g\frac{(m_2-m_3)^2}{m_2+m_3}-g(m_1-m_2-m_3) = 0,$$

leading to

$$\ddot{\xi}_1 = \frac{g\left[\frac{(m_2-m_3)^2}{m_2+m_3} + (m_1 - m_2 - m_3)\right]}{m_1 + m_2 + m_3 - \frac{(m_2-m_3)^2}{m_2+m_3}} = g\frac{m_1(m_2 + m_3) - 4m_2m_3}{m_1(m_2 + m_3) + 4m_2m_3}.$$

(6.11)

Then,

$$\ddot{\xi}_2 = \frac{m_2 - m_3}{m_2 + m_3}(g + \ddot{\xi}_1) = g\frac{m_2 - m_3}{m_2 + m_3}\left(1 + \frac{m_1(m_2 + m_3) - 4m_2m_3}{m_1(m_2 + m_3) + 4m_2m_3}\right)$$

$$= g\frac{2m_1(m_2 - m_3)}{m_1(m_2 + m_3) + 4m_2m_3}. \quad (6.11')$$

Therefore, the accelerations of the three bodies are:

$$a_1 = \ddot{\xi}_1 = g\frac{m_1(m_2 + m_3) - 4m_2m_3}{m_1(m_2 + m_3) + 4m_2m_3};$$

$$a_2 = \ddot{\xi}_2 = g\frac{2m_1(m_2 - m_3)}{m_1(m_2 + m_3) + 4m_2m_3}; \quad (6.12)$$

$$a_3 = -a_2 = g\frac{2m_1(m_3 - m_2)}{m_1(m_2 + m_3) + 4m_2m_3}.$$

At the end of this investigation we mention that $a_1 = \ddot{\xi}_1$ is determined with respect to the inertial frame xOy, while $a_2 = -a_3 = \ddot{\xi}_2$ is calculated relative to the pulley of radius r_2, which is a non-inertial frame.

3. Pendulum with horizontally oscillating point of suspension

The point of suspension of a gravitational pendulum of length l and mass m performs a horizontal motion, oscillating according to the law $X = a\cos\omega_0 t$ (see Fig.VI.3). Determine the equations of motion of the system, in both an inertial (IRF) and a non-inertial (NIRF) reference frames.

Solution

I. Inertial reference frame (IRF). Let us denote by θ the angle between the (ideal) rod and the vertical line, and choose the axes of the inertial frame (fixed with respect to the laboratory frame) as shown in Fig.VI.3. The oscillations of the point O obey the law $x_O = X = a\cos\omega_0 t$.

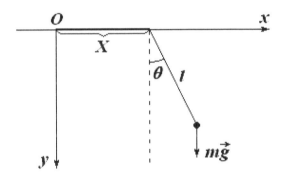

Fig.VI.3

There are two constraints, given by the equations
i) $f_1(x,y) = (x-X)^2 + y^2 - l^2 = 0$,
ii) $f_2(z) = z = 0$.
The system has $3-2 = 1$ degree of freedom. Let θ be the generalized coordinate. Then, we have

$$\begin{cases} x = X + l\sin\theta = a\cos\omega_0 t + l\sin\theta; \\ y = l\cos\theta, \end{cases}$$

and

$$\begin{cases} \dot{x} = -a\omega_0 \sin\omega_0 t + l\dot\theta\cos\theta; \\ \dot{y} = -l\dot\theta\sin\theta, \end{cases}$$

The kinetic energy then writes

$$T = \frac{1}{2}m(\dot{x}^2 + \dot{y}^2) = \frac{1}{2}m(a^2\omega_0^2 \sin^2\omega_0 t - 2al\omega_0\dot\theta \sin\omega_0 t \cos\theta + l^2\dot\theta^2).$$

In its turn, the potential (gravitational) energy is obtained by means of the standard procedure

$$dV = -dA = -\vec{G} \cdot d\vec{r} = -mg\,dy.$$

If we choose $V(y=0) = 0$, the last relation yields

$$V = -mgy = -mgl\cos\theta.$$

The Lagrangian therefore is

$$L = T - V = \frac{1}{2}m(a^2\omega_0^2\sin^2\omega_0 t - 2al\omega_0\dot\theta\sin\omega_0 t\cos\theta + l^2\dot\theta^2) + mgl\cos\theta. \tag{6.13}$$

Since the term $\frac{1}{2}ma^2\omega_0^2\sin^2\omega_0 t$ can be written as

$$\frac{1}{2}ma^2\omega_0^2\sin^2\omega_0 t = \frac{1}{2}ma^2\omega_0^2\frac{d}{dt}\left(\frac{t}{2} - \frac{\sin 2\omega_0 t}{4\omega_0}\right)$$

$$= \frac{d}{dt}\left[\frac{1}{8}ma^2\omega_0(2\omega_0 t - \sin 2\omega_0 t)\right] = \frac{dF(t)}{dt},$$

where $F(t) = \frac{1}{8}ma^2\omega_0(2\omega_0 t - \sin 2\omega_0 t)$ is a function of time only, this term can be omitted and the equivalent Lagrangian writes

$$L = \frac{1}{2}m(l^2\dot\theta^2 - 2al\omega_0\dot\theta\sin\omega_0 t\cos\theta) + mgl\cos\theta.$$

Performing the calculations demanded by the Lagrange equation

$$\frac{d}{dt}\left(\frac{\partial L}{\partial \dot\theta}\right) - \frac{\partial L}{\partial \theta} = 0, \tag{6.14}$$

we finally arrive at the desired equation

$$l\ddot\theta - a\omega_0^2\cos\omega_0 t\cos\theta + g\sin\theta = 0. \tag{6.15}$$

II. Non-inertial reference frame (NIRF). First of all, we shall show that in a NIRF the general form of the Lagrangian writes

$$L = \frac{1}{2}m|\vec{v}_r|^2 + \frac{1}{2}m|\vec{\omega}\times\vec{r}'|^2 + m\vec{v}_r\cdot(\vec{\omega}\times\vec{r}') - m\vec{a}_0\cdot\vec{r}' - V(\vec{r}'), \tag{6.16}$$

where the significance of the quantities $\vec{v}_r$, $\vec{\omega}$, $\vec{r}'$, $\vec{a}_0$ and $V(\vec{r}')$ shall be explained later on in this application.

Let us consider a particle of mass m and report its motion relative to two frames $S(Oxyz)$ and $S'(O'x'y'z')$, the first being inertial (*e.g.* fixed with respect to the Earth), and the second non-inertial (engaged in an accelerated motion with respect to S). Any motion in the Universe can be considered as a "composite" motion. The motion of the particle with respect to S is called *absolute*, and the motion of the same particle with respect to S' is called *relative*. If the particle is fixed with respect to S', then the motion of S' relative to S is named *transport motion*. As an intuitive example one can consider the motion of a car with respect to the Earth (NIRF), the last one being engaged in its motion around the Sun (IRF).

To find the Lagrangian (6.16) we shall study the motion of a particle P of mass m relative to both frames S and S' (see Fig.VI.4).

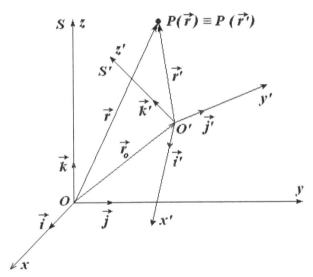

Fig.VI.4

Since
$$\vec{r} = \vec{r}_0 + \vec{r}\,',$$

the time derivative of this relation (in Newtonian mechanics, time intervals are the same in any reference frame) yields

$$\vec{v} = \dot{\vec{r}} = \dot{\vec{r}}_0 + \dot{\vec{r}}\,'. \tag{6.17}$$

Here $\vec{v} = \dot{\vec{r}}$ is called *absolute velocity*. Denoting $\vec{i}\,' = \vec{u}\,'_1$, $\vec{j}\,' = \vec{u}\,'_2$, $\vec{k}\,' = \vec{u}\,'_3$, we can write

$$\vec{r}\,' = x'_k \vec{u}\,'_k, \tag{6.18}$$

where Einstein's summation convention has been used. Therefore

$$\dot{\vec{r}}' = \dot{x}'_k \vec{u}'_k + x'_k \dot{\vec{u}}'_k = \vec{v}_r + x'_k \dot{\vec{u}}'_k. \tag{6.19}$$

Here $\vec{v}_r = \dot{x}'_k \vec{u}'_k$ is called *relative velocity*. On the other hand, let ω'_k be the components of $\dot{\vec{u}}'_k$ in the orthonormal basis $\vec{u}'_k$, i.e.

$$\dot{\vec{u}}'_k = \omega'_{ks} \vec{u}'_s. \tag{6.20}$$

The orthogonality condition $\vec{u}'_k \cdot \vec{u}'_s = \delta_{ks}$ then yields

$$\frac{d}{dt}(\vec{u}'_k \cdot \vec{u}'_s) = \dot{\vec{u}}'_k \cdot \vec{u}'_s + \vec{u}'_k \cdot \dot{\vec{u}}'_s = \omega'_{ks} + \omega'_{sk} = 0,$$

showing that ω'_{ks} ($k,s = \overline{1,3}$) are the components of an antisymmetric second rank tensor. If $\vec{\omega}\,(\omega'_k)$ is the axial vector associated with ω'_{sk}, then we can write

$$\omega'_{ks} = \varepsilon_{ksi} \omega'_i \quad (i,k,s = \overline{1,3}), \tag{6.21}$$

where ε_{ksi} is the Levi-Civita permutation symbol. Then

$$\dot{\vec{u}}'_k = \omega'_{ks} \vec{u}'_s = \varepsilon_{ksi} \omega'_i \vec{u}'_s = \omega'_i \vec{u}'_i \times \vec{u}'_k = \vec{\omega} \times \vec{u}'_k. \tag{6.22}$$

By means of (6.19) and (6.22), the absolute velocity given by (6.17) writes

$$\vec{v} = \vec{v}_0 + \vec{v}_r + x'_k \dot{\vec{u}}'_k = \vec{v}_0 + \vec{v}_r + x'_k \vec{\omega} \times \vec{u}'_k$$

$$= \vec{v}_0 + \vec{v}_r + \vec{\omega} \times \vec{r}', \tag{6.23}$$

where $\vec{v}_0 = \dot{\vec{r}}_0$ is the velocity of O' relative to O.

The absolute acceleration is obtained by taking the time derivative of (6.23). Recalling that $\vec{v}_r = \dot{x}'_k \vec{u}'_k$, we have:

$$\vec{a} = \frac{d\vec{v}}{dt} = \dot{\vec{v}}_0 + \ddot{x}'_k \vec{u}'_k + \dot{x}'_k \dot{\vec{u}}'_k + \dot{\vec{\omega}} \times \vec{r}' + \vec{\omega} \times \dot{\vec{r}}',$$

or, in view of (6.19) and (6.22),

$$\vec{a} = \vec{a}_0 + \vec{a}_r + \dot{\vec{\omega}} \times \vec{r}' + [\dot{x}'_k \vec{\omega} \times \vec{u}'_k + \vec{\omega} \times (\vec{v}_r + x'_k \dot{\vec{u}}'_k)]$$

$$= \vec{a}_0 + \vec{a}_r + \dot{\vec{\omega}} \times \vec{r}' + 2\vec{\omega} \times \vec{v}_r + \vec{\omega} \times (\vec{\omega} \times \vec{r}'), \tag{6.24}$$

where $\dot{\vec{v}}_0 = \vec{a}_0$ is the acceleration of O' with respect to O, and $\ddot{x}'\vec{u}'_k = \vec{a}_r$ the acceleration of the particle with respect to O', called *relative acceleration*. The term $\vec{\omega} \times (\vec{\omega} \times \vec{r}') = \vec{a}_{cp}$ is named *centripetal acceleration*, while $2\vec{\omega} \times \vec{v}_r = \vec{a}_c$ is the *Coriolis acceleration*.

If the particle is invariably attached to S' (in other words, the particle itself is a non-inertial frame), then $\vec{v}_r = 0$, $\vec{a}_r = 0$, and the last two relations yield

$$\vec{v} = \vec{v}_{tr} = \vec{v}_0 + \vec{\omega} \times \vec{r}'; \tag{6.25}$$

$$\vec{a} = \vec{a}_{tr} = \vec{a}_0 + \dot{\vec{\omega}} \times \vec{r}' + \vec{\omega} \times (\vec{\omega} \times \vec{r}'). \tag{6.26}$$

Here $\vec{v}_{tr}$ and $\vec{a}_{tr}$ are called *transport velocity* and *transport acceleration*, respectively. If $O \equiv O'$, then $\vec{r}_0 = 0$, $\vec{v}_0 = 0$, $\vec{a}_0 = 0$, $\vec{r}' = \vec{r}$, and we have

$$\vec{v} = \vec{\omega} \times \vec{r}, \tag{6.27}$$

$$\vec{a} = \dot{\vec{\omega}} \times \vec{r} + \vec{\omega} \times (\vec{\omega} \times \vec{r}). \tag{6.28}$$

With these notations, the absolute velocity and absolute acceleration can also be written as

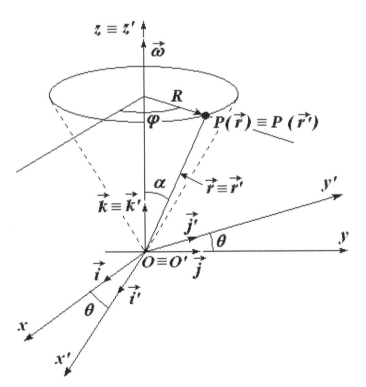

Fig.VI.5

$$\vec{v} = \vec{v}_r + \vec{v}_{tr}, \qquad (6.29)$$

$$\vec{a} = \vec{a}_r + \vec{a}_c + \vec{a}_{tr}. \qquad (6.30)$$

To determine the physical significance of the pseudovector $\vec{\omega}$, let us consider a special case: $O \equiv O'$, and $Oz \equiv Oz'$ as a fixed axis of rotation. Then (6.22) yields $\dot{\vec{k}} = \vec{\omega} \times \vec{k} = 0$, meaning that $\vec{\omega}$ and the axis of rotation are collinear: $\vec{\omega} = \omega \vec{k}$. On the other hand, (6.29) shows that the velocity $\vec{v}$ of the particle is orthogonal to the plane defined by $\vec{\omega}$ and $\vec{r}$ (see Fig.VI.5), its modulus being

$$v = |\vec{v}| = |\vec{\omega} \times \vec{r}| = \omega r \sin\alpha = \omega R.$$

As well-known, the velocity of a point engaged in a uniform circular motion is

$$v = \dot{\varphi} R = \omega R,$$

where $\dot{\varphi} = \omega$ is the angular velocity. The last two relations show that $\vec{\omega} = \vec{\omega}(t)$ is a pseudovector oriented along the axis of rotation, its magnitude being equal to the angular velocity $\dot{\varphi}$. It is called *instantaneous vector of rotation*.

Let us now turn back to our problem. In the IRF denoted by S, the fundamental equation of motion writes

$$\vec{F} = m\vec{a}, \qquad (6.31)$$

where $\vec{F} = -\nabla V(\vec{r})$ is the conservative force acting upon the particle. The Lagrangian in S then is

$$L = T - V = \frac{1}{2} m |\vec{v}|^2 - V(\vec{r}). \qquad (6.32)$$

To write the equation of motion of the particle in the non-inertial frame S', we have to express the Lagrangian L in terms of x'_i and $\dot{x}'_i$ ($i = \overline{1,3}$). In view of (6.23), we have

$$|\vec{v}|^2 = (\vec{v}_0 + \vec{v}_r + \vec{\omega} \times \vec{r}') \cdot (\vec{v}_0 + \vec{v}_r + \vec{\omega} \times \vec{r}')$$

$$= |\vec{v}_0|^2 + |\vec{v}_r|^2 + |\vec{\omega} \times \vec{r}'|^2$$

$$+ 2\vec{v}_0 \cdot \vec{v}_r + 2\vec{v}_0 \cdot (\vec{\omega} \times \vec{r}') + 2\vec{v}_r \cdot (\vec{\omega} \times \vec{r}').$$

In the non-inertial frame S' our Lagrangian L then writes

$$L = \frac{1}{2} m \left[|\vec{v}_0|^2 + |\vec{v}_r|^2 + |\vec{\omega} \times \vec{r}'|^2 \right.$$

$$+2\vec{v}_0 \cdot \vec{v}_r + 2\vec{v}_0 \cdot (\vec{\omega} \times \vec{r}\,') + 2\vec{v}_r \cdot (\vec{\omega} \times \vec{r}\,')\bigg] - V(\vec{r}\,').$$

The terms containing $\vec{v}_0$ can be transformed as follows:

$$|\vec{v}_0|^2 + 2\vec{v}_0 \cdot \vec{v}_r + 2\vec{v}_0 \cdot (\vec{\omega} \times \vec{r}\,') = |\vec{v}_0|^2 + 2\vec{v}_0 \cdot (\vec{v}_r + \vec{\omega} \times \vec{r}\,')$$

$$= |\vec{v}_0|^2 + 2\vec{v}_0 \cdot (\vec{v} - \vec{v}_0) = \vec{v}_0 \cdot (2\vec{v} - \vec{v}_0) = \vec{v}_0 \cdot \frac{d}{dt}(2\vec{r} - \vec{r}_0)$$

$$= \vec{v}_0 \cdot \frac{d}{dt}[2(\vec{r}_0 + \vec{r}\,') - \vec{r}_0] = \vec{v}_0 \cdot \frac{d}{dt}(\vec{r}_0 + 2\vec{r}\,')$$

$$= \frac{d}{dt}\left[\vec{v}_0 \cdot (\vec{r}_0 + 2\vec{r}\,')\right] - \vec{a}_0 \cdot (\vec{r}_0 + 2\vec{r}\,')$$

$$= \frac{d}{dt}\left[\vec{v}_0 \cdot (\vec{r}_0 + 2\vec{r}\,')\right] - \vec{a}_0 \cdot \vec{r}_0 - 2\vec{a}_0 \cdot \vec{r}\,'.$$

Since the expression $\vec{v}_0 \cdot (\vec{r}_0 + 2\vec{r}\,')$ depends on the coordinates and/or the time only (it cannot depend on velocity), while the quantity $\vec{a}_0 \cdot \vec{r}_0$ is also a function which depends only on the time (it can be written as a total derivative with respect to time of some function of time), the terms $\frac{d}{dt}\left[\vec{v}_0 \cdot (\vec{r}_0 + 2\vec{r}\,')\right]$ and $(-\vec{a}_0 \cdot \vec{r}_0)$ can be omitted in L and we are left with

$$L = \frac{1}{2}m|\vec{v}_r|^2 + \frac{1}{2}m|\vec{\omega} \times \vec{r}\,'|^2 + m\vec{v}_r \cdot (\vec{\omega} \times \vec{r}\,') - m\vec{a}_0 \cdot \vec{r}\,' - V(\vec{r}\,'), \quad (6.33)$$

which is precisely (6.16). Therefore, the proof is complete.

Let us now go further and use the Lagrangian formalism in order to write the equation of motion of the particle in a NIRF. To this end, it is more convenient to use vectors components in the Lagrangian (6.33), that is

$$L = \frac{1}{2}m\dot{x}'_i\dot{x}'_i + \frac{1}{2}m\varepsilon_{ijk}\varepsilon_{inl}\omega'_j\omega'_n x'_k x'_l + m\varepsilon_{ijk}\dot{x}'_i\omega'_j x'_k - ma'_{0i}x'_i - V(x'_k)$$

$$= \frac{1}{2}m\dot{x}'_i\dot{x}'_i + \frac{1}{2}m(\delta_{jn}\delta_{kl}$$

$$-\delta_{jl}\delta_{kn})\omega'_j\omega'_n x'_k x'_l + m\varepsilon_{ijk}\dot{x}'_i\omega'_j x'_k - ma'_{0i}x'_i - V(x'_k)$$

$$= \frac{1}{2}m\dot{x}'_i\dot{x}'_i + \frac{1}{2}m\omega'_j\omega'_j x'_k x'_k$$

$$-\frac{1}{2}m\omega'_j\omega'_k x'_k x'_j + m\varepsilon_{ijk}\dot{x}'_i\omega'_j x'_k - ma'_{0i}x'_i - V(x'_k)$$

$$= \frac{1}{2}m\dot{x}'_i\dot{x}'_i + \frac{1}{2}m\omega'_k\omega'_k x'_i x'_i$$

$$-\frac{1}{2}m(x'_i\omega'_i)(x'_k\omega'_k) + m\varepsilon_{ijk}\dot{x}'_i\omega'_j x'_k - ma'_{0i}x'_i - V(x'_k). \qquad (6.34)$$

We have:

$$\frac{\partial L}{\partial \dot{x}'_s} = m\dot{x}'_i\delta_{is} + m\varepsilon_{sjk}\omega'_j x'_k = m\dot{x}'_s + m\varepsilon_{sjk}\omega'_j x'_k;$$

$$\frac{d}{dt}\left(\frac{\partial L}{\partial \dot{x}'_s}\right) = m\ddot{x}'_s + m\varepsilon_{sjk}\dot{\omega}'_j x'_k + m\varepsilon_{sjk}\omega'_j \dot{x}'_k;$$

$$\frac{\partial L}{\partial x'_s} = m\omega'_k\omega'_k x'_i\delta_{is} - \frac{1}{2}m\omega'_i(x'_k\omega'_k)\delta_{is}$$

$$-\frac{1}{2}m\omega'_k(x'_i\omega'_i)\delta_{ks} + m\varepsilon_{ijs}\dot{x}'_i\omega'_j - ma'_{0s} - \frac{\partial V}{\partial x'_s}$$

$$= m\omega'_k\omega'_k x'_s - m(x'_i\omega'_i)\omega'_s + m\varepsilon_{ijs}\dot{x}'_i\omega'_j - ma'_{0s} - \frac{\partial V}{\partial x'_s}.$$

With these results, Lagrange equations of the second kind

$$\frac{d}{dt}\left(\frac{\partial L}{\partial \dot{x}'_s}\right) - \frac{\partial L}{\partial x'_s} = 0 \ (s = \overline{1,3}) \qquad (6.35)$$

become

$$m\ddot{x}'_s + m\varepsilon_{sjk}\dot{\omega}'_j x'_k + m\varepsilon_{sjk}\omega'_j \dot{x}'_k - m\omega'_k\omega'_k x'_s$$

$$+ m(x'_i\omega'_i)\omega'_s - m\varepsilon_{sij}\dot{x}'_i\omega'_j + ma'_{0s} + \frac{\partial V}{\partial x'_s}$$

$$= m\ddot{x}'_s + m(\dot{\vec{\omega}} \times \vec{r}')_s + 2m(\vec{\omega} \times \vec{v}_r)_s - m\omega'_k\omega'_k x'_s + m(x'_i\omega'_i)\omega'_s + ma'_{0s} + \frac{\partial V}{\partial x'_s}$$

$$= m\ddot{x}'_s + m(\dot{\vec{\omega}} \times \vec{r}')_s + 2m(\vec{\omega} \times \vec{v}_r)_s + m[\vec{\omega} \times (\vec{\omega} \times \vec{r}')]_s + ma'_{0s} + \frac{\partial V}{\partial x'_s} = 0.$$
$$(6.36)$$

Equation (6.36) is the x'_s-component of the vector equation

$$m\vec{a}_r + m\dot{\vec{\omega}} \times \vec{r}' + 2m\vec{\omega} \times \vec{v}_r + m\vec{\omega} \times (\vec{\omega} \times \vec{r}') + m\vec{a}_0 - \vec{F} = 0,$$

or, in a more eloquent form

$$m\vec{a}_r = \vec{F} - m\vec{a}_0 - m\dot{\vec{\omega}} \times \vec{r}' - m\vec{\omega} \times (\vec{\omega} \times \vec{r}') - 2m\vec{\omega} \times \vec{v}_r. \qquad (6.37)$$

Taking into account (6.26), we still have

$$m\vec{a}_r = \vec{F} - m\vec{a}_{tr} - m\vec{a}_c. \tag{6.38}$$

Introducing the notations

$$\vec{F}_{tr} = -m\vec{a}_{tr}; \quad \vec{F}_c = -m\vec{a}_c, \tag{6.39}$$

we finally obtain the equation of motion of the particle with respect to the non-inertial frame S'

$$m\vec{a}_r = \vec{F} + \vec{F}_{tr} + \vec{F}_c. \tag{6.40}$$

As it can be observed, the fundamental equation of motion of the particle *does not keep its form* when passing from the inertial frame S to the non-inertial frame S'. Together with the Newtonian force $\vec{F}$ appear two more forces $\vec{F}_{tr}$ and $\vec{F}_c$, called *inertial* or *apparent* forces. The inertial forces emerge as a result of the motion of S'. Indeed, if $\vec{a}_0 = 0$, $\vec{\omega} = 0$ (that is, if the frame S' becomes inertial), then $\vec{F}_{tr} = 0$, $\vec{F}_c = 0$, and the equation of motion gets its "inertial" form $m\vec{a}_r = m\vec{a} = \vec{F}$.

We should mention that, even if the inertial forces do not exist for an observer connected to S, they play the role of *real* forces for an observer fixed relative to S'. In such a frame (non-inertial) the inertial forces can be considered as being produced by some force fields which can be called (by analogy with the gravitational field) *inertial force fields*. As an example, the gravitational force in an inertial frame appears as an "effect" of the gravitational field, which is a potential (or, even, conservative) field. This idea lead Einstein to elaborating his general theory of relativity, by postulating the (local) equivalence between the gravitational field and the field of inertial forces.

Let us now come back to our problem and choose as the non-inertial frame S' a coordinate system connected to the point of suspension of the rod of pendulum, O' (see Fig.VI.6). This point oscillates with respect to O according to the law $X = a \cos\omega_0 t$, with acceleration $a_0 = |\vec{a}_0| = \ddot{X} = -a\omega_0^2 \cos\omega_0 t$. As one can see, in our case $\vec{\omega} = 0$, $\vec{v}_0 = (-a\omega_0 \sin\omega_0 t, 0, 0)$, $\vec{a}_0 = (-a\omega_0^2 \cos\omega_0 t, 0, 0)$, $\vec{r}' = (l\sin\theta, l\cos\theta, 0)$, and the Lagrangian (6.33) writes

$$L_{NIRF} \equiv L' = \frac{1}{2}m|\vec{v}_r|^2 - m\vec{a}_0 \cdot \vec{r}' - V(\vec{r}')$$

$$= \frac{1}{2}m|\vec{v} - \vec{v}_0|^2 - m\vec{a}_0 \cdot \vec{r}' - V(\vec{r}'). \tag{6.41}$$

Since
$$\vec{v} - \vec{v}_0 = (\dot{x} - v_{0x}, \dot{y}, 0) = (l\dot{\theta}\cos\theta, -l\dot{\theta}\sin\theta, 0)$$
and
$$V(\vec{r}') = V(y') = V(y) = -mgl\cos\theta,$$
we still have
$$L_{NIRF} \equiv L' = \frac{1}{2}ml^2\dot{\theta}^2 + mal\omega_0^2\cos\omega_0 t\sin\theta + mgl\cos\theta. \quad (6.42)$$

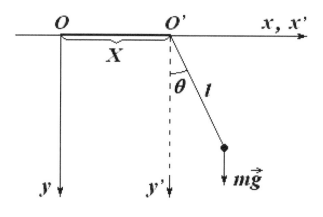

Fig.VI.6

The Lagrange equation of the second kind, written for L',
$$\frac{d}{dt}\left(\frac{\partial L'}{\partial \dot{\theta}}\right) - \frac{\partial L'}{\partial \theta} = 0, \quad (6.43)$$
gives
$$l\ddot{\theta} - a\omega_0^2\cos\omega_0 t\cos\theta + g\sin\theta = 0, \quad (6.44)$$
which is precisely equation (6.15) obtained while working in an inertial reference frame.

4. Problem of two identical coupled pendulums

Consider two identical simple pendulums, each of them with mass m and length l, connected by a spring of negligible mass. The spring is not tense when the pendulums are in equilibrium ($x_1 = x_2 = 0$ – see Fig.VI.7). The reader is asked to determine:
a) Proper frequencies of the system;
b) Solutions of the equations of motion, if at the initial moment $t_0 = 0$, $x_1(t_0) = a$, $x_2(t_0) = 0$, and $\dot{x}_1(t_0) = 0$, $\dot{x}_2(t_0) = 0$.

Solution

a) Recalling that both gravitational and elastic forces are potential forces, we deal with a *natural system*. According to the notations used in Fig.VI.7, the equations of constraints are:

i) $f_1(z_1) = z_1 = 0$;
ii) $f_2(x_1, y_1) = x_1^2 + y_1^2 - l^2 = 0$;
iii) $f_3(z_2) = z_2 = 0$;
iv) $f_4(x_2, y_2) = x_2^2 + y_2^2 - l^2 = 0$,

where each pendulum is reported to its own reference system. The x-axis is common for both systems, while the origins are taken in the two points of suspension. There is no restriction if we choose $z = 0$ as the plane of motion. For sufficiently small oscillations, the variation of coordinate y of the particles can be neglected, so that the motion can be considered as a unidimensional motion along the x-axis.

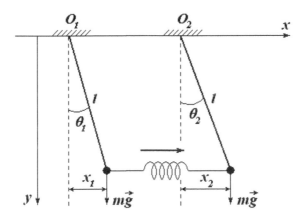

Fig.VI.7

Let x_1 and x_2 be the generalized coordinates associated with the $3 \cdot 2 - 4 = 2$ degrees of freedom of the system. The kinetic energy then writes

$$T = T_1 + T_2 = \frac{1}{2}m|\vec{v}_1|^2 + \frac{1}{2}m|\vec{v}_2|^2 = \frac{1}{2}m\dot{x}_1^2 + \frac{1}{2}m\dot{x}_2^2.$$

In its turn, the potential energy has two "components", one of gravitational nature V_g, and the other V_e due to elastic properties of the spring: $V = V_g + V_e$. By means of the usual procedure, we have

$$dV_g = -dA_g = -\vec{G}_1 \cdot d\vec{r}_1 - \vec{G}_2 \cdot d\vec{r}_2 = -mgdy_1 - mgdy_2,$$

or

$$V_g = -mgy_1 - mgy_2 + V_{g0}.$$

If we choose the plane xOz as a "reference level" for the gravitational potential energy, then $V_g(y_1 = 0, y_2 = 0) = 0$, and we may take the integration constant as $V_{g0} = 0$. This way,

$$V_g = -mgl\cos\theta_1 - mgl\cos\theta_2 = -mgl\left(\sqrt{1-\sin^2\theta_1} + \sqrt{1-\sin^2\theta_2}\right)$$

$$= -mgl\left(\sqrt{1-\left(\frac{x_1}{l}\right)^2} + \sqrt{1-\left(\frac{x_2}{l}\right)^2}\right) \simeq -2mgl + mg\frac{x_1^2}{2l} + mg\frac{x_2^2}{2l}.$$

Similarly,

$$dV_e = -dA_e = -\vec{F}_e \cdot d\vec{r} = -(-k\vec{r} \cdot d\vec{r}) = kx\,dx,$$

so that

$$V_e = \frac{1}{2}kx^2 + V_{e0}.$$

Taking the initial state (when the deformation of the spring is zero) as the reference level for V_e, one may choose $V_{e0} = 0$. Using again Fig.VI.7, we then have

$$V_e = \frac{1}{2}kx^2 \equiv \frac{1}{2}k(\Delta x)^2 = \frac{1}{2}k(x_2 - x_1)^2,$$

where Δx is the deformation of the spring. The total potential energy therefore is

$$V = mg\frac{x_1^2}{2l} + mg\frac{x_2^2}{2l} + \frac{1}{2}k(x_2 - x_1)^2,$$

where the constant $-2mgl$ has been dropped. The Lagrangian then writes

$$L = T - V = \frac{1}{2}m\dot{x}_1^2 + \frac{1}{2}m\dot{x}_2^2 - mg\frac{x_1^2}{2l} - mg\frac{x_2^2}{2l} - \frac{1}{2}k(x_2 - x_1)^2.$$

and Lagrange equations for x_1 and x_2

$$\frac{d}{dt}\left(\frac{\partial L}{\partial \dot{x}_j}\right) - \frac{\partial L}{\partial x_j} = 0 \quad (j = 1, 2)$$

yield the system

$$m\ddot{x}_1 = -mg\frac{x_1}{l} + k(x_2 - x_1), \tag{6.45}$$

$$m\ddot{x}_2 = -mg\frac{x_2}{l} - k(x_2 - x_1). \tag{6.46}$$

Assuming that the system performs only small oscillations, we search for solutions of the form

$$x_1 = A_1 \sin \omega t + B_1 \cos \omega t, \tag{6.47}$$

$$x_2 = A_2 \sin \omega t + B_2 \cos \omega t. \tag{6.48}$$

Introducing (6.47) and (6.48) into (6.45) and (6.46), one obtains

$$-m\omega^2 A_1 \sin \omega t - m\omega^2 B_1 \cos \omega t = -mg\frac{A_1}{l} \sin \omega t - mg\frac{B_1}{l} \cos \omega t$$

$$+kA_2 \sin \omega t + kB_2 \cos \omega t - kA_1 \sin \omega t - kB_1 \cos \omega t,$$

and

$$-m\omega^2 A_2 \sin \omega t - m\omega^2 B_2 \cos \omega t = -mg\frac{A_2}{l} \sin \omega t - mg\frac{B_2}{l} \cos \omega t$$

$$-kA_2 \sin \omega t - kB_2 \cos \omega t + kA_1 \sin \omega t + kB_1 \cos \omega t.$$

Identifying the coefficients of $\sin \omega t$ and $\cos \omega t$ of the two members of the above equations, we obtain

$$-m\omega^2 A_1 = -mg\frac{A_1}{l} + kA_2 - kA_1, \tag{6.49}$$

$$-m\omega^2 B_1 = -mg\frac{B_1}{l} + kB_2 - kB_1, \tag{6.50}$$

and

$$-m\omega^2 A_2 = -mg\frac{A_2}{l} - kA_2 + kA_1, \tag{6.51}$$

$$-m\omega^2 B_2 = -mg\frac{B_2}{l} - kB_2 + kB_1. \tag{6.52}$$

In order that the system (6.49) and (6.51) has at least one non-trivial solution, we must have

$$\begin{vmatrix} -m\omega^2 + \frac{mg}{l} + k & -k \\ -k & -m\omega^2 + \frac{mg}{l} + k \end{vmatrix} = 0,$$

which yields

$$\left(-m\omega^2 + \frac{mg}{l} + k\right)^2 = k^2,$$

or

$$-m\omega^2 + \frac{mg}{l} + k = \pm k.$$

We therefore have the following two solutions for ω^2:

$$\omega_{1,2}^2 = \frac{g}{l} + \frac{k}{m} \pm \frac{k}{m}, \qquad (6.53)$$

with

$$\omega_1^2 \equiv \omega_M^2 = \frac{g}{l} + 2\frac{k}{m}, \qquad (6.54)$$

and

$$\omega_2^2 \equiv \omega_m^2 = \frac{g}{l}. \qquad (6.55)$$

The reader can easily prove himself that the use of (6.50) and (6.52) leads to the same result.

Substituting ω_M^2 in (6.49) by its value given by (6.54), we have

$$-m\omega_M^2 A_1 = -mg\frac{A_1}{l} + kA_2 - kA_1,$$

that is

$$A_2 = -A_1. \qquad (6.56)$$

Using the same procedure in equation (6.50), one obtains

$$-m\frac{g}{l}B_1 - 2kB_1 = -m\frac{g}{l}B_1 + kB_2 - kB_1,$$

which means

$$B_2 = -B_1. \qquad (6.57)$$

Taking now ω_m^2 given by (6.55) and following the same simple calculations, we get

$$A_2 = A_1, \qquad (6.58)$$

and

$$B_2 = B_1. \qquad (6.59)$$

Since the system has two frequencies of oscillation, the solutions of the differential equations of motion (6.47) and (6.48) of the two coupled pendulums are written as a superposition of the two modes of oscillation, as

$$x_1 = C_1 \sin\omega_M t + C_2 \cos\omega_M t + C_3 \sin\omega_m t + C_4 \cos\omega_m t, \qquad (6.60)$$

and

$$x_2 = -C_1 \sin\omega_M t - C_2 \cos\omega_M t + C_3 \sin\omega_m t + C_4 \cos\omega_m t, \qquad (6.61)$$

where the constants of integration C_i ($i = \overline{1,4}$) are determined by means of initial conditions.

b) According to the statement of the problem, $x_1 = a$, $x_2 = 0$ at $t = 0$. This implies $\dot{x}_1(t = 0) = \dot{a} = 0$, $\dot{x}_2(t = 0) = 0$. Equations (6.60) and (6.61) yield the following expressions for velocities

$$\dot{x}_1 = \omega_M C_1 \cos \omega_M t - \omega_M C_2 \sin \omega_M t + \omega_m C_3 \cos \omega_m t - \omega_m C_4 \sin \omega_m t,$$

and

$$\dot{x}_2 = -\omega_M C_1 \cos \omega_M t + \omega_M C_2 \sin \omega_M t + \omega_m C_3 \cos \omega_m t - \omega_m C_4 \sin \omega_m t.$$

Imposing the initial conditions, we have

$$x_1(t = 0) = a = C_2 + C_4, \tag{6.62}$$

$$x_2(t = 0) = 0 = -C_2 + C_4, \tag{6.63}$$

$$\dot{x}_1(t = 0) = 0 = \omega_M C_1 + \omega_m C_3, \tag{6.64}$$

$$\dot{x}_2(t = 0) = 0 = -\omega_M C_1 + \omega_m C_3. \tag{6.65}$$

It then follows $C_2 = C_4 = a/2$, and $C_1 = C_3 = 0$, and the solutions become

$$x_1 = \frac{a}{2}(\cos \omega_M t + \cos \omega_m t), \tag{6.66}$$

$$x_2 = \frac{a}{2}(\cos \omega_m t - \cos \omega_M t), \tag{6.67}$$

or

$$x_1 = a \cos\left(\frac{\omega_M + \omega_m}{2}\right)t \cos\left(\frac{\omega_M - \omega_m}{2}\right)t$$

$$= A_{mod}(t) \cos\left(\frac{\omega_M + \omega_m}{2}\right)t, \tag{6.68}$$

$$x_2 = a \sin\left(\frac{\omega_M + \omega_m}{2}\right)t \sin\left(\frac{\omega_M - \omega_m}{2}\right)t$$

$$= B_{mod}(t) \sin\left(\frac{\omega_M + \omega_m}{2}\right)t, \tag{6.69}$$

where A_{mod} and B_{mod} denote the amplitude of the two modes of oscillation.

As one can see, there appears a phenomenon of *beating* between the two pendulums. The beating frequency is

$$\omega_{beat} = 2\omega_{mod} = \omega_M - \omega_m. \tag{6.70}$$

5. Problem of two different coupled pendulums

Two simple pendulums of masses m_1, m_2 and lengths l_1, l_2 are coupled by a spring with spring constant k, mounted at the distance h with respect to their suspension points O_1 and O_2 (see Fig.VI.8). Supposing that the spring is not tensioned when pendulums are in equilibrium ($\theta_1 = \theta_2 = 0$), write the equations of motion of the system and determine their solutions in case of small oscillations. The motion takes place in a vertical plane.

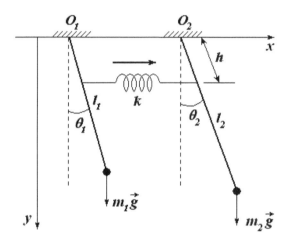

Fig.VI.8

Solution

Since both the gravitational and elastic forces are potential force fields, we have to do with a natural system. The constraints are given by

    i) $f_1(z_1) = z_1 = 0$;
    ii) $f_2(x_1, y_1) = x_1^2 + y_1^2 - l_1^2 = 0$;
    iii) $f_3(z_2) = z_2 = 0$;
    iv) $f_4(x_2, y_2) = x_2^2 + y_2^2 - l_2^2 = 0$,

where with each pendulum is associated a reference system, with origins at the suspension points O_1 and O_2, the axis Ox being a common axis. The system has $3 \cdot 2 - 4 = 2$ degrees of freedom. Let θ_1 and θ_2 be the associated generalized coordinates. The two rods are supposed to be ideal (rigid, and massless). Since

$$x_1 = l_1 \sin \theta_1; \quad y_1 = l_1 \cos \theta_1,$$

$$x_2 = l_2 \sin \theta_2; \quad y_2 = l_2 \cos \theta_2,$$

the kinetic energy of the system is

$$T = T_1 + T_2 = \frac{1}{2}m_1|\vec{v}_1|^2 + \frac{1}{2}m_2|\vec{v}_2|^2 = \frac{1}{2}(m_1 l_1^2 \dot{\theta}_1^2 + m_2 l_2^2 \dot{\theta}_2^2),$$

while the potential energy writes

$$V = V_g + V_e,$$

where V_g and V_e are the gravitational and elastic potential energies, respectively. To find V_g, we apply the usual method (see previous problems):

$$dV_g = -dA_g = -\vec{G}_1 \cdot d\vec{r}_1 - \vec{G}_2 \cdot d\vec{r}_2 = -m_1 g\, dy_1 - m_2 g\, dy_2,$$

so that

$$V_g = -m_1 g y_1 - m_2 g y_2 + V_{g0}.$$

If we choose the plane xOz as a reference level for the gravitational potential energy, then $V_g(y_1 = 0, y_2 = 0) = 0$, and we may take the integration constant as $V_{g0} = 0$. Therefore,

$$V_g = -m_1 g l_1 \cos\theta_1 - m_2 g l_2 \cos\theta_2.$$

We still have

$$dV_e = -dA_e = -\vec{F}_e \cdot d\vec{r} = -(-k\vec{x} \cdot d\vec{x}) = kx\, dx$$

so that

$$V_e = \frac{1}{2}kx^2 + V_{e0}.$$

The integration constant V_{e0} may be taken as zero, if we conveniently choose the initial state of the spring ($x = 0$) as the "reference level" for V_e. Thus,

$$V_e = \frac{1}{2}kx^2 = \frac{1}{2}(\Delta x)^2,$$

where Δx is, obviously, the deformation of the spring (see Fig.VI.8). As one observes,

$$|\Delta x| = |x_2 - x_1| = h|\sin\theta_2 - \sin\theta_1|,$$

and we still have

$$V_e = \frac{1}{2}kh^2\left[4\sin^2\left(\frac{\theta_2 - \theta_1}{2}\right)\cos^2\left(\frac{\theta_2 + \theta_1}{2}\right)\right].$$

Suppose we restrict the pendulum's oscillations to small angles ($< 4°$). Then we can approximate

$$\sin\left(\frac{\theta_2 - \theta_1}{2}\right) = \frac{\theta_2 - \theta_1}{2} + \mathcal{O}\left(\frac{\theta_2 - \theta_1}{2}\right)^3 \simeq \frac{\theta_2 - \theta_1}{2};$$

$$\cos\left(\frac{\theta_1 + \theta_2}{2}\right) = 1 - \frac{1}{2}\left(\frac{\theta_1 + \theta_2}{2}\right)^2 + \mathcal{O}\left(\frac{\theta_1 + \theta_2}{2}\right)^4 \simeq 1,$$

and V_e becomes

$$V_e = \frac{1}{2}kh^2(\theta_2 - \theta_1)^2.$$

Using the same approximation for $\cos\theta_1$ and $\cos\theta_2$ in V_g, we have

$$V_g = -g\left[m_1 l_1\left(1 - \frac{\theta_1^2}{2}\right) + m_2 l_2\left(1 - \frac{\theta_2^2}{2}\right)\right].$$

The Lagrangian of the system then writes

$$L = T - V = \frac{1}{2}\left(m_1 l_1^2 \dot\theta_1^2 + m_2 l_2^2 \dot\theta_2^2\right)$$

$$-\frac{1}{2}g\left(m_1 l_1 \theta_1^2 + m_2 l_2 \theta_2^2\right) - \frac{1}{2}kh^2(\theta_2 - \theta_1)^2, \tag{6.71}$$

where the constant term $g(m_1 l_1 + m_2 l_2)$ has been dropped.

Once the Lagrangian is known, the Lagrange equations of the second kind

$$\frac{d}{dt}\left(\frac{\partial L}{\partial \dot\theta_i}\right) - \frac{\partial L}{\partial \theta_i} = 0 \quad (i = 1, 2)$$

yield

$$m_1 l_1^2 \ddot\theta_1 + g m_1 l_1 \theta_1 - kh^2(\theta_2 - \theta_1) = 0; \tag{6.72}$$

$$m_2 l_2^2 \ddot\theta_2 + g m_2 l_2 \theta_2 + kh^2(\theta_2 - \theta_1) = 0. \tag{6.73}$$

Let us now turn back to the Lagrangian written for small oscillations (6.71) and introduce the following substitutions: $\xi_1 = \sqrt{m_1}\, l_1 \theta_1$, and $\xi_2 = \sqrt{m_2}\, l_2 \theta_2$. Obviously, we also have $\dot\xi_1 = \sqrt{m_1}\, l_1 \dot\theta_1$, and $\dot\xi_2 = \sqrt{m_2}\, l_2 \dot\theta_2$. With these notations, the Lagrangian (6.71) writes

$$L = \frac{1}{2}(\dot\xi_1^2 + \dot\xi_2^2) - \frac{1}{2}\left(\frac{g}{l_1}\xi_1^2 + \frac{g}{l_2}\xi_2^2\right) - \frac{1}{2}kh^2\left(\frac{\xi_2}{l_2\sqrt{m_2}} - \frac{\xi_1}{l_1\sqrt{m_1}}\right)^2$$

$$= \frac{1}{2}(\dot\xi_1^2 + \dot\xi_2^2) - \frac{1}{2}\left[\left(\frac{g}{l_1} + \frac{kh^2}{m_1 l_1^2}\right)\xi_1^2 + \left(\frac{g}{l_2} + \frac{kh^2}{m_2 l_2^2}\right)\xi_2^2\right]$$

$$+\frac{kh^2}{l_1 l_2 \sqrt{m_1 m_2}} \xi_1 \xi_2 = \frac{1}{2}(\dot\xi_1^2 + \dot\xi_2^2) - \frac{1}{2}(\omega_1^2 \xi_1^2 + \omega_2^2 \xi_2^2) + \kappa \xi_1 \xi_2, \quad (6.74)$$

where the following notations have been used:

$$\omega_{1,2}^2 = \frac{g}{l_{1,2}} + \frac{kh^2}{m_{1,2} l_{1,2}^2}; \quad (6.75)$$

$$\kappa = \frac{kh^2}{l_1 l_2 \sqrt{m_1 m_2}}. \quad (6.76)$$

Consequently, the problem reduces to the study of small oscillations of a system described by the Lagrangian (6.74). Let us write this Lagrangian in *normal coordinates*[1]. To this end, it is convenient to make a change of generalized coordinates, such as

$$(\xi_1, \xi_2) \to (\eta_1, \eta_2), \quad (6.77)$$

so that the new Lagrangian remains quadratic in the new generalized velocities, and, in addition, becomes quadratic in the new generalized coordinates (in other words, the coefficient of the mixed term vanishes). This can be accomplished by taking (6.77) as a plane rotation:

$$\begin{cases} \xi_1 = \eta_1 \cos\varphi - \eta_2 \sin\varphi; \\ \xi_2 = \eta_1 \sin\varphi + \eta_2 \cos\varphi, \end{cases} \quad (6.78)$$

where φ is a constant. As one can see

$$\dot\xi_1^2 + \dot\xi_2^2 = \dot\eta_1^2 + \dot\eta_2^2,$$

which means that the Lagrangian (6.74) remains quadratic in the new generalized velocities $\dot\eta_1$ and $\dot\eta_2$. The other two terms yield:

$$\kappa \xi_1 \xi_2 = \kappa(\eta_1 \cos\varphi - \eta_2 \sin\varphi)(\eta_1 \sin\varphi + \eta_2 \cos\varphi)$$
$$= \kappa \sin\varphi \cos\varphi (\eta_1^2 - \eta_2^2) + \kappa \eta_1 \eta_2 (\cos^2\varphi - \sin^2\varphi),$$

and

$$-\frac{1}{2}(\omega_1^2 \xi_1^2 + \omega_2^2 \xi_2^2)$$
$$= -\frac{1}{2}\left[\omega_1^2 (\eta_1 \cos\varphi - \eta_2 \sin\varphi)^2 + \omega_2^2 (\eta_1 \sin\varphi + \eta_2 \cos\varphi)^2\right]$$

---

[1] A set of coordinates for a coupled system such that each equation of motion involve only one of these coordinates.

$$= -\frac{1}{2}\Big[\eta_1^2\big(\omega_1^2\cos^2\varphi + \omega_2^2\sin^2\varphi\big) + \eta_2^2\big(\omega_1^2\sin^2\varphi + \omega_2^2\cos^2\varphi\big)$$

$$+ 2\eta_1\eta_2\sin\varphi\cos\varphi\big(\omega_2^2 - \omega_1^2\big)\Big].$$

In order that the Lagrangian becomes quadratic in the new variables, the mixed term must be zero, that is

$$\eta_1\eta_2\Big[\kappa\big(\cos^2\varphi - \sin^2\varphi\big) - \sin\varphi\cos\varphi\big(\omega_2^2 - \omega_1^2\big)\Big] = 0,$$

or

$$\cot 2\varphi = \frac{\omega_2^2 - \omega_1^2}{2\kappa}. \tag{6.79}$$

Therefore, if we choose φ according to (6.79), the new generalized coordinates η_1 and η_2 are *normal coordinates*, and the Lagrangian (6.74) writes

$$L = \frac{1}{2}\big(\dot\eta_1^2 + \dot\eta_2^2\big) - \frac{1}{2}\Big[\eta_1^2\big(\omega_1^2\cos^2\varphi + \omega_2^2\sin^2\varphi - 2\kappa\sin\varphi\cos\varphi\big)$$

$$+ \eta_2^2\big(\omega_1^2\sin^2\varphi + \omega_2^2\cos^2\varphi + 2\kappa\sin\varphi\cos\varphi\big)\Big].$$

Let us denote

$$\overline{\omega}_1^2 = \omega_1^2\cos^2\varphi + \omega_2^2\sin^2\varphi - 2\kappa\sin\varphi\cos\varphi;$$

$$\overline{\omega}_2^2 = \omega_1^2\sin^2\varphi + \omega_2^2\cos^2\varphi + 2\kappa\sin\varphi\cos\varphi.$$

Then

$$\overline{\omega}_1^2 = \frac{1}{2}\big(\omega_1^2 + \omega_2^2\big) - \sqrt{\kappa^2 + \frac{(\omega_2^2 - \omega_1^2)}{4}}; \tag{6.80}$$

$$\overline{\omega}_2^2 = \frac{1}{2}\big(\omega_1^2 + \omega_2^2\big) + \sqrt{\kappa^2 + \frac{(\omega_2^2 - \omega_1^2)}{4}}. \tag{6.81}$$

To obtain (6.80) and (6.81) we used (6.79). Denoting

$$\cot 2\varphi = \frac{\omega_2^2 - \omega_1^2}{2\kappa} = a, \tag{6.82}$$

we have

$$\frac{1 - 2\sin^2\varphi}{2\sin\varphi\cos\varphi} = a \Leftrightarrow 1 + 4\sin^4\varphi - 4\sin^2\varphi = 4a^2\sin^2\varphi\big(1 - \sin^2\varphi\big),$$

with the physically acceptable solution

$$\sin^2 \varphi = \frac{1}{2}\left(1 - \frac{a}{\sqrt{1+a^2}}\right). \qquad (6.83)$$

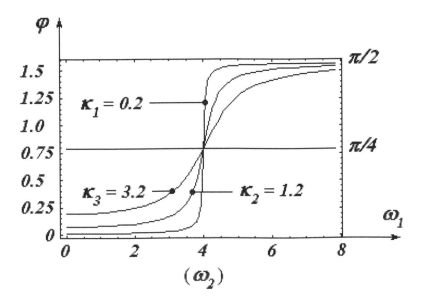

Fig.VI.9

In this case

$$-2\kappa \sin \varphi \cos \varphi = -\kappa \sin 2\varphi = -\kappa \frac{\cos 2\varphi}{\cot 2\varphi}$$

$$= -\kappa \frac{\cos^2 \varphi - \sin^2 \varphi}{a} = -\frac{\kappa}{\sqrt{1+a^2}}.$$

The rest of calculations are very simple and they remain up to the reader. So, the new Lagrangian finally writes

$$L = \frac{1}{2}\left(\dot{\eta}_1^2 + \dot{\eta}_2^2\right) - \frac{1}{2}\left(\overline{\omega}_1^2 \eta_1^2 + \overline{\omega}_2^2 \eta_2^2\right). \qquad (6.84)$$

Let us now investigate the dependence of φ on the frequencies ω_1 and ω_2. Fig.VI.9 shows this dependence for ω_1, and ω_2 fixed. As one can see, φ varies between 0 and $\pi/2$, the interval $\Delta \omega_1$ corresponding to transition from $\varphi \simeq 0$ to $\varphi \simeq \pi/2$ being of order κ/ω_2. Therefore, the smaller is κ, the narrower is this interval. Fig.VI.9 shows this dependence for three different values of κ, namely $\kappa_1 = 0.2$, $\kappa_2 =$

1.2, $\kappa_3 = 3.2$. This has been done by means of the *Mathematica* software package, as follows:

```
Off[General::spell]
Remove["Global`*"];
Unprotect[In, Out];
Clear[In, Out];

k1 := 0.2;
k2 := 1.2;
k3 := 3.2;

Omega2 := 4;

f1 = Plot[Pi/4, {x, 0, 2.5*Pi}];
f2 = Plot[ArcSin[
   Sqrt[(1*(1 - (Omega2^2 - x^2)/(2*k1*Sqrt[1 + ((Omega2^2 -
x^2)/(2*k1))^2])))/2]],
   {x, 0, 2.5*Pi}, Frame -> True, PlotRange -> All];
f3 = Plot[ArcSin[
   Sqrt[(1*(1 - (Omega2^2 - x^2)/(2*k2*Sqrt[1 + ((Omega2^2 -
x^2)/(2*k2))^2])))/2]],
   {x, 0, 2.5*Pi}, Frame -> True, PlotRange -> All];
f4 = Plot[ArcSin[
   Sqrt[(1*(1 - (Omega2^2 - x^2)/(2*k3*Sqrt[1 + ((Omega2^2 -
x^2)/(2*k3))^2])))/2]],
   {x, 0, 2.5*Pi}, Frame -> True, PlotRange -> All];

Show[f1, f2, f3, f4];
```

The interval of variation of φ should be between 0 and $\pi/2$, but practically these values can only be approximated. As an example, take $\omega_2 = 4\,rad/s$ and $\kappa = 0.01\,(rad/s)^2$ and calculate

$$\lim_{\omega_1 \to 0} \varphi = \lim_{\omega_1 \to 0} \left[\arcsin \sqrt{\frac{1}{2}\left(1 - \frac{\frac{\omega_2^2 - \omega_1^2}{2\kappa}}{\sqrt{1 + \left(\frac{\omega_2^2 - \omega_1^2}{2\kappa}\right)^2}}\right)}\right]$$

$$= \arcsin \sqrt{\frac{1}{2}\left(1 - \frac{\omega_2^2}{\sqrt{4\kappa^2 + \omega_2^4}}\right)}$$

$$= \arcsin\sqrt{\frac{1}{2}\left(1 - \frac{16}{\sqrt{4.10^{-4} + 4^4}}\right)} = 0.000625,$$

which can be approximated by zero. In general, no matter what the individual values of ω_2 and κ are, but satisfy the condition $\frac{4\kappa^2}{\omega_2^4} \to 0$, we have

$$\lim_{\omega_1 \to 0} \varphi = \arcsin\sqrt{\frac{1}{2}\left(1 - \frac{\omega_2^2}{\omega_2^2\sqrt{1 + \frac{4\kappa^2}{\omega_2^4}}}\right)}\Bigg|_{\frac{4\kappa^2}{\omega_2^4} \to 0}$$

$$\simeq \arcsin\sqrt{\frac{1}{2}\left(1 - \frac{\omega_2^2}{\omega_2^2}\right)} = \arcsin 0 = 0.$$

In the same way one can calculate the limit

$$\lim_{\omega_1 \to \infty} \varphi.$$

Simple calculations show that, for the same values of ω_2 and κ, this limit is $\frac{\pi}{2} = 1.5708$. It can be easily verified that, for $\omega_1 = 87$, the expression $\arcsin\sqrt{\frac{1}{2}\left(1 - \frac{\frac{\omega_2^2 - \omega_1^2}{2\kappa}}{\sqrt{1 + \left(\frac{\omega_2^2 - \omega_1^2}{2\kappa}\right)^2}}\right)}$ already has the value $\pi/2$. This value maintains, no matter how much the increase of ω_1 is. This result can also be easily verified by means of a straight calculation.

In case of a weak coupling ($\kappa \ll |\omega_2^2 - \omega_1^2|$), the normal oscillations are "localized", that is, for $\omega_1 < \omega_2$, the relation (6.83) yields

$$\varphi\bigg|_{\frac{\omega_2^2 - \omega_1^2}{\kappa} \gg 1} = \arcsin\left[\sqrt{\frac{1}{2}\left(1 - \frac{\frac{\omega_2^2 - \omega_1^2}{2\kappa}}{\sqrt{1 + \left(\frac{\omega_2^2 - \omega_1^2}{2\kappa}\right)^2}}\right)}\right]_{\frac{\omega_2^2 - \omega_1^2}{\kappa} \gg 1}$$

$$\simeq \arcsin\left[\sqrt{\frac{1}{2}\left(1 - \frac{\frac{\omega_2^2 - \omega_1^2}{2\kappa}}{\sqrt{\left(\frac{\omega_2^2 - \omega_1^2}{2\kappa}\right)^2}}\right)}\right] = \arcsin 0 = 0,$$

and (6.78) leads to
$$\xi_1 \simeq \eta_1; \quad \xi_2 \simeq \eta_2. \tag{6.85}$$

If $\omega_1 > \omega_2$, we have:

$$\varphi = \arcsin\left[\sqrt{\frac{1}{2}\left(1 - \frac{\frac{\omega_2^2-\omega_1^2}{2\kappa}}{\sqrt{1+\left(\frac{\omega_2^2-\omega_1^2}{2\kappa}\right)^2}}\right)}\right] \begin{cases} |\omega_2^2 - \omega_1^2| \gg k \\ \omega_1 > \omega_2 \end{cases}$$

$$\simeq \arcsin\sqrt{1} = \frac{\pi}{2},$$

and (6.78) yield
$$\xi_1 \simeq -\eta_2, \quad \xi_2 \simeq \eta_1. \tag{6.86}$$

In case of a strong coupling, that is $|\omega_2^2 - \omega_1^2| \ll \kappa$, the oscillations are not localized. Indeed,

$$\varphi_{\frac{|\omega_2^2-\omega_1^2|}{\kappa} \ll 1} = \arcsin\left[\sqrt{\frac{1}{2}\left(1 - \frac{\frac{\omega_2^2-\omega_1^2}{2\kappa}}{\sqrt{1+\left(\frac{\omega_2^2-\omega_1^2}{2\kappa}\right)^2}}\right)}\right]_{\frac{|\omega_2^2-\omega_1^2|}{\kappa} \ll 1}$$

$$= \arcsin\left[\sqrt{\frac{1}{2}\left(1 - \frac{\frac{\omega_2^2-\omega_1^2}{2\kappa}}{\sqrt{1+\left(\frac{\omega_2^2-\omega_1^2}{2\kappa}\right)^2}}\right)}\right]_{\frac{|\omega_2^2-\omega_1^2|}{\kappa} \to 0}$$

$$= \arcsin\frac{1}{\sqrt{2}} = \frac{\pi}{4},$$

and (6.78) lead to
$$\xi_1 \simeq \frac{1}{\sqrt{2}}(\eta_1 - \eta_2); \quad \xi_2 \simeq \frac{1}{\sqrt{2}}(\eta_1 + \eta_2). \tag{6.87}$$

Relations (6.80) and (6.81) show the dependence of the normal frequencies on the parameters ω_1, ω_2, and κ. This dependence is graphically illustrated in Fig.VI.10. Here ω_2 is fixed ($\omega_2 = 4\,rad/s$), and the coupling is relatively weak ($\kappa = 8(rad/s)^2$). The graphic representation has been performed using the same software package, by means of the following set of instructions:

```
Off[General::spell]
Remove["Global`*"];
Unprotect[In, Out];
Clear[In, Out];

k := 8;
Omega2 := 4;

g1 = Sqrt[(Omega2^2 + x^2)/2 - Sqrt[k^2 + (Omega2^2 - x^2)^2/4]];
g2 = Sqrt[(Omega2^2 + x^2)/2 + Sqrt[k^2 + (Omega2^2 - x^2)^2/4]];
g3 = Sqrt[Abs[(Omega2^2 + x^2)/2 - Sqrt[k^2 + (Omega2^2 –
    x^2)^2/4]]];

h1 = Plot[g1, {x, k/Omega2, 3*Pi}, Frame -> True, PlotRange -> All];
h2 = Plot[g2, {x, 0, 3*Pi}, Frame -> True, PlotRange -> All];
h3 = Plot[a, {x, 0, 3*Pi}, Frame -> True, PlotRange -> All];
h4 = Plot[x, {x, 0, 3*Pi}, Frame -> True, PlotRange -> All];
h5 = Plot[g3, {x, 0, 3*Pi}, Frame -> True, PlotRange -> All];

Show[h1, h2, h3, h4];
Show[h1, h2, h3, h4, h5];
```

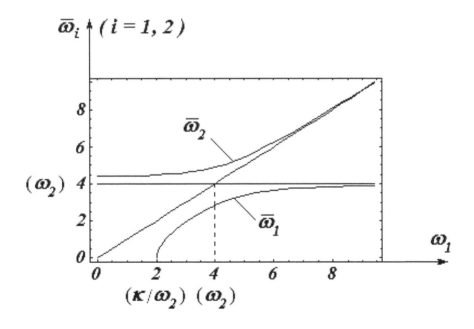

Fig.VI.10

As one observes, $\overline{\omega}_1 < \min(\omega_1, \omega_2)$, and $\overline{\omega}_2 > \max(\omega_1, \omega_2)$. If the coupling is weak (κ is small enough), except for the domain of degeneracy $|\omega_2^2 - \omega_1^2| \simeq \kappa$, the normal frequencies practically coincide with ω_1 and ω_2, respectively. Indeed, if κ can be neglected as compared to $\frac{\omega_2^2 - \omega_1^2}{2}$, then

$$\sqrt{\kappa^2 + \frac{(\omega_2^2 - \omega_1^2)^2}{4}} \simeq \frac{\omega_2^2 - \omega_1^2}{2},$$

and (6.80), (6.81) yield

$$\overline{\omega}_{1,2}^2 = \frac{\omega_1^2}{2} + \frac{\omega_2^2}{2} \mp \left(\frac{\omega_2^2}{2} - \frac{\omega_1^2}{2}\right),$$

namely $\overline{\omega}_1 = \omega_1$, and $\overline{\omega}_2 = \omega_2$.

Finally, for frequencies ω_1 very small (see further), one of the normal frequencies (namely $\overline{\omega}_1$) becomes imaginary, which means that the system is not stable anymore. Indeed, in order to be imaginary, that is

$$\overline{\omega}_1^2 = \frac{\omega_1^2}{2} + \frac{\omega_2^2}{2} - \sqrt{\kappa^2 + \frac{(\omega_2^2 - \omega_1^2)^2}{4}} < 0,$$

we must have

$$\left(\frac{\omega_1^2 + \omega_2^2}{2}\right)^2 < \kappa^2 + \frac{(\omega_2^2 - \omega_1^2)^2}{4},$$

or

$$\omega_1 < \frac{\kappa}{\omega_2},$$

which explains what we mean by "very small frequencies". This can be easily seen in Fig.VI.11, where ω_2 and $\sqrt{|\overline{\omega}_1^2|} = \overline{\overline{\omega}}_1$ are represented in terms of ω_1. As observed, $\overline{\overline{\omega}}_1$ presents two branches: the shortest branch, situated on the left from $\omega_1 = \frac{\kappa}{\omega_2}$ appears as an image in a mirror situated on $O\omega_1$ axis and face downwards of the graphic of $\overline{\omega}_1^2 = \overline{\omega}_1^2(\omega_1)$ "affected by the radical" on the interval $\left[0, \frac{\kappa}{\omega_2}\right]$.

Let us explain what "affected by the radical" means. Function $\overline{\omega}_1 = \overline{\omega}_1(\omega_1)$ cannot have a graphic representation on the interval $[0, \kappa/\omega_2]$ as a real function, because within this interval $\overline{\omega}_1 = \sqrt{\overline{\omega}_1^2}$ does not exist. In the above defined interval, only $\overline{\omega}_1^2$ can be graphically represented. But the mirror image (which exists as an effect of the modulus operation) of the function $\overline{\omega}_1^2 = \overline{\omega}_1^2(\omega_1)$ is not exactly the branch situated on the left of the value κ/ω_2 of the graph $\overline{\overline{\omega}}_1 = \sqrt{|\overline{\omega}_1^2|}$. The difference is due to the radical function. These observations are emphasized by Fig.VI.12, where this difference is even more obvious.

Indeed, the two branches on the lef the value κ/ω_2 of the functions $\overline{\omega}_1^2 = \overline{\omega}_1^2(\omega_1)$ and $\overline{\overline{\omega}}_1 = \sqrt{|\overline{\omega}_1^2|}$ are not as an object and its image in a horizontal mirror, situated on $O\omega_1$ axis.

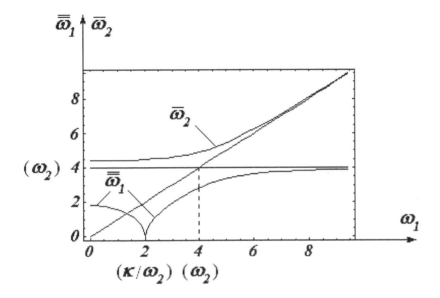

Fig.VI.11

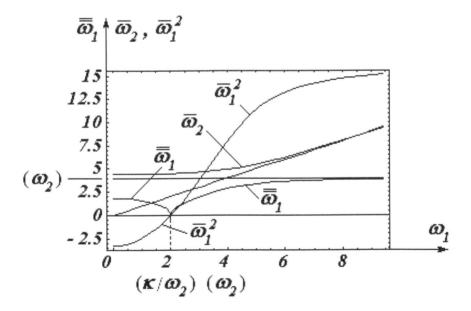

Fig.VI.12

The law of motion is given by
$$\begin{cases} \eta_1 = A_1 \cos(\overline{\omega}_1 t + \psi_1), \\ \eta_2 = A_2 \cos(\overline{\omega}_2 t + \psi_2). \end{cases} \quad (6.88)$$

Here the arbitrary constants A_1, A_2, ψ_1, ψ_2 are determined by means of the initial conditions. In the plane of normal coordinates, which is rotated by the angle φ with respect to the plane of the initial coordinates, the trajectory covers densely the rectangle delimited by $-A_1 \leq \eta_1 \leq A_1$ and $-A_2 \leq \eta_2 \leq A_2$ (see Fig.VI.13).

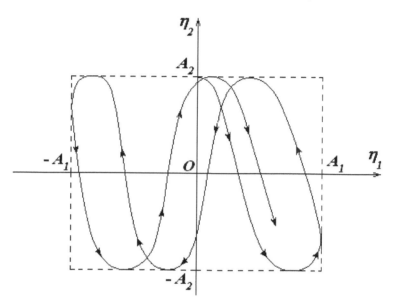

Fig.VI.13

If $\overline{\omega}_1$ and $\overline{\omega}_2$ are not commensurable, the trajectory is not closed, meaning that the motion is not periodical. Nevertheless, the projection of the representative point on both axes is periodical. But, if the normal frequencies are commensurable,
$$\frac{\overline{\omega}_1}{\overline{\omega}_2} = \frac{n_1}{n_2} \quad (n_1, n_2 \in \mathbb{N}),$$
the trajectories become closed and appear the so-called *Lissajous figures*. The motion is now periodical, with the period
$$\tau = 2\pi \frac{n_1}{\overline{\omega}_1} = 2\pi \frac{n_2}{\overline{\omega}_2}. \quad (6.89)$$

Under these conditions, $\xi_1(t)$ and $\xi_2(t)$ write:
$$\begin{cases} \xi_1(t) = A_1 \cos(\overline{\omega}_1 t + \psi_1) \cos\varphi - A_2 \cos(\overline{\omega}_2 t + \psi_2) \sin\varphi, \\ \xi_2(t) = A_1 \cos(\overline{\omega}_1 t + \psi_1) \sin\varphi - A_2 \cos(\overline{\omega}_2 t + \psi_2) \cos\varphi. \end{cases} \quad (6.90)$$

The trajectory in the plane $\xi_1 O \xi_2$ is the same, but the axis $(-A_1, A_1)$ of the rectangle are rotated by the angle φ with respect to $O\xi_1$ axis (see Fig.VI.14).

Let us now turn back to our problem and first discuss the case (a) $l_1 = l_2 = l$. The relations (6.75) and (6.76) then become:

$$\omega_{1,2}^2 = \frac{g}{l} + \left(\frac{h}{l}\right)^2 \frac{k}{m_{1,2}} \; ; \tag{6.75'}$$

$$\kappa = \left(\frac{h}{l}\right)^2 \frac{k}{\sqrt{m_1 m_2}}. \tag{6.76'}$$

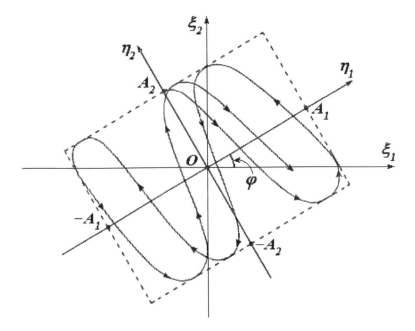

Fig.VI.14

Denoting $\omega_0^2 = \frac{g}{l}$ and $\omega_{e1,2}^2 = \frac{k}{m_{1,2}}$, we still have:

$$\omega_{1,2}^2 = \omega_0^2 + \left(\frac{h}{l}\right)^2 \omega_{e1,2}^2, \tag{6.75''}$$

$$\kappa = \left(\frac{h}{l}\right)^2 \omega_{e1}\omega_{e2}. \tag{6.76''}$$

With these notations, the normal frequencies are:

$$\overline{\omega}_1^2 = \frac{\omega_1^2 + \omega_2^2}{2} - \sqrt{\kappa^2 + \frac{(\omega_2^2 - \omega_1^2)^2}{4}}$$

$$= \frac{2\omega_0^2 + \left(\frac{h}{l}\right)^2(\omega_{e1}^2 + \omega_{e2}^2)}{2} - \frac{\left(\frac{h}{l}\right)^2(\omega_{e1}^2 + \omega_{e2}^2)}{2} = \omega_0^2,$$

$$\overline{\omega}_2^2 = \frac{\omega_1^2 + \omega_2^2}{2} + \sqrt{\kappa^2 + \frac{(\omega_2^2 - \omega_1^2)^2}{4}}$$

$$= \omega_0^2 + \left(\frac{h}{l}\right)^2(\omega_{e1}^2 + \omega_{e2}^2).$$

and (6.79) yields

$$\cot 2\varphi = \frac{\omega_2^2 - \omega_1^2}{2\kappa} = \frac{\left(\frac{h}{l}\right)^2(\omega_{e2}^2 - \omega_{e1}^2)}{2\left(\frac{h}{l}\right)^2 \omega_{e1}\omega_{e2}} = \frac{1}{2}\left(\frac{\omega_{e2}}{\omega_{e1}} - \frac{\omega_{e1}}{\omega_{e2}}\right).$$

In this case, the condition for weak coupling $\kappa \ll |\omega_2^2 - \omega_1^2|$ becomes

$$\left(\frac{h}{l}\right)^2 \omega_{e1}\omega_{e2} \ll \left|\left(\frac{h}{l}\right)^2(\omega_{e2}^2 - \omega_{e1}^2)\right|,$$

that is

$$\omega_{e1}\omega_{e2} \ll |\omega_{e2}^2 - \omega_{e1}^2|. \tag{6.91}$$

As we have seen while discussing the general case, there are two possible situations for the weak coupling:

i) $\omega_1 < \omega_2$ which, in our case, means $\omega_{e1} < \omega_{e2}$, that is $m_1 > m_2$, and (6.91) becomes

$$\frac{k}{\sqrt{m_1 m_2}} \ll \frac{k}{m_2} - \frac{k}{m_1},$$

or

$$m_1 \gg m_2\left(3 - \frac{m_2}{m_1}\right),$$

in agreement with our hypothesis $m_1 > m_2$. Consequently, if the coupling between the two pendulums is weak and $m_1 \gg m_2$, then according to the general analysis $\varphi \simeq 0$ and oscillations are localized.

ii) $\omega_1 > \omega_2$, meaning $\omega_{e1} > \omega_{e2}$, that is $m_1 < m_2$. Then (6.91) leads to

$$\frac{k}{\sqrt{m_1 m_2}} \ll \frac{k}{m_1} - \frac{k}{m_2},$$

or

$$m_2 \gg m_1\left(3 - \frac{m_1}{m_2}\right),$$

which emphasizes our condition $m_1 < m_2$. Therefore, if the coupling is weak and $m_1 \ll m_2$, the general theory says that $\varphi \simeq \frac{\pi}{2}$, and oscillations are also localized.

In both cases ($\varphi \simeq 0$ and $\varphi \simeq \pi/2$), the pendulum of mass m_1 oscillates with normal frequency $\overline{\omega}_1 = \omega_0$, while the other one, as the case may be, with normal frequency

$$\overline{\omega}_2 = \sqrt{\omega_0^2 + \left(\frac{h}{l}\right)^2 (\omega_{e1}^2 + \omega_{e2}^2)} \simeq \sqrt{\omega_0^2 + \left(\frac{h}{l}\omega_{e2}\right)^2}, \quad (m_1 \gg m_2),$$

↗
↘

$$\overline{\omega}_2 = \sqrt{\omega_0^2 + \left(\frac{h}{l}\right)^2 (\omega_{e1}^2 + \omega_{e2}^2)} \simeq \sqrt{\omega_0^2 + \left(\frac{h}{l}\omega_{e1}\right)^2}, \quad (m_1 \ll m_2).$$

According to our general analysis, we then have:

$$\theta_1 \simeq \overline{\theta}_1 = \overline{A}_1 \cos(\overline{\omega}_1 t + \psi_1) = \overline{A}_1 \cos(\omega_0 t + \psi_1), \tag{6.92}$$

and

$$\theta_2 \simeq \overline{\theta}_2 = \overline{A}_2 \cos(\overline{\omega}_2 t + \psi_2)$$

$$= \begin{cases} \overline{A}_2 \cos\left(\sqrt{\omega_0^2 + \left(\frac{h}{l}\omega_{e2}\right)^2}\, t + \psi_2\right), & (m_1 \gg m_2), \\ \\ \overline{A}_2 \cos\left(\sqrt{\omega_0^2 + \left(\frac{h}{l}\omega_{e1}\right)^2}\, t + \psi_2\right), & (m_1 \ll m_2). \end{cases} \tag{6.93}$$

If $m_1 = m_2 = m$, then $\omega_{e1} = \omega_{e2} = \omega_e$, and

$$\cot 2\varphi = \frac{\omega_2^2 - \omega_1^2}{2\kappa} = \frac{1}{2}\left(\frac{\omega_{e2}}{\omega_{e1}} - \frac{\omega_{e1}}{\omega_{e2}}\right) = 0,$$

that is $\varphi = \pi/4$. This means that oscillations are not localized and we have

$$\begin{cases} \theta_1 = A_1 \cos(\overline{\omega}_1 t + \psi_1) - A_2 \cos(\overline{\omega}_2 t + \psi_2), \\ \theta_2 = A_1 \cos(\overline{\omega}_1 t + \psi_1) + A_2 \cos(\overline{\omega}_2 t + \psi_2), \end{cases} \tag{6.94}$$

where $\overline{\omega}_1 = \omega_0$ and $\overline{\omega}_2 = \sqrt{\omega_0^2 + 2\left(\frac{h}{l}\omega_e\right)^2}$, with $\omega_e = \sqrt{\frac{k}{m}}$. Indeed, the two normal modes of oscillation are associated with the normal coordinates

$$\begin{cases} \overline{\theta}_1 = \overline{A}_1 \cos(\overline{\omega}_1 t + \psi_1), \\ \overline{\theta}_2 = \overline{A}_2 \cos(\overline{\omega}_2 t + \psi_2), \end{cases}$$

while θ_1 and θ_2 are given by [see (6.78)]

$$\begin{cases} \theta_1 = \overline{\theta}_1 \cos\varphi - \overline{\theta}_2 \sin\varphi, \\ \theta_2 = \overline{\theta}_1 \sin\varphi + \overline{\theta}_2 \cos\varphi, \end{cases}$$

with $\varphi = \pi/4$. This investigation leads to (6.94), where $A_1 = \frac{\overline{A}_1}{\sqrt{2}}$, and $A_2 = \frac{\overline{A}_2}{\sqrt{2}}$. The constants A_1, A_2, ψ_1 and ψ_2 are determined by means of the initial conditions.

Let us now investigate three situations that are particularly important in this case, as follows:

1) The pendulums are initially displaced on the same side of the vertical axis, $\theta_1(0) = \theta_2(0) = \theta_0$, and let them oscillate freely, $\dot{\theta}_1(0) = 0$, $\dot{\theta}_2(0) = 0$. Since

$$\begin{cases} \dot{\theta}_1 = -\overline{\omega}_1 A_1 \sin(\overline{\omega}_1 t + \psi_1) + \overline{\omega}_2 A_2 \sin(\overline{\omega}_2 t + \psi_2), \\ \dot{\theta}_2 = -\overline{\omega}_1 A_1 \sin(\overline{\omega}_1 t + \psi_1) - \overline{\omega}_2 A_2 \sin(\overline{\omega}_2 t + \psi_2), \end{cases}$$

we have

$$\begin{cases} \theta_0 = A_1 \cos\psi_1 - A_2 \cos\psi_2, \\ \theta_0 = A_1 \cos\psi_1 + A_2 \cos\psi_2, \\ 0 = -\overline{\omega}_1 A_1 \sin\psi_1 + \overline{\omega}_2 A_2 \sin\psi_2, \\ 0 = -\overline{\omega}_1 A_1 \sin\psi_1 - \overline{\omega}_2 A_2 \sin\psi_2. \end{cases}$$

Some very simple algebraic manipulation lead to: $\psi_1 = 0$, $\psi_2 = 0$, $A_1 = \theta_0$, $A_2 = 0$. With these values of the constants, (6.94) becomes

$$\begin{cases} \theta_1 = \theta_0 \cos(\overline{\omega}_1 t) = \theta_0 \cos(\omega_0 t), \\ \theta_2 = \theta_0 \cos(\overline{\omega}_1 t) = \theta_0 \cos(\omega_0 t), \end{cases} \quad (6.95)$$

meaning that the two pendulums oscillate in phase, with frequency ω_0.

2) The two pendulums are initially displaced on one side and the other of the vertical axis, $\theta_1(0) = -\theta_2(0) = \theta_0$, and let them oscillate freely, $\dot{\theta}_1(0) = 0$, $\dot{\theta}_2(0) = 0$. We then have:

$$\begin{cases} +\theta_0 = A_1 \cos\psi_1 - A_2 \cos\psi_2, \\ -\theta_0 = A_1 \cos\psi_1 + A_2 \cos\psi_2, \\ 0 = -\overline{\omega}_1 A_1 \sin\psi_1 + \overline{\omega}_2 A_2 \sin\psi_2, \\ 0 = -\overline{\omega}_1 A_1 \sin\psi_1 - \overline{\omega}_2 A_2 \sin\psi_2, \end{cases}$$

which yield $\psi_1 = 0$, $\psi_2 = 0$, $A_1 = 0$, $A_2 = -\theta_0$, and (6.94) take the form

$$\begin{cases} \theta_1 = \theta_0 \cos(\overline{\omega}_2 t) = \theta_0 \cos\left(\sqrt{\omega_0^2 + 2\left(\frac{h}{l}\omega_e\right)^2}\, t\right), \\ \theta_2 = -\theta_0 \cos(\overline{\omega}_2 t) = \theta_0 \cos(\overline{\omega}_2 t + \pi) \\ \quad = \theta_0 \cos\left(\sqrt{\omega_0^2 + 2\left(\frac{h}{l}\omega_e\right)^2}\, t + \pi\right). \end{cases} \quad (6.96)$$

This result shows that the pendulums oscillate with the same frequency, $\overline{\omega}_2 = \sqrt{\omega_0^2 + 2\left(\frac{h}{l}\omega_e\right)^2} = \sqrt{\frac{g}{l} + \frac{2k}{m}\left(\frac{h}{l}\right)^2}$, but in opposition of phase.

3) Only one pendulum is displaced (say, the one denoted by number 1), the other one keeping its equilibrium position, $\theta_1(0) = \theta_0$, $\theta_2(0) = 0$, and let them oscillate without initial velocity, $\dot{\theta}_1(0) = 0$, $\dot{\theta}_2(0) = 0$. Then, we have:

$$\begin{cases} \theta_0 = A_1 \cos \psi_1 - A_2 \cos \psi_2, \\ 0 = A_1 \cos \psi_1 + A_2 \cos \psi_2, \\ 0 = -\overline{\omega}_1 A_1 \sin \psi_1 + \overline{\omega}_2 A_2 \sin \psi_2, \\ 0 = -\overline{\omega}_1 A_1 \sin \psi_1 - \overline{\omega}_2 A_2 \sin \psi_2. \end{cases}$$

One easily finds $\psi_1 = 0$, $\psi_2 = 0$, $A_1 = -A_2 = \theta_o/2$, and (6.94) yield

$$\begin{cases} \theta_1 = \frac{\theta_0}{2}[\cos(\overline{\omega}_1 t) + \cos(\overline{\omega}_2 t)] \\ \quad = \frac{\theta_0}{2}\left[\cos(\omega_0 t) + \cos\left(\sqrt{\omega_0^2 + 2\left(\frac{h}{l}\omega_e\right)^2}\, t\right)\right], \\ \theta_2 = \frac{\theta_0}{2}[\cos(\overline{\omega}_1 t) - \cos(\overline{\omega}_2 t)] \\ \quad = \frac{\theta_0}{2}\left[\cos(\omega_0 t) - \cos\left(\sqrt{\omega_0^2 + 2\left(\frac{h}{l}\omega_e\right)^2}\, t\right)\right]. \end{cases} \quad (6.97)$$

By means of the well-known trigonometric relations

$$\cos x - \cos y = -2 \sin \frac{x-y}{2} \sin \frac{x+y}{2},$$

$$\cos x + \cos y = 2 \cos \frac{x-y}{2} \cos \frac{x+y}{2},$$

we still have

$$\begin{cases} \theta_1(t) = \theta_0 \cos\left(\omega_o - \sqrt{\omega_0^2 + 2\left(\omega_e \frac{h}{l}\right)^2}\right)\frac{t}{2} \\ \qquad \times \cos\left(\omega_o + \sqrt{\omega_0^2 + 2\left(\omega_e \frac{h}{l}\right)^2}\right)\frac{t}{2}, \\ \theta_2(t) = -\theta_0 \sin\left(\omega_o - \sqrt{\omega_0^2 + 2\left(\omega_e \frac{h}{l}\right)^2}\right)\frac{t}{2} \\ \qquad \times \sin\left(\omega_o + \sqrt{\omega_0^2 + 2\left(\omega_e \frac{h}{l}\right)^2}\right)\frac{t}{2}. \end{cases} \qquad (6.98)$$

Since $h < l$, we still have

$$\sqrt{\omega_0^2 + 2\left(\omega_e \frac{h}{l}\right)^2} = \omega_0 \sqrt{1 + 2\left(\frac{\omega_e}{\omega_0}\right)^2 \left(\frac{h}{l}\right)^2}$$

$$= \omega_0 \left[1 + \left(\frac{\omega_e}{\omega_0}\right)^2 \left(\frac{h}{l}\right)^2 + \mathcal{O}\left(\frac{\omega_e}{\omega_0}\frac{h}{l}\right)^4\right]_{\omega_e < \omega_0} \simeq \omega_0 \left[1 + \left(\frac{\omega_e}{\omega_0}\right)^2 \left(\frac{h}{l}\right)^2\right].$$

In other words, if $\omega_e < \omega_0$, which is equivalent to $k < \frac{mg}{l}$ (relatively weak coupling), the frequency $\bar{\omega}_2$ becomes

$$\bar{\omega}_2 = \sqrt{\omega_0^2 + 2\left(\omega_e \frac{h}{l}\right)^2} \simeq \omega_0 \left[1 + \left(\frac{\omega_e}{\omega_0}\right)^2 \left(\frac{h}{l}\right)^2\right] = \omega_0 + 2\gamma,$$

with $\gamma = \frac{\omega_e^2}{\omega_0}\left(\frac{h}{l}\right)^2 \ll \omega_0$. In this case, θ_1 and θ_2 are given by

$$\begin{cases} \theta_1(t) = \theta_0 \cos\gamma t \,\cos(\omega_0 + \gamma)t \simeq \theta_0 \cos\omega_0 t \,\cos\gamma t = \theta_1^{mod}(t) \cos\omega_0 t, \\ \theta_2(t) = \theta_0 \sin\gamma t \,\sin(\omega_0 + \gamma)t \simeq \theta_0 \sin\omega_0 t \,\sin\gamma t = \theta_2^{mod}(t) \sin\omega_0 t, \end{cases}$$
(6.99)

where $\theta_1^{mod}(t)$ and $\theta_2^{mod}(t)$ denote the amplitudes of the two modulated oscillations. As one can see, there appears the phenomenon of *beatings*, as shown in Fig.VI.15. Here, the following values have been arbitrary chosen: $\theta_0 = 1$, $\omega_0 = 12$, $\gamma = 1$.

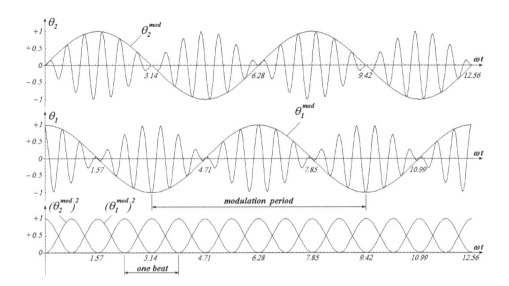

Fig.VI.15

b) $l_1 \neq l_2$ (e.g. $l_1 > l_2$). It is convenient for our investigation to find the *characteristic equation*. To this end, we use the equations (6.72) and (6.73), written as

$$\ddot{\theta}_1 + \left(\frac{g}{l_1} + \frac{kh^2}{m_1 l_1^2}\right)\theta_1 - \frac{kh^2}{m_1 l_1^2}\theta_2 = 0, \qquad (6.72')$$

$$\ddot{\theta}_2 + \left(\frac{g}{l_2} + \frac{kh^2}{m_2 l_2^2}\right)\theta_2 - \frac{kh^2}{m_2 l_2^2}\theta_1 = 0. \qquad (6.73')$$

We search for solutions of the form $\theta_i = A_i \sin \overline{\omega} t + B_i \cos \overline{\omega} t$ ($i = 1, 2$), in order to determine the frequencies $\overline{\omega}$ of the two normal modes of oscillations of the system. Since

$$\ddot{\theta}_i = -\overline{\omega}^2 A_i \sin \overline{\omega} t - \overline{\omega}^2 B_i \cos \overline{\omega} t \quad (i = 1, 2),$$

equations (6.72') and (6.73') write

$$-\overline{\omega}^2 A_1 \sin \overline{\omega} t - \overline{\omega}^2 B_1 \cos \overline{\omega} t + \left(\frac{g}{l_1} + \frac{kh^2}{m_1 l_1^2}\right)(A_1 \sin \overline{\omega} t + B_1 \cos \overline{\omega} t)$$

$$-\frac{kh^2}{m_1 l_1^2}(A_2 \sin \overline{\omega} t + B_2 \cos \overline{\omega} t) = 0,$$

$$-\overline{\omega}^2 A_2 \sin \overline{\omega} t - \overline{\omega}^2 B_2 \cos \overline{\omega} t + \left(\frac{g}{l_2} + \frac{kh^2}{m_2 l_2^2}\right)(A_2 \sin \overline{\omega} t + B_2 \cos \overline{\omega} t)$$

$$-\frac{kh^2}{m_2 l_2^2}(A_1 \sin\overline{\omega} t + B_1 \cos\overline{\omega} t) = 0.$$

Identifying the coefficients of $\sin\overline{\omega} t$ and $\cos\overline{\omega} t$, we have

$$\left[-\overline{\omega}^2 + \left(\frac{g}{l_1} + \frac{kh^2}{m_1 l_1^2}\right)\right] A_1 - \frac{kh^2}{m_1 l_1^2} A_2 = 0,$$

$$\left[-\overline{\omega}^2 + \left(\frac{g}{l_1} + \frac{kh^2}{m_1 l_1^2}\right)\right] B_1 - \frac{kh^2}{m_1 l_1^2} B_2 = 0,$$

and

$$\left[-\overline{\omega}^2 + \left(\frac{g}{l_2} + \frac{kh^2}{m_2 l_2^2}\right)\right] A_2 - \frac{kh^2}{m_2 l_2^2} A_1 = 0,$$

$$\left[-\overline{\omega}^2 + \left(\frac{g}{l_2} + \frac{kh^2}{m_2 l_2^2}\right)\right] B_2 - \frac{kh^2}{m_2 l_2^2} B_1 = 0.$$

The system of two algebraic equations in A_1 and A_2 (or B_1 and B_2) has non-trivial solution if the characteristic determinant is zero

$$\begin{vmatrix} -\overline{\omega}^2 + \left(\frac{g}{l_1} + \frac{kh^2}{m_1 l_1^2}\right) & -\frac{kh^2}{m_1 l_1^2} \\ -\frac{kh^2}{m_2 l_2^2} & -\overline{\omega}^2 + \left(\frac{g}{l_2} + \frac{kh^2}{m_2 l_2^2}\right) \end{vmatrix} = 0,$$

which yields the characteristic equation

$$\left[-\overline{\omega}^2 + \left(\frac{g}{l_1} + \frac{kh^2}{m_1 l_1^2}\right)\right]\left[-\overline{\omega}^2 + \left(\frac{g}{l_2} + \frac{kh^2}{m_2 l_2^2}\right)\right] - \frac{k^2 h^4}{m_1 m_2 l_1^2 l_2^2} = 0.$$
(6.100)

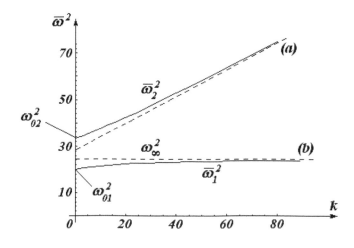

Fig.VI.16

In the plane $(k,\overline{\omega}^2)$ this represents a hyperbola whose branches corresponding to the values with physical signification (we must not forget that $k \in [0,\infty)$) are represented in Fig.VI.16. Here the following numerical values have been chosen: $l_1 = 0.5$, $l_2 = 0.3$, $m_1 = 0.2$, $m_2 = 0.3$, $h = 0.1$, $g = 10$.

By means of notations (6.74) and (6.75), Eq. (6.100) can be written as
$$(\overline{\omega}^2 - \omega_1^2)(\overline{\omega}^2 - \omega_2^2) - \kappa^2 = 0,$$
with its solutions (already determined by a different procedure, see (6.80) and (6.81)):
$$\overline{\omega}_1^2 = \frac{1}{2}(\omega_1^2 + \omega_2^2) - \sqrt{\kappa^2 + \frac{(\omega_2^2 - \omega_1^2)^2}{4}},$$
and
$$\overline{\omega}_2^2 = \frac{1}{2}(\omega_1^2 + \omega_2^2) + \sqrt{\kappa^2 + \frac{(\omega_2^2 - \omega_1^2)^2}{4}}.$$

For k very small ($k \to 0$; the spring is very weak), the normal frequencies $\overline{\omega}_1^2$ and $\overline{\omega}_2^2$ tend to the frequencies of free pendulums, $\omega_{01}^2 = g/l_1$, and $\omega_{02}^2 = g/l_2$. Indeed, for $k \to 0$, equation (6.100) becomes $(\overline{\omega}^2 - \omega_{01}^2)(\overline{\omega}^2 - \omega_{02}^2) = 0$, with obvious solutions $\overline{\omega}_1^2 = \omega_{01}^2$ and $\overline{\omega}_2^2 = \omega_{02}^2$.

For k very big ($k \to \infty$; the spring is very strong) the normal frequency $\overline{\omega}_2^2$ tends to infinity as

$$\left(\frac{h}{l_1}\right)^2 \omega_{e1}^2 + \left(\frac{h}{l_2}\right)^2 \omega_{e2}^2 \qquad (6.101)$$

(asymptote (a) of Fig.VI.16). Indeed, let us take

$$\alpha \equiv \lim_{k\to\infty} \frac{\overline{\omega}_2^2}{k} = \lim_{k\to\infty} \frac{\frac{1}{2}(\omega_1^2 + \omega_2^2) + \sqrt{\kappa^2 + (\omega_2^2 - \omega_1^2)^2/4}}{k}$$

$$= \lim_{k\to\infty} \frac{\omega_{01}^2 + \left(\frac{h}{l_1}\right)^2 \frac{k}{m_1} + \omega_{02}^2 + \left(\frac{h}{l_2}\right)^2 \frac{k}{m_2}}{2k}$$

$$+ \lim_{k\to\infty} \frac{\sqrt{\left(\frac{h^2}{l_1 l_2}\right)^2 \frac{k^2}{m_1 m_2} + \frac{1}{4}\left[\omega_{02}^2 + \left(\frac{h}{l_2}\right)^2 \frac{k}{m_2} - \omega_{01}^2 - \left(\frac{h}{l_1}\right)^2 \frac{k}{m_1}\right]^2}}{k}$$

$$= \left(\frac{h}{l_1}\right)^2 \frac{1}{2m_1} + \left(\frac{h}{l_2}\right)^2 \frac{1}{2m_2}$$

$$+ \sqrt{\left(\frac{h^2}{l_1 l_2}\right)^2 \frac{1}{m_1 m_2} + \left[\left(\frac{h}{l_2}\right)^2 \frac{1}{2m_2} - \left(\frac{h}{l_1}\right)^2 \frac{1}{2m_1}\right]^2}$$

$$= \left(\frac{h}{l_1}\right)^2 \frac{1}{2m_1} + \left(\frac{h}{l_2}\right)^2 \frac{1}{2m_2} + \sqrt{\left[\left(\frac{h}{l_2}\right)^2 \frac{1}{2m_2} + \left(\frac{h}{l_1}\right)^2 \frac{1}{2m_1}\right]^2}$$

$$= \left(\frac{h}{l_1}\right)^2 \frac{1}{m_1} + \left(\frac{h}{l_2}\right)^2 \frac{1}{m_2},$$

and we finally have

$$\overline{\omega}_{2(k\to\infty)}^2 = k\alpha = \left(\frac{h}{l_1}\right)^2 \frac{k}{m_1} + \left(\frac{h}{l_2}\right)^2 \frac{k}{m_2} = \left(\frac{h}{l_1}\right)^2 \omega_{e1}^2 + \left(\frac{h}{l_2}\right)^2 \omega_{e2}^2,$$

which is precisely the quantity (6.101).

Let us next calculate $\overline{\omega}_1^2$, subject to the same condition $k \to \infty$. Dividing (6.100) by k, then taking the limit, we successively have:

$$\frac{1}{k}\left\{\left[\overline{\omega}^2 - \frac{g}{l_1} - \left(\frac{h}{l_1}\right)^2 \frac{k}{m_1}\right]\left[\overline{\omega}^2 - \frac{g}{l_2} - \left(\frac{h}{l_2}\right)^2 \frac{k}{m_2}\right]\right.$$

$$\left. - \left(\frac{h^2}{l_1 l_2}\right)^2 \frac{k^2}{m_1 m_2}\right\} = 0,$$

or

$$\frac{1}{k}\left\{\overline{\omega}^4 - \overline{\omega}^2\left[\frac{g}{l_1} + \frac{g}{l_2} + \left(\frac{h}{l_1}\right)^2 \frac{k}{m_1} + \left(\frac{h}{l_2}\right)^2 \frac{k}{m_2}\right]\right.$$

$$+ \frac{g}{l_1}\left[\frac{g}{l_2} + \left(\frac{h}{l_2}\right)^2 \frac{k}{m_2}\right] + \frac{g}{l_2}\left[\frac{g}{l_1} + \left(\frac{h}{l_1}\right)^2 \frac{k}{m_1}\right]$$

$$\left. + \left(\frac{h^2}{l_1 l_2}\right) \frac{k^2}{m_1 m_2} - \left(\frac{h^2}{l_1 l_2}\right) \frac{k^2}{m_1 m_2}\right\} = 0,$$

or, still,

$$\frac{\overline{\omega}^4}{k} - \frac{\overline{\omega}^2}{k}\left(\frac{g}{l_1} + \frac{g}{l_2}\right) - \overline{\omega}^2\left[\left(\frac{h}{l_1}\right)^2 \frac{1}{m_1} + \left(\frac{h}{l_2}\right)^2 \frac{1}{m_2}\right]$$

$$+\frac{g}{l_1}\left(\frac{h}{l_2}\right)^2\frac{1}{m_2}+\frac{g}{l_2}\left(\frac{h}{l_1}\right)^2\frac{1}{m_1}+\frac{2}{k}\frac{g^2}{l_1 l_2}=0.$$

For $k \to \infty$, this becomes

$$-\overline{\omega}_\infty^2\left[\left(\frac{h}{l_1}\right)^2\frac{1}{m_1}+\left(\frac{h}{l_2}\right)^2\frac{1}{m_2}\right]+\frac{g}{l_1}\left(\frac{h}{l_2}\right)^2\frac{1}{m_2}+\frac{g}{l_2}\left(\frac{h}{l_1}\right)^2\frac{1}{m_1}=0. \tag{6.102}$$

which yields

$$\overline{\omega}_\infty^2 = \frac{\frac{g}{l_1}\left(\frac{h}{l_2}\right)^2\frac{1}{m_2}+\frac{g}{l_2}\left(\frac{h}{l_1}\right)^2\frac{1}{m_1}}{\left(\frac{h}{l_1}\right)^2\frac{1}{m_1}+\left(\frac{h}{l_2}\right)^2\frac{1}{m_2}} = g\frac{m_1 l_1 + m_2 l_2}{m_1 l_1^2 + m_2 l_2^2} \equiv \omega_\infty^2. \tag{6.103}$$

Therefore, $\overline{\omega}_1^2$ tends to ω_∞^2 given by (6.103) (asymptote (b) shown in Fig.VI.16), when k tends to infinity. This is the frequency of a pendulum with two masses m_1 and m_2, situated on the same ideal rod, at the distances l_1 and $l_2 < l_1$ with respect to the point of suspension O (see Fig.VI.17).

Let us now prove, by means of the analytical formalism, that formula (6.103) expresses the frequency of the motion of pendulum represented in Fig.VI.17. Since both bodies are on the same rod, the system has one degree of freedom, and let θ be the associated generalized coordinate. Since

$$x_i = l_i \sin\theta; \quad y_i = l_i \cos\theta \ (i=1,2),$$

and

$$\dot{x}_i = l_i \dot\theta \cos\theta; \quad \dot{y}_i = -l_i \dot\theta \sin\theta \ (i=1,2),$$

the kinetic energy writes

$$T = T_1 + T_2 = \frac{1}{2}m_1|\vec{v}_1|^2 + \frac{1}{2}m_2|\vec{v}_2|^2 = \frac{1}{2}\dot\theta^2(m_1 l_1^2 + m_2 l_2^2), \tag{6.104}$$

while the potential energy is obtained by the usual procedure

$$dV = -dA = -\vec{G}_1 \cdot d\vec{r}_1 - \vec{G}_2 \cdot d\vec{r}_2 = -m_1 g\, dy_1 - m_2 g\, dy_2.$$

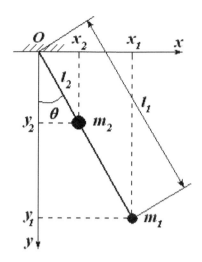

Fig.VI.17

Choosing as the reference level the plane xOz, meaning $y_1 = 0$, $y_2 = 0$, the constant of integration can be assumed to be zero. Then,

$$V = -m_1 g y_1 - m_2 g y_2 = -g \cos\theta (m_1 l_1 + m_2 l_2), \qquad (6.105)$$

and the Lagrangian writes

$$L = T - V = \frac{1}{2}\dot{\theta}^2 (m_1 l_1^2 + m_2 l_2^2) + g(m_1 l_1 + m_2 l_2)\cos\theta. \qquad (6.106)$$

The Lagrange equation of the second kind

$$\frac{d}{dt}\left(\frac{\partial L}{\partial \dot{\theta}}\right) - \frac{\partial L}{\partial \theta} = 0, \qquad (6.107)$$

then easily yields the differential equation of motion

$$\ddot{\theta}(m_1 l_1^2 + m_2 l_2^2) + g(m_1 l_1 + m_2 l_2)\sin\theta = 0. \qquad (6.108)$$

In the limit of small oscillations ($\sin\theta \simeq \theta$), if one denotes

$$\Omega = \sqrt{g \frac{m_1 l_1 + m_2 l_2}{m_1 l_1^2 + m_2 l_2^2}} \quad (\equiv \omega_\infty), \qquad (6.109)$$

then equation (6.108) takes the expected form

$$\ddot{\theta} + \Omega^2 \theta = 0. \qquad (6.110)$$

Consequently, our system behaves like a simple pendulum, the oscillation frequency being given by (6.109).

6. Problem of three identical coupled pendulums

Consider three identical simple pendulums, of mass m and length l, connected by two springs of negligible mass, situated at distance h with respect to the point of suspension (see Fig.VI.18). The springs are not tensioned at the state of equilibrium ($\theta_1 = \theta_2 = \theta_3 = 0$). Write the equations of motion of the system and find the solutions for the case of small oscillations.

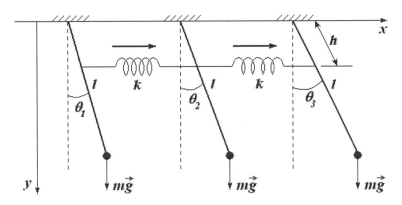

Fig.VI.18

Solution

Since both gravitational and elastic force fields are conservative, here we have again a natural system. Supposing that the motion takes place in the xOy plane, the system is subject to the following constraints:

i) $f_1(z_1) = z_1 = 0$,
ii) $f_2(x_1, y_1) = x_1^2 + y_1^2 - l^2 = 0$,
iii) $f_3(z_2) = z_2 = 0$,
iv) $f_4(x_2, y_2) = x_2^2 + y_2^2 - l^2 = 0$,
v) $f_5(z_3) = z_3 = 0$,
vi) $f_6(x_3, y_3) = x_3^2 + y_3^2 - l^2 = 0$,

where each pendulum is reported to its own reference frame, with the common x-axis. The system has $3 \cdot 3 - 6 = 3$ degrees of freedom, and let $\theta_1, \theta_2, \theta_3$ (see Fig.VI.18) be the associated generalized coordinates.

The kinetic energy is

$$T = T_1 + T_2 + T_3 = \frac{1}{2}m|\vec{v}_1|^2 + \frac{1}{2}m|\vec{v}_2|^2 + \frac{1}{2}m|\vec{v}_3|^2$$

$$= \frac{1}{2}m(\dot{x}_1^2 + \dot{y}_1^2) + \frac{1}{2}m(\dot{x}_2^2 + \dot{y}_2^2) + \frac{1}{2}m(\dot{x}_3^2 + \dot{y}_3^2).$$

Since $x_i = l\sin\theta_i$, $y_i = l\cos\theta_i$, $(i = \overline{1,3})$, we still have

$$T = \frac{1}{2}ml^2(\dot{\theta}_1^2 + \dot{\theta}_2^2 + \dot{\theta}_3^2).$$

In its turn, the potential energy has two components, of gravitational and elastic nature: $V = V_g + V_e$. The gravitational energy is obtained by the usual procedure

$$dV_g = -dA_g = -\vec{G}_1 \cdot d\vec{r}_1 - \vec{G}_2 \cdot d\vec{r}_2 - \vec{G}_3 \cdot d\vec{r}_3 = -mg(dy_1 + dy_2 + dy_3),$$

and, by integration

$$V_g = -mg(y_1 + y_2 + y_3) + V_{g0}.$$

A convenient choice of the reference frame

$$V_g(y_1 = 0, y_2 = 0, y_3 = 0) = 0,$$

allows us to take $V_{g0} = 0$. Then,

$$V = -mgl(\cos\theta_1 + \cos\theta_2 + \cos\theta_3).$$

In its turn, the elementary elastic energy

$$dV_e = -dA_e = -\vec{F}_e \cdot d\vec{r} = kx\,dx,$$

leads to

$$V_e = \frac{1}{2}kx^2 + V_{e0}.$$

Taking as the "reference level" for V_e the initial state when the deformation of the two strings is zero, $V_e(x = 0) = 0$, the integration constant V_{e0} can also be taken to be zero. Under these assumptions, the variables x_1 and x_2 stand for the deformations Δx_1 and Δx_2 of the springs, and we are left with

$$V_e = V_{e1} + V_{e2} = \frac{1}{2}kx_1^2 + \frac{1}{2}kx_2^2 = \frac{1}{2}k(\Delta x_1)^2 + \frac{1}{2}k(\Delta x_2)^2.$$

According to Fig.VI.18, we can write

$$\Delta x_1 = x_2 - x_1 = h(\sin\theta_2 - \sin\theta_1);$$

$$\Delta x_2 = x_3 - x_2 = h(\sin\theta_3 - \sin\theta_2).$$

Since

$$\sin a - \sin b = 2\sin\left(\frac{a-b}{2}\right)\cos\left(\frac{a+b}{2}\right),$$

we still have

$$V_e = \frac{1}{2}kh^2\left[4\sin^2\left(\frac{\theta_2-\theta_1}{2}\right)\cos^2\left(\frac{\theta_1+\theta_2}{2}\right)\right]$$

$$+\frac{1}{2}kh^2\left[4\sin^2\left(\frac{\theta_3-\theta_2}{2}\right)\cos^2\left(\frac{\theta_2+\theta_3}{2}\right)\right].$$

Using the hypothesis of small oscillations, we can write

$$\sin\left(\frac{\theta_2-\theta_1}{2}\right) = \frac{\theta_2-\theta_1}{2} + \mathcal{O}\left(\frac{\theta_2-\theta_1}{2}\right)^3 \simeq \frac{\theta_2-\theta_1}{2},$$

$$\cos\left(\frac{\theta_1+\theta_2}{2}\right) = 1 - \frac{1}{2}\left(\frac{\theta_1+\theta_2}{2}\right)^2 + \mathcal{O}\left(\frac{\theta_1+\theta_2}{2}\right)^4 \simeq 1,$$

and, similarly,

$$\sin\left(\frac{\theta_3-\theta_2}{2}\right) \simeq \frac{\theta_3-\theta_2}{2}; \quad \cos\left(\frac{\theta_2+\theta_3}{2}\right) \simeq 1.$$

Under these assumptions, V_e becomes

$$V_e = \frac{1}{2}kh^2\left[(\theta_2-\theta_1)^2 + (\theta_3-\theta_2)^2\right],$$

and the total potential energy writes

$$V = V_g + V_e = -mgl(\cos\theta_1 + \cos\theta_2 + \cos\theta_3)$$

$$+\frac{1}{2}kh^2\left[(\theta_2-\theta_1)^2 + (\theta_3-\theta_2)^2\right].$$

The Lagrangian of the system then is

$$L = T - V = \frac{1}{2}ml^2(\dot{\theta}_1^2 + \dot{\theta}_2^2 + \dot{\theta}_3^2) + mgl(\cos\theta_1 + \cos\theta_2 + \cos\theta_3)$$

$$-\frac{1}{2}kh^2\left[(\theta_2-\theta_1)^2+(\theta_3-\theta_2)^2\right],$$

and Lagrange equations of the second kind

$$\frac{d}{dt}\left(\frac{\partial L}{\partial \dot{\theta}_i}\right)-\frac{\partial L}{\partial \theta_i}=0 \quad (i=\overline{1,3})$$

yield

$$ml^2\ddot{\theta}_1+mgl\sin\theta_1-kh^2(\theta_2-\theta_1)=0,$$
$$ml^2\ddot{\theta}_2+mgl\sin\theta_2+2kh^2\theta_2-kh^2(\theta_1+\theta_3)=0,$$
$$ml^2\ddot{\theta}_3+mgl\sin\theta_3+kh^2(\theta_3-\theta_2)=0,$$

or, if the oscillations are presumably small ($\sin\theta_i \simeq \theta_i$, $i=\overline{1,3}$),

$$\ddot{\theta}_1+\left(\frac{g}{l}+\frac{kh^2}{ml^2}\right)\theta_1-\frac{kh^2}{ml^2}\theta_2=0, \tag{6.111}$$

$$\ddot{\theta}_2+\left(\frac{g}{l}+\frac{2kh^2}{ml^2}\right)\theta_2-\frac{kh^2}{ml^2}(\theta_1+\theta_3)=0, \tag{6.112}$$

$$\ddot{\theta}_3+\left(\frac{g}{l}+\frac{kh^2}{ml^2}\right)\theta_3-\frac{kh^2}{ml^2}\theta_2=0. \tag{6.113}$$

As usual in the case of small oscillations, we are looking for solutions of the form

$$\theta_i = A_i\sin\omega t + B_i\cos\omega t \quad (i=\overline{1,3}). \tag{6.114}$$

Introducing these solutions into (6.111)-(6.113), then identifying the coefficients of $\sin\omega t$ and $\cos\omega t$ in each equation, one obtains the following algebraic system for the coefficients A_i, B_i ($i=\overline{1,3}$):

$$\left(-\omega^2+\frac{g}{l}+\frac{kh^2}{ml^2}\right)A_1-\frac{kh^2}{ml^2}A_2=0, \tag{6.115}$$

$$\left(-\omega^2+\frac{g}{l}+\frac{kh^2}{ml^2}\right)B_1-\frac{kh^2}{ml^2}B_2=0, \tag{6.116}$$

$$-\frac{kh^2}{ml^2}A_1+\left(-\omega^2+\frac{g}{l}+\frac{2kh^2}{ml^2}\right)A_2-\frac{kh^2}{ml^2}A_3=0, \tag{6.117}$$

$$-\frac{kh^2}{ml^2}B_1+\left(-\omega^2+\frac{g}{l}+\frac{2kh^2}{ml^2}\right)B_2-\frac{kh^2}{ml^2}B_3=0, \tag{6.118}$$

$$-\frac{kh^2}{ml^2}A_2 + \left(-\omega^2 + \frac{g}{l} + \frac{kh^2}{ml^2}\right)A_3 = 0, \qquad (6.119)$$

$$-\frac{kh^2}{ml^2}B_2 + \left(-\omega^2 + \frac{g}{l} + \frac{kh^2}{ml^2}\right)B_3 = 0. \qquad (6.120)$$

To have at least one non-trivial solution, the system of equations (6.110), (6.117), and (6.119) must obey the condition

$$\begin{vmatrix} -\omega^2 + \frac{g}{l} + \frac{kh^2}{ml^2} & -\frac{kh^2}{ml^2} & 0 \\ -\frac{kh^2}{ml^2} & -\omega^2 + \frac{g}{l} + \frac{2kh^2}{ml^2} & -\frac{kh^2}{ml^2} \\ 0 & -\frac{kh^2}{ml^2} & -\omega^2 + \frac{g}{l} + \frac{kh^2}{ml^2} \end{vmatrix} = 0,$$

or

$$\left(-\omega^2 + \frac{g}{l} + \frac{kh^2}{ml^2}\right)\left[\left(-\omega^2 + \frac{g}{l} + \frac{2kh^2}{ml^2}\right)\left(-\omega^2 + \frac{g}{l} + \frac{kh^2}{ml^2}\right)\right.$$

$$\left. -2\left(\frac{kh^2}{ml^2}\right)^2\right] = 0.$$

This equation is satisfied if

$$-\omega^2 + \frac{g}{l} + \frac{kh^2}{ml^2} = 0$$

with the solution

$$\omega_I^2 = \frac{g}{l} + \frac{kh^2}{ml^2}, \qquad (6.121)$$

or if

$$\left(-\omega^2 + \frac{g}{l} + \frac{2kh^2}{ml^2}\right)\left(-\omega^2 + \frac{g}{l} + \frac{kh^2}{ml^2}\right) = 2\left(\frac{kh^2}{ml^2}\right)^2,$$

which can also be written as

$$\omega^4 - 2\omega^2\left(\frac{g}{l} + \frac{3kh^2}{2ml^2}\right)$$

$$+ \left(\frac{g}{l} + \frac{kh^2}{ml^2}\right)\left(\frac{g}{l} + \frac{2kh^2}{ml^2}\right) - 2\left(\frac{kh^2}{ml^2}\right)^2 = 0,$$

with the solutions

$$\omega_{II,III}^2 = \frac{g}{l} + \frac{3kh^2}{2ml^2}$$

$$\pm\sqrt{\left(\frac{g}{l}+\frac{3kh^2}{2ml^2}\right)^2 - \left(\frac{g}{l}+\frac{kh^2}{ml^2}\right)\left(\frac{g}{l}+\frac{2kh^2}{ml^2}\right) + 2\left(\frac{kh^2}{ml^2}\right)^2}$$

$$= \frac{g}{l} + \frac{3kh^2}{2ml^2} \pm \frac{3kh^2}{2ml^2},$$

that is

$$\omega_{II}^2 \equiv \omega_M^2 = \frac{g}{l} + \frac{3kh^2}{ml^2}, \tag{6.122}$$

and

$$\omega_{III}^2 \equiv \omega_m^2 = \frac{g}{l}. \tag{6.123}$$

The reader is advised to prove by himself that the remaining equations (6.116), (6,118), and (6.120) in the unknowns B_1, B_2, and B_3 lead to the same frequencies.

Changing the solution notations, the three normal frequencies of the system of three coupled pendulums are

$$\omega_1 = \sqrt{\frac{g}{l}}, \quad \omega_2 = \sqrt{\frac{g}{l} + \frac{kh^2}{ml^2}}, \quad \omega_3 = \sqrt{\frac{g}{l} + \frac{3kh^2}{ml^2}}. \tag{6.124}$$

Introducing ω_1^2 into (6.115), we obtain

$$A_2 = A_1. \tag{6.125}$$

Using the same procedure in (6.116), we arrive at

$$B_2 = B_1. \tag{6.126}$$

By means of (6.125) and (6.117), with ω_1^2 given by (6.124), we arrive at

$$A_3 = A_2 = A_1. \tag{6.127}$$

In the same way, making use of (6.118), (6.124), and (6.126), we have

$$B_3 = B_2 = B_1. \tag{6.128}$$

Resuming the same operation, but this time with ω_2^2 given by (6.124), and introducing it into (6.115), we obtain

$$A_2 = 0. \tag{6.129}$$

Next, introduce ω_2^2 into (6.116) and get

$$B_2 = 0, \tag{6.130}$$

while (6.117), in view of (6.129), gives

$$A_3 = -A_1, \tag{6.131}$$

and (6.118), by means of (6.130), yields

$$B_3 = -B_1. \tag{6.132}$$

Finally, using ω_3^2 displayed by (6.124), and the relations (6.115), (6.116), and (6.117) as well, we successively obtain

$$A_2 = -2A_1; \tag{6.133}$$

$$B_2 = -2B_1; \tag{6.134}$$

$$A_3 = A_1. \tag{6.135}$$

Using (6.134) and (6.118), one also obtains

$$B_3 = B_1. \tag{6.136}$$

The remaining relations (6.119) and (6.120) are identically satisfied by all three solutions, and we leave this up to the reader.

Since the system has three oscillation frequencies, the solutions of the differential equations of motion of the three coupled pendulums are written as a superposition of the three modes of oscillations. In view if (6.114) and (6.125)-(6.136), we then have:

$$\begin{cases} \theta_1 = A\sin\omega_1 t + B\cos\omega_1 t + C\sin\omega_2 t + D\cos\omega_2 t \\ \quad + E\sin\omega_3 t + F\cos\omega_3 t, \\ \theta_2 = A\sin\omega_1 t + B\cos\omega_1 t - 2E\sin\omega_3 t - 2F\cos\omega_3 t, \\ \theta_3 = \theta_1 = A\sin\omega_1 t + B\cos\omega_1 t - C\sin\omega_2 t - D\cos\omega_2 t \\ \quad + E\sin\omega_3 t + F\cos\omega_3 t, \end{cases} \tag{6.137}$$

or

$$\begin{cases} \theta_1 = K_1\sin(\omega_1 t + \alpha_1) + K_2\sin(\omega_2 t + \alpha_2) \\ \quad + K_3\sin(\omega_3 t + \alpha_3), \\ \theta_2 = K_1\sin(\omega_1 t + \alpha_1) - 2K_3\sin(\omega_3 t + \alpha_3), \\ \theta_3 = K_1\sin(\omega_1 t + \alpha_1) - K_2\sin(\omega_2 t + \alpha_2) \\ \quad + K_3\sin(\omega_3 t + \alpha_3), \end{cases} \tag{6.138}$$

or, still, in a matrix form

$$\begin{pmatrix} \theta_1 \\ \theta_2 \\ \theta_3 \end{pmatrix} = K_1 \begin{pmatrix} +1 \\ +1 \\ +1 \end{pmatrix} \sin\left(\sqrt{\frac{g}{l}}t + \alpha_1\right)$$

$$+K_2 \begin{pmatrix} +1 \\ 0 \\ -1 \end{pmatrix} \sin\left(\sqrt{\frac{g}{l} + \frac{kh^2}{ml^2}}\, t + \alpha_2\right) \qquad (6.139)$$

$$+K_3 \begin{pmatrix} +1 \\ -2 \\ +1 \end{pmatrix} \sin\left(\sqrt{\frac{g}{l} + \frac{3kh^2}{ml^2}}\, t + \alpha_3\right),$$

where the integration constants A, B, C, D, E, F as well as K_i, α_i, ($i = \overline{1,3}$) are determined by means of the initial conditions:

$$\theta_i(t=0) = \theta_{0i}, \quad \dot{\theta}_i(t=0) = \dot{\theta}_{0i} \quad (i = \overline{1,3}). \qquad (6.140)$$

7. Problem of double gravitational pendulum

Study the motion of a double gravitational pendulum composed by two bodies (particles) of masses m_1, m_2, and lengths of the rods l_1, l_2 (Fig.VI.19). The rods are supposed to be massless and the connections frictionless.

Solution

Let us first identify the constraints acting on the system. Since the motion takes place in a plane, say xOy, we can write:
  i) $f_1(z_1) = z_1 = 0$,
  ii) $f_2(z_2) = z_2 = 0$,
  iii) $f_3(x_1, y_1) = x_1^2 + y_1^2 - l_1^2 = 0$,
  iv) $f_4(x_1, x_2, y_1, y_2) = (x_2 - x_1)^2 + (y_2 - y_1)^2 - l_2^2 = 0$.
Therefore, the system is submitted to four holonomic (bilateral, scleronomous, and finite) constraints, and has $3 \cdot 2 - 4 = 2$ degrees of freedom. Suppose that θ_1 and θ_2 are the generalized coordinates associated with the two degrees of freedom (see Fig.VI.19). Since the only applied forces are the gravitational forces, this is a natural system. To write the Lagrangian, we need to determine its potential and kinetic energies.

Using the same procedure as in the case of the simple pendulum, we can write

$$dV = -dA = -\vec{G}_1 \cdot d\vec{r}_1 - \vec{G}_2 \cdot d\vec{r}_2 = -m_1 g\, dx_1 - m_2 g\, dx_2,$$

so that
$$V = -g(m_1 x_1 + m_2 x_2) + V_0.$$

The integration constant V_0 can be conveniently chosen by defining the reference level for the potential energy. In our case, if this level corresponds to the plane yOz, we may take $V_0 = 0$. Indeed,

$$V(x_1 = 0, x_2 = 0) = 0 \Rightarrow V_0 = 0.$$

Since $x_1 = l_1 \cos\theta_1$, $x_2 = l_1 \cos\theta_1 + l_2 \cos\theta_2$, we have

$$V = -g(m_1 x_1 + m_2 x_2) = -gl_1(m_1+m_2)\cos\theta_1 - m_2 g l_2 \cos\theta_2. \quad (6.141)$$

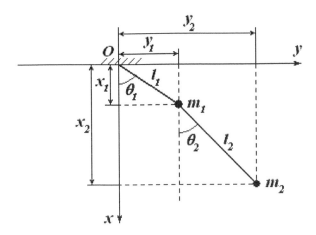

Fig.VI.19

We also observe that

$$x_1 = l_1 \cos\theta_1, \ y_1 = l_1 \sin\theta_1,$$

$$x_2 = l_1 \cos\theta_1 + l_2 \cos\theta_2, \ y_2 = l_1 \sin\theta_1 + l_2 \sin\theta_2,$$

which means

$$\dot{x}_1 = -l_1 \dot\theta_1 \sin\theta_1, \ \dot{y}_1 = l_1 \dot\theta_1 \cos\theta_1,$$

$$\dot{x}_2 = -l_1 \dot\theta_1 \sin\theta_1 - l_2 \dot\theta_2 \sin\theta_2, \ \dot{y}_2 = l_1 \dot\theta_1 \cos\theta_1 + l_2 \dot\theta_2 \cos\theta_2,$$

and the kinetic energy writes

$$T = T_1 + T_2 = \frac{1}{2} m_1 |\vec{v}_1|^2 + \frac{1}{2} m_2 |\vec{v}_2|^2$$

$$= \frac{1}{2}\left[m_1(\dot{x}_1^2 + \dot{y}_1^2) + m_2(\dot{x}_2^2 + \dot{y}_2^2)\right]$$

$$= \frac{1}{2} l_1^2 \dot\theta_1^2 (m_1 + m_2) + \frac{1}{2} m_2 l_2^2 \dot\theta_2^2 + m_2 l_1 l_2 \dot\theta_1 \dot\theta_2 \cos(\theta_2 - \theta_1). \quad (6.142)$$

The Lagrangian therefore is:

$$L = T - V = \frac{1}{2}l_1^2\dot\theta_1^2(m_1+m_2) + \frac{1}{2}m_2 l_2^2\dot\theta_2^2 + m_2 l_1 l_2 \dot\theta_1\dot\theta_2 \cos(\theta_2-\theta_1)$$

$$+ gl_1(m_1+m_2)\cos\theta_1 + m_2 g l_2 \cos\theta_2. \qquad (6.143)$$

As one can see, the Lagrangian does not explicitly depend on time. This means that the Lagrange equations of the second kind admit the energy first integral

$$E_{tot} = E_{cin} + E_{pot} = \dot\theta_1 \frac{\partial L}{\partial \dot\theta_1} + \dot\theta_2 \frac{\partial L}{\partial \dot\theta_2} - L = const. \qquad (6.144)$$

The Lagrange equations of the second kind, associated with the generalized coordinates θ_1 and θ_2, are

$$\frac{d}{dt}\left(\frac{\partial L}{\partial \dot\theta_i}\right) - \frac{\partial L}{\partial \theta_i} = 0 \quad (i=1,2). \qquad (6.145)$$

Since

$$\frac{\partial L}{\partial \dot\theta_1} = l_1^2 \dot\theta_1 (m_1+m_2) + m_2 l_1 l_2 \dot\theta_2 \cos(\theta_2-\theta_1);$$

$$\frac{\partial L}{\partial \dot\theta_2} = m_2 l_2^2 \dot\theta_2 + m_2 l_1 l_2 \dot\theta_1 \cos(\theta_2-\theta_1);$$

$$\frac{d}{dt}\left(\frac{\partial L}{\partial \dot\theta_1}\right) = l_1^2 \ddot\theta_1 (m_1+m_2) + m_2 l_1 l_2 \ddot\theta_2 \cos(\theta_2-\theta_1)$$

$$- m_2 l_1 l_2 \dot\theta_2^2 \sin(\theta_2-\theta_1) + m_2 l_1 l_2 \dot\theta_1 \dot\theta_2 \sin(\theta_2-\theta_1);$$

$$\frac{d}{dt}\left(\frac{\partial L}{\partial \dot\theta_2}\right) = m_2 l_2^2 \ddot\theta_2 + m_2 l_1 l_2 \ddot\theta_1 \cos(\theta_2-\theta_1)$$

$$+ m_2 l_1 l_2 \dot\theta_1^2 \sin(\theta_2-\theta_1) - m_2 l_1 l_2 \dot\theta_1 \dot\theta_2 \sin(\theta_2-\theta_1);$$

$$\frac{\partial L}{\partial \theta_1} = m_2 l_1 l_2 \dot\theta_1 \dot\theta_2 \sin(\theta_2-\theta_1) - (m_1+m_2) g l_1 \sin\theta_1;$$

$$\frac{\partial L}{\partial \theta_2} = -m_2 l_1 l_2 \dot\theta_1 \dot\theta_2 \sin(\theta_2-\theta_1) - m_2 g l_2 \sin\theta_2,$$

the Lagrange equations (6.145) write

$$l_1^2 \ddot\theta_1(m_1+m_2) + m_2 l_1 l_2 \ddot\theta_2 \cos(\theta_2-\theta_1) - m_2 l_1 l_2 \dot\theta_2^2 \sin(\theta_2-\theta_1)$$

$$+ (m_1+m_2) g l_1 \sin\theta_1 = 0, \qquad (6.146)$$

and
$$m_2 l_2^2 \ddot{\theta}_2 + m_2 l_1 l_2 \ddot{\theta}_1 \cos(\theta_2 - \theta_1)$$
$$+ m_2 l_1 l_2 \dot{\theta}_1^2 \sin(\theta_2 - \theta_1) + m_2 g l_2 \sin \theta_2 = 0. \qquad (6.146')$$

This way, we are left with the differential equations of motion of the system. Since their general analytical solution is not possible, one usually apply to the most important limit cases. Most frequently is studied the case of small oscillations, $\cos(\theta_2 - \theta_1) \simeq 1$ and $\cos \theta_i \simeq 1 - \frac{\theta_i^2}{2}$, $(i = 1, 2)$. In this case, the Lagrangian (6.143) becomes

$$L = \frac{1}{2} l_1^2 \dot{\theta}_1^2 (m_1 + m_2) + \frac{1}{2} m_2 l_2^2 \dot{\theta}_2^2 + m_2 l_1 l_2 \dot{\theta}_1 \dot{\theta}_2$$
$$- \frac{1}{2} g l_1 (m_1 + m_2) \theta_1^2 - \frac{1}{2} m_2 g l_2 \theta_2^2, \qquad (6.147)$$

where the constant terms $g l_1 (m_1 + m_2)$ and $g m_2 l_2$, in agreement with the general analytical formalism, have been omitted.

In the following approach we shall restrict out investigation to the particular case $l_1 = l_2 = l$. Let us define as coordinates of small oscillations the quantities $\xi_1 = \sqrt{m_1} l \theta_1$ and $\xi_2 = \sqrt{m_2} l (\theta_1 + \theta_2)$. The Lagrangian then writes

$$L = \frac{1}{2}(\dot{\xi}_1^2 + \dot{\xi}_2^2) - \frac{1}{2} \frac{g}{l} \left[\left(1 + 2\frac{m_2}{m_1}\right) \xi_1^2 + \xi_2^2 \right] + \sqrt{\frac{m_2}{m_1}} \frac{g}{l} \xi_1 \xi_2. \qquad (6.148)$$

This is precisely the Lagrangian (6.74) used in Problem No.5, with

$$\omega_1^2 = \frac{g}{l} \left(1 + 2\frac{m_2}{m_1}\right), \quad \omega_2^2 = \frac{g}{l}, \qquad (6.149)$$

and

$$\kappa = \frac{g}{l} \sqrt{\frac{m_2}{m_1}}. \qquad (6.150)$$

The frequencies of the normal modes are obtained by means of the formulas (6.80) and (6.81) obtained in the same Problem No.5. The results are:

$$\bar{\omega}_1^2 = \frac{1}{2}(\omega_1^2 + \omega_2^2) - \sqrt{\kappa^2 + (\omega_2^2 - \omega_1^2)/4} = \frac{g}{l}\left[1 + \frac{m_2}{m_1}\left(1 - \sqrt{1 + \frac{m_1}{m_2}}\right)\right];$$
$$(6.151)$$

$$\overline{\omega}_2^2 = \frac{1}{2}(\omega_1^2+\omega_2^2)+\sqrt{\kappa^2+(\omega_2^2-\omega_1^2)/4} = \frac{g}{l}\left[1+\frac{m_2}{m_1}\left(1+\sqrt{1+\frac{m_1}{m_2}}\right)\right].$$
(6.152)

To the general analysis with respect to ω_1 and κ (with ω_2 fixed) developed in Problem No.5, now corresponds the behavior of the ratio $\frac{m_2}{m_1}$. Here we shall consider only the most interesting limit cases: (1) $\frac{m_2}{m_1} \ll 1$, (2) $\frac{m_2}{m_1} \gg 1$, and (3) $m_1 = m_2$. What we meant by "weak coupling" (localized oscillations) in Problem No.5, $\kappa \ll |\omega_2^2-\omega_1^2|$, is now represented by

$$\frac{g}{l}\sqrt{\frac{m_2}{m_1}} \ll \left|2\frac{g}{l}\frac{m_2}{m_1}\right| \iff \sqrt{\frac{m_2}{m_1}} \gg 1 \iff \frac{m_2}{m_1} \gg 1,$$

while the condition of "strong coupling" (non-localized oscillations), $\kappa \gg |\omega_2^2-\omega_1^2|$, now becomes

$$\frac{g}{l}\sqrt{\frac{m_2}{m_1}} \gg \left|2\frac{g}{l}\frac{m_2}{m_1}\right| \iff \sqrt{\frac{m_2}{m_1}} \ll 1 \iff \frac{m_2}{m_1} \ll 1.$$

Let us now analyze the above mentioned three cases.

(1) If $\frac{m_2}{m_1} \ll 1$, oscillations are not localized, and according to the general analysis developed in Problem No.5, we have

$$\xi_1 \simeq \frac{\eta_1-\eta_2}{\sqrt{2}}, \quad \xi_2 \simeq \frac{\eta_1+\eta_2}{\sqrt{2}}, \qquad (6.153)$$

where η_i ($i=1,2$) are normal coordinates, so that we may write

$$\eta_i = A_i\cos(\overline{\omega}_i t+\psi_i) \quad (i=1,2), \qquad (6.154)$$

where the constants A_i, ψ_i ($i=1,2$) are determined by means of initial conditions. Indeed, for $\cot 2\varphi = \frac{\omega_2^2-\omega_1^2}{2\kappa}$ we have

$$\cot 2\varphi = \frac{-2\frac{g}{l}\frac{m_2}{m_1}}{2\frac{g}{l}\sqrt{\frac{m_2}{m_1}}} = -\sqrt{\frac{m_2}{m_1}} \quad (\to 0 \text{ through negative values})$$

meaning that $\varphi \to \frac{\pi}{4}$ through values greater than $\frac{\pi}{4}$, and the relations (6.78) with actual notations

$$\begin{cases} \xi_1 = \eta_1\cos\varphi - \eta_2\sin\varphi, \\ \xi_2 = \eta_1\sin\varphi + \eta_2\cos\varphi, \end{cases}$$

easily lead to (6.153). Up to terms of the first order in $\sqrt{m_2/m_1}$, the quantities $\overline{\omega}_i$ ($i=1,2$) interfering in (6.154) are given by

$$\overline{\omega}_{1,2} = \omega_0 \left(1 \mp \frac{1}{2}\sqrt{\frac{m_2}{m_1}}\right) = \omega_0 \mp \gamma, \qquad (6.155)$$

with $\gamma = \omega_0 \frac{1}{2}\sqrt{\frac{m_2}{m_1}} \ll \omega_0$, and $\omega_0^2 = \frac{g}{l}$. Indeed, denoting $z = \sqrt{\frac{m_2}{m_1}}$, the frequency $\overline{\omega}_1^2$ defined by (6.151) writes

$$\overline{\omega}_1 = \omega_0 \sqrt{1 + z^2 \left(1 - \sqrt{1 + \frac{1}{z^2}}\right)}.$$

Using approximation $\sqrt{1 \pm \zeta} \simeq 1 \pm \frac{\zeta}{2}$, we still have

$$\sqrt{1 + z^2 \left(1 - \sqrt{1 + \frac{1}{z^2}}\right)} = \sqrt{1 + z^2 - z\sqrt{z^2 + 1}}$$

$$\simeq \sqrt{1 + z^2 - z\left(1 + \frac{z^2}{2}\right)} \simeq \sqrt{1 - z} \simeq 1 - \frac{z}{2},$$

so that

$$\overline{\omega}_1 \simeq \omega_0 \left(1 - \frac{z}{2}\right) = \omega_0 \left(1 - \frac{1}{2}\sqrt{\frac{m_2}{m_1}}\right). \qquad (6.156)$$

Proceeding in the same manner, we also obtain

$$\overline{\omega}_2 \simeq \omega_0 \left(1 + \frac{z}{2}\right) = \omega_0 \left(1 + \frac{1}{2}\sqrt{\frac{m_2}{m_1}}\right). \qquad (6.157)$$

Suppose that initially the pendulums are at rest, $\dot{\theta}_1(t_0 = 0) = 0$, $\dot{\theta}_2(t_0 = 0) = 0$, and only the pendulum of mass m_2 is displaced from equilibrium, $\theta_1(t_0 = 0) = 0$, $\theta_2(t_0 = 0) = \theta_0$. Then

$$\theta_1 = \frac{\xi_1}{l\sqrt{m_1}} \simeq \frac{\eta_1 - \eta_2}{l\sqrt{2m_1}}$$

$$= \frac{1}{l\sqrt{2m_1}}\left[A_1 \cos(\overline{\omega}_1 t + \psi_1) - A_2 \cos(\overline{\omega}_2 t + \psi_2)\right], \qquad (6.158)$$

$$\theta_2 = \frac{\xi_2}{l\sqrt{m_2}} - \frac{\xi_1}{l\sqrt{m_1}} \simeq \frac{\eta_1 + \eta_2}{l\sqrt{2m_2}} - \frac{\eta_1 - \eta_2}{l\sqrt{2m_1}}$$

$$= \left(\frac{1}{l\sqrt{2m_2}} - \frac{1}{l\sqrt{2m_1}}\right) A_1 \cos(\overline{\omega}_1 t + \psi_1) \quad (6.159)$$

$$+ \left(\frac{1}{l\sqrt{2m_2}} + \frac{1}{l\sqrt{2m_1}}\right) A_2 \cos(\overline{\omega}_2 t + \psi_2),$$

where the arbitrary constants A_1, A_2, ψ_1, and ψ_2 are determined by means of the initial conditions. Since

$$\dot{\theta}_1 = \frac{1}{l\sqrt{2m_1}} \left[\overline{\omega}_2 A_2 \sin(\overline{\omega}_2 t + \psi_2) - \overline{\omega}_1 A_1 \sin(\overline{\omega}_1 t + \psi_1)\right],$$

$$\dot{\theta}_2 = \left(\frac{\overline{\omega}_1}{l\sqrt{2m_1}} - \frac{\overline{\omega}_1}{l\sqrt{2m_2}}\right) A_1 \sin(\overline{\omega}_1 t + \psi_1)$$

$$- \left(\frac{\overline{\omega}_2}{l\sqrt{2m_1}} + \frac{\overline{\omega}_2}{l\sqrt{2m_2}}\right) A_2 \sin(\overline{\omega}_2 t + \psi_2),$$

the use of initial conditions leads to the following system of four equations in the above mentioned four constants:

$$\begin{cases} 0 = \frac{1}{l\sqrt{2m_1}}(A_1 \cos\psi_1 - A_2 \cos\psi_2), \\ \theta_0 = \left(\frac{1}{l\sqrt{2m_2}} - \frac{1}{l\sqrt{2m_1}}\right) A_1 \cos\psi_1 + \left(\frac{1}{l\sqrt{2m_2}} + \frac{1}{l\sqrt{2m_1}}\right) A_2 \cos\psi_2, \\ 0 = \frac{1}{l\sqrt{2m_1}}(\overline{\omega}_2 A_2 \sin\psi_2 - \overline{\omega}_1 A_1 \sin\psi_1), \\ 0 = \left(\frac{\overline{\omega}_1}{l\sqrt{2m_1}} - \frac{\overline{\omega}_1}{l\sqrt{2m_2}}\right) A_1 \sin\psi_1 - \left(\frac{\overline{\omega}_2}{l\sqrt{2m_2}} + \frac{\overline{\omega}_2}{l\sqrt{2m_1}}\right) A_2 \sin\psi_2. \end{cases}$$

The third equation yields $\overline{\omega}_1 A_1 \sin\psi_1 = \overline{\omega}_2 A_2 \sin\psi_2$, and, consequently, the fourth equation leads to $\psi_1 = 0, \psi_2 = 0$, so that the first gives $A_1 = A_2$. Finally, the second equation says that $A_1 = A_2 = \frac{\theta_0 l \sqrt{m_2}}{\sqrt{2}}$. Then we are left with

$$\theta_1 = \frac{\xi_1}{l\sqrt{m_1}} \simeq \frac{\eta_1 - \eta_2}{l\sqrt{2m_1}} = \frac{\theta_0}{2}\sqrt{\frac{m_2}{m_1}}(\cos\overline{\omega}_1 t - \cos\overline{\omega}_2 t);$$

$$\theta_2 = \left(\frac{\theta_0}{2} - \frac{\theta_0}{2}\sqrt{\frac{m_2}{m_1}}\right)\cos\overline{\omega}_1 t + \left(\frac{\theta_0}{2} + \frac{\theta_0}{2}\sqrt{\frac{m_2}{m_1}}\right)\cos\overline{\omega}_2 t$$

$$= \frac{\theta_0}{2}(\cos\overline{\omega}_1 t + \cos\overline{\omega}_2 t) + \frac{\theta_0}{2}\sqrt{\frac{m_2}{m_1}}(\cos\overline{\omega}_2 t - \cos\overline{\omega}_1 t).$$

In view of the trigonometric identities

$$\begin{aligned} \cos x + \cos y &= 2\cos\tfrac{x-y}{2}\cos\tfrac{x+y}{2}, \\ \cos x - \cos y &= -2\sin\tfrac{x-y}{2}\cos\tfrac{x+y}{2}, \end{aligned} \quad (6.160)$$

the angular coordinates θ_1 and θ_2 become:

$$\theta_1 = \theta_0 \sqrt{\frac{m_2}{m_1}} \sin \frac{(\overline{\omega}_2 - \overline{\omega}_1)t}{2} \sin \frac{(\overline{\omega}_2 + \overline{\omega}_1)t}{2}$$

$$= \theta_0 \sqrt{\frac{m_2}{m_1}} \sin\left(\frac{\omega_0 t}{2} \sqrt{\frac{m_2}{m_1}}\right) \sin \omega_0 t = \theta_0 \sqrt{\frac{m_2}{m_1}} \sin \omega_0 t \sin \gamma t,$$

$$\theta_2 = \theta_0 \cos \omega_0 t \cos \gamma t + \theta_0 \sqrt{\frac{m_2}{m_1}} \cos \omega_0 t \cos \gamma t$$

$$= \theta_0 \left(1 + \sqrt{\frac{m_2}{m_1}}\right) \cos \omega_0 t \cos \gamma t \simeq \theta_0 \cos \omega_0 t \cos \gamma t.$$

To conclude, the obtained solution

$$\begin{cases} \theta_1 = \theta_0 \sqrt{\frac{m_2}{m_1}} \sin \omega_0 t \sin \gamma t, \\ \theta_2 = \theta_0 \cos \omega_0 t \cos \gamma t, \end{cases} \tag{6.161}$$

shows that the pendulums oscillate "alternatively": each of them reaches its amplitude, when the other one is at equilibrium. At the same time, one can see that the amplitude of the pendulum of mass m_1 is $\sqrt{m_1/m_2}$ times smaller than the amplitude of the pendulum of mass m_2. This result can also be interpreted as being of the "beatings" type: we have oscillations with frequency ω_0 and amplitude $\theta_1^{mod}(t) = \theta_0 \sqrt{\frac{m_2}{m_1}} \sin \gamma t$, on the one hand, and oscillations with amplitude $\theta_2^{mod}(t) = \theta_0 \cos \gamma t$, on the other. As seen, the change with time of the first amplitude is much slower than of the second one.

(2) If $\frac{m_2}{m_1} \gg 1$, the oscillations are localized and, since $\omega_1 > \omega_2$, according to the general analysis developed in Problem No.5, we have $\varphi \simeq \frac{\pi}{2}$, so that

$$\xi_1 \simeq -\eta_2, \quad \xi_2 \simeq \eta_1.$$

Indeed, we have:

$$\cot 2\varphi = \frac{\omega_2^2 - \omega_1^2}{2\kappa} = \frac{-2\frac{g}{l}\frac{m_2}{m_1}}{2\frac{g}{l}\sqrt{\frac{m_2}{m_1}}} = -\sqrt{\frac{m_2}{m_1}} \;(\to -\infty),$$

which means that $\varphi \simeq \frac{\pi}{2}$, and

$$\begin{cases} \xi_1 = \eta_1 \cos \varphi - \eta_2 \sin \varphi, \\ \xi_2 = \eta_1 \sin \varphi + \eta_2 \cos \varphi, \end{cases}$$

lead to $\xi_1 \simeq -\eta_2 = -A_2 \cos(\overline{\omega}_2 t + \psi_2)$, and $\xi_2 \simeq \eta_1 = A_1 \cos(\overline{\omega}_1 t + \psi_1)$, which completes the proof.

Working within the same first-order approximation for $\sqrt{\frac{m_1}{m_2}} \ll 1$, we obtain

$$\overline{\omega}_1 \simeq \frac{\omega_0}{\sqrt{2}}, \quad \overline{\omega}_2 \simeq \omega_0 \sqrt{\frac{2m_2}{m_1}} \quad \left(\omega_0^2 = \frac{g}{l}\right).$$

These results can be verified as follows. Denoting $z = \sqrt{\frac{m_1}{m_2}}$, we have

$$\overline{\omega}_1 = \sqrt{\frac{g}{l}} \sqrt{1 + \frac{m_2}{m_1}\left(1 - \sqrt{1 + \frac{m_1}{m_2}}\right)}$$

$$= \omega_0 \sqrt{1 + \frac{1}{z^2}\left(1 - \sqrt{1 + z^2}\right)} \simeq \omega_0 \sqrt{1 + \frac{1}{z^2}\left(1 - 1 - \frac{z^2}{2}\right)} = \frac{\omega_0}{\sqrt{2}},$$

as well as

$$\overline{\omega}_2 = \omega_0 \sqrt{1 + \frac{1}{z^2}\left(1 + \sqrt{1 + z^2}\right)}$$

$$\simeq \omega_0 \sqrt{1 + \frac{1}{z^2}\left(1 + 1 + \frac{z^2}{2}\right)} = \omega_0 \frac{\sqrt{2}}{z} = \omega_0 \sqrt{\frac{2m_2}{m_1}}.$$

Suppose, this time, that initially the pendulums are at rest, $\dot{\theta}_1(t_0 = 0) = 0$, $\dot{\theta}_2(t_0 = 0) = 0$, and only the pendulum of mass m_1 is displaced from its equilibrium position, $\theta_1(t_0 = 0) = \theta_0$, $\theta_2(t_0 = 0) = 0$. Then, we have:

$$\theta_1 = \frac{\xi_1}{l\sqrt{m_1}} \simeq -\frac{\eta_2}{l\sqrt{m_1}} = -\frac{A_2}{l\sqrt{m_1}} \cos(\overline{\omega}_2 t + \psi_2),$$

$$\theta_2 = \frac{\xi_2}{l\sqrt{m_2}} - \frac{\xi_1}{l\sqrt{m_1}} \simeq \frac{\eta_1}{l\sqrt{m_2}} + \frac{\eta_2}{l\sqrt{m_1}}$$

$$= \frac{A_1}{l\sqrt{m_2}} \cos(\overline{\omega}_1 t + \psi_1) + \frac{A_2}{l\sqrt{m_1}} \cos(\overline{\omega}_2 t + \psi_2).$$

We still have

$$\dot{\theta}_1 = \frac{\overline{\omega}_2 A_2}{l\sqrt{m_1}} \sin(\overline{\omega}_2 t + \psi_2),$$

$$\dot{\theta}_2 = -\frac{\overline{\omega}_1 A_1}{l\sqrt{m_2}}\sin(\overline{\omega}_1 t + \psi_1) - \frac{\overline{\omega}_2 A_2}{l\sqrt{m_1}}\sin(\overline{\omega}_2 t + \psi_2).$$

Imposing the initial conditions, we are left with the following system of four algebraic equations in four unknowns A_1, A_2, ψ_1 and ψ_2:

$$\begin{cases} \theta_0 = -\frac{A_2}{l\sqrt{m_1}}\cos\psi_2, \\ 0 = \frac{A_1}{l\sqrt{m_2}}\cos\psi_1 + \frac{A_2}{l\sqrt{m_1}}\cos\psi_2, \\ 0 = \frac{\overline{\omega}_2 A_2}{l\sqrt{m_1}}\sin\psi_2, \\ 0 = -\frac{\overline{\omega}_1 A_1}{l\sqrt{m_2}}\sin\psi_1 - \frac{\overline{\omega}_2 A_2}{l\sqrt{m_1}}\sin\psi_2. \end{cases}$$

The third equation yields $\psi_2 = 0$, and, as a result, the last equation gives $\psi_1 = 0$. Then the first two equations result in $A_2 = -l\theta_0\sqrt{m_1}$, $A_1 = l\theta_0\sqrt{m_2}$. Therefore,

$$\theta_1 = \theta_0 \cos\overline{\omega}_2 t = \theta_0 \cos\left(\omega_0 \sqrt{\frac{2m_2}{m_1}} t\right); \quad (6.162)$$

$$\theta_2 = \theta_0 \cos\overline{\omega}_1 t - \theta_0 \cos\overline{\omega}_2 t = \theta_0 \cos\frac{\omega_0}{\sqrt{2}} t - \theta_0 \cos\left(\omega_0\sqrt{\frac{2m_2}{m_1}} t\right)$$

$$= 2\theta_0 \sin\frac{\omega_0\left(\sqrt{\frac{2m_2}{m_1}} - \frac{1}{\sqrt{2}}\right)}{2} t \, \sin\frac{\omega_0\left(\sqrt{\frac{2m_2}{m_1}} + \frac{1}{\sqrt{2}}\right)}{2} t \quad (6.163)$$

$$\simeq 2\theta_0 \sin^2\left(\omega_0\sqrt{\frac{m_2}{2m_1}} t\right) = \theta_0 \left[1 - \cos\left(\omega_0\sqrt{\frac{2m_2}{m_1}} t\right)\right].$$

(3) If $m_1 = m_2 = m$, by means of $\cot 2\varphi = \frac{\omega_2^2 - \omega_1^2}{2\kappa} = a = -1$, where a is a notation, we have

$$\sin^2\varphi = \frac{1}{2}\left(1 - \frac{a}{\sqrt{1+a^2}}\right) = \frac{1+\sqrt{2}}{2\sqrt{2}}$$

$$\implies \sin\varphi = \sqrt{\frac{1+\sqrt{2}}{2\sqrt{2}}} \implies \varphi = \frac{3\pi}{8},$$

and

$$\cos^2\varphi = \frac{1}{2}\left(1 + \frac{a}{\sqrt{1+a^2}}\right) = \frac{1}{2}\left(1 - \frac{1}{\sqrt{2}}\right) = \frac{\sqrt{2}-1}{2\sqrt{2}}$$

$$\implies \cos\varphi = \sqrt{\frac{\sqrt{2}-1}{2\sqrt{2}}} \implies \varphi = \frac{3\pi}{8}.$$

Then ξ_1, ξ_2 write

$$\xi_1 = \eta_1 \cos\varphi - \eta_2 \sin\varphi = \sqrt{\frac{\sqrt{2}-1}{2\sqrt{2}}}\eta_1 - \sqrt{\frac{\sqrt{2}+1}{2\sqrt{2}}}\eta_2,$$

$$\xi_2 = \eta_1 \sin\varphi + \eta_2 \cos\varphi = \sqrt{\frac{\sqrt{2}+1}{2\sqrt{2}}}\eta_1 + \sqrt{\frac{\sqrt{2}-1}{2\sqrt{2}}}\eta_2.$$

We also have:

$$\theta_1 = \frac{\xi_1}{l\sqrt{m}} = \frac{1}{l\sqrt{m}}\left(\sqrt{\frac{\sqrt{2}-1}{2\sqrt{2}}}\eta_1 - \sqrt{\frac{\sqrt{2}+1}{2\sqrt{2}}}\eta_2\right)$$

$$= \frac{1}{l\sqrt{m}}\left[\sqrt{\frac{\sqrt{2}-1}{2\sqrt{2}}}A_1 \cos(\overline{\omega}_1 t + \psi_1) - \sqrt{\frac{\sqrt{2}+1}{2\sqrt{2}}}A_2 \cos(\overline{\omega}_2 t + \psi_2)\right];$$

(6.164)

$$\theta_2 = \frac{1}{l\sqrt{m}}(\xi_2 - \xi_1) = \frac{1}{l\sqrt{m}}\left(\sqrt{\frac{\sqrt{2}+1}{2\sqrt{2}}}\eta_1 + \sqrt{\frac{\sqrt{2}-1}{2\sqrt{2}}}\eta_2\right)$$

$$- \frac{1}{l\sqrt{m}}\left(\sqrt{\frac{\sqrt{2}-1}{2\sqrt{2}}}\eta_1 - \sqrt{\frac{\sqrt{2}+1}{2\sqrt{2}}}\eta_2\right)$$

$$= \frac{1}{l\sqrt{m}}\left(\sqrt{\frac{\sqrt{2}+1}{2\sqrt{2}}} - \sqrt{\frac{\sqrt{2}-1}{2\sqrt{2}}}\right)A_1 \cos(\overline{\omega}_1 t + \psi_1) \quad (6.165)$$

$$+ \frac{1}{l\sqrt{m}}\left(\sqrt{\frac{\sqrt{2}-1}{2\sqrt{2}}} + \sqrt{\frac{\sqrt{2}+1}{2\sqrt{2}}}\right)A_2 \cos(\overline{\omega}_2 t + \psi_2),$$

as well as

$$\dot{\theta}_1 = \frac{1}{l\sqrt{m}}\left[\sqrt{\frac{\sqrt{2}+1}{2\sqrt{2}}}A_2\overline{\omega}_2 \sin(\overline{\omega}_2 t + \psi_2)\right.$$

$$\left. - \sqrt{\frac{\sqrt{2}-1}{2\sqrt{2}}}A_1\overline{\omega}_1 \sin(\overline{\omega}_1 t + \psi_1)\right],$$

$$\dot{\theta}_2 = \frac{1}{l\sqrt{m}}\left(\sqrt{\frac{\sqrt{2}-1}{2\sqrt{2}}} - \sqrt{\frac{\sqrt{2}+1}{2\sqrt{2}}}\right)A_1\overline{\omega}_1 \sin(\overline{\omega}_1 t + \psi_1)$$

$$-\frac{1}{l\sqrt{m}}\left(\sqrt{\frac{\sqrt{2}-1}{2\sqrt{2}}}+\sqrt{\frac{\sqrt{2}+1}{2\sqrt{2}}}\right)A_2\bar{\omega}_2\sin(\bar{\omega}_2 t+\psi_2).$$

The initial conditions $\theta_1(t_0 = 0) = 0$, $\theta_2(t_0 = 0) = \theta_0$, $\dot{\theta}_1(t_0 = 0) = 0$, and $\dot{\theta}_2(t_0 = 0) = 0$ then allow one to write the following four algebraic equations for the unknowns A_1, A_2, ψ_1, and ψ_2:

$$0 = \sqrt{\frac{\sqrt{2}-1}{2\sqrt{2}}}A_1\cos\psi_1 - \sqrt{\frac{\sqrt{2}+1}{2\sqrt{2}}}A_2\cos\psi_2,$$

$$l\theta_0\sqrt{m} = \left(\sqrt{\frac{\sqrt{2}+1}{2\sqrt{2}}} - \sqrt{\frac{\sqrt{2}-1}{2\sqrt{2}}}\right)A_1\cos\psi_1$$
$$+ \left(\sqrt{\frac{\sqrt{2}-1}{2\sqrt{2}}} + \sqrt{\frac{\sqrt{2}+1}{2\sqrt{2}}}\right)A_2\cos\psi_2,$$

$$0 = \sqrt{\frac{\sqrt{2}+1}{2\sqrt{2}}}A_2\bar{\omega}_2\sin\psi_2 - \sqrt{\frac{\sqrt{2}-1}{2\sqrt{2}}}A_1\bar{\omega}_1\sin\psi_1,$$

$$0 = \left(\sqrt{\frac{\sqrt{2}-1}{2\sqrt{2}}} - \sqrt{\frac{\sqrt{2}+1}{2\sqrt{2}}}\right)A_1\bar{\omega}_1\sin\psi_1$$
$$- \left(\sqrt{\frac{\sqrt{2}-1}{2\sqrt{2}}} + \sqrt{\frac{\sqrt{2}+1}{2\sqrt{2}}}\right)A_2\bar{\omega}_2\sin\psi_2.$$

The last two equations give $\psi_1 = 0$ and $\psi_2 = 0$. With these values, the first equation leads to

$$A_2 = \sqrt{\frac{\sqrt{2}-1}{\sqrt{2}+1}}A_1,$$

and, finally, the second relation shows that

$$A_1 = \frac{l\theta_0\sqrt{m}}{\sqrt{\frac{\sqrt{2}+1}{2\sqrt{2}}} + \frac{\sqrt{2}-1}{\sqrt{4+2\sqrt{2}}}} = \frac{l\theta_0\sqrt{m}}{\frac{1}{\sqrt{4-2\sqrt{2}}} + \frac{\sqrt{2}-1}{\sqrt{4+2\sqrt{2}}}},$$

so that
$$A_2 = \frac{l\theta_0\sqrt{m}\sqrt{\frac{\sqrt{2}-1}{\sqrt{2}+1}}}{\frac{1}{\sqrt{4-2\sqrt{2}}}+\frac{\sqrt{2}-1}{\sqrt{4+2\sqrt{2}}}} = \frac{l\theta_0\sqrt{m}}{\frac{\sqrt{2}+1}{\sqrt{4-2\sqrt{2}}}+\frac{1}{\sqrt{4+2\sqrt{2}}}}.$$

With these results, θ_1 given by (6.164) writes

$$\theta_1 = \frac{1}{l\sqrt{m}}\left[\sqrt{\frac{\sqrt{2}-1}{2\sqrt{2}}}\left(\frac{l\theta_0\sqrt{m}}{\frac{1}{\sqrt{4-2\sqrt{2}}}+\frac{\sqrt{2}-1}{\sqrt{4+2\sqrt{2}}}}\right)\cos\overline{\omega}_1 t\right.$$

$$\left.-\sqrt{\frac{\sqrt{2}+1}{2\sqrt{2}}}\left(\frac{l\theta_0\sqrt{m}}{\frac{\sqrt{2}+1}{\sqrt{4-2\sqrt{2}}}+\frac{1}{\sqrt{4+2\sqrt{2}}}}\right)\cos\overline{\omega}_2 t\right]$$

$$= \frac{\theta_0}{2\sqrt{2}}(\cos\overline{\omega}_1 t - \cos\overline{\omega}_2 t),$$

while (6.165) gives

$$\theta_2 = \frac{1}{l\sqrt{m}}\left(\sqrt{\frac{\sqrt{2}+1}{2\sqrt{2}}}-\sqrt{\frac{\sqrt{2}-1}{2\sqrt{2}}}\right)\left(\frac{l\theta_0\sqrt{m}}{\frac{1}{\sqrt{4-2\sqrt{2}}}+\frac{\sqrt{2}-1}{\sqrt{4+2\sqrt{2}}}}\right)\cos\overline{\omega}_1 t$$

$$+\frac{1}{l\sqrt{m}}\left(\sqrt{\frac{\sqrt{2}-1}{2\sqrt{2}}}+\sqrt{\frac{\sqrt{2}+1}{2\sqrt{2}}}\right)\left(\frac{l\theta_0\sqrt{m}}{\frac{\sqrt{2}+1}{\sqrt{4-2\sqrt{2}}}+\frac{1}{\sqrt{4+2\sqrt{2}}}}\right)\cos\overline{\omega}_2 t$$

$$= \frac{\theta_0}{2}(\cos\overline{\omega}_1 t + \cos\overline{\omega}_2 t).$$

Recalling our initial assumptions ($m_1 = m_2 = m$, $l_1 = l_2 = l$), we also have
$$\overline{\omega}_1^2 = \frac{g}{l}(2-\sqrt{2}); \quad \overline{\omega}_2^2 = \frac{g}{l}(2+\sqrt{2}),$$

and the two angular variables finally write

$$\theta_1(t) = \frac{\theta_0}{2\sqrt{2}}\left[\cos\left(\sqrt{2-\sqrt{2}}\,\omega_0 t\right)-\cos\left(\sqrt{2+\sqrt{2}}\,\omega_0 t\right)\right], \quad (6.166)$$

$$\theta_2(t) = \frac{\theta_0}{2}\left[\cos\left(\sqrt{2-\sqrt{2}}\,\omega_0 t\right)+\cos\left(\sqrt{2+\sqrt{2}}\,\omega_0 t\right)\right]. \quad (6.167)$$

Observation. This last case ($m_1 = m_2 = m$, $l_1 = l_2 = l$) can also be approached independently from the general considerations exposed in Problem No.5. With these assumptions, equations (6.146) and (6.146') become

$$2\ddot{\theta}_1 + \ddot{\theta}_2 \cos(\theta_2 - \theta_1) - \dot{\theta}_2^2 \sin(\theta_2 - \theta_1) + 2\frac{g}{l}\sin\theta_1 = 0, \qquad (6.168)$$

$$\ddot{\theta}_2 + \ddot{\theta}_1 \cos(\theta_2 - \theta_1) + \dot{\theta}_1^2 \sin(\theta_2 - \theta_1) + \frac{g}{l}\sin\theta_2 = 0. \qquad (6.168')$$

Furthermore, if the oscillations are small, $\cos(\theta_2 - \theta_1) \simeq 1$, $\sin(\theta_2 - \theta_1) \simeq 0$, $\sin\theta_1 \simeq \theta_1$, $\sin\theta_2 \simeq \theta_2$, and the last two equations receive a simpler form

$$2\ddot{\theta}_1 + \ddot{\theta}_2 + 2\frac{g}{l}\theta_1 = 0; \qquad (6.169)$$

$$\ddot{\theta}_2 + \ddot{\theta}_1 + \frac{g}{l}\theta_2 = 0. \qquad (6.169')$$

We are looking for solutions of the form $\theta_1 = A_1 e^{i\omega t}$, $\theta_2 = A_2 e^{i\omega t}$, leading to the following system

$$\begin{cases} 2\left(\frac{g}{l} - \omega^2\right) A_1 - \omega^2 A_2 = 0, \\ -\omega^2 A_1 + \left(\frac{g}{l} - \omega^2\right) A_2 = 0. \end{cases}$$

This system admits non-trivial solution for the amplitudes A_1 and A_2 only if

$$\begin{vmatrix} 2\left(\frac{g}{l} - \omega^2\right) & -\omega^2 \\ -\omega^2 & \left(\frac{g}{l} - \omega^2\right) \end{vmatrix} = 0, \qquad (6.170)$$

that is

$$\omega^4 - 2\left(\frac{g}{l} - \omega^2\right)^2 = 0.$$

Denoting $\omega^2 = r$, $\frac{g}{l} = \omega_0^2$, one obtains the following equation of the second degree in r

$$r^2 - 4\omega_0^2 r + 2\omega_0^4 = 0,$$

with the solutions

$$r_{1,2} = \omega_0^2 (2 \pm \sqrt{2}),$$

or, if we return to the old variable

$$\omega_1 = \omega_0 \sqrt{2 - \sqrt{2}}, \qquad \omega_2 = \omega_0 \sqrt{2 + \sqrt{2}}. \qquad (6.171)$$

These are the two *normal frequencies* of oscillation. It then follows that the system possesses two *normal modes of oscillation*, corresponding to its two degrees of freedom.

As known, the choice of the generalized coordinates is - up to some extent - arbitrary. This fact allows us to represent the system as an assembly of linear harmonic oscillators, with one degree of freedom for each of them. In other words, to each normal frequency one can associate a periodically time-varying generalized coordinate. These coordinates are called *normal*.

Suppose that our system oscillates with one of the normal frequencies, say ω_1. Then, since $\theta_1 = A_1 e^{i\omega_1 t}$, $\theta_2 = A_2 e^{i\omega_1 t}$, equation (6.169) becomes

$$(-2\omega_1^2 + 2\omega_0^2)\theta_1 = \omega_1^2 \theta_2.$$

Recalling that $\omega_1^2 = (2 - \sqrt{2})\omega_0^2$, we still have

$$(-2 + 2\sqrt{2})\theta_1 = (2 - \sqrt{2})\theta_2,$$

or

$$\frac{\theta_2}{\theta_1} = \sqrt{2}. \tag{6.172}$$

Calculations for the second normal frequency are similar, and the result is $\theta_2/\theta_1 = -\sqrt{2}$. Therefore, for the two normal modes of oscillation, the relations between the two generalized coordinates are

$$\theta_2 = \theta_1 \sqrt{2} \quad (\text{for } \omega = \omega_1),$$

$$\theta_2 = -\theta_1 \sqrt{2} \quad (\text{for } \omega = \omega_2).$$

In the first case the modes are called *symmetric*, and in the second - *antisymmetric*.

If we choose as normal coordinates the quantities $\eta_s = \theta_2 + \sqrt{2}\theta_1$, $\eta_{as} = \theta_2 - \sqrt{2}\theta_1$, where the indices "s" and "as" stand for "symmetric" and "antisymmetric", and place our discussion within the same particular case ($m_1 = m_2 = m$, $l_1 = l_2 = l$, $\cos(\theta_2 - \theta_1) \simeq 1$, $\cos\theta_i \simeq 1 - \frac{\theta_i^2}{2}$ ($i = 1, 2$)), the Lagrangian (6.143) writes

$$L = ml^2 \dot{\theta}_1^2 + \frac{1}{2}ml^2 \dot{\theta}_2^2 + ml^2 \dot{\theta}_1 \dot{\theta}_2 - mgl\theta_1^2 - \frac{1}{2}mgl\theta_2^2$$

$$= ml^2 \left(\frac{\dot{\eta}_s - \dot{\eta}_{as}}{2\sqrt{2}}\right)^2 + \frac{1}{2}ml^2 \left(\frac{\dot{\eta}_s + \dot{\eta}_{as}}{2}\right)^2$$

$$+ ml^2 \left(\frac{\dot{\eta}_s - \dot{\eta}_{as}}{2\sqrt{2}}\right)\left(\frac{\dot{\eta}_s + \dot{\eta}_{as}}{2}\right)$$

$$-mgl\left(\frac{\eta_s - \eta_{as}}{2\sqrt{2}}\right)^2 - \frac{1}{2}mgl\left(\frac{\eta_s + \eta_{as}}{2}\right)^2$$

$$= \frac{1}{4\sqrt{2}}ml^2\left[\dot{\eta}_s^2(\sqrt{2}+1) + \dot{\eta}_{as}^2(\sqrt{2}-1)\right] - \frac{mgl}{4}(\eta_s^2 + \eta_{as}^2).$$

By choosing as normal coordinates

$$\eta_1 = \eta_s\sqrt{\frac{\sqrt{2}+1}{2\sqrt{2}}ml^2};$$

$$\eta_2 = \eta_{as}\sqrt{\frac{\sqrt{2}-1}{2\sqrt{2}}ml^2},$$

the Lagrangian writes

$$L = \frac{1}{2}(\dot{\eta}_1^2 + \dot{\eta}_2^2) - \frac{g}{l}\left(\frac{\eta_1^2}{2+\sqrt{2}} + \frac{\eta_2^2}{2-\sqrt{2}}\right)$$

$$= \frac{1}{2}(\dot{\eta}_1^2 + \dot{\eta}_2^2) - \frac{1}{2}\omega_0^2\left[(2-\sqrt{2})\eta_1^2 + (2+\sqrt{2})\eta_2^2\right]$$

$$= \frac{1}{2}(\dot{\eta}_1^2 + \dot{\eta}_2^2) - \frac{1}{2}(\omega_1^2\eta_1^2 + \omega_2^2\eta_2^2)$$

$$\equiv \frac{1}{2}(\dot{\eta}_1^2 + \dot{\eta}_2^2) - \frac{1}{2}(\overline{\omega}_1^2\eta_1^2 + \overline{\omega}_2^2\eta_2^2),$$

and we meet again the Lagrangian (6.147) for the third case ($m_1 = m_2 = m$; $l_1 = l_2 = l$). To find $\theta_i = \theta_i(t)$ ($i = 1, 2$) we use the relations

$$\eta_1 = l\eta_s\sqrt{m}\sqrt{\frac{\sqrt{2}+1}{2\sqrt{2}}} = l\sqrt{m}\sqrt{\frac{\sqrt{2}+1}{2\sqrt{2}}}(\theta_2 + \sqrt{2}\theta_1)$$

$$= A_1\cos(\omega_1 t + \psi_1); \qquad (6.173)$$

$$\eta_2 = l\eta_{as}\sqrt{m}\sqrt{\frac{\sqrt{2}-1}{2\sqrt{2}}} = l\sqrt{m}\sqrt{\frac{\sqrt{2}-1}{2\sqrt{2}}}(\theta_2 - \sqrt{2}\theta_1)$$

$$= A_2\cos(\omega_2 t + \psi_2). \qquad (6.174)$$

Using the same initial conditions, $\theta_1(t_0 = 0) = 0$, $\theta_2(t_0 = 0) = \theta_0$, $\dot{\theta}_1(t_0 = 0) = 0$, $\dot{\theta}_2(t_0 = 0) = 0$, the constants A_1, A_2, ψ_1, ψ_2 are found as solutions of the following algebraic system

$$\begin{cases} l\theta_0\sqrt{m}\sqrt{\frac{\sqrt{2}+1}{2\sqrt{2}}} = A_1\cos\psi_1, \\ l\theta_0\sqrt{m}\sqrt{\frac{\sqrt{2}-1}{2\sqrt{2}}} = A_2\cos\psi_2, \\ 0 = -\omega_1 A_1\sin\psi_1, \\ 0 = -\omega_2 A_2\sin\psi_2. \end{cases}$$

The last two equations give $\psi_1 = \psi_2 = 0$, while the other two equations yield

$$A_1 = l\theta_0\sqrt{m}\sqrt{\frac{\sqrt{2}+1}{2\sqrt{2}}}; \quad A_2 = l\theta_0\sqrt{m}\sqrt{\frac{\sqrt{2}-1}{2\sqrt{2}}}$$

and we finally obtain

$$\theta_1(t) = \frac{\theta_0}{2\sqrt{2}}\left[\cos\left(\sqrt{2-\sqrt{2}}\,\omega_0 t\right) - \cos\left(\sqrt{2+\sqrt{2}}\,\omega_0 t\right)\right];$$

$$\theta_2(t) = \frac{\theta_0}{2}\left[\cos\left(\sqrt{2-\sqrt{2}}\,\omega_0 t\right) + \cos\left(\sqrt{2+\sqrt{2}}\,\omega_0 t\right)\right],$$

which are precisely solutions (6.166) and (6.167) obtained by the first method.

CHAPTER VII

PROBLEMS OF EQUILIBRIUM AND SMALL OSCILLATIONS

Problem 1. A particle (material point) moves without friction inside a pipe of elliptic shape and constant cross section, rotating about its major axis with constant angular velocity ω (see Fig.VII.1). The mass m of the particle, the gravitational acceleration g, and the semi-axes $a > b$ of the ellipse are given. Determine:

a) The equilibrium positions of the particle;
b) Stability of the equilibrium positions;
c) Period of small oscillations of the particle about the stable positions of equilibrium.

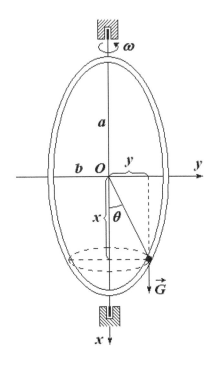

Fig.VII.1

Solution

Obviously, if the ellipse does not rotate, position given by $\theta = 0$ shall be a position of stable equilibrium, while $\theta = \pi$ would be a position of unstable equilibrium. If the ellipse rotates, then depending on the value of ω, position $\theta = 0$ could become a position of unstable equilibrium, since upon the particle act both the centrifugal and gravitational forces, the first tending to move away the particle from the position of stable equilibrium, and the second tending to bring the particle back to this position. If ω exceeds a certain critical value ω_{cr}, then the position $\theta = 0$ becomes a position of unstable equilibrium, while the situation $\omega < \omega_{cr}$ corresponds to a stable equilibrium. In its turn, position given by $\theta = \pi$ is always a position of unstable equilibrium, no matter how big or small ω is taken. It is also possible to appear one or more intermediate states of equilibrium for $\theta \in [0, \pi]$. For $\omega \to \infty$, the centrifugal force is far superior to the force of gravity, and $\theta = \pi/2$ corresponds to a position of stable equilibrium. Using Lagrange equations of the second kind formalism, let us prove all these intuitive observations.

Since this application implies a centrifugal force, which is a force of inertia, we shall investigate the problem from two different points of view, associated with inertial and non-inertial approaches.

(A) Inertial reference frame (IRF)

a) As we know, there is no inertial force (such as the centrifugal force) in an IRF. From the point of view of the IRF, the only applied force is the force of gravity $\vec{G}$. Since the gravitational force derives from a potential, our system is a natural one. The gravitational field being conservative, the time-independent potential is precisely the potential energy of the system. The equilibrium positions of the particle are given by the extremum (extrema) of the potential energy. According to analytical approach,

$$dV = -dA,$$

where

$$dA = \vec{G} \cdot d\vec{r}$$

is the elementary mechanical work done by the gravitational force field. The choice of the coordinate system (see Fig.VII.1) allows us to write

$$dV = -mg\,dx$$

so that

$$V = -mgx + V_0.$$

The most convenient choice for V_0 is $V_0 = 0$. This condition is fulfilled by taking the yOz-plane as the reference level for the potential energy

$$V(x=0) = 0 \Rightarrow V_0 = 0.$$

The standard form equation of the ellipse

$$\frac{x^2}{a^2} + \frac{y^2}{b^2} = 1$$

allows one the parametrization

$$\begin{cases} x = a \cos\theta; \\ y = b \sin\theta, \end{cases}$$

and the potential energy of the particle becomes

$$V = -mga\cos\theta. \tag{7.1}$$

Equating to zero the first derivative of $V = V(\theta)$, one obtains

$$\sin\theta = 0$$

with "physical" solutions

$$\theta_1 = 0, \quad \theta_2 = \pi. \tag{7.2}$$

The stability of these equilibrium positions is determined by performing the second derivative

$$\frac{d^2V}{d\theta^2} = mga\cos\theta,$$

which means that $\theta_1 = 0$ is a position of stable equilibrium ($\frac{d^2V}{d\theta^2}|_{\theta=0} > 0$), while $\theta_2 = \pi$ corresponds to a position of unstable equilibrium ($\frac{d^2V}{d\theta^2}|_{\theta=\pi} < 0$).

These results can be intuitively anticipated if the pipe is at rest. But, if the pipe rotates, this fact has to be considered under our assumption that the frame is inertial. To solve the problem, we have to draw our attention to the "true" value of the potential energy of the particle, due to the action of both the gravity and the motion of rotation. This is called *effective potential energy*.

To determine the effective potential energy, let us first write the Lagrangian of the problem. To the single degree of freedom of the particle we shall associate the generalized coordinate $q = \theta$, so that $L = T - V(\theta)$. Since the particle is subjected to a composite motion (a motion of translation along the ellipse, in the xOy plane, and a motion of rotation with constant angular velocity ω about the x-axis), the kinetic energy writes

$$T = T_{tr} + T_{rot} = \frac{1}{2}m\vec{v}_{tr}^{\,2} + \frac{1}{2}m\vec{v}_{rot}^{\,2} = \frac{1}{2}m(\dot{x}^2 + \dot{y}^2) + \frac{1}{2}m\omega^2 y^2$$

$$= \frac{1}{2}m\dot{\theta}^2(a^2 \sin^2\theta + b^2 \cos^2\theta) + \frac{1}{2}m\omega^2 b^2 \sin^2\theta.$$

The Lagrangian then becomes

$$L = \frac{1}{2}m\dot{\theta}^2(a^2 \sin^2\theta + b^2 \cos^2\theta) + \frac{1}{2}m\omega^2 b^2 \sin^2\theta + mga\cos\theta. \quad (7.3)$$

As one can see, the Lagrangian does not explicitly depend on time. Consequently, Lagrange equation of the second kind admits the first integral

$$\dot{\theta}\frac{\partial L}{\partial \dot{\theta}} - L = const.,$$

expressing conservation of the total energy of the system

$$const. = E_{tot} = E_{cin} + E_{pot} = \dot{\theta}\frac{\partial L}{\partial \dot{\theta}} - L$$

$$= \frac{1}{2}m\dot{\theta}^2\left(a^2 \sin^2\theta + b^2 \cos^2\theta\right) - \frac{1}{2}m\omega^2 b^2 \sin^2\theta - mga\cos\theta.$$

Therefore, the total energy consists of two groups of terms, one involving the squared generalized velocity, which is the kinetic energy

$$E_{cin} = \frac{1}{2}m\dot{\theta}^2\left(a^2 \sin^2\theta + b^2 \cos^2\theta\right), \quad (7.4)$$

and the other written in terms of ω and θ, namely

$$E_{pot} = -\frac{1}{2}m\omega^2 b^2 \sin^2\theta - mga\cos\theta. \quad (7.4')$$

We have to emphasize that, within the Lagrangian formalism, the kinetic energy is expressed in terms of the generalized velocity (in our case $\dot{\theta}$). Therefore, we may call the quantity given by (7.4) the "true/genuine" kinetic energy. On the other hand, as seen from (7.4'), E_{pot} contains the supplementary term $-\frac{1}{2}m\omega^2 b^2 \sin^2\theta$, in addition to the term $V(\theta) = -mga\cos\theta$, meaning that (7.4') is *the effective potential energy*. We want to draw attention to the reader that the term containing ω^2 is *not* connected to the kinetic energy, even if ω is, in its turn, the time derivative of an angular coordinate (which is not the generalized coordinate of the problem, but it belongs to the uniform circular motion of the material point about the x-axis, i.e., it is an angular coordinate in a plane orthogonal to the x-axis).

Equating to zero the first derivative of E_{pot}, one obtains

$$\frac{dE_{pot}}{d\theta} = -m\omega^2 b^2 \sin\theta\cos\theta + mga\cos\theta = 0,$$

giving the equilibrium positions of the particle

$$\begin{cases} \sin\theta = 0, \\ ga - \omega^2 b^2 \cos\theta = 0. \end{cases}$$

The first equation yields the already known positions of equilibrium

$$\theta_1 = 0, \quad \theta_2 = \pi,$$

while the second equation leads to a new position of equilibrium, given by

$$\theta_3 = \arccos\left(\frac{ga}{\omega^2 b^2}\right). \tag{7.5}$$

Using notation

$$\omega_{cr} = \frac{\sqrt{ga}}{b}, \tag{7.6}$$

we still have

$$\theta_3 = \arccos\left(\frac{\omega_{cr}}{\omega}\right)^2. \tag{7.6'}$$

b) To investigate the stability of the determined equilibrium positions, one must study the sign of the second derivative of the effective potential energy at these points. We have:

$$\frac{d^2 E_{pot}}{d\theta^2} = -m\omega^2 b^2 \cos^2\theta + m\omega^2 b^2 \sin^2\theta + mga\cos\theta$$

$$= m\omega^2 b^2 (1 - 2\cos^2\theta) + mga\cos\theta,$$

leading to
  i)

$$\left.\frac{d^2 E_{pot}}{d\theta^2}\right|_{\theta=0} = -m\omega^2 b^2 + mga = mga\left(1 - \frac{\omega^2 b^2}{ga}\right)$$

$$= mga\left[1 - \left(\frac{\omega}{\omega_{cr}}\right)^2\right] \quad \begin{matrix} \nearrow > 0, \ \omega < \omega_{cr}, & \text{(stable equilibrium);} \\ \searrow < 0, \ \omega > \omega_{cr}, & \text{(unstable equilibrium);} \end{matrix}$$

ii)

$$\left.\frac{d^2 E_{pot}}{d\theta^2}\right|_{\theta=\pi} = -m\omega^2 b^2 - mga < 0, \quad \forall \ \omega \quad \text{(unstable equilibrium);}$$

iii)

$$\left.\frac{d^2 E_{pot}}{d\theta^2}\right|_{\theta=\arccos\left(\frac{\omega_{cr}}{\omega}\right)^2} = m\omega^2 b^2\left(1 - 2\left(\frac{ag}{\omega^2 b^2}\right)^2\right) + mga\frac{ga}{\omega^2 b^2}$$

$$= m\omega^2 b^2\left[1 - \left(\frac{\omega_{cr}}{\omega}\right)^4\right] \quad \begin{matrix} \nearrow > 0, \ \omega > \omega_{cr}, & \text{(stable equilibrium);} \\ \searrow < 0, \ \omega < \omega_{cr}, & \text{(unstable equilibrium).} \end{matrix}$$

Therefore, the equilibrium position $\theta_1 = 0$ can be of both stable equilibrium (for $\omega < \omega_{cr}$) and unstable equilibrium (for $\omega > \omega_{cr}$), position $\theta_2 = \pi$ is always a position of unstable equilibrium, while the intermediate position $\theta_3 = \arccos\left(\frac{\omega_{cr}}{\omega}\right)^2$ can be, in its turn, a position of both stable equilibrium (for $\omega > \omega_{cr}$), and unstable equilibrium (for $\omega < \omega_{cr}$).

If $\omega = \omega_{cr}$, that is $\omega^2 b^2 = ga$, then $\theta_3 = \arccos 1 = 0$, and

$$\left.\frac{d^2 E_{pot}}{d\theta^2}\right|_{\theta=\arccos\left(\frac{\omega_{cr}}{\omega(=\omega_{cr})}\right)^2} = m\omega^2 b^2\left[1 - \left(\frac{\omega_{cr}}{\omega(=\omega_{cr})}\right)^4\right] = 0.$$

To determine the stability of equilibrium in this case ($\theta_3 = 0$, for $\omega = \omega_{cr}$), we have to go "further" with derivatives. Since

$$\frac{d^3 E_{pot}}{d\theta^3} = \frac{d}{d\theta}[m\omega^2 b^2(1 - 2\cos^2\theta) + mga\cos\theta]$$

$$= mga\sin\theta\left(\frac{4\omega^2 b^2}{ga}\cos\theta - 1\right) = mga\sin\theta\left[\left(\frac{2\omega}{\omega_{cr}}\right)^2\cos\theta - 1\right],$$

we have

$$\left.\frac{d^3 E_{pot}}{d\theta^3}\right|_{\theta=\arccos\left(\frac{\omega_{cr}}{\omega(=\omega_{cr})}\right)^2} = \left\{mga\sin\theta\left[\left(\frac{2\omega}{\omega_{cr}}\right)^2\cos\theta - 1\right]\right\}_{\omega=\omega_{cr}}$$

$$= mga\sqrt{1 - \left(\frac{\omega_{cr}}{\omega(=\omega_{cr})}\right)^4} \left[\left(2\frac{\omega(=\omega_{cr})}{\omega_{cr}}\right)^2 \left(\frac{\omega_{cr}}{\omega(=\omega_{cr})}\right)^2 - 1\right] = 0.$$

This result shows that we have to make one step further and calculate the fourth derivative. Since

$$\frac{d^4 E_{pot}}{d\theta^4} = \frac{d}{d\theta}\left\{mga\sin\theta\left[\left(\frac{2\omega}{\omega_{cr}}\right)^2 \cos\theta - 1\right]\right\}$$

$$= mga\left(\frac{2\omega}{\omega_{cr}}\right)^2 (2\cos^2\theta - 1) - mga\cos\theta,$$

we are left with

$$\left.\frac{d^4 E_{pot}}{d\theta^4}\right|_{\theta=\arccos\left(\frac{\omega_{cr}}{\omega(=\omega_{cr})}\right)^2}$$

$$= \left[mga\left(\frac{2\omega}{\omega_{cr}}\right)^2 (2\cos^2\theta - 1) - mga\cos\theta\right]_{\omega=\omega_{cr}} = 3mga > 0.$$

Since in this special case ($\omega = \omega_{cr}$) the fourth derivative at the point $\theta_3 = 0$ is strictly positive, the equilibrium position $\theta_3 = 0$ is also a position of stable equilibrium, but the small oscillations about this position are not harmonic anymore.

c) Obviously, small oscillations can take place only about a position of stable equilibrium. According to our investigation, there exist two positions with this property in our case: $\theta_1 = 0$ (if $\omega < \omega_{cr}$), and $\theta_3 = \arccos\left(\frac{\omega_{cr}}{\omega}\right)^2$ (if $\omega > \omega_{cr}$). The most important question arising in a problem of small oscillations is to determine the periods of the normal modes of oscillation. Since our application involves a single degree of freedom, this part of investigation is not very difficult.

There are several procedures used to determine the period(s) of oscillation of a physical system performing small oscillations. If the system possesses more than one degree of freedom, the Newtonian formalism shows to be difficult, or even useless, while the Lagrangian approach offers a simple and elegant way, by solving the characteristic equation of the system. (This is an equation of order n in ω^2, with real and positive solutions, n being the number of degrees of freedom).

Let us first briefly discuss the solution to the problem for systems with one degree of freedom, from the Newtonian point of view. This approach is based on the analogy with the simplest mechanical system performing linear (monodimensional) harmonic oscillations, which is a body (particle) of mass m, connected to a spring of elastic constant k. As known, the period of small oscillations of such a system is

$$\tau = 2\pi \sqrt{\frac{m}{k}}. \tag{7.7}$$

In fact, this is one of the simplest mechanical systems, acted by an elastic-type force $\vec{F} = -k\vec{x}$. Since the field of the elastic forces is a potential field, we have

$$dV = -dA = -\vec{F} \cdot d\vec{x} = k\vec{x} \cdot d\vec{x} = k\,x\,dx,$$

which yields

$$V = k \int x\,dx = \frac{1}{2}kx^2 + V_0.$$

If we choose $x = 0$ as the "reference level" (corresponding to the undeformed spring) for the potential energy, then we can write

$$V = \frac{kx^2}{2}. \tag{7.8}$$

Here are two simple, Newtonian procedures to determine the period of linear harmonic small oscillations. The first one - which we may call "dynamical" - consists in determination of both the resulting force acting on the system, and the resulting acceleration of the system. If the resulting force is elastic, being of the form $\vec{F} = -K\vec{\xi}$, where $\vec{\xi}$ is elongation of the motion, then it allows one to determine the "elastic constant" K (analogous to k interfering in $\vec{F} = -k\vec{x}$). In its turn, the resultant acceleration - by means of the fundamental equation of dynamics $\vec{F} = M\vec{a} = M\ddot{\vec{\xi}}$ - enables determination of "the mass" M, which is analogous to mass m of the body connected to the spring with elastic constant k. Then, by analogy with (7.7), the period of the small oscillations of the system writes

$$\tau = 2\pi \sqrt{\frac{M}{K}}. \tag{7.9}$$

Another simple Newtonian method - let us call it "energetic" - implies determination of both kinetic and potential energies of the system performing small oscillations. If these quantities are expressed as

$$E_c = \frac{1}{2}M\dot{\xi}^2, \qquad E_p = \frac{1}{2}K\xi^2,$$

where ξ has the same significance as above, then by analogy with (7.7), the period of small oscillations is also given by (7.9).

Based on the Lagrangian formalism, we shall now develop a simple and efficient method of determination the oscillation period of any physical system with one degree of freedom, performing small, linear, harmonic oscillations. Working again on the above example, one can remember the potential given by (7.8), while the kinetic energy associated with the single degree of freedom is

$$T = \frac{1}{2}m\dot{x}^2.$$

The Lagrangian of the system therefore is

$$L = T - V = \frac{1}{2}m\dot{x}^2 - \frac{1}{2}kx^2.$$

The corresponding Lagrange equation of the second kind

$$\frac{d}{dt}\left(\frac{\partial L}{\partial \dot{x}}\right) - \frac{\partial L}{\partial x} = 0$$

then leads to the well-known equation of the harmonic oscillator

$$m\ddot{x} + kx = 0,$$

or

$$\ddot{x} + \omega_0^2 x = 0,$$

where $\omega_0^2 = k/m$. As we know (see Chap.III), the solution of this equation could have four versions, with the period given by

$$\tau = \frac{2\pi}{\omega_0} = 2\pi\sqrt{\frac{m}{k}}.$$

Let us consider a physical system with a single degree of freedom, whose motion is unknown so far. Denote by ξ the associated generalized coordinate, and take the Lagrangian of the system of the form

$$L = A\dot{\xi}^2 - B\xi^2, \tag{7.10}$$

where A and B are two real, non-zero constants, both having the same sign. The Lagrange equation

$$\frac{d}{dt}\left(\frac{\partial L}{\partial \dot{\xi}}\right) - \frac{\partial L}{\partial \xi} = 0$$

then yields
$$A\ddot{\xi} + B\xi = 0,$$
or
$$\ddot{\xi} + \Omega_0^2 \xi = 0,$$
with $\Omega_0^2 = B/A$. As one can see, we have obtained the same equation of the linear harmonic oscillator, the oscillation period being

$$\tau = \frac{2\pi}{\Omega_0} = 2\pi\sqrt{\frac{A}{B}}. \qquad (7.11)$$

Consequently, if a physical system with a single degree of freedom has a Lagrangian of the form (7.10), then the system performs an oscillatory harmonic motion, with the period given by (7.11). Such a Lagrangian is called *Lagrangian of small oscillations* associated with a physical system with a single degree of freedom.

Returning now to our problem, in order to determine the period of small oscillations of the particle moving without friction inside the rotating pipe, we first have to write the suitable Lagrangian. To this end, we shall use the Lagrangian given by (7.3):

$$L = \frac{1}{2}m\dot{\theta}^2(a^2 \sin^2\theta + b^2 \cos^2\theta) + \frac{1}{2}m\omega^2 b^2 \sin^2\theta + mga\cos\theta.$$

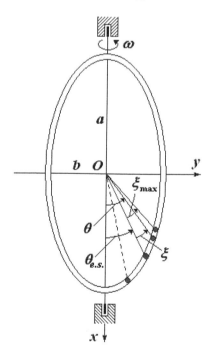

Fig.VII.2

To go further with our investigation, let us suppose that the particle of mass m performs small oscillations about a position of stable equilibrium denoted by $\theta_{se} = const.$ (see Fig.7.2). The elongation ξ of the oscillating motion then is

$$\xi = \theta - \theta_{se}. \tag{7.12}$$

Taking into account the expression of the Lagrangian of small oscillations (7.10), we have to consider the quadratic terms in the small quantity ξ, meaning that the elongation of small oscillations satisfies the relation

$$\mathcal{O}(\xi^3) = 0.$$

The law of the oscillating motion can be written as

$$\xi = \xi_{max} \sin(\Omega_0 t + \varphi_0),$$

where the constants ξ_{max} and φ_0 can be determined by means of initial conditions: $\xi_0 = \xi(t=0)$, and $\dot{\xi}_0 = \dot{\xi}(t=0)$.

There are two ways of writing the Lagrangian of small oscillations in our particular case:

1. Using (7.12), the angle θ is expressed in terms of the new variable ξ

$$\theta = \theta_{se} + \xi,$$

this value is introduced into the Lagrangian (7.3), then expand L in Maclaurin series in terms of ξ about the point $\xi = 0$, keeping the terms up to the second power.

2. Expand the Lagrangian (7.3) in Taylor series in terms of θ about the value $\theta = \theta_{se}$ and keep the terms up to the second power of $(\theta - \theta_{se}) = \xi$.

We shall use the second procedure in our investigation. Observing that L can be written as

$$L = E_{cin} - E_{pot},$$

the expansion in Taylor series of L implies the series expansion of E_{cin} and E_{pot} in terms of θ, about the value $\theta = \theta_{se}$. Therefore,

$$E_{pot}(\theta) = E_{pot}(\theta_{se}) + \frac{1}{1!} \frac{dE_{pot}}{d\theta}\bigg|_{\theta=\theta_{se}} (\theta - \theta_{se})$$

$$+ \frac{1}{2!} \frac{d^2 E_{pot}}{d\theta^2}\bigg|_{\theta=\theta_{se}} (\theta - \theta_{se})^2 + \mathcal{O}\left[(\theta - \theta_{se})^3\right]$$

$$\simeq E_{pot}(\theta_{se}) + \frac{1}{2!}\frac{d^2 E_{pot}}{d\theta^2}\bigg|_{\theta=\theta_{se}}\xi^2 \equiv E_{pot}(\xi),$$

where we have used the property of the equilibrium positions

$$\frac{dE_{pot}}{d\theta}\bigg|_{\theta=\theta_{se}} = 0,$$

for any kind of equilibrium (stable or unstable).

We also have

$$E_{cin}(\theta,\dot{\theta}) = \frac{1}{2}m\dot{\theta}^2(a^2\sin^2\theta + b^2\cos^2\theta) = \frac{1}{2}m\dot{\xi}^2(a^2\sin^2\theta + b^2\cos^2\theta)$$

$$= E_{cin}(\theta,\dot{\xi}) = E_{cin}(\theta_{se},\dot{\xi}) + \frac{1}{1!}\frac{dE_{cin}}{d\theta}\bigg|_{\theta=\theta_{se}}(\theta-\theta_{se})$$

$$+\frac{1}{2!}\frac{d^2 E_{cin}}{d\theta^2}\bigg|_{\theta=\theta_{se}}(\theta-\theta_{se})^2 + \mathcal{O}\left[(\theta-\theta_{se})^3\right] \simeq E_{cin}(\theta_{se},\dot{\xi}).$$

Here we used the fact that the term

$$\frac{1}{1!}\frac{dE_{cin}}{d\theta}\bigg|_{\theta=\theta_{se}}(\theta-\theta_{se})$$

$$= \frac{1}{1!}\frac{d}{d\theta}\left[\frac{1}{2}\dot{\xi}^2(a^2\sin^2\theta+b^2\cos^2\theta)\right]\bigg|_{\theta=\theta_{se}}\xi$$

contains the factor $\dot{\xi}^2\xi$, which is of the third power in ξ. Therefore

$$E_{cin}(\theta,\dot{\theta}) \simeq E_{cin}(\theta_{se},\dot{\xi}) = \frac{1}{2}m\dot{\xi}^2(a^2\sin^2\theta_{se}+b^2\cos^2\theta_{se}) \equiv E_{cin}(\dot{\xi}).$$

The term $E_{pot}(\theta_{se})$ is a constant and, according to analytical formalism, can be omitted in the Lagrangian of the small oscillations L_{so}. Consequently,

$$L_{so} = E_{cin}(\theta,\dot{\theta}) - E_{pot}(\theta) = E_{cin}(\dot{\xi}) - E_{pot}(\xi)$$

$$= \frac{1}{2}m\dot{\xi}^2(a^2\sin^2\theta_{se}+b^2\cos^2\theta_{se})$$

$$-\frac{1}{2!}\frac{d^2 E_{pot}}{d\theta^2}\bigg|_{\theta=\theta_{se}}\xi^2 \equiv L(\xi,\dot{\xi}) = A\dot{\xi}^2 - B\xi^2,$$

with

$$A = \frac{1}{2}m(a^2\sin^2\theta_{se}+b^2\cos^2\theta_{se}) = const. > 0,$$

and
$$B = \frac{1}{2!}\frac{d^2 E_{pot}}{d\theta^2}\bigg|_{\theta=\theta_{se}} = const. > 0.$$

In view of (7.11), the period of small oscillations of the system is

$$\tau = 2\pi \sqrt{\frac{m(a^2 \sin^2 \theta_{se} + b^2 \cos^2 \theta_{se})}{\frac{d^2 E_{pot}}{d\theta^2}\bigg|_{\theta=\theta_{se}}}}$$

$$= 2\pi \sqrt{\frac{m(a^2 \sin^2 \theta_{se} + b^2 \cos^2 \theta_{se})}{[m\omega^2 b^2(1 - 2\cos^2\theta) + mga\cos\theta]\bigg|_{\theta=\theta_{se}}}}$$

$$= 2\pi \sqrt{\frac{\frac{1}{\omega^2}\frac{a^2}{b^2}\sin^2\theta_{se} + \frac{1}{\omega^2}\cos^2\theta_{se}}{1 - 2\cos^2\theta_{se} + \left(\frac{\omega_{cr}}{\omega}\right)^2 \cos\theta_{se}}}$$

$$= \frac{2\pi}{\omega}\sqrt{\frac{\left(1 - \frac{a^2}{b^2}\right)\cos^2\theta_{se} + \frac{a^2}{b^2}}{1 - 2\cos^2\theta_{se} + \left(\frac{\omega_{cr}}{\omega}\right)^2 \cos\theta_{se}}}$$

$$= \frac{2\pi}{\omega\sqrt{1-e^2}}\sqrt{\frac{1 - e^2\cos^2\theta_{se}}{1 - 2\cos^2\theta_{se} + \left(\frac{\omega_{cr}}{\omega}\right)^2 \cos\theta_{se}}}, \quad (7.13)$$

where
$$e = \frac{c}{a} = \frac{\sqrt{a^2 - b^2}}{a} = \sqrt{1 - \left(\frac{b}{a}\right)^2}$$

is the ellipse eccentricity.

We are now able to calculate the oscillation period for both positions of stable equilibrium $\theta_1 = 0$ (for $\omega < \omega_{cr}$) and $\theta_3 = \arccos\left(\frac{\omega_{cr}}{\omega}\right)^2$ (for $\omega > \omega_{cr}$), as follows:

$$\tau_1 = \tau|_{\theta_{se}=0} = \frac{2\pi}{\omega\sqrt{1-e^2}}\sqrt{\frac{1-e^2}{\left(\frac{\omega_{cr}}{\omega}\right)^2 - 1}}$$

$$= \frac{2\pi}{\sqrt{\omega_{cr}^2 - \omega^2}} \quad (\omega < \omega_{cr}), \quad (7.14)$$

and
$$\tau_2 = \tau|_{\theta_{se}=\arccos\left(\frac{\omega_{cr}}{\omega}\right)^2}$$

$$= \frac{2\pi}{\omega\sqrt{1-e^2}}\sqrt{\frac{1-e^2\cos^2\theta_{se}}{1-2\cos^2\theta_{se}+\left(\frac{\omega_{cr}}{\omega}\right)^2\cos\theta_{se}}}$$

$$= \frac{2\pi}{\omega\sqrt{1-e^2}}\sqrt{\frac{1-e^2\left(\frac{\omega_{cr}}{\omega}\right)^4}{1-\left(\frac{\omega_{cr}}{\omega}\right)^4}} \quad (\omega > \omega_{cr}). \tag{7.15}$$

If $\omega \to \infty$, then $\theta_3 = \arccos\left(\frac{\omega_{cr}}{\omega}\right)^2$, as a position of stable equilibrium for $\omega > \omega_{cr}$, gives rise to a new position of stable equilibrium, namely:

$$\theta_4 = \lim_{\omega\to\infty}\theta_3 = \lim_{\omega\to\infty}\left[\arccos\left(\frac{\omega_{cr}}{\omega}\right)^2\right] = \arccos 0 = \frac{\pi}{2}, \tag{7.16}$$

in complete agreement with our initial qualitative analysis. The period of small oscillations about this position of equilibrium is

$$\tau_3 = \tau|_{\theta_{se}=\frac{\pi}{2}} = \frac{2\pi}{\omega}\frac{1}{\sqrt{1-e^2}} = \tau_0\frac{1}{\sqrt{1-e^2}}, \tag{7.17}$$

where notation $\tau_0 = 2\pi/\omega$ is obvious. If the ellipse degenerates into a circle ($e = 0$), we are left with

$$\tau_3 = \tau|_{\theta_{se}=\frac{\pi}{2}} = \tau_0. \tag{7.18}$$

(B) Non-inertial reference frame (NIRF)

a) In order to determine the equilibrium positions of the particle of mass m we shall use the same requirement of cancellation of the first derivative of potential energy. As a non-inertial reference system, one can consider a frame invariably connected to the moving particle. Unlike the inertial frame, for an observer situated in the non-inertial system not only the gravitational force, but also the centrifugal force $\vec{F}_{cf}$ has to be considered as an applied force. Therefore, the potential of the resultant of applied forces satisfies the relation

$$dV = -dA = -(\vec{G}\cdot d\vec{r} + \vec{F}_{cf}\cdot d\vec{r}), \tag{7.19}$$

where $\vec{F}_{cf}$ is orthogonal to the rotation axis and points radially outwards (see Fig.VII.3).

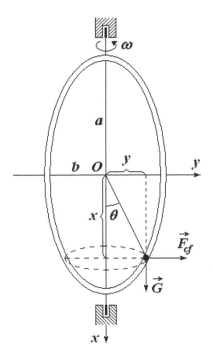

Fig. VII.3

To facilitate the calculation of the scalar products in (7.19), let us write the vectors $\vec{G}$, $\vec{F}_{cf}$, and $d\vec{r}$ in terms of their components. According to Fig. VII.3, we have:

$$\vec{G} = (mg, 0, 0), \quad \vec{F}_{cf} = (0, m\omega^2 y, 0), \quad d\vec{r} = (dx, dy, dz),$$

and (7.19) becomes

$$dV = -mg\, dx - m\omega^2 y\, dy,$$

or, by integration,

$$V = -mgx - \frac{1}{2}m\omega^2 y^2 + V_0. \tag{7.20}$$

Since the potential does not explicitly depend on time, it coincides with the potential energy of the system. The most convenient choice of the arbitrary constant of integration V_0 is

$$V(x=0, y=0) = V_0 = 0,$$

and the potential energy of the body (particle) becomes

$$V = V(x, y) = -mgx - \frac{1}{2}m\omega^2 y^2 = -mga\cos\theta - \frac{1}{2}m\omega^2 b^2 \sin^2\theta$$

$$= V(\theta) \equiv E_{pot}(\theta).$$

This result has been already obtained in our approach (A) to the problem (study in an inertial frame), but this time (working in NIRF) the formula of the potential energy was found in a more simple way. This observation is valid in general, for more complicated applications, but everything depends on the ability of the researcher. From now on, the reasoning and calculations shall follow the way developed within the inertial frame.

Problem 2. A particle P of mass m and electric charge q moves without friction inside a pipe of elliptic shape and constant cross section, rotating in vacuum about its major axis with constant angular velocity ω. A point charge q' is fixed at the lowest point of the ellipse (see Fig.VII.4). Determine:

a) The equilibrium positions of the particle with charge q;
b) Stability of these positions of equilibrium;
c) Period of small oscillations of the particle about the stable positions of equilibrium.

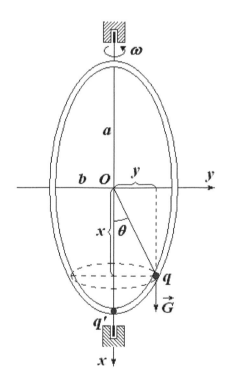

Fig.VII.4

Solution.
a) This problem is similar to the previous one, so that we shall assume the results of our preceding investigation. This time, in addition to the gravitational force, we have to consider the electrostatic force of interaction between charges q and q'. Before going further, we have to mention that the magnetic effects due to the motion of particle with charge q are neglected. We also specify that our reference frame is the laboratory frame, which is an inertial reference system.

Since the electrostatic and gravitational fields are conservative, our system (*i.e.* the particle of mass m and charge q) is a natural system. The kinetic energy is the same as in Problem 1

$$T = \frac{1}{2}m\dot{\theta}^2(a^2\sin^2\theta + b^2\cos^2\theta) + \frac{1}{2}m\omega^2 b^2\sin^2\theta,$$

but to the gravitational potential energy

$$V_g = -mga\cos\theta$$

a supplementary electrostatic term has to be added

$$V_{el} = \frac{1}{4\pi\varepsilon_0}\frac{qq'}{\sqrt{(a-x)^2+y^2}} = \frac{1}{4\pi\varepsilon_0}\frac{qq'}{\sqrt{4a^2\sin^4\frac{\theta}{2}+b^2\sin^2\theta}}$$

$$= \frac{qq'(4\pi\varepsilon_0)^{-1}}{\sqrt{4a^2\sin^4\frac{\theta}{2}+4b^2\sin^2\frac{\theta}{2}\cos^2\frac{\theta}{2}}} = \frac{qq'(4\pi\varepsilon_0)^{-1}}{2|\sin\frac{\theta}{2}|\sqrt{(a^2-b^2)\sin^2\frac{\theta}{2}+b^2}}$$

$$= \frac{qq'}{8\pi\varepsilon_0}\frac{1}{\sin\frac{\theta}{2}\sqrt{c^2\sin^2\frac{\theta}{2}+b^2}},$$

where we used the fact that $\sin\frac{\theta}{2} > 0$ on the interval $\theta \in (0,\pi]$.

The Lagrangian of our problem therefore is

$$L = \frac{1}{2}m\dot{\theta}^2(a^2\sin^2\theta + b^2\cos^2\theta) + \frac{1}{2}m\omega^2 b^2\sin^2\theta$$

$$+ mga\cos\theta - \frac{qq'}{8\pi\varepsilon_0}\frac{1}{\sin\frac{\theta}{2}\sqrt{c^2\sin^2\frac{\theta}{2}+b^2}}.$$

Since L does not explicitly depend on time, the Lagrange equations of the second kind admit the first integral

$$\dot{\theta}\frac{\partial L}{\partial \dot{\theta}} - L = const.,$$

where the constant is the total energy of the particle E_{tot}

$$const. = E_{tot} = E_{cin} + E_{pot} = \dot\theta \frac{\partial L}{\partial \dot\theta} - L = \frac{1}{2}m\dot\theta^2(a^2\sin^2\theta + b^2\cos^2\theta)$$

$$-\frac{1}{2}m\omega^2 b^2 \sin^2\theta - mga\cos\theta + \frac{qq'}{8\pi\varepsilon_0}\frac{1}{\sin\frac{\theta}{2}\sqrt{c^2\sin^2\frac{\theta}{2}+b^2}}.$$

By identification, we then have

$$E_{cin} = \frac{1}{2}m\dot\theta^2(a^2\sin^2\theta + b^2\cos^2\theta), \qquad (7.21)$$

and

$$E_{pot} = -\frac{1}{2}m\omega^2 b^2 \sin^2\theta - mga\cos\theta + \frac{qq'}{8\pi\varepsilon_0}\frac{1}{\sin\frac{\theta}{2}\sqrt{c^2\sin^2\frac{\theta}{2}+b^2}}.$$
(7.21')

It is worthwhile to mention, as we previously did, that within the analytical formalism, the kinetic energy can be expressed only in terms of the squared generalized velocities.

To find the equilibrium positions, one takes the first derivative of the effective potential energy $E_{pot}(\theta)$:

$$\frac{dE_{pot}}{d\theta} = -m\omega^2 b^2 \sin\theta\cos\theta + mga\sin\theta$$

$$-\frac{qq'}{8\pi\varepsilon_0}\frac{\frac{1}{2}\cos\frac{\theta}{2}\sqrt{c^2\sin^2\frac{\theta}{2}+b^2} + \frac{c^2\sin^2\frac{\theta}{2}\cos\frac{\theta}{2}}{2\sqrt{c^2\sin^2\frac{\theta}{2}+b^2}}}{\sin^2\frac{\theta}{2}(c^2\sin^2\frac{\theta}{2}+b^2)}$$

$$= -m\omega^2 b^2 \sin\theta\cos\theta + mga\sin\theta$$

$$-\frac{qq'}{8\pi\varepsilon_0}\frac{2c^2\sin^2\frac{\theta}{2}\cos\frac{\theta}{2} + b^2\cos\frac{\theta}{2}}{2\sin^2\frac{\theta}{2}(c^2\sin^2\frac{\theta}{2}+b^2)^{3/2}}.$$

The equilibrium positions of the system are obtained as solutions of the equation $\frac{dE_{pot}}{d\theta} = 0$, that is

$$\sin\theta\left[m\omega^2 b^2 \cos\theta - mga + \frac{qq'}{8\pi\varepsilon_0}\frac{2c^2\sin\frac{\theta}{2} + b^2\csc\frac{\theta}{2}}{4\sin^2\frac{\theta}{2}(c^2\sin^2\frac{\theta}{2}+b^2)^{3/2}}\right] = 0,$$

or, equivalently

$$\begin{cases} \sin\theta = 0; \\ m\omega^2 b^2 \cos\theta - mga + \frac{qq'}{8\pi\varepsilon_0}\frac{2c^2\sin\frac{\theta}{2}+b^2\csc\frac{\theta}{2}}{4\sin^2\frac{\theta}{2}(c^2\sin^2\frac{\theta}{2}+b^2)^{3/2}} = 0. \end{cases}$$

The first equation gives two positions of equilibrium

$$\theta_1^{(1)} = 0, \qquad \theta_2^{(1)} = \pi,$$

but only the second is acceptable from the physical point of view. Denoting $k = 4\pi\varepsilon_o$, the second equation can also be written as

$$8k\, m \sin^3 \frac{\theta}{2} \left(c^2 \sin^2 \frac{\theta}{2} + b^2\right)^{3/2} (\omega^2 b^2 \cos\theta - ga)$$

$$+ qq' \left(2c^2 \sin^2 \frac{\theta}{2} + b^2\right) = 0. \qquad (7.22)$$

Using formula $\cos\theta = 1 - 2\sin^2 \frac{\theta}{2}$ and squaring the last equation, we still have

$$64k^2 m^2 \sin^6 \frac{\theta}{2} \left(c^2 \sin^2 \frac{\theta}{2} + b^2\right)^3 \left[ga - \omega^2 b^2 \left(1 - 2\sin^2 \frac{\theta}{2}\right)\right]^2$$

$$- (qq')^2 \left(2c^2 \sin^2 \frac{\theta}{2} + b^2\right)^2 = 0. \qquad (7.23)$$

This is a trigonometric equation for the unknown variable $0 < \theta \leq \pi$. Its real solutions $\theta_i^{(2)}$ ($i = 1, 2, ...$) give new equilibrium positions. As one can see, it is not possible to solve this equation analytically, so that we have to follow a numerical procedure. Even so, due to the multitude of values which can be taken by the constant quantities a, b, q, q', m, and ω, it is a very difficult task to solve the equation.

We first observe that the solutions of equation (7.23) is not affected by the sign of the electric charges. Since discussion of the general case is very complicated, we shall focus our attention upon analysis in terms of m, q, and ω. To this purpose, we shall use the software *Mathematica*, specialized in analytical and numerical calculations. *Mathematica* furnishes a simple and convenient way to solve this problem, using the graphic representation of ω, considered as an independent variable (or, better, as a parameter) in terms of θ (taken as an dependent variable). This dependence is implicitly given by means of equation (7.23). For the calculations to follow, we chose the following values for the interfering constants: $k = 4\pi\varepsilon_0 = 10^{-9} F/m$, $a = 0.2\,m$, $b = 0.15\,m$, $q = 10^{-3} C$, $q' = 5.10^{-3} C$.

1) $m = 1\,\mu g = 10^{-6} kg$

The command lines allowing the graphic representation of interdependence between ω and θ are:

```
Remove["Global`*"];
Unprotect[In, Out];
Clear[In, Out];

k = 10^-9/9;
a = 0.2;
b = 0.15;
c = Sqrt[a^2-b^2];
g = 9.8;
m = 10^-6;
q = 10^-3;
q1 = 5*10^-3;

NSolve[64*k^2*m^2*(Sin[Theta/2]^2)^3*(c^2*(Sin[Theta/2]^2)+b^2)^3*
(g*a - ω^2*b^2*(1 – 2*(Sin[Theta/2]^2)))^2 - (q*q1)^2*(2*c^2*
(Sin[Theta/2]^2) + b^2)^2 == 0, Theta];

Plot[Evaluate[ Theta /. %], {ω, 10^6, 3*10^7}]
```

The dependence $\theta = \theta(\omega)$ is graphically represented in Fig.VII.5. As can be observed, there is a minimum value of the angular velocity $\omega = \omega_{min}$, below which equation has no real solutions. This value can be determined observing that $\theta^{(2)} \to \pi$ for $\omega \to \omega_{min}$. Solving equation (7.23) for $\theta = \pi$, one finds $\omega_{min} = 1340475.662 \; rad \cdot s^{-1}$.

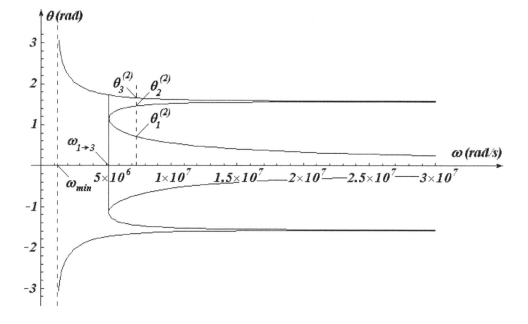

Fig.VII.5

One also observes that until a value of ω denoted $\omega_{1\to 3}$, the variation of θ in terms of ω is univocal, but beginning with $\omega_{1\to 3}$ the graph begins to split in three branches. This shows that within interval $\omega \in (\omega_{min}, \omega_{1\to 3})$ equation (7.23) has a single root and, consequently, there exists a single equilibrium position, while for $\omega \in (\omega_{1\to 3}, \infty)$ equation (7.23) has three solutions, corresponding to three equilibrium positions of the system, determined by three values of angle θ, denoted in Fig.VII.5 by $\theta_1^{(2)}$, $\theta_2^{(2)}$ and $\theta_3^{(2)}$.

In addition, as shown in Fig.VII.6, in the (theoretical) limit $\omega \to \infty$, the equilibrium value $\theta_1^{(2)}$ of θ tends to a minimum value $\theta_{1min}^{(2)}$, determined by the equilibrium between gravitational and electrostatic forces (if the problem is analyzed in an inertial frame), and between gravitational, electrostatic and centrifugal forces (if the problem is approached in a noninertial frame, invariably connected to the charged body, rotating about x-axis with angular velocity ω). In the same limit $\omega \to \infty$, the other two equilibrium values $\theta_2^{(2)}$ and $\theta_3^{(2)}$ tend to the same limit $\lim_{\omega\to\infty}\theta_2^{(2)} = \lim_{\omega\to\infty}\theta_3^{(2)} = \frac{\pi}{2}$.

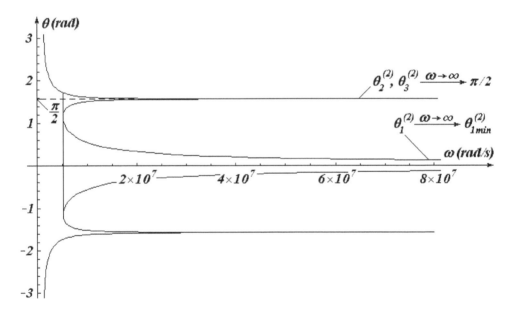

Fig.VII.6

Observation. The vertical line connecting the branches of the graph $\theta = \theta(\omega)$ near the value $\omega_{1\to 3}$ in Fig.VII.5 (also appearing in some forthcoming drawings, like Fig.VII.10 and/or Fig.VII.14) are due to certain specific errors in graphic representations by means of instruction "Plot" of *Mathematica*. This kind of error is generated when the first argument in "Plot" does not evaluate to a machine-size real number at the indicated point[1]. The cause of this error can often be determined by temporarily assigning the indicated value to the plot variable, evaluating the first argument of the plotting function, and observing whether or not the result is a real number.

Separation of the two distinct domains, corresponding to one and three solutions for equation (7.23), respectively, is displayed even more clearly in Fig.VII.7. As can be observed, for $\omega < \omega_{1\to3}$ equation (7.23) has a single solution, $\theta_1^{(2)}$, while for $\omega > \omega_{1\to3}$ there exist three solutions $\theta_1^{(2)}$, $\theta_2^{(2)}$ and $\theta_3^{(2)}$. Not to mention the equilibrium position corresponding to the analytical solution $\theta_2^{(1)} = \pi$ of equation (7.23), in addition to the equilibrium positions determined by the numerical solutions of (7.23).

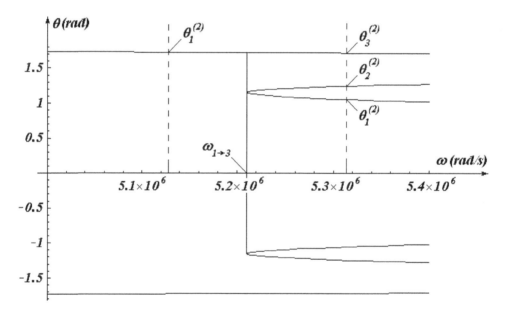

Fig.VII.7

2) $m = 1\,g = 10^{-3}kg$

In the same way, for ω_{min} one obtains the value $\omega_{min} = 42389,561$ $rad \cdot s^{-1}$; therefore, if $\omega < \omega_{min}$, the system has no extra equilibrium position (in addition to the analytic solution $\theta_2^{(1)} = \pi$). This fact has a graphic representation in Fig.VII.8, where dependence $\theta = \theta(\omega)$ is

---

[1] The "indicated point" is given in every line error that appears in *Mathematica* during the calculations (i.e., during the time interval between the instant of launching the "Plot" command and the instant when the graphical representation is depicted).

given for a large scale variation of ω. As one can see, there are no major qualitative differences between the last two cases.

The analysis for the case $m = 10^{-3} kg$ is performed as for the previous case $m = 10^{-6} kg$. The only clear-cut difference refers to the order of magnitude of ω corresponding to the number of equilibrium positions of the system. This can be easily observed by comparing the graphic representations of the two cases (Fig.VII.6 with Fig.VII.8, or Fig.VII.7 with Fig.VII.9 and Fig.VII.10).

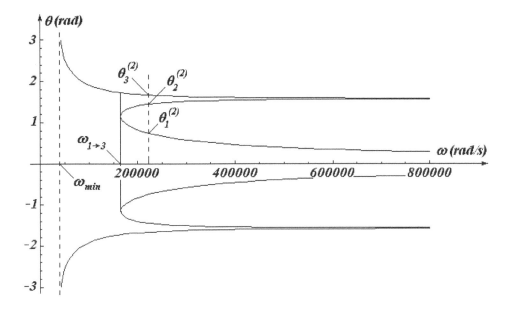

Fig.VII.8

We realize that an increase of three times of the order of magnitude of mass produces a two-order of magnitude reduction of the angular velocity corresponding to separation of the two domains, associated with one or three equilibrium positions of the system, respectively.

Fig.VII.9 shows a graphic representation of the dependence $\theta = \theta(\omega)$ for a narrow interval of variation of angular velocity ω, situated to the left of $\omega_{1\to 3}$, that is within the domain in which the system possess a single equilibrium position, $\theta_1^{(2)}$.

The same dependence is displayed in Fig.VII.10 and Fig.VII.11, but for an interval of variation of ω situated to the right of $\omega_{1\to 3}$, that is within the domain where the body of mass m and charge q has three equilibrium positions. This has been done to give even a more clear explanation regarding displacement of the three equilibrium positions to the limit values discussed at point 1). Indeed, comparison between Fig.VII.10 and Fig.VII.11 shows that, together with increase of ω, the "separation/distance" between the equilibrium positions $\theta_2^{(2)}$ and $\theta_3^{(2)}$ (both tending to the same limit for $\omega \to \infty$) decreases, while the "distance" between the "pair" $(\theta_2^{(1)}, \theta_3^{(2)})$ and $\theta_1^{(2)}$ increases.

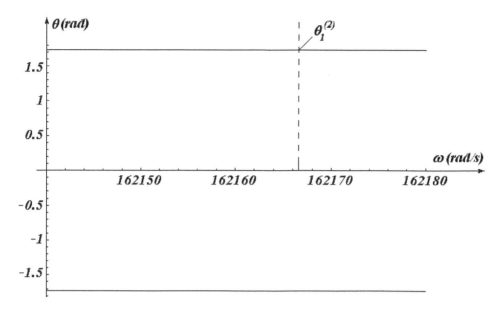

Fig.VII.9

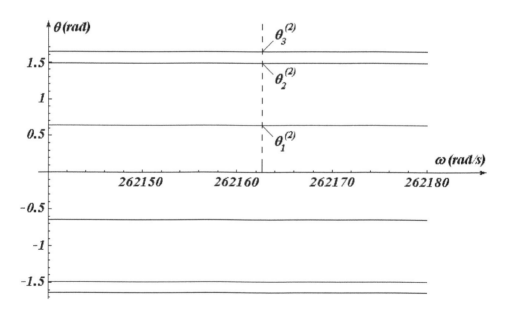

Fig.VII.10

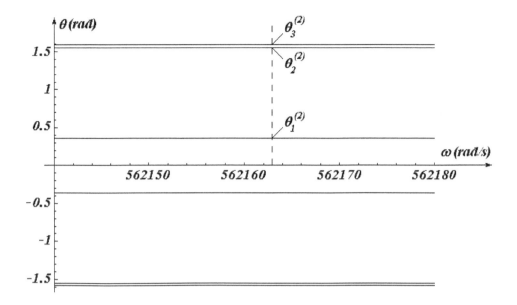

Fig.VII.11

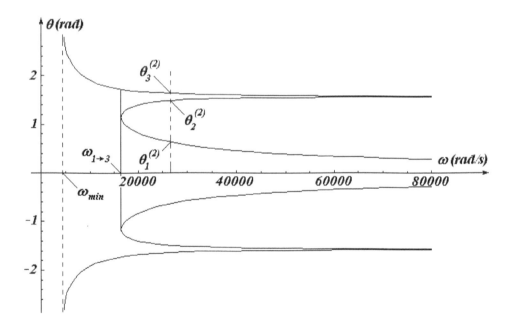

Fig.VII.12

3) $m = 100\,g = 10^{-1}kg$

In this case, the graphic representation of $\theta = \theta(\omega)$ is given in Fig. VII.12. It can be seen that the qualitative dependence of θ on ω keeps the same properties as in the previous two cases. The minimum value of ω now is $\omega_{min} = 4238,946\,rad \cdot s^{-1}$. As in the preceding case, the main difference as regarded the first case consists in diminution of the order of magnitude of angular velocity w, with respect to the mass increase. So, to an increase of five orders of magnitude (as compared to the first case), and two orders of magnitude (as compared to the second case) of the mass, corresponds a decrease of three orders of magnitude of angular velocity (with respect to the first case), and one order of magnitude (as compared to the second case), respectively.

b) To determine stability of the above discussed equilibrium positions we must investigate the sign of the second derivative of the effective potential energy in these points. We have:

$$\frac{d^2 E_{pot}}{d\theta^2} = -m\omega^2 b^2 \cos^2\theta + m\omega^2 b^2 \sin^2\theta + mga\cos\theta - \frac{qq'}{8\pi\varepsilon_0}$$

$$\times \frac{2\left(2c^2 \sin\frac{\theta}{2}\cos^2\frac{\theta}{2} - c^2 \sin^3\frac{\theta}{2} - \frac{1}{2}b^2 \sin\frac{\theta}{2}\right)\sin^2\frac{\theta}{2}\left(c^2 \sin^2\frac{\theta}{2} + b^2\right)^{3/2}}{4\sin^4\frac{\theta}{2}\left(c^2 \sin^2\frac{\theta}{2} + b^2\right)^3}$$

$$+\frac{qq'}{8\pi\varepsilon_0} \frac{2\left(2c^2 \sin^2\frac{\theta}{2}\cos\frac{\theta}{2} + b^2 \cos\frac{\theta}{2}\right)\left[\sin\frac{\theta}{2}\cos\frac{\theta}{2}\left(c^2\sin^2\frac{\theta}{2}+b^2\right)^{3/2}\right]}{4\sin^4\frac{\theta}{2}\left(c^2\sin^2\frac{\theta}{2}+b^2\right)^3}$$

$$+\frac{qq'}{8\pi\varepsilon_0}$$

$$\times \frac{2\left(2c^2 \sin^2\frac{\theta}{2}\cos\frac{\theta}{2}+b^2\cos\frac{\theta}{2}\right)\left[\frac{3}{2}c^2 \sin^3\frac{\theta}{2}\cos\frac{\theta}{2}\left(c^2\sin^2\frac{\theta}{2}+b^2\right)^{1/2}\right]}{4\sin^4\frac{\theta}{2}\left(c^2\sin^2\frac{\theta}{2}+b^2\right)^3}$$

$$= -m\omega^2 b^2 \cos^2\theta + m\omega^2 b^2 \sin^2\theta + mga\cos\theta + \frac{qq'}{4\pi\varepsilon_0}\frac{\sin^{-3}\frac{\theta}{2}}{8\left(c^2\sin^2\frac{\theta}{2}+b^2\right)^{5/2}}$$

$$\times \left[2c^4 \sin^4\frac{\theta}{2}\left(3 - 2\sin^2\frac{\theta}{2}\right) + c^2 b^2 \sin^2\frac{\theta}{2}\left(5 - 2\sin^2\frac{\theta}{2}\right) + b^4\left(2 - \sin^2\frac{\theta}{2}\right)\right].$$

Denoting

$$f(\theta) = \frac{\sin^{-3}\frac{\theta}{2}}{8\left(c^2\sin^2\frac{\theta}{2}+b^2\right)^{5/2}}$$

$$\times \left[2c^4 \sin^4\frac{\theta}{2}\left(3 - 2\sin^2\frac{\theta}{2}\right) + c^2 b^2 \sin^2\frac{\theta}{2}\left(5 - 2\sin^2\frac{\theta}{2}\right) + b^4\left(2 - \sin^2\frac{\theta}{2}\right)\right],$$

the second derivative with respect to θ of the potential energy becomes:

$$\frac{d^2 E_{pot}}{d\theta^2} = -m\omega^2 b^2 \cos^2\theta + m\omega^2 b^2 \sin^2\theta + mga\cos\theta + \frac{qq'}{4\pi\varepsilon_0}f(\theta)$$

$$= -m\omega^2 b^2 (1 - 2\sin^2\theta) + mga\cos\theta + k^{-1}qq' f(\theta). \quad (7.24)$$

To investigate the sign of this expression, we shall appeal to soft *Mathematica*. As can be observed, the sign of this expression significally depends on both mass and angular velocity, for given values of the other quantities. So, for $m = 1\,g$, if $\omega_{min} < \omega < \omega_e^* = 585310,223\,rad\cdot s^{-1}$, equation

$$\frac{d^2 E_{pot}}{d\theta^2} = -m\omega^2 b^2 (1 - 2\sin^2\theta) + mga\cos\theta + k^{-1}qq' f(\theta) = 0 \quad (7.25)$$

has a single positive real solution, therefore the function $F(\theta) = \frac{d^2 E_{pot}}{d\theta^2}$ changes its sigh once in interval $\theta \in \left(0, \pi\right]$, its graph being shown in Fig.VII.13. The positive, real solution of equation (7.25), corresponding to $\omega = 200000\,rad\cdot s^{-1}$ (this value of ω is arbitrarily chosen, being contained in the above-mentioned interval of variation) can be determined by the command line

**NSolve[(q*q1/k)*(1/x^3)*(1/(8*(c^2*x^2+b^2)^(5/2)))*(2*c^4*x^4*
(3-2*x^2)+c^2*b^2*x^2*(5-2*x^2)+b^4*(2-x^2))-m* ⍵^2*b^2*
(1-8*x^2(1-x^2))+m*g*a*(1-2*x^2)== 0, x]**

The physically acceptable solution of the equation in x is $x_0 = 0,9299$, corresponding to an angle of value $\theta_0 = 2,388\,rad \simeq 0,76\,\pi\,rad$. Next command lines allow one to give a graphic representation of $F(\theta)$ in the interval $\theta \in (0, \pi]$, for the arbitrary values $m = 1\,g$ and $\omega = 2 \times 10^5\,rad\cdot s^{-1}$.

**Remove["Global`*"];
Unprotect[In, Out];
Clear[In, Out];**

**k = 10^-9/9;
a = 0.2;
b = 0.15;
c = Sqrt[a^2-b^2];
g = 9.8;
m = 0.001;
q = 10^-3;
q1 = 5*10^-3;
⍵ = 2*10^5;**

**Plot[(q*q1/k)*(1/Sin[th/2]^3)*(1/(8*(c^2*Sin[th/2]^2+b^2)^(5/2)))*(2*c^4*
Sin[th/2]^4*(3-2*Sin[th/2]^2)+c^2*b^2*Sin[th/2]^2*(5-2*Sin[th/2]^2)+b^4*
(2-Sin[th/2]^2))-m* ⍵^2*b^2*(1-2*Sin[th]^2)+m*g*a*Cos[th],{th,0,Pi}]**

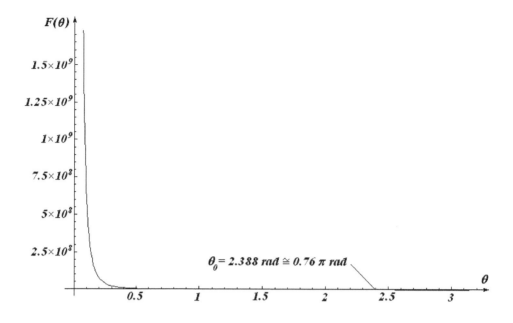

Fig.VII.13

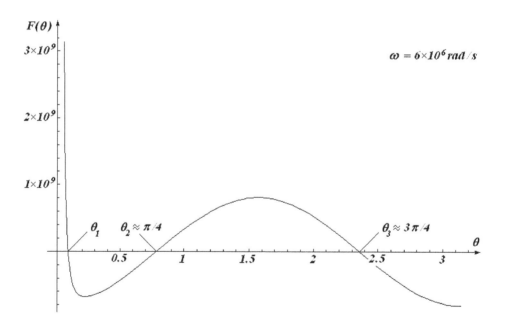

Fig.VII.14

According to Fig.VII.13, the equilibrium positions (physically acceptable) obtained as solutions of equation (7.23) situated within the interval $\theta \in (0, \theta_0)$ are positions of *stable equilibrium*, while those belonging to the interval $\theta \in (\theta_0, \pi]$ are positions of *unstable equilibrium*.

For $\omega > \omega^*$, equations (7.25) admits three real, positive, physically acceptable solutions. In other words, in the interval $\theta \in (\theta_0, \pi]$ function $F(\theta)$ changes its sign three times (see Fig.VII.14, which gives a graphic representation of $F(\theta)$ for a conveniently chosen $\omega = 6 \times 10^6 rad \cdot s^{-1}$).

Consequently, the equilibrium positions (obtained as solutions of equation (7.23)) satisfying relation $\theta \in (0, \theta_1) \cup (\theta_2, \theta_3)$ are positions of *stable equilibrium* (the second derivative of the effective potential energy is positive), while those situated within the interval $\theta \in (\theta_1, \theta_2) \cup (\theta_1, \pi]$ are positions of *unstable equilibrium*. The numerical analysis shows that the smallest solution of (7.25) diminishes with the increase of ω, tending to zero when $\omega \to \infty$ (see Fig.VII.17). Therefore, at high angular velocities the stable equilibrium positions are contained *only* within the interval $\theta \in (\theta_2, \theta_3) \approx \left(\frac{\pi}{4}, \frac{3\pi}{4}\right)$.

c) Small oscillations can occur, obviously, only about positions of stable equilibrium. Since we have already obtained the kinetic energy in the frame of Problem 1, the same formula

$$\tau = 2\pi \sqrt{\frac{m(a^2 \sin^2 \theta_{se} + b^2 \cos^2 \theta_{se})}{(d^2 E_{pot}/d\theta^2)|_{\theta=\theta_{se}}}},$$

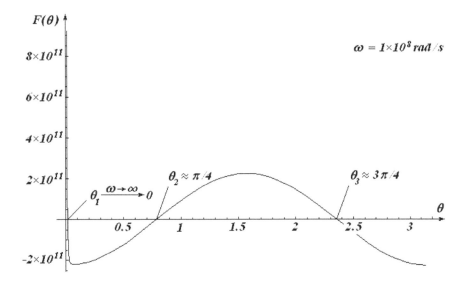

Fig.VII.15

can be used to find the oscillation period of the system about the position of stable equilibrium, where the values of θ_{se} are replaced with the expressions determined at the points a) and b) of this problem.

Particular case: $a = b = R$ (the ellipse degenerates into a circle of radius R). In this case

$$V_{el} = \frac{1}{4\pi\varepsilon_0} \frac{qq'}{2R\sin\frac{\theta}{2}},$$

and the Lagrangian becomes

$$L = \frac{1}{2}mR^2\dot{\theta}^2 + \frac{1}{2}m\omega^2 R^2 \sin^2\theta + mgR\cos\theta - \frac{1}{4\pi\varepsilon_0}\frac{qq'}{2R\sin\frac{\theta}{2}}.$$

This formula shows that the effective potential energy writes [see (7.21)]:

$$E_{pot} = -\frac{1}{2}m\omega^2 R^2 \sin^2\theta - mgR\cos\theta + \frac{1}{4\pi\varepsilon_0}\frac{qq'}{2R\sin\frac{\theta}{2}}.$$

The equilibrium positions of the system are then obtained as

$$\frac{dE_{pot}}{d\theta} = \frac{d}{d\theta}\left[-\frac{1}{2}m\omega^2 R^2 \sin^2\theta - mgR\cos\theta + \frac{1}{4\pi\varepsilon_0}\frac{qq'}{2R\sin\frac{\theta}{2}}\right]$$

$$= -m\omega^2 R^2 \sin\theta\cos\theta + mgR\sin\theta - \frac{1}{4\pi\varepsilon_0}\frac{qq'\cos\frac{\theta}{2}}{4R\sin^2\frac{\theta}{2}} = 0,$$

or

$$\frac{\cos\frac{\theta}{2}}{16\pi\varepsilon_0 R \sin^2\frac{\theta}{2}}\left[-32\pi\varepsilon_0 m\omega^2 R^3 \sin^3\frac{\theta}{2}\right.$$

$$\left.+64\pi\varepsilon_0 m\omega^2 R^3 \sin^5\frac{\theta}{2} + 32\pi\varepsilon_0 mgR^2 \sin^3\frac{\theta}{2} - qq'\right] = 0,$$

which is equivalent to the following system of two equations

$$\begin{cases} \cos\frac{\theta}{2} = 0; \\ 32\pi m\varepsilon_0 R^2 \sin^3\frac{\theta}{2}(g - \omega^2 R\cos\theta) - qq' = 0. \end{cases} \quad (7.26)$$

The physically accepted solution of the first equation is $\theta_1 = \pi$, while the second equation has no analytical solution (which is exact), while a numerical solution is almost always approximate. To solve it, we shall use again the soft *Mathematica*, but this time we shall focus our attention upon only one case, namely: $m = 1\,g$, $R = 0.2\,m$, $q = 10^{-3}\,C$, $q' = 5\times 10^{-3}\,C$. The following command lines allow one to give the graphic representation of $\theta = \theta(\omega)$, which is implicitly expressed by the equation $(7.26)_2$, as shown in Fig.VII.16.

```
Remove["Global`*"];
Unprotect[In, Out];
Clear[In, Out];

k = 10^-9/9;
R = 0.2;
g = 9.8;
m = 0.001;
q = 10^-3;
q1 = 5*10^-3;

NSolve[8*k*m*R^2*Sin[th/2]^3*(g-omega^2*R*Cos[th])-q*q1== 0, th];
g1=Plot[Evaluate[th /. %],{omega, 20000, 1000000}];
g2=Plot[Pi/2,{omega, 20000, 1000000}];

Show[g1,g2]
```

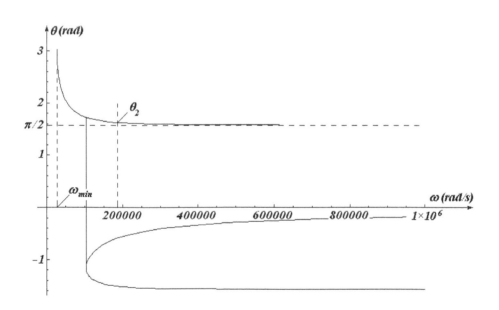

Fig.VII.16

The minimum value of angular velocity $\omega_{min} = 26516,503\,rad \cdot s^{-1}$ (i.e., under which equation $(7.26)_2$ has no real solution) can be determined by means of the command lines:

```
8*k*m*R^2*Sin[Pi/2]^3*(g-ω^2*R*Cos[Pi])-q*q1  //N
-5.× 10⁻⁶+3.55556× 10⁻¹⁴(9.8+0.2 ω²)

NSolve[-5.`*^-6+3.555555555555556`*^-14(9.8`+0.2` ω²), ω]
{{ω→-26516.503370542654`}, {ω→26516.503370542654`}}
```

As can be observed in Fig.VII.16, for any value $\omega > \omega_{min}$ of angular velocity, equation $(7.26)_2$ has a solution which is physically acceptable, namely $\pi/2 < \theta_2 < \pi$. Consequently, putting our results all together, the system has two equilibrium positions, given by the analytical solution $\theta_1 = \pi$, and, respectively, by the numeric solution $\pi/2 < \theta_2 < \pi$. The stability of these positions can be studied by means of the second derivative with respect to θ of the potential energy:

$$\frac{d^2 E_{pot}}{d\theta^2} = \frac{d}{d\theta}\left[-m\omega^2 R^2 \sin\theta\cos\theta + mgR\sin\theta - \frac{1}{4\pi\varepsilon_0}\frac{qq'\cos\frac{\theta}{2}}{4R\sin^2\frac{\theta}{2}}\right]$$

$$= m\omega^2 R^2(1 - 2\cos^2\theta) + mgR\cos\theta + \frac{qq'}{4\pi\varepsilon_0}\frac{1+\cos^2\frac{\theta}{2}}{8R\sin^3\frac{\theta}{2}}.$$

In order to investigate the sign of this expression within the interval $0 < \theta \leq \pi$, we consider the graphic representation of function

$$D(\theta) \equiv \frac{d^2 E_{pot}}{d\theta^2} = m\omega^2 R^2(1 - 2\cos^2\theta) + mgR\cos\theta + \frac{qq'}{4\pi\varepsilon_0}\frac{1+\cos^2\frac{\theta}{2}}{8R\sin^3\frac{\theta}{2}}.$$

In this case also exists a certain value of the angular velocity ω, namely $\omega_c^* = 379162,149\,rad\cdot s^{-1}$ below which equation $D(\theta) = 0$ has a single physically accepted solution; in other words, for $\omega < \omega_c^*$, the function $D(\theta)$ changes only once its sign in the interval $0 < \theta \leq \pi$ (i.e., the graph of $D(\theta)$ passes only once through zero). For example, taking $\omega = 10^5\,rad\cdot s^{-1}$, equation $D(\theta) = 0$ has only one physically accepted solution (see Fig.VII.17), which is

$$\theta_0 = 2,405\,rad \simeq 0,766\,\pi\,rad.$$

Any equilibrium position $0 < \theta \leq \pi$, physically accepted, of equation $(7.26)_2$ is a position of *stable equilibrium* if $\theta \in (0, \theta_0)$, and, respectively, of *unstable equilibrium*, if $\theta \in (\theta_0, \pi]$.

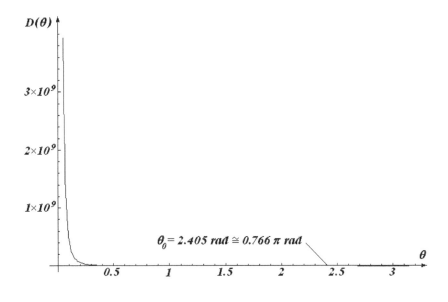

Fig.VII.17

If $\omega > \omega_c^*$, then equation $D(\theta) = \frac{d^2 E_{pot}}{d\theta^2} = 0$ has three real solutions, all of them bearing physical significance for our problem. For example, if we set $\omega = 5 \times 10^6 \, rad \cdot s^{-1}$, then the graphical representation of dependence $D = D(\theta)$ looks like in Fig.VII.18, which puts into evidence two subintervals for each positive - and, respectively, negative - values of angle θ, as solution of the equation $D(\theta) = 0$. As observed in Fig.VII.18, if $\theta \in (0, \theta_1) \bigcup (\theta_2, \theta_3)$, then the considered equilibrium position is a position of *stable equilibrium*, while if $\theta \in (\theta_1, \theta_2) \bigcup (\theta_3, \pi]$, we have a position of *unstable equilibrium*.

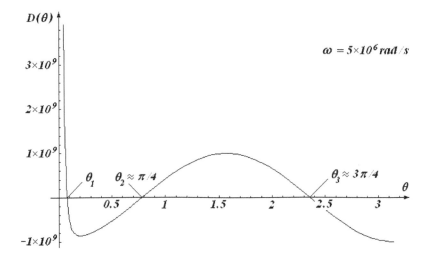

Fig.VII.18

The numerical analysis shows that in the limit $\omega \to \infty$, $\theta_1 \to 0$ (see Fig.VII.19). Therefore, for very large angular velocities, as much as for the ellipse case, the positions of stable equilibrium are comprised only in the interval $\theta \in (\theta_2, \theta_3) \simeq \left(\frac{\pi}{4}, \frac{3\pi}{4}\right)$.

To determine the period of small oscillations, one uses the formula (previously obtained for ellipse, by setting $a = b = R$)

$$\tau = 2\pi R \sqrt{\frac{m}{\left.\frac{d^2 E_{pot}}{d\theta^2}\right|_{\theta=\theta_{se}}}},$$

where the values of θ_{se} corresponding to the stable equilibrium states are replaced in terms of the new configuration (circle of radius R).

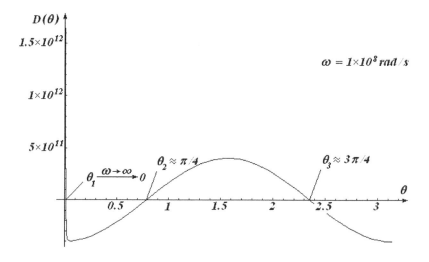

Fig.VII.19

Observation. If, in addition to the above conditions, the circle is at rest, our problem becomes considerably simpler. Setting $\omega = 0$ and resuming the previous steps, we have:

$$L = \frac{1}{2}mR^2\dot{\theta}^2 + mgR\cos\theta - \frac{1}{4\pi\varepsilon_0}\frac{qq'}{2R\sin\frac{\theta}{2}},$$

it which case the effective potential energy is

$$E_{pot} = -mgR\cos\theta + \frac{1}{4\pi\varepsilon_0}\frac{qq'}{2R\sin\frac{\theta}{2}}.$$

The equilibrium positions of the system are found by means of equation

$$\frac{dE_{pot}}{d\theta} = mgR\sin\theta - \frac{1}{4\pi\varepsilon_0}\frac{qq'\cos\frac{\theta}{2}}{4R\sin^2\frac{\theta}{2}} = 0$$

or
$$\frac{\cos\frac{\theta}{2}}{16\pi\varepsilon_0 R \sin^2\frac{\theta}{2}} \left(32\pi\varepsilon_0 mgR^2 \sin^3\frac{\theta}{2} - qq'\right) = 0,$$

which is equivalent to the system of two equations

$$\begin{cases} \cos\dfrac{\theta}{2} = 0; \\ 32\pi\varepsilon_0 mgR^2 \sin^3\dfrac{\theta}{2} - qq' = 0. \end{cases} \quad (7.27)$$

The only physically acceptable solution of the first equation is $\theta_1 = \pi$. Equation $(7.27)_2$ furnishes a second equilibrium position, given by

$$\theta_2 = 2\arcsin\left(\frac{qq'}{32\pi\varepsilon_0 mgR^2}\right)^{1/3},$$

provided that the charges have the same sign, and $qq' < 32\pi\varepsilon_0 mgR^2$.

The type of equilibrium (stable or unstable) is shown by the sign of the second-order derivative of E_{pot} at the points θ_1 and θ_2: "plus" for stable equilibrium, and "minus" for unstable equilibrium. We have:

$$\frac{d^2 E_{pot}}{d\theta^2} = \frac{d}{d\theta}\left[mgR\sin\theta - \frac{1}{4\pi\varepsilon_0}\frac{qq'\cos\frac{\theta}{2}}{4R\sin^2\frac{\theta}{2}}\right]$$

$$= mgR\cos\theta + \frac{qq'}{32\pi\varepsilon_0 R}\frac{\sin^2\frac{\theta}{2} + 2\cos^2\frac{\theta}{2}}{\sin^3\frac{\theta}{2}}$$

$$= mgR\left(1 - 2\sin^2\frac{\theta}{2}\right) + \frac{qq'}{32\pi\varepsilon_0 R}\frac{2 - \sin^2\frac{\theta}{2}}{\sin^3\frac{\theta}{2}},$$

leading to

$$\left.\frac{d^2 E_{pot}}{d\theta^2}\right|_{\theta=\pi} = -mgR + \frac{qq'}{32\pi\varepsilon_0 R}, \quad (7.28)$$

and

$$\left.\frac{d^2 E_{pot}}{d\theta^2}\right|_{\theta=\theta_2} = mgR\left(1 - 2\sin^2\frac{\theta_2}{2}\right) + \frac{qq'}{32\pi\varepsilon_0 R}\frac{2 - \sin^2\frac{\theta_2}{2}}{\sin^3\frac{\theta_2}{2}}$$

$$= mgR\left[1 - 2\left(\frac{qq'}{32\pi\varepsilon_0 mgR^2}\right)^{2/3}\right] + \frac{qq'}{32\pi\varepsilon_0 R}\frac{2 - \left(\frac{qq'}{32\pi\varepsilon_0 mgR^2}\right)^{2/3}}{\frac{qq'}{32\pi\varepsilon_0 mgR^2}}$$

$$= mgR\left[1 - 2\left(\frac{qq'}{32\pi\varepsilon_0 mgR^2}\right)^{2/3}\right] + mgR\left[2 - \left(\frac{qq'}{32\pi\varepsilon_0 mgR^2}\right)^{2/3}\right]$$

$$= 3mgR - 3mgR\left(\frac{qq'}{32\pi\varepsilon_0 mgR^2}\right)^{2/3} = 3mgR\cos^2\frac{\theta_2}{2} > 0. \quad (7.29)$$

According to (7.28), if charges have different signs, or have the same sign and $qq' < 32\pi\varepsilon_0 mgR^2$, then $\theta_1 = \pi$ is a position of *unstable equilibrium*, while if charges have the same sign and $qq' > 32\pi\varepsilon_0 mgR^2$, then $\theta_1 = \pi$ is a position of *stable equilibrium*. Relation (7.29) shows that the equilibrium position given by

$$\theta = \theta_2 = 2\arcsin\left(\frac{qq'}{32\pi\varepsilon_0 mgR^2}\right)^{1/3},$$

if exists, is a position of *stable equilibrium*.

The period of small oscillations about the equilibrium positions is

$$\tau = 2\pi R\sqrt{\frac{m}{\left.\frac{d^2 E_{pot}}{d\theta^2}\right|_{\theta=\theta_{se}}}}.$$

According to our previous results, if charges have the same sign and $qq' > 32\pi\varepsilon_0 mgR^2$, then about position of stable equilibrium $\theta_1 = \pi$ occur small oscillations with period [see (7.28)]

$$\tau = 2\pi R\sqrt{\frac{m}{\left.\frac{d^2 E_{pot}}{d\theta^2}\right|_{\theta=\pi}}} = 2\pi R\sqrt{\frac{m}{-mgR + \frac{qq'}{32\pi\varepsilon_0 R}}}$$

$$= 2\pi R\sqrt{\frac{32\pi\varepsilon_0 mR}{qq' - 32\pi\varepsilon_0 mgR^2}} = 2\pi\sqrt{\frac{R}{g}}\left(\frac{qq'}{32\pi\varepsilon_0 mgR^2} - 1\right)^{-1/2}$$

$$= \tau_0\left(\frac{qq'}{32\pi\varepsilon_0 mgR^2} - 1\right)^{-1/2},$$

where

$$\tau_0 = 2\pi\sqrt{\frac{R}{g}}.$$

If charges have the same sign and $qq' < 32\pi\varepsilon_0 mgR^2$, then about the position of stable equilibrium $\theta_2 = 2\arcsin\left(qq'/32\pi\varepsilon_0 mgR^2\right)^{1/3}$ take place small oscillations with period [see (7.29)]

$$\tau = 2\pi R\sqrt{\frac{m}{\left.\frac{d^2 E_{pot}}{d\theta^2}\right|_{\theta=\theta_2}}} = 2\pi R\sqrt{\frac{1}{3gR - 3gR\left(\frac{qq'}{32\pi\varepsilon_0 mgR^2}\right)^{2/3}}}$$

$$= 2\pi\sqrt{\frac{R}{g}}\left\{3\left[1-\left(\frac{qq'}{32\pi\varepsilon_0 mgR^2}\right)^{2/3}\right]\right\}^{-1/2}$$

$$= \tau_0\left\{3\left[1-\left(\frac{qq'}{32\pi\varepsilon_0 mgR^2}\right)^{2/3}\right]\right\}^{-1/2}.$$

Observation. If, in particular, $qq' = 32\pi\varepsilon_0 mgR^2$, the equilibrium position given by $\theta_2 = 2\arcsin(qq'/32\pi\varepsilon_0 mgR^2)^{1/3}$ shall coincide with that given by $\theta_1 = \pi$. In addition, we have

$$\left.\frac{d^2 E_{pot}}{d\theta^2}\right|_{\theta=\theta_2=\pi} = \left(3mgR\cos^2\frac{\theta_2}{2}\right)_{\theta_2=\pi} = 0,$$

therefore, to determine the stability of the equilibrium position $\theta_2 = \pi$ in this special case ($qq' = 32\pi\varepsilon_0 mgR^2$) we must calculate the derivatives of higher order of the effective potential energy, more precisely, up to a non-zero derivative for $\theta = \pi$. We have:

$$\frac{d^3 E_{pot}}{d\theta^3} = \frac{d}{d\theta}\left(\frac{d^2 E_{pot}}{d\theta^2}\right)$$

$$= \frac{d}{d\theta}\left[mgR(1-2\sin^2\frac{\theta}{2}) + \frac{qq'}{32\pi\varepsilon_0 R}\frac{2-\sin^2\frac{\theta}{2}}{\sin^3\frac{\theta}{2}}\right]$$

$$= -2mgR\sin\frac{\theta}{2}\cos\frac{\theta}{2}$$

$$+\frac{qq'}{32\pi\varepsilon_0 R}\frac{-\sin^4\frac{\theta}{2}\cos\frac{\theta}{2} - \frac{3}{2}\sin^2\frac{\theta}{2}\cos\frac{\theta}{2}\left(2-\sin^2\frac{\theta}{2}\right)}{\sin^6\frac{\theta}{2}}$$

$$= -mgR\sin\theta - \frac{qq'}{64\pi\varepsilon_0 R}\frac{\left(6-\sin^2\frac{\theta}{2}\right)\cos\frac{\theta}{2}}{\sin^4\frac{\theta}{2}},$$

which yields

$$\left.\frac{d^3 E_{pot}}{d\theta^3}\right|_{\theta=\pi} = 0.$$

This means that we also have to calculate

$$\frac{d^4 E_{pot}}{d\theta^4} = \frac{d}{d\theta}\left(\frac{d^3 E_{pot}}{d\theta^3}\right)$$

$$= -\frac{d}{d\theta}\left[mgR\sin\theta + \frac{qq'}{64\pi\varepsilon_0 R}\frac{\left(6-\sin^2\frac{\theta}{2}\right)\cos\frac{\theta}{2}}{\sin^4\frac{\theta}{2}}\right]$$

$$= -mgR\cos\theta - \frac{qq'}{64\pi\varepsilon_0 R} \frac{\left(-\frac{1}{2}\sin\frac{\theta}{2}(6-\sin^2\frac{\theta}{2}) - \sin\frac{\theta}{2}\cos^2\frac{\theta}{2}\right)\sin^4\frac{\theta}{2}}{\sin^8\frac{\theta}{2}}$$

$$+ \frac{qq'}{64\pi\varepsilon_0 R} \frac{2\sin^3\frac{\theta}{2}\cos^2\frac{\theta}{2}(6-\sin^2\frac{\theta}{2})}{\sin^8\frac{\theta}{2}}$$

$$= -mgR\cos\theta + \frac{qq'}{64\pi\varepsilon_0 R} \frac{6-\sin^2\frac{\theta}{2} + 2\cos^2\frac{\theta}{2}}{2\sin^3\frac{\theta}{2}}$$

$$+ \frac{qq'}{64\pi\varepsilon_0 R} \frac{2\cos^2\frac{\theta}{2}(6-\sin^2\frac{\theta}{2})}{\sin^5\frac{\theta}{2}}.$$

Setting now $\theta = \pi$ and $qq' = 32\pi\varepsilon_0 mgR^2$, we still have

$$\left.\frac{d^4 E_{pot}}{d\theta^4}\right|_{\substack{\theta=\pi \\ qq'=32\pi\varepsilon_0 mgR^2}}$$

$$= \left(-mgR\cos\theta + \frac{qq'}{64\pi\varepsilon_0 R} \frac{6-\sin^2\frac{\theta}{2} + 2\cos^2\frac{\theta}{2}}{2\sin^3\frac{\theta}{2}}\right.$$

$$\left.+ \frac{qq'}{64\pi\varepsilon_0 R} \frac{2\cos^2\frac{\theta}{2}(6-\sin^2\frac{\theta}{2})}{\sin^5\frac{\theta}{2}}\right)_{\substack{\theta=\pi \\ qq'=32\pi\varepsilon_0 mgR^2}}$$

$$= mgR + \frac{5}{4}\left(\frac{qq'}{32\pi\varepsilon_0 R}\right)_{qq'=32\pi\varepsilon_0 mgR^2} = \frac{9}{4}mgR > 0,$$

Consequently, in this special case ($qq' = 32\pi\varepsilon_0 mgR^2$) the equilibrium position for $\theta = \pi$ is a position of *stable equilibrium*, but oscillations about it *are no longer harmonic*.

CHAPTER VIII

PROBLEMS SOLVED BY MEANS OF THE HAMILTONIAN FORMALISM

Problem 1

An *annular pendulum* is a physical pendulum consisting of a homogeneous circular crown of radii r and R. Using the Hamiltonian technique, determine the equation of motion of such a pendulum, and the period of small oscillations about a fixed point O situated on the interior circle (see Fig. VIII.1).

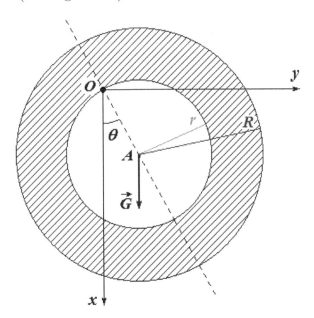

Fig. VIII.1

Solution

Let d be the thickness of the circular crown, and μ its mass density. Then the mass of the crown is

$$m = \pi \mu d (R^2 - r^2). \tag{8.1}$$

Obviously, the system has only one degree of freedom. Denoting by θ the associated generalized coordinate, we are now prepared to write Hamilton's function. Since the system is conservative, the Hamiltonian is expressed in terms of the kinetic and potential energies.

The kinetic energy of the pendulum is

$$T = \frac{1}{2}I\omega^2 = \frac{1}{2}I\dot{\theta}^2, \qquad (8.2)$$

where I is the moment of inertia of the circular crown with respect to z-axis. To determine this quantity, we appeal to Steiner's theorem

$$I \equiv I_O = I_A + mr^2, \qquad (8.3)$$

where I_A is the moment of inertia of the circular crown with respect to the axis passing through its centre of symmetry A, being parallel to Oz. According to the definition,

$$I_A = \int_{(D)} (x'^2 + y'^2)\, dm, \qquad (8.4)$$

where x' and y' are considered with respect to a reference frame with its origin at A, and (D) is the domain represented by all points of the circular crown.

Due to cylindrical symmetry of the crown, the integral (8.4) can be easier calculated in cylindrical coordinates. Thus, we have:

$$I_A = \int_D \rho^2\, dm = \int_D \rho^2\, \mu\, \rho\, \delta\rho\, dz\, d\varphi = \mu \int_r^R \rho^3\, d\rho \int_0^d dz \int_0^{2\pi} d\varphi$$

$$= \mu \frac{R^4 - r^4}{4}\, d\, 2\pi = \frac{1}{2}\pi\, d\, \mu(R^4 - r^4) = \frac{m}{2}(R^2 + r^2). \qquad (8.5)$$

In this case, equation (8.3) leads to

$$I = I_A + mr^2 = \frac{m}{2}(R^2 + 3r^2), \qquad (8.6)$$

and the kinetic energy of the annular pendulum writes

$$T = \frac{1}{2}I\dot{\theta}^2 = \frac{m}{4}(R^2 + 3r^2)\dot{\theta}^2. \qquad (8.7)$$

Since the only applied force is the force of gravity $\vec{G} = (mg, 0, 0)$, which is a conservative force, and taking into account the symmetry of the body, the potential energy of the system can be easily determined by means of
$$\vec{G} = -\mathrm{grad} V,$$
that is
$$dV = -\vec{G} \cdot d\vec{r} = -mg\, dx,$$
so that
$$V = -mgx + V_0.$$
A convenient choice for the reference level of the potential energy, $V(x = 0) = 0$, yields
$$V = -mgx = -mgr\cos\theta. \tag{8.8}$$

The Lagrangian of the system then is
$$L = T - V = \frac{m}{4}(R^2 + 3r^2)\dot{\theta}^2 + mgr\cos\theta, \tag{8.9}$$
allowing to calculate the generalized momentum associated with θ
$$p_\theta = \frac{\partial L}{\partial \dot{\theta}} = \frac{m}{2}(R^2 + 3r^2)\dot{\theta}, \tag{8.10}$$
and the Hamiltonian writes
$$H = p_\theta \dot{\theta} - L = \frac{m}{4}(R^2 + 3r^2)\dot{\theta}^2 - mgr\cos\theta. \tag{8.11}$$
Since
$$\dot{\theta} = \frac{2p_\theta}{m(R^2 + 3r^2)},$$
we still have
$$H = H(\theta, p_\theta) = \frac{p_\theta^2}{m(R^2 + 3r^2)} - mgr\cos\theta. \tag{8.12}$$

Hamilton's canonical equations then yield:
$$\begin{cases} \dot{\theta} = \dfrac{\partial H}{\partial p_\theta} = \dfrac{2p_\theta}{m(R^2 + 3r^2)}; \\ \dot{p}_\theta = -\dfrac{\partial H}{\partial \theta} = -mgr\sin\theta. \end{cases} \tag{8.13}$$

Taking the time derivative of $(8.13)_1$, and using $(8.13)_2$, we have

$$\ddot{\theta} + \frac{2gr}{R^2 + 3r^2} \sin\theta = 0, \qquad (8.14)$$

which is the differential equation of motion. For small oscillations ($\sin\theta \approx \theta$), equation (8.14) becomes

$$\ddot{\theta} + \frac{2gr}{R^2 + 3r^2}\theta = 0. \qquad (8.15)$$

Denoting

$$\omega_0^2 = \frac{2gr}{R^2 + 3r^2}, \qquad (8.16)$$

we are left with

$$\ddot{\theta} + \omega_0^2 \theta = 0, \qquad (8.17)$$

which is the well-known differential equation of the linear harmonic oscillator, whose period is

$$T_0 = \frac{2\pi}{\omega_0} = 2\pi \sqrt{\frac{R^2 + 3r^2}{2gr}}. \qquad (8.18)$$

The last formula can be used to determine the gravitational acceleration g at the place of the Earth where the experiment is performed:

$$g = 2\pi^2 \frac{R^2 + 3r^2}{r} \frac{1}{T_0^2}. \qquad (8.19)$$

Suppose, in this respect, that the radii r and R, and the period of small oscillations T_0 are determined with high precision, while the conditions of isochronocity are fulfilled as well as possible. In this case, the *relative error* in determination - with the help of annular pendulum - of the gravitational acceleration g, can be estimated by means of the *logarithmic finite difference method*. So, we have:

$$\ln g = \ln(2\pi^2) + \ln(R^2 + 3r^2) - \ln r - \ln T_0^2.$$

Differentiating this relation, then going to finite variations, we have:

$$\frac{\Delta g}{g} = \frac{2R^2}{R^2 + 3r^2}\frac{\Delta R}{R} - \frac{R^2 - 3r^2}{R^2 + 3r^2}\frac{\Delta r}{r} - 2\frac{\Delta T_0}{T_0}. \qquad (8.20)$$

Denoting by $\delta g = \frac{\Delta g}{g}$ the relative error on g, and by δr, δR, δT_0 the relative errors committed in determination of the corresponding quantities, we finally obtain

$$\delta g = \frac{2R^2}{R^2 + 3r^2}\delta R + \frac{|R^2 - 3r^2|}{R^2 + 3r^2}\delta r + 2\delta T_0, \tag{8.21}$$

where, in order to consider the maximum relative error affecting the determination of g, all terms in (8.20) have been taken with plus sign. To conclude, the determination of g has to be written as

$$g = \left(2\pi^2 \frac{R^2 + 3r^2}{r}\frac{1}{T_0^2} \pm \Delta g\right) m \cdot s^{-2}, \tag{8.22}$$

where $\Delta g = g\, \delta g$, or

$$g = \left(2\pi^2 \frac{R^2 + 3r^2}{r}\frac{1}{T_0^2}\right) m \cdot s^{-2} \pm \delta g \cdot 100\%, \tag{8.23}$$

where δg is given by (8.21). Here δR, δr and δT_0 are usually given by the characteristics of the measuring instruments. For example, if the radii of the circular crown are determined by the slide rule, then the absolute error affecting the determination of r and R is about $0.1\,mm$, while if the time interval is measured by a digital chronometer, the absolute error on time could be of $0.01\,s$, or even smaller. In general, the errors on distances and on time are dictated by both the precision of the employed instruments, and the method of measurement. If, for instance, $r = 10\,cm$, $R = 41\,cm$, and $T_0 = 2s$, in which case we have to do with a *second pendulum*[1], then, according to (8.19), we have

$$g = 2\pi^2 \frac{R^2 + 3r^2}{r}\frac{1}{T_0^2} = 9.8\, m \cdot s^{-2}. \tag{8.24}$$

To conclude, in agreement with (8.22), the result regarding determination of g must be written as

$$g = (9.8 \pm 0.1) m \cdot s^{-2}, \tag{8.25}$$

or, equivalently,

$$g = 9.8\, m \cdot s^{-2} \pm 1.11\%, \tag{8.26}$$

---

[1] A second pendulum is a pendulum whose period is precisely two seconds: one second for a swing in one direction, and one second for the return swing.

where we used (8.21) to write

$$\delta g = 1.11 \times 10^{-2}. \tag{8.27}$$

The relative error on determination of g is dictated by the relative error on T_0, which is five times bigger than that on r, and more than twenty times bigger than that on R:

$$\delta T_0 = 5 \cdot 10^{-3}, \quad \delta r \simeq 10^{-3}, \quad \delta R \simeq 2.44 \cdot 10^{-4}.$$

Problem 2

Using the Hamiltonian formalism, write the differential equation of motion of a homogeneous circular cylinder of radius a and mass m, rolling without sliding inside a fixed cylinder of radius R, as shown in Fig.VIII.2. Determine the period of small oscillations of the cylinder about its position of stable equilibrium.

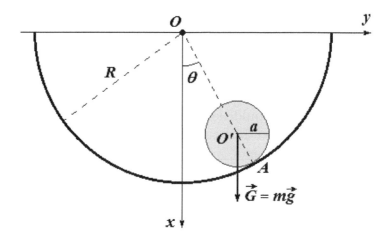

Fig.VIII.2

Solution

The motion of rolling without sliding of the cylinder or radius a on the inner surface of the fixed cylinder of radius R is a motion about the instantaneous axis of rotation, which coincides with the contact line of the two cylinders. This motion can be decomposed in two motions, as follows: a motion of pure rotation of the cylinder of radius a about its own axis of rotation (passing through O' and parallel to both generatrix line and instantaneous axis of rotation - see Fig.VIII.2), and a motion of "translation" of the symmetry centre O' along a circle arc with all points at the same distance relative to the inner surface of the fixed cylinder. In fact, the second motion is also a motion of rotation, but of the cylinder of radius a "as a whole" about z-axis.

The kinetic energy associated to this motion can be determined considering that the cylinder mass is concentrated at point O', which rotates about z-axis on a circle of radius $R - a$. The motion of the moving cylinder is subject to two constraints:

i) $z_{O'} = const.$,
ii) $x_{O'}^2 + y_{O'}^2 - (R - a)^2 = 0$,

meaning that the system has one degree of freedom. Denoting by θ the angle between vertical and the straight line connecting the centers of cylinders, the linear velocity of the cylinder of radius a moving "as a whole" about z-axis is $|\vec{v}_{O'}| = \dot{\theta}(R - a)$, and the associated kinetic energy writes

$$T_{O'} = \frac{1}{2}m|\vec{v}_{O'}|^2 = \frac{1}{2}m(R-a)^2\dot{\theta}^2. \quad (8.28)$$

The total kinetic energy of the moving cylinder then writes as a sum of the kinetic energy of the body "as a whole", given by (8.28), and the kinetic energy of the moving cylinder about its symmetry axis:

$$T \equiv T_O = T_{O'} + T' = \frac{1}{2}m|\vec{v}_{O'}|^2 + \frac{1}{2}I_{z'z'}\omega^2, \quad (8.29)$$

where $I_{z'z'}$ is the moment of inertia of the moving cylinder with respect to its symmetry axis

$$I_{z'z'} = \int_D r^2\, dm = \int_0^a r^2 \rho\, h\, 2\pi r\, dr = \frac{1}{2}ma^2. \quad (8.30)$$

Here D is the set of all material points that belong to the moving cylinder, h is its length, and ρ its mass density. Relation (8.29) represents nothing else but König's second theorem for the compound motion of the moving cylinder.

The relation between ω and θ can be found at least in two different ways: (i) by means of velocity analysis, and (ii) by means of estimation of angles swept by various axes of the moving cylinder in different reference frames. The simplest way takes into account the fact that the centre of inertia of the moving cylinder is situated on its symmetry axis (passing through O'), while its linear velocity is

$$v = |\vec{v}_{O'}| = (R-a)\dot{\theta}.$$

The angular velocity of the motion of rotation about the instantaneous axis of rotation, which coincides with the contact lines of cylinders (passing through A and being parallel to z-axis) then is

$$\Omega = \frac{v}{a} = \frac{R-a}{a}\dot{\theta}.$$

Since $\omega = -\Omega$ (rotation of O' with respect to A, with angular velocity $|\vec{\Omega}| = \Omega = \omega$ is equivalent to rotation of A with respect to O' with angular velocity $|\vec{\omega}| = |-\vec{\Omega}| = \Omega = \omega$), it follows that

$$\omega = |\vec{\omega}| = \frac{R-a}{a}\dot{\theta},$$

so that

$$T' = \frac{1}{2}I_{z'z'}\omega^2 = \frac{1}{4}m(R-a)^2\dot{\theta}^2$$

and, therefore,

$$T \equiv T_O = T_{O'} + T' = \frac{3}{4}m(R-a)^2\dot{\theta}^2. \tag{8.31}$$

The Lagrangian then writes

$$L = T - V = \frac{3}{4}m(R-a)^2\dot{\theta}^2 - V. \tag{8.32}$$

Since the only force applied to the moving cylinder is the force of gravity $\vec{G} = m\vec{g}$, and due to cylindrical symmetry of the body, we have

$$dV = -\vec{G} \cdot d\vec{r} = -mg\,dx$$

or

$$V(x) = -mgx + V_0. \tag{8.33}$$

A convenient choice for the reference level of the potential energy, that is $V(x=0) = 0$, leads to $V(x) = -mgx$, or

$$V(\theta) = -mg(R-a)\cos\theta, \tag{8.34}$$

and the Lagrangian (8.32) takes the form

$$L = \frac{3}{4}m(R-a)^2\dot{\theta}^2 + mg(R-a)\cos\theta. \tag{8.35}$$

The generalized momentum p_θ, canonically conjugated with θ, is

$$p_\theta = \frac{\partial L}{\partial \dot\theta} = \frac{3}{2}m(R-a)^2\dot\theta, \tag{8.36}$$

and the Hamiltonian writes

$$H(\theta,\dot\theta) = \dot\theta p_\theta - L = \frac{3}{4}m(R-a)^2\dot\theta^2 - mg(R-a)\cos\theta. \tag{8.37}$$

To express H in terms of θ and p_θ, one uses (8.36). The result is:

$$H(\theta,p_\theta) = \frac{p_\theta^2}{3m(R-a)^2} - mg(R-a)\cos\theta. \tag{8.38}$$

Hamilton's canonical equations then yield

$$\begin{cases} \dot\theta = \dfrac{\partial H}{\partial p_\theta} = \dfrac{2p_\theta}{3m(R-a)^2}; \\ \dot p_\theta = -\dfrac{\partial H}{\partial \theta} = -mg(R-a)\sin\theta. \end{cases} \tag{8.39}$$

Taking the time derivative of $(8.39)_1$ and using $(8.39)_2$, we obtain

$$\ddot\theta + \frac{2g}{3(R-a)}\sin\theta = 0. \tag{8.40}$$

Denoting

$$\omega_0^2 = \frac{2g}{3(R-a)},$$

we finally have

$$\ddot\theta + \omega_0^2\sin\theta = 0. \tag{8.41}$$

This is the differential equation of motion of the cylinder of mass m and radius a "as a whole" relative to the reference frame $Oxyz$. For small angular amplitudes ($\sin\theta \simeq \theta$), equation (8.41) turns into a well-known equation of the linear harmonic oscillator. The period of small oscillations about the position of stable equilibrium is

$$T_0 = \frac{2\pi}{\omega_0} = 2\pi\sqrt{\frac{3(R-a)}{2g}}. \tag{8.42}$$

This formula can be used to determine the gravitational acceleration g:

$$g = \frac{6\pi^2(R-a)}{T_0^2}. \tag{8.43}$$

Problem 3

Write the Hamiltonian of a material point (particle) of mass m, situated in a conservative force field, in Cartesian, cylindrical, plane-polar, spherical, parabolic, and elliptic coordinates.

Solution

i) Cartesian coordinates x, y, z

Let us suppose that the particle moves freely in the conservative field (there are no constraints) characterized by the potential energy $V(x, y, z)$. The Lagrangian then writes

$$L = T - V = \frac{1}{2}m(v_x^2 + v_y^2 + v_z^2) - V(x, y, z)$$

$$= \frac{1}{2}m(\dot{x}^2 + \dot{y}^2 + \dot{z}^2) - V(x, y, z). \qquad (8.44)$$

The generalized momenta, canonically conjugated to the generalized coordinates x, y, z are:

$$\begin{cases} p_x = \dfrac{\partial L}{\partial \dot{x}} = m\dot{x} \ ; \\ p_y = \dfrac{\partial L}{\partial \dot{y}} = m\dot{y} \ ; \\ p_z = \dfrac{\partial L}{\partial \dot{z}} = m\dot{z} \ . \end{cases} \qquad (8.45)$$

The Hamiltonian then writes

$$H = p_i \dot{q}_i - L = \frac{1}{2}m(\dot{x}^2 + \dot{y}^2 + \dot{z}^2) + V(x, y, z). \qquad (8.46)$$

Expressing $\dot{x}, \dot{y}, \dot{z}$ in terms of p_x, p_y, p_z,

$$\begin{cases} \dot{x} = \dfrac{p_x}{m}, \\ \dot{y} = \dfrac{p_y}{m}, \\ \dot{z} = \dfrac{p_z}{m}, \end{cases} \qquad (8.47)$$

and introducing these quantities into (8.46), we finally have

$$H = \frac{1}{2m}(p_x^2 + p_y^2 + p_z^2) + V(x, y, z). \qquad (8.48)$$

ii) Cylindrical coordinates ρ, φ, z

In this case, the coordinate transformations are:

$$\begin{cases} x = \rho \cos \varphi; \\ y = \rho \sin \varphi; \\ z = z. \end{cases} \quad (8.49)$$

Then

$$\begin{cases} \dot{x} = \dot{\rho} \cos \varphi - \rho \dot{\varphi} \sin \varphi; \\ \dot{y} = \dot{\rho} \sin \varphi + \rho \dot{\varphi} \cos \varphi; \\ \dot{z} = \dot{z}, \end{cases} \quad (8.50)$$

and the squared velocity is

$$v^2 = \dot{x}^2 + \dot{y}^2 + \dot{z}^2 = \dot{\rho}^2 + \rho^2 \dot{\varphi}^2 + \dot{z}^2,$$

which allows us to write the kinetic energy

$$T(\rho, \dot{\rho}, \dot{\varphi}, \dot{z}) = \frac{1}{2} m (\dot{\rho}^2 + \rho^2 \dot{\varphi}^2 + \dot{z}^2),$$

and the Lagrangian is

$$L = T - V = \frac{1}{2} m (\dot{\rho}^2 + \rho^2 \dot{\varphi}^2 + \dot{z}^2) - V(\rho, \varphi, z). \quad (8.51)$$

The associated conjugate momenta are:

$$\begin{cases} p_\rho = \dfrac{\partial L}{\partial \dot{\rho}} = m \dot{\rho}; \\ p_\varphi = \dfrac{\partial L}{\partial \dot{\varphi}} = m \rho^2 \dot{\varphi}; \\ p_z = \dfrac{\partial L}{\partial \dot{z}} = m \dot{z}. \end{cases} \quad (8.52)$$

According to definition, the Hamiltonian is

$$H = \sum_{i=1}^{3} p_i \dot{q}_i - L = p_\rho \dot{\rho} + p_\varphi \dot{\varphi} + p_z \dot{z} - L$$

$$= \frac{1}{2} m (\dot{\rho}^2 + \rho^2 \dot{\varphi}^2 + \dot{z}^2) + V(\rho, \varphi, z). \quad (8.53)$$

Expressing $\dot{\rho}$, $\dot{\varphi}$, $\dot{z}$ in terms of generalized momenta

$$\begin{cases} \dot{\rho} = \dfrac{1}{m} p_\rho; \\ \dot{\varphi} = \dfrac{1}{m \rho^2} p_\varphi; \\ \dot{z} = \dfrac{1}{m} p_z, \end{cases} \quad (8.54)$$

and introducing these quantities into (8.53), we finally obtain

$$H = \frac{1}{2m}\left(p_\rho^2 + \frac{1}{\rho^2}p_\varphi^2 + p_z^2\right) + V(\rho, \varphi, z). \tag{8.55}$$

iii) Plane-polar coordinates ρ, θ

The plane-polar coordinates are obtained from cylindrical coordinates by setting $z = 0$ and $\varphi \to \theta$. The result is:

$$H = \frac{1}{2m}\left(p_\rho^2 + \frac{1}{\rho^2}p_\theta^2\right) + V(\rho, \theta). \tag{8.56}$$

iv) Spherical coordinates r, θ, φ

Using relations between Cartesian and spherical coordinates

$$\begin{cases} x = r\sin\theta\cos\varphi; \\ y = r\sin\theta\sin\varphi; \\ z = r\cos\theta, \end{cases} \tag{8.57}$$

we have

$$\begin{cases} \dot{x} = \dot{r}\sin\theta\cos\varphi + r\dot\theta\cos\theta\cos\varphi - r\dot\varphi\sin\theta\sin\varphi; \\ \dot{y} = \dot{r}\sin\theta\sin\varphi + r\dot\theta\cos\theta\sin\varphi + r\dot\varphi\sin\theta\sin\varphi; \\ \dot{z} = \dot{r}\cos\theta - r\dot\theta\sin\theta. \end{cases} \tag{8.58}$$

The kinetic energy writes

$$T(r, \theta, \varphi) = \frac{1}{2}m(\dot r^2 + r^2\dot\theta^2 + \dot\varphi^2 r^2 \sin^2\theta),$$

and the Lagrangian is

$$L = T - V = \frac{1}{2}m(\dot r^2 + r^2\dot\theta^2 + \dot\varphi^2 r^2 \sin^2\theta) - V(r, \theta, \varphi). \tag{8.59}$$

The generalized momenta are

$$\begin{cases} p_r = \dfrac{\partial L}{\partial \dot r} = m\dot r; \\ p_\theta = \dfrac{\partial L}{\partial \dot\theta} = mr^2\dot\theta; \\ p_\varphi = \dfrac{\partial L}{\partial \dot\varphi} = mr^2\dot\varphi\sin^2\theta. \end{cases} \tag{8.60}$$

The Hamiltonian then is

$$H = \sum_{i=1}^{3} p_i \dot{q}_i - L = p_r \dot{r} + p_\theta \dot{\theta} + p_\varphi \dot{\varphi} - L$$

$$= \frac{1}{2} m(\dot{r}^2 + r^2 \dot{\theta}^2 + \dot{\varphi}^2 r^2 \sin^2 \theta) + V(r, \theta, \varphi), \tag{8.61}$$

or, in terms of generalized coordinates and generalized momenta

$$H = \frac{1}{2m}\left(p_r^2 + \frac{1}{r^2} p_\theta^2 + \frac{1}{r^2 \sin^2 \theta} p_\varphi^2 \right) + V(r, \theta, \varphi). \tag{8.62}$$

v) Parabolic coordinates ξ, η, φ

The parabolic coordinates ξ, η, φ can be defined starting from cylindrical coordinates ρ, φ, z, as follows:

$$\begin{cases} z = \frac{1}{2}(\xi - \eta); \\ \rho = \sqrt{\xi \eta}. \end{cases} \tag{8.63}$$

The coordinate φ is the same in both coordinate systems, cylindrical and parabolic. Coordinates ξ and η can take any value between 0 and ∞. Eliminating η from (8.63), we have

$$\xi^2 - 2\xi z - \rho^2 = 0, \tag{8.64}$$

with the solutions

$$\xi_{1,2} = z \pm \sqrt{z^2 + \rho^2}.$$

Setting $\xi = const.$, this means

$$z \pm \sqrt{z^2 + \rho^2} = const. = C,$$

or

$$z = \frac{1}{2C}(C^2 - \rho^2), \tag{8.65}$$

which is a family of paraboloids of revolution, with z-axis as axis of symmetry. Proceeding in the same manner and eliminating ξ from (8.63), we obtain

$$\eta_{1,2} = -z \pm \sqrt{z^2 + \rho^2},$$

so that by taking $\eta = const. = C_1$, equation

$$z = \frac{1}{2C_1}(\rho^2 - C_1^2) \qquad (8.66)$$

represents, in its turn, a family of paraboloids of revolution, with the same axis of symmetry.

Observation. Performing the change of variable

$$r = \sqrt{z^2 + \rho^2}, \qquad (8.67)$$

we can write relations (8.63) in a more convenient form. Observing that $r = \frac{1}{2}(\xi + \eta)$ and using $z = \frac{1}{2}(\xi - \eta)$, we obtain

$$\begin{cases} \xi = r + z; \\ \eta = r - z. \end{cases} \qquad (8.68)$$

To write the Lagrangian of a heavy particle in parabolic coordinates ξ, η, φ, we use the Lagrangian expressed in cylindrical coordinates (8.51)

$$L = \frac{1}{2}m(\dot{\rho}^2 + \rho^2\dot{\varphi}^2 + \dot{z}^2) - V(\rho, \varphi, z).$$

Since

$$\begin{cases} \rho = \sqrt{\xi\eta}; \\ \varphi = \varphi; \\ z = \frac{1}{2}(\xi - \eta), \end{cases}$$

we have:

$$\begin{cases} \dot{\rho} = \frac{1}{2\sqrt{\xi\eta}}(\dot{\xi}\eta + \xi\dot{\eta}); \\ \dot{\varphi} = \dot{\varphi}; \\ \dot{z} = \frac{1}{2}(\dot{\xi} - \dot{\eta}). \end{cases} \qquad (8.69)$$

Then we can write

$$\dot{\rho}^2 + \rho^2\dot{\varphi}^2 + \dot{z}^2 = \frac{1}{4\xi\eta}(\dot{\xi} + \xi\dot{\eta})^2 + \xi\eta\dot{\varphi}^2 + \frac{1}{4}(\dot{\xi} - \dot{\eta})^2$$

$$= \frac{1}{4}(\xi + \eta)\left(\frac{\dot{\xi}^2}{\xi} + \frac{\dot{\eta}^2}{\eta}\right) + \xi\eta\dot{\varphi}^2. \qquad (8.70)$$

Making allowance for (8.70), the Lagrangian (8.51) writes

$$L = \frac{m}{8}(\xi+\eta)\left(\frac{\dot{\xi}^2}{\xi}+\frac{\dot{\eta}^2}{\eta}\right)+\frac{m}{2}\xi\eta\dot{\varphi}^2-V(\xi,\eta,\varphi). \qquad (8.71)$$

The associated momenta then are

$$\begin{cases} p_\xi = \dfrac{\partial L}{\partial \dot{\xi}} = \dfrac{m}{4\xi}(\xi+\eta)\dot{\xi}; \\ p_\eta = \dfrac{\partial L}{\partial \dot{\eta}} = \dfrac{m}{4\eta}(\xi+\eta)\dot{\eta}; \\ p_\varphi = \dfrac{\partial L}{\partial \dot{\varphi}} = m\xi\eta\dot{\varphi}. \end{cases} \qquad (8.72)$$

The Hamiltonian is

$$H = \sum_{i=1}^{3} p_i \dot{q}_i - L = p_\xi \dot{\xi} + p_\eta \dot{\eta} + p_\varphi \dot{\varphi} - L$$

$$= \frac{m}{4\xi}(\xi+\eta)\dot{\xi}^2 + \frac{m}{4\eta}(\xi+\eta)\dot{\eta}^2 + m\xi\eta\dot{\varphi}^2 - \frac{m}{8\xi}(\xi+\eta)\dot{\xi}^2$$

$$- \frac{m}{8\eta}(\xi+\eta)\dot{\eta}^2 - \frac{m}{2}\xi\eta\dot{\varphi}^2 + V(\xi,\eta,\varphi)$$

$$= \frac{m}{8}(\xi+\eta)\left(\frac{\dot{\xi}^2}{\xi}+\frac{\dot{\eta}^2}{\eta}\right)+\frac{m}{2}\xi\eta\dot{\varphi}^2+V(\xi,\eta,\varphi). \qquad (8.73)$$

In view of (8.72), we have

$$\begin{cases} \dot{\xi} = \dfrac{4\xi p_\xi}{m(\xi+\eta)}; \\ \dot{\eta} = \dfrac{4\eta p_\eta}{m(\xi+\eta)}; \\ \dot{\varphi} = \dfrac{p_\varphi}{m\xi\eta}, \end{cases} \qquad (8.74)$$

and, introducing these expressions into (8.73), we are left with

$$H = \frac{m}{8}(\xi+\eta)\left[\frac{16\xi p_\xi^2}{m^2(\xi+\eta)^2}+\frac{16\eta p_\eta^2}{m^2(\xi+\eta)^2}\right]$$

$$+\frac{p_\varphi^2}{2m\xi\eta}+V(\xi,\eta,\varphi) = \frac{2}{m}\frac{\xi p_\xi^2+\eta p_\eta^2}{\xi+\eta}+\frac{p_\varphi^2}{2m\xi\eta}+V(\xi,\eta,\varphi). \qquad (8.75)$$

vi) Elliptic coordinates λ, μ, φ

Transition from cylindrical to elliptic coordinates λ, μ, φ is performed by means of relations

$$\begin{cases} \rho = \sigma\sqrt{(\lambda^2 - 1)(1 - \mu^2)}; \\ \varphi = \varphi; \\ z = \sigma\lambda\mu, \end{cases} \tag{8.76}$$

where σ is a constant called *transformation parameter*. The elliptic coordinate λ takes all possible values from 1 to ∞, while μ can take all values from -1 to $+1$.

Before going further, we observe that the relations (8.76) can be written in a more explicit geometric form. To this end, let us denote by d_1 and d_2 the distances, determined on z-axis, between the origin of the coordinate system and the points P_1 and P_2, defined by $z_1 = \sigma$ and $z_2 = -\sigma$, respectively. In this case, the lines $\lambda = const.$ display the family of ellipsoids

$$\frac{z^2}{\sigma^2\lambda^2} + \frac{\rho^2}{\sigma^2(\lambda^2 - 1)} = 1$$

with foci at P_1 and P_2, while the lines $\mu = const.$ represent the family of hyperboloids

$$\frac{z^2}{\sigma^2\mu^2} - \frac{\rho^2}{\sigma^2(1 - \mu^2)} = 1,$$

with the same foci. The two segments d_1 and d_2 are given by

$$\begin{cases} d_1 = \sqrt{(z - \sigma)^2 + \rho^2}; \\ d_2 = \sqrt{(z + \sigma)^2 + \rho^2}. \end{cases} \tag{8.77}$$

Using (8.76), we still have

$$\begin{cases} d_1 = \sigma(\lambda - \mu); \\ d_2 = \sigma(\lambda + \mu), \end{cases} \tag{8.78}$$

which yields

$$\begin{cases} \lambda = \dfrac{d_1 + d_2}{2\sigma}; \\ \mu = \dfrac{d_2 - d_1}{2\sigma}. \end{cases} \tag{8.79}$$

Let us now write the Lagrangian of a heavy particle moving in a conservative force field in elliptic coordinates. In this respect, we have to introduce (8.76) into (8.51). Observing that

$$\begin{cases} \dot{\rho} = \dfrac{\sigma}{2\sqrt{(\lambda^2 - 1)(1 - \mu^2)}} \left[2\lambda\dot{\lambda}(1 - \mu^2) - 2\mu\dot{\mu}(\lambda^2 - 1) \right]; \\ \dot{\varphi} = \dot{\varphi}; \\ \dot{z} = \sigma(\dot{\lambda}\mu + \lambda\dot{\mu}), \end{cases} \tag{8.80}$$

and, also,

$$\dot{\rho}^2 + \rho^2\dot{\varphi}^2 + \dot{z}^2 = \frac{\sigma^2}{(\lambda^2-1)(1-\mu^2)}\left[\lambda\dot{\lambda}(1-\mu^2) - \mu\dot{\mu}(\lambda^2-1)\right]^2$$
$$+\sigma^2(\lambda^2-1)(1-\mu^2)\dot{\varphi}^2 + \sigma^2(\dot{\lambda}\mu + \lambda\dot{\mu})$$
$$= \sigma^2(\lambda^2-\mu^2)\left(\frac{\dot{\lambda}^2}{\lambda^2-1} - \frac{\dot{\mu}^2}{1-\mu^2}\right) + \sigma^2(\lambda^2-1)(1-\mu^2)\dot{\varphi}^2,$$

the Lagrangian becomes

$$L = \frac{m\sigma^2}{2}(\lambda^2-\mu^2)\left(\frac{\dot{\lambda}^2}{\lambda^2-1} - \frac{\dot{\mu}^2}{1-\mu^2}\right)$$
$$+ \frac{m\sigma^2}{2}(\lambda^2-1)(1-\mu^2)\dot{\varphi}^2 - V(\lambda,\mu,\varphi). \tag{8.81}$$

The generalized momenta, canonically conjugated to the generalized coordinates λ, μ, φ, are:

$$\begin{cases} p_\lambda = \dfrac{\partial L}{\partial \dot{\lambda}} = m\sigma^2(\lambda^2-\mu^2)\dfrac{\dot{\lambda}}{\lambda^2-1}; \\ p_\mu = \dfrac{\partial L}{\partial \dot{\mu}} = m\sigma^2(\lambda^2-\mu^2)\dfrac{\dot{\mu}}{\mu^2-1}; \\ p_\varphi = \dfrac{\partial L}{\partial \dot{\varphi}} = m\sigma^2(\lambda^2-1)(1-\mu^2)\dot{\varphi}. \end{cases} \tag{8.82}$$

The Hamiltonian, according to its definition, then is:

$$H = \sum_{i=1}^{3} p_i \dot{q}_i - L = \dot{\lambda} p_\lambda + \dot{\mu} p_\mu + \dot{\varphi} p_\varphi - L$$
$$= \frac{m\sigma^2}{2}(\lambda^2-\mu^2)\left(\frac{\dot{\lambda}^2}{\lambda^2-1} - \frac{\dot{\mu}^2}{1-\mu^2}\right)$$
$$+ \frac{m\sigma^2}{2}(\lambda^2-1)(1-\mu^2)\dot{\varphi}^2 + V(\lambda,\mu,\varphi). \tag{8.83}$$

The last step is to write the Hamiltonian in terms of generalized coordinates λ, μ, φ and generalized momenta p_λ, p_μ, p_φ. Since

$$\begin{cases} \dot{\lambda} = \dfrac{(\lambda^2-1)p_\lambda}{m\sigma^2(\lambda^2-\mu^2)}; \\ \dot{\mu} = \dfrac{(1-\mu^2)p_\mu}{m\sigma^2(\lambda^2-\mu^2)}; \\ \dot{\varphi} = \dfrac{p_\varphi}{m\sigma^2(\lambda^2-1)(1-\mu^2)}, \end{cases} \tag{8.84}$$

the Hamiltonian finally writes:

$$H = \frac{1}{2m\sigma^2(\lambda^2 - \mu^2)}\left[(\lambda^2 - 1)p_\lambda^2 + (1 - \mu^2)p_\mu^2\right.$$

$$\left. + \left(\frac{1}{\lambda^2 - 1} + \frac{1}{1 - \mu^2}\right)p_\varphi^2\right] + V(\lambda, \mu, \varphi). \tag{8.85}$$

Problem 4

Let $f(q, p, t)$ and $g(q, p, t)$ be two arbitrary functions of canonical variables $q_1, q_2, ..., q_n$ and $p_1, p_2, ..., p_n$. The expression

$$\{f, g\} = \frac{\partial f}{\partial q_i}\frac{\partial g}{\partial p_i} - \frac{\partial g}{\partial q_i}\frac{\partial f}{\partial p_i} \tag{8.86}$$

is called the *Poisson bracket of f and g*. If $\vec{r} = x_i\vec{u}_i$, $\vec{p} = m\dot{\vec{r}}$, and $\vec{l} = \vec{r} \times \vec{p}$ are the radius vector, the linear and the angular momenta, respectively, of a particle of mass m with respect to the Cartesian frame $Oxyz$ of versors $\vec{u}_i$ ($i = \overline{1, 3}$), prove the following relations:

(i) $\{l_i, x_j\} = \varepsilon_{ijk}x_k$;
(ii) $\{l_i, p_j\} = \varepsilon_{ijk}p_k$;
(iii) $\{l_i, l_j\} = \varepsilon_{ijk}l_k$;
(iv) $\{\vec{l}, r^2\} = 0$;
(v) $\{\vec{l}, p^2\} = 0$.

Solution

In order to solve the problem, we remind the reader the main properties of the Poisson brackets, together with some useful relations involving the completely antisymmetric, third rank pseudotensor ε_{ijk} (the Levi-Civita symbol). So, one can easily prove that
1°. $\{C, f\} = 0 \quad C = const.$;
2°. $\{Cf, g\} = C\{f, g\}$;
3°. $\{f, g\} = -\{g, f\}$;
4°. $\{-f, g\} = -\{f, g\} = \{g, f\}$;
5°. $\{f_1 + f_2, g\} = \{f_1, g\} + \{f_2, g\}$;
6°. $\{f_1 f_2, g\} = f_1\{f_2, g\} + f_2\{f_1, g\}$;
7°. $\mathcal{D}\{f, g\} = \{\mathcal{D}f, g\} + \{f, \mathcal{D}g\}$, where $\mathcal{D}$ is any scalar or vector differential operator, like $\frac{\partial}{\partial t}$, $\frac{\partial}{\partial x}$, ∇, etc.;
8°. $\{q_i, q_k\} = \{p_i, p_k\} = 0$, $\forall\ i \neq k$;

9°. $\{q_i, p_k\} = \delta_{ik}$;
10°. $\{f, \{g, h\}\} + \{h, \{f, g\}\} + \{g, \{h, f\}\} = 0$ (Jacobi's identity);
11°. $\{f, g(y_1, y_2, ..., y_n)\} = \sum_{k=1}^{n} \frac{\partial g}{\partial y_k} \{f, y_k\}$,
as well as

12°.
$$\varepsilon_{ijk}\varepsilon_{lmn} = \begin{vmatrix} \delta_{il} & \delta_{im} & \delta_{in} \\ \delta_{jl} & \delta_{jm} & \delta_{jn} \\ \delta_{kl} & \delta_{km} & \delta_{kn} \end{vmatrix};$$

13°. $\varepsilon_{ijk}\varepsilon_{imn} = \delta_{jm}\delta_{kn} - \delta_{jn}\delta_{km}$;
14°. $\varepsilon_{ijk}\varepsilon_{ijl} = 2\delta_{kl}$;
15°. $\varepsilon_{ijk}\varepsilon_{ijk} = 3!$;
16°. $\vec{u}_i \times \vec{u}_j = \varepsilon_{ijk}\vec{u}_k$ $(i, j, k = \overline{1,3})$;
17°. $\vec{u}_i = \frac{1}{2}\varepsilon_{ijk}\vec{u}_j \times \vec{u}_k$ $(i, j, k = \overline{1,3})$;
18°. $\vec{u}_i \cdot (\vec{u}_j \times \vec{u}_k) = \varepsilon_{ijk}$;
19°. $\varepsilon_{ijk}\delta_{ij} = \varepsilon_{ijk}\delta_{ik} = \varepsilon_{ijk}\delta_{jk} = 0$.

Let us now proceed to solve our exercises (i)-(v):

(i) Since $\vec{l} = \vec{r} \times \vec{p} = \varepsilon_{ijk}x_j p_k \vec{u}_i$, we have $l_i = \varepsilon_{ijk}x_j p_k$, therefore

$$\{l_i, x_j\} = \{\varepsilon_{ilk}x_l p_k, x_j\} = \varepsilon_{ilk}\{x_l p_k, x_j\}$$

$$= \varepsilon_{ilk}x_l\{p_k, x_j\} + \varepsilon_{ilk}p_k\{x_l, x_j\}$$

$$= -\varepsilon_{ilk}x_l\delta_{jk} = \varepsilon_{ijl}x_l = \varepsilon_{ijk}x_k.$$

(ii)
$$\{l_i, p_j\} = \{\varepsilon_{ilk}x_l p_k, p_j\} = \varepsilon_{ilk}\{x_l p_k, p_j\}$$

$$= \varepsilon_{ilk}x_l\{p_k, p_j\} + \varepsilon_{ilk}p_k\{x_l, p_j\}$$

$$= \varepsilon_{ilk}p_k\{x_l, p_j\} = \varepsilon_{ilk}p_k\delta_{lj} = \varepsilon_{ijk}p_k.$$

(iii)
$$\{l_i, l_j\} = \{\varepsilon_{ilk}x_l p_k, \varepsilon_{jmn}x_m p_n\} = \varepsilon_{ilk}\varepsilon_{jmn}\{x_l p_k, x_m p_n\}$$

$$= \varepsilon_{ilk}\varepsilon_{jmn}(x_l\{p_k, x_m p_n\} + p_k\{x_l, x_m p_n\})$$

$$= \varepsilon_{ilk}\varepsilon_{jmn}\left[x_l\Big(x_m\{p_k, p_n\} + p_n\{p_k, x_m\}\Big)\right.$$

$$\left.+ p_k\Big(x_m\{x_l, p_n\} + p_n\{x_l, x_m\}\Big)\right]$$

$$= \varepsilon_{ilk}\varepsilon_{jmn}(-x_l p_n \delta_{km} + p_k x_m \delta_{ln})$$

$$= -\varepsilon_{ilk}\varepsilon_{jkn}x_l p_n + \varepsilon_{ilk}\varepsilon_{jml}p_k x_m$$

$$= \varepsilon_{kil}\varepsilon_{kjn}x_l p_n - \varepsilon_{lik}\varepsilon_{ljm}p_k x_m$$

$$= (\delta_{ij}\delta_{ln} - \delta_{in}\delta_{lj})x_l p_n - (\delta_{ij}\delta_{km} - \delta_{im}\delta_{kj})x_m p_k$$

$$= \delta_{ij}x_l p_l - x_j p_i - \delta_{ij} x_k p_k + x_i p_j = x_i p_j - x_j p_i \stackrel{not.}{=} A_{ij} = \varepsilon_{ijk} l_k,$$

where the pseudovector $\vec{l}$ is connected with the antisymmetric tensor A_{ik} by relation (see Appendix B)

$$l_k = \frac{1}{2}\varepsilon_{kij} A_{ij} = \frac{1}{2}\varepsilon_{kij}(x_i p_j - x_j p_i) = \frac{1}{2} 2\varepsilon_{kij} x_i p_j = (\vec{r}\times\vec{p})_k,$$

which means that the pseudovector $\vec{l}$ is precisely the angular momentum.

(iv)
$$\{\vec{l}, \vec{r}^2\} = \vec{i}\{l_x, \vec{r}^2\} + \vec{j}\{l_y, \vec{r}^2\} + \vec{k}\{l_z, \vec{r}^2\}.$$

The first Poisson bracket on the r.h.s. yields

$$\{l_x, \vec{r}^2\} = \{\varepsilon_{1jk} x_j p_k, x_l x_l\} = \varepsilon_{1jk}\{x_j p_k, x_l x_l\}$$

$$= \varepsilon_{1jk}\big(x_j\{p_k, x_l x_l\} + p_k\{x_j, x_l x_l\}\big) = \varepsilon_{1jk} x_j \{p_k, x_l x_l\}$$

$$= 2\varepsilon_{1jk} x_j x_l\{p_k, x_l\} = -2\varepsilon_{1jk} x_j x_l \delta_{kl} = -2\varepsilon_{1jk} x_j x_k = 0,$$

because $x_j x_k = x_k x_j$ is a symmetric tensor in the pair of indices j and k, and ε_{1jk} is antisymmetric in the same indices. Indeed, if S_{ij} and A_{ij} are symmetric and, respectively, antisymmetric tensors, we have:

$$A_{ij} S_{ij} = \frac{1}{2}(A_{ij} S_{ij} + A_{ij} S_{ij}) = \frac{1}{2}(A_{ij} S_{ij} + A_{ji} S_{ji})$$

$$= \frac{1}{2}(A_{ij} S_{ij} - A_{ij} S_{ij}) = 0.$$

Similarly, it can be shown that the other two Poisson brackets $\{l_y, \vec{r}^2\}$ and $\{l_z, \vec{r}^2\}$ are also zero. Consequently,

$$\{\vec{l}, \vec{r}^2\} = 0.$$

(v)
$$\{\vec{l}, \vec{p}^2\} = \vec{i}\{l_x, \vec{p}^2\} + \vec{j}\{l_y, \vec{p}^2\} + \vec{k}\{l_z, \vec{p}^2\}.$$

Using the previous procedure, let us calculate the first Poisson bracket on the r.h.s. We have:

$$\{l_x, \vec{p}^2\} = \{\varepsilon_{1jk} x_j p_k, p_l p_l\} = \varepsilon_{1jk}\big(x_j\{p_k, p_l p_l\} + p_k\{x_j, p_l p_l\}\big)$$

$$= \varepsilon_{1jk} p_k\{x_j, p_l p_l\} = 2\varepsilon_{1jk} p_k p_j = 0,$$

for the same reason as that displayed in the exercise (iv). In an analogous way, it can be proved that $\{l_y, \vec{p}^2\} = 0$, and $\{l_z, \vec{p}^2\} = 0$, which finally yields
$$\{\vec{l}, \vec{p}^2\} = 0.$$

Problem 5

Using the results obtained in Problem 4 above, calculate the Poisson brackets among the Hamiltonian of a free particle moving in a potential force field $\vec{F} = -\operatorname{grad} V(\vec{r})$ and the following functions, generically denoted by $f(q, p, t)$:

(i) $f(q, p, t) = q_j = x_j \quad (j = 1, 2, 3)$;
(ii) $f(q, p, t) = p_j \quad (j = 1, 2, 3)$;
(iii) $f(q, p, t) = l_j \quad (j = 1, 2, 3)$;
(iv) $f(q, p, t) = \vec{r} = x_j \vec{u}_j \quad (j = 1, 2, 3)$;
(v) $f(q, p, t) = \vec{p} = p_j \vec{u}_j \quad (j = 1, 2, 3)$;
(vi) $f(q, p, t) = \vec{r} \cdot \vec{p} = x_j p_j \quad (j = 1, 2, 3)$;
(vii) $f(q, p, t) = \vec{r} \times \vec{p} = \varepsilon_{ijk} x_j p_k \vec{u}_i$;
(viii) $f(q, p, t) = \vec{r}^2 = x_j x_j \quad (j = 1, 2, 3)$;
(ix) $f(q, p, t) = \vec{p}^2 = p_j p_j \quad (j = 1, 2, 3)$;
(x) $f(q, p, t) = \vec{l}^{\,2} = l_j l_j \quad (j = 1, 2, 3)$;

Solution

As well-known, the Hamiltonian of a free particle moving in a potential force field with the potential $V = V(\vec{r})$ equals the total mechanical energy of the particle (being a conserved quantity):

$$H \equiv E = T + V = \frac{\vec{p}^2}{2m} + V(\vec{r}).$$

Taking into account the properties of the Poisson brackets, we then have:

(i)
$$\{H, x_j\} = \left\{\frac{\vec{p}^2}{2m} + V(\vec{r}), x_j\right\}$$
$$= \frac{1}{2m}\{\vec{p}^2, x_j\} + \{V(\vec{r}), x_j\} = \frac{1}{2m}\{p_k p_k, x_j\}$$
$$= \frac{1}{2m}(p_k\{p_k, x_j\} + \{p_k, x_j\}p_k) = \frac{p_k}{m}\{p_k, x_j\}$$
$$= -\frac{p_k}{m}\delta_{kj} = -\frac{p_j}{m} = -v_j = -\dot{x}_j,$$

where we used the fact that $V(\vec{r})$ does not depend on velocity, while the variables x_k and p_k are independent. Therefore,

$$\dot{x}_j = \{x_j, H\}.$$

(ii)

$$\{H, p_j\} = \left\{\frac{\vec{p}^2}{2m} + V(\vec{r}), p_j\right\} = \frac{1}{2m}\{\vec{p}^2, p_j\} + \{V(\vec{r}), p_j\}$$

$$= \frac{\partial V}{\partial x_i}\frac{\partial p_j}{\partial p_i} - \frac{\partial V}{\partial p_i}\frac{\partial p_j}{\partial x_i} = \frac{\partial V}{\partial x_i}\delta_{ij}$$

$$= \frac{\partial V}{\partial x_j} = [\mathrm{grad} V(\vec{r})]_j = -F_j = -\dot{p}_j,$$

so that

$$\dot{p}_j = \{p_j, H\}.$$

(iii)

$$\{H, l_j\} = \{H, \varepsilon_{jsm} x_s p_m\} = \varepsilon_{jsm}\{H, x_s p_m\}$$

$$= \varepsilon_{jsm} x_s \{H, p_m\} + \varepsilon_{jsm} p_m \{H, x_s\}$$

$$= -\varepsilon_{jsm} x_s \dot{p}_m - \varepsilon_{jsm} \dot{x}_s p_m = -\frac{d}{dt}(\varepsilon_{jsm} x_s p_m) = -\frac{d}{dt} l_j = -\dot{l}_j$$

which can be written as

$$\dot{l}_j = \{l_j, H\}.$$

Observation. According to the angular momentum theorem for the particle of mass m

$$\frac{d\vec{l}}{dt} = \vec{\mathcal{M}} = \vec{r} \times \vec{F},$$

our result can also be expressed as

$$\{l_j, H\} = (\vec{r} \times \vec{F})_j = \mathcal{M}_j.$$

Indeed:

$$\{l_j, H\} = \left\{l_j, \frac{\vec{p}^2}{2m} + V(\vec{r})\right\} = \frac{1}{2m}\{l_j, p_k p_k\} + \{l_j, V(\vec{r})\}$$

$$= \frac{1}{m} p_k \{l_j, p_k\} + \frac{\partial l_j}{\partial x_i}\frac{\partial V}{\partial p_i} - \frac{\partial l_j}{\partial p_i}\frac{\partial V}{\partial x_i}$$

$$= \frac{1}{m}\varepsilon_{jks}p_k p_s - \frac{\partial}{\partial p_i}(\varepsilon_{jks}x_k p_s)\frac{\partial V}{\partial x_i}$$

$$= -\varepsilon_{jki}x_k \frac{\partial V}{\partial x_i} = (\vec{r} \times \vec{F})_j = \mathcal{M}_j.$$

If the force is of central-type, $\vec{F} = f(r)\vec{r}$, then $\{l_j, H\} = 0$. This shows that, if l_j does not explicitly depend on time $\left(\frac{\partial l_j}{\partial t} = 0\right)$, then $l_j = const.$ is a first integral of the Hamilton's canonical equations.

Let us show that this result is also valid for a particle of mass m and charge e, moving in the external static electromagnetic field $\vec{E} = -\mathrm{grad}\Phi(\vec{r})$, $\vec{B} = \mathrm{curl}\vec{A}(\vec{r})$. In this case the Hamiltonian writes

$$H = \frac{1}{2m}(p_k - eA_k)(p_k - eA_k) + e\Phi,$$

while the projection on x_j-axis of the angular momentum is

$$l_j = \varepsilon_{jsn}x_s(p_n - eA_n),$$

so that

$$\{l_j, H\} = \left\{\varepsilon_{jsn}x_s(p_n - eA_n), \frac{1}{2m}(p_k - eA_k)(p_k - eA_k) + e\Phi\right\}$$

$$= \varepsilon_{jsn}\left\{x_s(p_n - eA_n), \frac{1}{2m}(p_k - eA_k)(p_k - eA_k)\right\}$$

$$+ \varepsilon_{jsn}\left\{x_s(p_n - eA_n), e\Phi\right\}$$

$$= \frac{1}{m}\varepsilon_{jsn}(p_k - eA_k)\{x_s(p_n - eA_n), p_k - eA_k\} + e\varepsilon_{jsn}x_s\{p_n - eA_n, \Phi\}$$

$$+ e\varepsilon_{jsn}(p_n - eA_n)\{x_s, \Phi\}$$

$$= \frac{1}{m}\varepsilon_{jsn}(p_k - eA_k)(p_n - eA_n)\{x_s, p_k - eA_k\}$$

$$+ \frac{1}{m}\varepsilon_{jsn}x_s(p_k - eA_k)\{p_n - eA_n, p_k - eA_k\} + e\varepsilon_{jsn}x_s\{p_n - eA_n, \Phi\}$$

$$= \frac{1}{m}\varepsilon_{jsn}(p_s - eA_s)(p_n - eA_n) - \frac{e}{m}\varepsilon_{jsn}x_s(p_k - eA_k)\frac{\partial A_n}{\partial x_k}$$

$$+ \frac{e}{m}\varepsilon_{jsn}x_s(p_k - eA_k)\frac{\partial A_k}{\partial x_n} - e\varepsilon_{jsn}x_s\frac{\partial \Phi}{\partial x_n}$$

$$= \varepsilon_{jsn}x_s\left[-e\frac{\partial \Phi}{\partial x_n} + e\frac{p_k - eA_k}{m}\left(\frac{\partial A_k}{\partial x_n} - \frac{\partial A_n}{\partial x_k}\right)\right]$$

$$= \varepsilon_{jsn}x_s\left[-e\frac{\partial \Phi}{\partial x_n} + ev_k\left(\frac{\partial A_k}{\partial x_n} - \frac{\partial A_n}{\partial x_k}\right)\right]$$

$$= \varepsilon_{jsn}x_s\left[eE_n + e(\vec{v}\times\vec{B})_n\right] = \varepsilon_{jsn}x_s F_n = (\vec{r}\times\vec{F})_j,$$

which completes the proof. Here we used the formula

$$v_k\left(\frac{\partial A_k}{\partial x_n} - \frac{\partial A_n}{\partial x_k}\right) \equiv v_k T_{nk}(= -v_k T_{kn}) = v_k \varepsilon_{nks}B_s$$

$$= \varepsilon_{nks}v_k B_s = (\vec{v}\times\vec{B})_n,$$

where, according to (B.91), we wrote

$$B_s = \frac{1}{2}\varepsilon_{snk}\left(\frac{\partial A_k}{\partial x_n} - \frac{\partial A_n}{\partial x_k}\right) = \varepsilon_{snk}\frac{\partial A_k}{\partial x_n} = (\text{curl}\vec{A})_s.$$

(iv)
$$\{H,\vec{r}\} = \{H,x_j\vec{u}_j\} = \{H,x_j\}\vec{u}_j + \{H,\vec{u}_j\}x_j$$
$$= \{H,x_j\}\vec{u}_j = -\dot{x}_j\vec{u}_j = -\dot{\vec{r}},$$

where we used the fact that the versors of the Cartesian coordinate axes are constant quantities, as well as the result obtained at the point (i). Consequently,

$$\dot{\vec{r}} = \{\vec{r},H\}.$$

(v)
$$\{H,\vec{p}\} = \{H,p_j\vec{u}_j\} = \{H,p_j\}\vec{u}_j + \{H,\vec{u}_j\}p_j$$
$$= \{H,p_j\}\vec{u}_j = -\dot{p}_j\vec{u}_j = -\dot{\vec{p}},$$

or
$$\dot{\vec{p}} = \{\vec{p},H\},$$

where we used the result obtained at point (ii), and the fact that $\vec{u}_j$ for Cartesian coordinate systems are constant quantities, as well.

(vi)
$$\{H,\vec{r}\cdot\vec{p}\} = \{H,x_j p_j\} = \{H,x_j\}p_j + \{H,p_j\}x_j$$
$$= -\dot{x}_j p_j - \dot{p}_j x_j = -\frac{d}{dt}(x_j p_j)$$
$$= -\frac{d}{dt}(\vec{r}\cdot\vec{p}),$$

or
$$\frac{d}{dt}(\vec{r}\cdot\vec{p}) = \{\vec{r}\cdot\vec{p},H\},$$

where (i) and (ii) have been used.
(vii)
$$\{H, \vec{l}\} = \{H, l_j \vec{u}_j\} = \{H, l_j\}\vec{u}_j + \{H, \vec{u}_j\}l_j = -\dot{l}_j\vec{u}_j = -\dot{\vec{l}},$$

or
$$\dot{\vec{l}} = \{\vec{l}, H\},$$

where the point (iii) has been taken into account.
(viii)
$$\{H, \vec{r}^2\} = \{H, x_j x_j\} = 2x_j\{H, x_j\} = -2x_j\dot{x}_j = -2\vec{r}\cdot\dot{\vec{r}},$$

or
$$\frac{d}{dt}(\vec{r}^2) = \{\vec{r}^2, H\},$$

where the point (i) has been used.
(ix)
$$\{H, \vec{p}^2\} = \{H, p_j p_j\} = 2p_j\{H, p_j\} = -2p_j\dot{p}_j,$$

or
$$\frac{d}{dt}(\vec{p}^2) = \{\vec{p}^2, H\},$$

where the result (ii) has been considered.
(x)
$$\{H, \vec{l}^{\,2}\} = \{H, l_j l_j\} = 2l_j\{H, l_j\} = -2l_j\dot{l}_j = -2\vec{l}\cdot\dot{\vec{l}} = -\frac{d}{dt}(\vec{l}^{\,2}),$$

or
$$\frac{d}{dt}(\vec{l}^{\,2}) = \{\vec{l}^{\,2}, H\},$$

where the point (iii) has been used.

Observation. All considered cases lead to results of the form

$$\frac{df}{dt} = \{f, H\},$$

where, seriatim, as function f have been chosen: x_j, p_j, l_j, $\vec{r}$, $\vec{p}$, $\vec{r}\cdot\vec{p}$, $\vec{l}$, $\vec{r}^2$, $\vec{p}^2$, and $\vec{l}^{\,2}$. In all these cases, f does not explicitly depend on time. In general, for $f = f(q, p, t)$, we have

$$\frac{df}{dt} = \frac{\partial f}{\partial t} + \dot{q}_j \frac{\partial f}{\partial q_j} + \dot{p}_j \frac{\partial f}{\partial p_j}$$

$$= \frac{\partial f}{\partial t} + \frac{\partial H}{\partial p_j}\frac{\partial f}{\partial q_j} - \frac{\partial H}{\partial q_j}\frac{\partial f}{\partial p_j} = \frac{\partial f}{\partial t} + \{f, H\}.$$

Problem 6.
Show that the temporal average values of the total kinetic and potential energies of a stable and spatially bounded system are given by the relations
$$\overline{E_c} = \frac{n}{n+2}H,$$
and, respectively,
$$\overline{V} = \frac{2}{n+2}H,$$
where H is the Hamiltonian of the system, while n is Euler's degree (order) of homogeneity of the potential energy.

Solution.
Consider a stable and spatially limited (bounded) physical system composed by N particles, each particle P_k having mass m_k ($k = \overline{1, N}$), acted by the conservative potential forces

$$\vec{F}_{kj} = -\text{grad}_{kj} V_{kj} \quad (\text{no summation};\quad k, j = \overline{1, N}), \qquad (8.87)$$

where V_{kj} is the potential energy of interaction between particles P_k and P_j, while grad_{kj} stands for the partial derivative with respect to the components of the vector $\vec{r}_{kj} = \vec{r}_k - \vec{r}_j$ (see Fig. VIII.3).

By definition, the quantity

$$G = \sum_{k=1}^{N} \vec{r}_k \cdot \vec{p}_k, \qquad (8.88)$$

where $\vec{p}_k = m_k \vec{v}_k = m_k \dot{\vec{x}}_k$ is the linear momentum of the particle P_k, is called the *scalar virial of the system*. Let us calculate the total derivative with respect to time of the virial G. We have:

$$\frac{dG}{dt} = \frac{d}{dt}\sum_{k=1}^{N} \vec{r}_k \cdot \vec{p}_k = \sum_{k=1}^{N} \frac{d}{dt}(\vec{r}_k \cdot \vec{p}_k)$$

$$= \sum_{k=1}^{N} \frac{d\vec{r}_k}{dt} \cdot \vec{p}_k + \sum_{k=1}^{N} \vec{r}_k \cdot \frac{d\vec{p}_k}{dt}$$

$$= \sum_{k=1}^{N} \frac{1}{m_k} \vec{p}_k \cdot \vec{p}_k + \sum_{k=1}^{N} \vec{r}_k \cdot \vec{F}_k = 2E_c + \sum_{k=1}^{N} \vec{r}_k \cdot \vec{F}_k,$$

where $E_c = \sum_{k=1}^{N} \frac{1}{m_k} \vec{p}_k^{\,2}$ is the total kinetic energy of the system of particles.

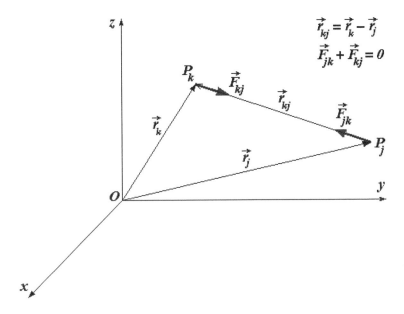

Fig.VIII.3

Since

$$\frac{\partial E_c}{\partial p_j} = \sum_{k=1}^{N} \frac{1}{m_k} p_k \delta_{kj} = \frac{p_j}{m_j} \quad \text{(no summation)},$$

we can write

$$2E_c = \sum_{k=1}^{N} \frac{1}{m_k} \vec{p}_k \cdot \vec{p}_k = \sum_{k=1}^{N} \frac{1}{m_k} p_k p_k = \sum_{k=1}^{N} p_k \frac{\partial E_c}{\partial p_k}, \qquad (8.89)$$

expressing the fact that the total kinetic energy of the system of particles is a Eulerian homogeneous function of the second degree. Here we remind the reader that the function $f(x_1, x_2, ..., x_n)$ is called homogeneous of degree s with respect to the independent variables $x_1, x_2, ..., x_n$, if

$$f(tx_1, tx_2, ..., tx_n) = t^s f(x_1, x_2, ..., x_n) \quad (t \in \mathbb{R}^*)$$

in which case is valid Euler's theorem regarding homogeneous functions of degree s:

$$\sum_{k=1}^{n} x_k \frac{\partial f}{\partial x_k} = sf. \qquad (8.90)$$

We have therefore shown that

$$\frac{dG}{dt} = 2E_c + \sum_{k=1}^{N} \vec{r}_k \cdot \vec{F}_k, \qquad (8.91)$$

where $\vec{F}_k$ is the total force acting on the particle P_k,

$$\vec{F}_k = \sum_{\substack{j=1 \\ j \neq k}}^{N} \vec{F}_{kj} \qquad (k = 1, 2, ..., N). \qquad (8.92)$$

Since forces $\vec{F}_{kj}$ are conservative, deriving from the potential $V_{kj}(|\vec{r}_{kj}|)$, that is

$$\vec{F}_{kj} = -\nabla_{\vec{r}_{kj}} V_{kj}(|\vec{r}_{kj}|), \qquad (8.93)$$

we can calculate the second term of the r.h.s. of Eq. (8.91). Indeed,

$$\sum_{k=1}^{N} \vec{r}_k \cdot \vec{F}_k = \sum_{k=1}^{N} \vec{r}_k \cdot \sum_{\substack{j=1 \\ j \neq k}}^{N} \vec{F}_{kj} = \sum_{k=1}^{N} \sum_{\substack{j=1 \\ j \neq k}}^{N} \vec{r}_k \cdot \vec{F}_{kj}$$

$$= \sum_{k=1}^{N} \sum_{\substack{j=1 \\ j<k}}^{N} \vec{r}_k \cdot \vec{F}_{kj} + \sum_{k=1}^{N} \sum_{\substack{j=1 \\ j>k}}^{N} \vec{r}_k \cdot \vec{F}_{kj}$$

$$= \sum_{k=1}^{N} \sum_{\substack{j=1 \\ j<k}}^{N} \vec{r}_k \cdot \vec{F}_{kj} + \sum_{k=1}^{N} \sum_{\substack{j=1 \\ j<k}}^{N} \vec{r}_j \cdot \vec{F}_{jk}$$

$$= \sum_{k=1}^{N} \sum_{\substack{j=1 \\ j<k}}^{N} \vec{r}_k \cdot \vec{F}_{kj} - \sum_{k=1}^{N} \sum_{\substack{j=1 \\ j<k}}^{N} \vec{r}_j \cdot \vec{F}_{kj}$$

$$= \sum_{k=1}^{N} \sum_{\substack{j=1 \\ j<k}}^{N} (\vec{r}_k - \vec{r}_j) \cdot \vec{F}_{kj} = \sum_{k=1}^{N} \sum_{\substack{j=1 \\ j<k}}^{N} \vec{r}_{kj} \cdot \vec{F}_{kj}, \qquad (8.94)$$

where we have used the action and reaction principle of Newtonian mechanics
$$\vec{F}_{kj} + \vec{F}_{jk} = 0. \tag{8.95}$$

Therefore,
$$\sum_{k=1}^{N} \vec{r}_k \cdot \vec{F}_k = \sum_{k=1}^{N} \sum_{\substack{j=1 \\ j<k}}^{N} \vec{r}_{kj} \cdot \vec{F}_{kj}$$

$$= -\sum_{k=1}^{N} \sum_{\substack{j=1 \\ j<k}}^{N} \vec{r}_{kj} \cdot \nabla_{\vec{r}_{kj}} V_{kj}(r_{kj}) = -\sum_{k=1}^{N} \sum_{\substack{j=1 \\ j<k}}^{N} \vec{r}_{kj} \cdot \frac{\partial V_{kj}}{\partial \vec{r}_{kj}} = -nV, \tag{8.96}$$

where we used (8.93), as well as the fact that the total potential energy V of the system is a homogeneous function of degree n. In view of (8.96), equation (8.91) yields

$$\frac{dG}{dt} = 2E_c - nV. \tag{8.97}$$

Let us now show that the temporal average (mean) value of the total time derivative of a bounded function is zero. To this end, we appeal to the definition of the temporal average value of some function $f(t)$

$$\overline{f(t)} = \lim_{\tau \to \infty} \frac{1}{\tau} \int_0^{\tau} f(t)\,dt. \tag{8.98}$$

If as function $f(t)$ we take the time derivative of the virial of a stable and spatially bounded system of particles, we have:

$$\overline{\frac{dG}{dt}} = \lim_{\tau \to \infty} \frac{1}{\tau} \int_0^{\tau} \frac{dG}{dt}\,dt = \lim_{\tau \to \infty} \frac{G(\tau) - G(0)}{\tau} = 0, \tag{8.99}$$

since for a stable and spatially bounded system of particles both $G(\tau)$ and $G(0)$ are bounded quantities, because the virial depends only on radius-vectors and momenta of the particles which, both, are bounded quantities. In view of (8.99), from (8.97) we therefore have

$$2\overline{E_c} = n\overline{V}. \tag{8.100}$$

As well-known, the total energy of a stable and spatially bounded system of particles, acted by conservative forces is a first integral of the motion and coincides with the Hamiltonian H of the system

$$E = E_c + V = H = const. \tag{8.101}$$

The mean value of this relation is

$$\overline{E_c} + \overline{V} = \overline{H} = H. \qquad (8.102)$$

Relations (8.100) and (8.102) form a system of two equations with two unknowns ($\overline{E_c}$ and $\overline{V}$), with the following solutions:

$$\begin{cases} \overline{E_c} = \dfrac{n}{n+2} H; \\ \overline{V} = \dfrac{2}{n+2} H, \end{cases} \qquad (8.103)$$

and the problem is solved.

Relation (8.100) is called *the virial theorem* and can be found under different forms, such as the so-called *variational form*, developed in 1945 by Paul Ledoux, and the *tensor form*, due to Eugene Parker, Subrahmanyan Chandrasekhar, and Enrico Fermi (1953). Our result (8.103) is valid only if the total potential energy of the system of particles

$$V = \sum_{k=1}^{N} \sum_{\substack{j=1 \\ j<k}}^{N} V_{kj}(r_{kj}) \qquad (8.104)$$

is a homogeneous function of degree n. The most common form of such a function is

$$V_{kj}(r_{kj}) = a r_{kj}^n, \qquad (8.105)$$

where a is a non-zero constant. According to (8.105), the energy of interaction between two particles is proportional to the nth power of the distance between the particles. A special case is furnished by the interaction conservative gravitational and electrostatic forces, characterized by $n = -1$, which leads to

$$2\overline{E_c} = -\overline{V}, \qquad (8.106)$$

called *Lagrange's identity*.

The virial theorem has several important applications, like: study of stability of cosmological systems (Henri Poincaré, 1911), determination of Chandrasekhar limit (1953), investigations upon hypothesis of existence of dark matter (Fritz Zwicky, 1933), etc. We also mention the efficiency of the virial theorem in evaluation of the mean value of kinetic energy of stable and spatially bounded physical systems, even for very complex ones, like systems studied by statistical physics.

In this last case, the temporal mean value of the total kinetic energy of a complex system is closely related to the system temperature, through the medium of equipartition of energy theorem. Nevertheless, the virial theorem does not depend on the notion of temperature (a purely statistical quantity), but can also be used for systems which are not in thermodynamic equilibrium.

Observation. The scalar virial of a mechanical system (containing constant masses m_k, $k = \overline{1,N}$) can also be defined in terms of the moment of inertia I of the system. Indeed,

$$I = \sum_{k=1}^{N} m_k \vec{r}_k^2 = \sum_{k=1}^{N} m_k r_k^2,$$

so that

$$\frac{1}{2}\frac{dI}{dt} = \sum_{k=1}^{N} m_k \frac{d\vec{r}_k}{dt} \cdot \vec{r}_k = \sum_{k=1}^{N} \vec{r}_k \cdot \vec{p}_k = G.$$

Problem 7.

Determine the Hamiltonian of a double coplanar pendulum (see Fig.VIII.4) and write the corresponding equations of motion.

Solution

This system was studied within the frame of Lagrangian formalism (see Problem 7, Chap.VI). If θ_1 and θ_2 are the generalized coordinates associated with the two degrees of freedom of the system, then the Lagrangian writes [see (6.143)]

$$L = \frac{1}{2}l_1^2\dot{\theta}_1^2(m_1 + m_2) + \frac{1}{2}m_2 l_2^2 \dot{\theta}_2^2 + m_2 l_1 l_2 \dot{\theta}_1 \dot{\theta}_2 \cos(\theta_2 - \theta_1)$$

$$+ g l_1 (m_1 + m_2) \cos\theta_1 + m_2 g l_2 \cos\theta_2. \qquad (8.107)$$

The generalized momenta conjugated with θ_1 and θ_2 are:

$$p_1 = \frac{\partial L}{\partial \dot{\theta}_1} = l_1^2 \dot{\theta}_1 (m_1 + m_2) + m_2 l_1 l_2 \dot{\theta}_2 \cos(\theta_2 - \theta_1), \qquad (8.108)$$

and

$$p_2 = \frac{\partial L}{\partial \dot{\theta}_2} = m_2 l_2^2 \dot{\theta}_2 + m_2 l_1 l_2 \dot{\theta}_1 \cos(\theta_2 - \theta_1). \qquad (8.109)$$

The Hamiltonian therefore is

$$H = p_1 \dot{\theta}_1 + p_2 \dot{\theta}_2 - L = \frac{1}{2}l_1^2 \dot{\theta}_1^2 (m_1 + m_2) + \frac{1}{2}m_2 l_2^2 \dot{\theta}_2^2$$

$$+m_2l_1l_2\dot\theta_1\dot\theta_2\cos(\theta_2-\theta_1)-(m_1+m_2)gl_1\cos\theta_1-m_2gl_2\cos\theta_2. \quad (8.110)$$

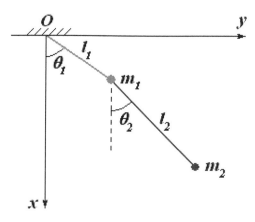

Fig. VIII.4

To express H in terms of canonical variables θ_1, θ_2, p_1, and p_2, one applies the usual procedure. Using (8.108), we have

$$l_1\dot\theta_1 = \frac{p_1}{l_1(m_1+m_2)} - \frac{m_2l_2\dot\theta_2\cos(\theta_2-\theta_1)}{m_1+m_2} \quad (8.111)$$

and (8.109) yields

$$p_2 = m_2l_2^2\dot\theta_2 + m_2l_2\cos(\theta_2-\theta_1)\left[\frac{p_1}{l_1(m_1+m_2)} - \frac{m_2l_2\dot\theta_2\cos(\theta_2-\theta_1)}{m_1+m_2}\right],$$

or

$$p_2l_1(m_1+m_2) = m_2l_1(m_1+m_2)l_2^2\dot\theta_2 + m_2l_2p_1\cos(\theta_2-\theta_1)$$
$$-m_2^2l_1l_2^2\dot\theta_2\cos^2(\theta_2-\theta_1),$$

which gives

$$\dot\theta_2 = \frac{p_2l_1(m_1+m_2) - m_2l_2p_1\cos(\theta_2-\theta_1)}{m_2l_1l_2^2[m_1+m_2-m_2\cos^2(\theta_2-\theta_1)]}. \quad (8.112)$$

With this result, (8.111) becomes

$$\dot\theta_1 = \frac{p_1}{l_1^2(m_1+m_2)}$$

296

$$-\frac{m_2 l_2 \cos(\theta_2 - \theta_1)}{l_1(m_1 + m_2)} \frac{p_2 l_1 (m_1 + m_2)}{m_2 l_1 l_2^2 [m_1 + m_2 - m_2 \cos^2(\theta_2 - \theta_1)]}$$

$$+\frac{m_2 l_2 \cos(\theta_2 - \theta_1)}{l_1(m_1 + m_2)} \frac{m_2 l_2 p_1 \cos(\theta_2 - \theta_1)}{m_2 l_1 l_2^2 [m_1 + m_2 - m_2 \cos^2(\theta_2 - \theta_1)]},$$

or

$$\dot{\theta}_1 = \frac{p_1}{l_1^2(m_1 + m_2)} + \frac{m_2 p_1 \cos^2(\theta_2 - \theta_1)}{l_1^2(m_1 + m_2)[m_1 + m_2 - m_2 \cos^2(\theta_2 - \theta_1)]}$$

$$-\frac{p_2 \cos(\theta_2 - \theta_1)}{l_1 l_2 [m_1 + m_2 - m_2 \cos^2(\theta_2 - \theta_1)]},$$

and, still

$$\dot{\theta}_1 = \frac{p_1}{l_1^2 [m_1 + m_2 - m_2 \cos^2(\theta_2 - \theta_1)]}$$

$$-\frac{p_2 \cos(\theta_2 - \theta_1)}{l_1 l_2 [m_1 + m_2 - m_2 \cos^2(\theta_2 - \theta_1)]}. \quad (8.113)$$

Introducing (8.112) and (8.113) into (8.110), we have

$$H = \frac{1}{2} l_1^2 (m_1 + m_2) \left\{ \frac{p_1 l_2 - p_2 l_1 \cos(\theta_2 - \theta_1)}{l_1^2 l_2 [m_1 + m_2 \sin^2(\theta_2 - \theta_1)]} \right\}^2$$

$$+\frac{1}{2} m_2 l_2^2 \left\{ \frac{p_2 l_1 (m_1 + m_2) - m_2 l_2 p_1 \cos(\theta_2 - \theta_1)}{m_2 l_1 l_2^2 [m_1 + m_2 \sin^2(\theta_2 - \theta_1)]} \right\}^2$$

$$+m_2 l_1 l_2 \cos(\theta_2 - \theta_1)$$

$$\times \frac{[p_1 l_2 - p_2 l_1 \cos(\theta_2 - \theta_1)] [p_2 l_1 (m_1 + m_2) - m_2 l_2 p_1 \cos(\theta_2 - \theta_1)]}{m_2 l_1^3 l_2^3 [m_1 + m_2 \sin^2(\theta_2 - \theta_1)]^2}$$

$$-(m_1 + m_2) g l_1 \cos \theta_1 - m_2 g l_2 \cos \theta_2,$$

or, finally,

$$H = \frac{1}{2} \frac{p_1^2}{l_1^2 [m_1 + m_2 \sin^2(\theta_2 - \theta_1)]}$$

$$+\frac{1}{2} \frac{m_1 + m_2}{m_2} \frac{p_2^2}{l_2^2 [m_1 + m_2 \sin^2(\theta_2 - \theta_1)]}$$

$$-\frac{p_1 p_2 \cos(\theta_2 - \theta_1)}{l_1 l_2 [m_1 + m_2 \sin^2(\theta_2 - \theta_1)]}$$

$$-(m_1 + m_2)gl_1 \cos\theta_1 - m_2 g l_2 \cos\theta_2. \qquad (8.114)$$

Hamilton's canonical equations

$$\begin{cases} \dot{\theta}_1 = \dfrac{\partial H}{\partial p_1}; & \dot{\theta}_2 = \dfrac{\partial H}{\partial p_2}; \\ \dot{p}_1 = -\dfrac{\partial H}{\partial \theta_1}; & \dot{p}_2 = -\dfrac{\partial H}{\partial \theta_2}, \end{cases}$$

then yield

$$\begin{cases} \dot{\theta}_1 = \dfrac{p_1}{l_1^2 \left[m_1 + m_2 \sin^2(\theta_2 - \theta_1)\right]} - \dfrac{p_2 \cos(\theta_2 - \theta_1)}{l_1 l_2 \left[m_1 + m_2 \sin^2(\theta_2 - \theta_1)\right]}, \\ \dot{\theta}_2 = \dfrac{m_1 + m_2}{m_2} \dfrac{p_2}{l_2^2 \left[m_1 + m_2 \sin^2(\theta_2 - \theta_1)\right]} \\ \qquad - \dfrac{p_1 \cos(\theta_2 - \theta_1)}{l_1 l_2 \left[m_1 + m_2 \sin^2(\theta_2 - \theta_1)\right]}, \\ \dot{p}_1 = -\dfrac{2 m_2 \sin(\theta_2 - \theta_1)\cos(\theta_2 - \theta_1)}{\left[m_1 + m_2 \sin^2(\theta_2 - \theta_1)\right]^2} \\ \qquad \times \left[\dfrac{p_1^2}{2 l_1^2} + \dfrac{m_1 + m_2}{2 m_2}\dfrac{p_2^2}{l_2^2} - \dfrac{p_1 p_2 \cos(\theta_2 - \theta_1)}{l_1 l_2}\right] \\ \qquad + \dfrac{p_1 p_2 \sin(\theta_2 - \theta_1)}{l_1 l_2 \left[m_1 + m_2 \sin^2(\theta_2 - \theta_1)\right]} - (m_1 + m_2) g l_1 \sin\theta_1, \\ \dot{p}_2 = \dfrac{2 m_2 \sin(\theta_2 - \theta_1)\cos(\theta_2 - \theta_1)}{\left[m_1 + m_2 \sin^2(\theta_2 - \theta_1)\right]^2} \\ \qquad \times \left[\dfrac{p_1^2}{2 l_1^2} + \dfrac{m_1 + m_2}{2 m_2}\dfrac{p_2^2}{l_2^2} - \dfrac{p_1 p_2 \cos(\theta_2 - \theta_1)}{l_1 l_2}\right] \\ \qquad - \dfrac{p_1 p_2 \sin(\theta_2 - \theta_1)}{l_1 l_2 \left[m_1 + m_2 \sin^2(\theta_2 - \theta_1)\right]} - m_2 g l_2 \sin\theta_2. \end{cases}$$

$$(8.115)$$

The differential equations of motion are obtained by taking the time derivative of the first two equations (8.115), where $\dot{p}_1$ and $\dot{p}_2$ are replaced by the last two equations (8.115). Since $\dot{p}_1$ and $\dot{p}_2$ are expressed in terms of p_1 and p_2, these quantities has to be expressed, in their turn, as functions of generalized coordinates and generalized velocities by means of (8.108) and (8.109). After some straightforward calculations we obtain the following results:

$$\ddot{\theta}_1 = \dfrac{m_2 \sin(\theta_2 - \theta_1)\cos(\theta_2 - \theta_1)}{m_1 + m_2 \sin^2(\theta_2 - \theta_1)} \dot{\theta}_1^2$$

$$+\frac{l_2}{l_1}\frac{m_2\sin(\theta_2-\theta_1)}{m_1+m_2\sin^2(\theta_2-\theta_1)}\dot{\theta}_2^2$$

$$-\frac{g}{l_1}\frac{m_1\sin\theta_1}{m_1+m_2\sin^2(\theta_2-\theta_1)}+\frac{g}{l_1}\frac{m_2\cos\theta_2\sin(\theta_2-\theta_1)}{m_1+m_2\sin^2(\theta_2-\theta_1)},$$

$$\ddot{\theta}_2=\frac{l_1}{l_2}\frac{(m_1+m_2)\sin(\theta_1-\theta_2)}{m_1+m_2\sin^2(\theta_1-\theta_2)}\dot{\theta}_1^2$$

$$+\frac{m_2\sin(\theta_1-\theta_2)\cos(\theta_1-\theta_2)}{m_1+m_2\sin^2(\theta_1-\theta_2)}\dot{\theta}_2^2$$

$$+\frac{g}{l_2}\frac{(m_1+m_2)\cos\theta_1\sin(\theta_1-\theta_2)}{m_1+m_2\sin^2(\theta_1-\theta_2)}. \qquad (8.116)$$

Relations (8.116) are the second order, differential equations of motion of the analyzed physical system. They are written in the canonical form from mathematical point of view, but not in the Hamiltonian sense. One can easily prove that equations (8.116) are precisely the equations (6.146) and (6.146'), obtained by means of the Lagrangian formalism, which are

$$\begin{cases} l_1^2\ddot{\theta}_1(m_1+m_2)+m_2l_1l_2\ddot{\theta}_2\cos(\theta_2-\theta_1) \\ -m_2l_1l_2\dot{\theta}_2^2\sin(\theta_2-\theta_1)+(m_1+m_2)gl_1\sin\theta_1=0, \\ \\ m_2l_2^2\ddot{\theta}_2+m_2l_1l_2\ddot{\theta}_1\cos(\theta_2-\theta_1) \\ +m_2l_1l_2\dot{\theta}_1^2\sin(\theta_2-\theta_1)+m_2gl_2\sin\theta_2=0. \end{cases}$$

As one can see, the study of double gravitational plane pendulum by means of the Hamiltonian formalism is more exhausting than the investigation of the same problem by using the Lagrangian technique. Of course, this is not a rule, in many other cases the situation being the other way round. We also mention that in some applications both Lagrangian and Hamiltonian procedures prove to be difficult and clumsy (or even useless), and the only efficient approach is the Hamilton-Jacobi formalism.

Problem 8

Using the Hamiltonian formalism, write the differential equations of motion of a particle of mass m, sliding without friction on a fixed sphere of radius R (spherical pendulum - see Fig.VIII.5).

Solution

One can choose the Cartesian coordinate system $Oxyz \equiv Ox_1x_2x_3$ as reference frame, with the origin O at the centre of the sphere, and z-axis oriented along the force of gravity acting on the particle $\vec{G} = m\vec{g}$ (see Fig.VIII.5). The motion of the particle is subject to a single constraint, namely

$$f(x, y, z) = x^2 + y^2 + z^2 - R^2 = 0, \tag{8.117}$$

which means that the system has $3N-s = 3-1 = 2$ degrees of freedom. Due to spherical symmetry of the problem, it is more convenient to work in spherical coordinates r, θ, φ. In such a system, the velocity of a free particle writes

$$\vec{v} = v_r \vec{u}_r + v_\theta \vec{u}_\theta + v_\varphi \vec{u}_\varphi = \dot{r}\vec{u}_r + r\dot{\theta}\vec{u}_\theta + r\dot{\varphi}\sin\theta\vec{u}_\varphi,$$

so that the kinetic energy of the particle is

$$T = \frac{1}{2}m|\vec{v}|^2 = \frac{1}{2}m(\dot{r}^2 + r^2\dot{\theta}^2 + \dot{\varphi}^2 r^2 \sin^2\theta). \tag{8.118}$$

Since in our case $r = R = const.$, we still have

$$T = \frac{1}{2}mR^2(\dot{\theta}^2 + \dot{\varphi}^2 \sin^2\theta). \tag{8.119}$$

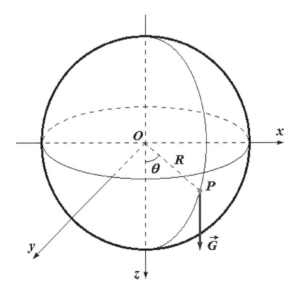

Fig.VIII.5

In its turn, the potential energy (our system is conservative) is obtained by means of the usual formula

$$dV = -\vec{G} \cdot d\vec{r} = -mg\,dz = -mgR\,d(\cos\theta) = mgR\sin\theta\,d\theta, \quad (8.120)$$

or

$$V = -mgR\cos\theta + V_0,$$

where the integration constant V_0 is determined by a suitable choice of the reference level for V. Taking the xy-plane as reference height for the potential energy, that is

$$V\left(\theta = \frac{\pi}{2}\right) = V_0 = 0,$$

we finally have

$$V = -mgR\cos\theta. \quad (8.121)$$

The Lagrangian therefore is

$$L = T - V = \frac{1}{2}mR^2(\dot{\theta}^2 + \dot{\varphi}^2\sin^2\theta) + mgR\cos\theta. \quad (8.122)$$

This allows one to write the generalized momenta, associated with the generalized coordinates θ and φ:

$$\begin{cases} p_\theta = \dfrac{\partial L}{\partial \dot{\theta}} = mR^2\dot{\theta}, \\ p_\varphi = \dfrac{\partial L}{\partial \dot{\varphi}} = mR^2\dot{\varphi}\sin^2\theta. \end{cases} \quad (8.123)$$

The Hamiltonian then is

$$H = \dot{\theta}p_\theta + \dot{\varphi}p_\varphi - L = \frac{1}{2}mR^2(\dot{\theta}^2 + \dot{\varphi}^2\sin^2\theta) - mgR\cos\theta,$$

or, in terms of generalized coordinates and generalized momenta

$$H(q,p) = H(\theta, \varphi, p_\theta, p_\varphi) = \frac{1}{2}\frac{p_\theta^2}{mR^2} + \frac{1}{2}\frac{p_\varphi^2}{mR^2\sin^2\theta} - mgR\cos\theta. \quad (8.124)$$

Hamilton's canonical equations are

$$\begin{cases} \dot{\theta} = \dfrac{\partial H}{\partial p_\theta} = \dfrac{p_\theta}{mR^2}, \\ \dot{\varphi} = \dfrac{\partial H}{\partial p_\varphi} = \dfrac{p_\varphi}{mR^2\sin^2\theta}, \\ \dot{p}_\theta = -\dfrac{\partial H}{\partial \theta} = -\dfrac{1}{2}\dfrac{p_\varphi^2}{mR^2}\dfrac{\sin 2\theta}{\sin^4\theta} - mgR\sin\theta, \\ \dot{p}_\varphi = -\dfrac{\partial H}{\partial \varphi} = 0. \end{cases} \quad (8.125)$$

Taking the time derivative of $(8.125)_{1,2}$ and using $(8.125)_{3,4}$, we arrive at the differential equations of motion:

$$\begin{cases} \ddot{\theta} = \dfrac{\dot{p}_\theta}{mR^2} = -\left(\dfrac{p_\varphi}{mR^2}\right)^2 \dfrac{\cos\theta}{\sin^3\theta} - \dfrac{g\sin\theta}{R}, \\ \ddot{\varphi} = \dfrac{\dot{p}_\varphi mR^2\sin^2\theta - 2mR^2\sin\theta\cos\theta\,\dot{\theta}\,p_\varphi}{m^2R^4\sin^4\theta} = -\dfrac{2p_\varphi\dot{\theta}\cos\theta}{mR^2\sin^3\theta}. \end{cases}$$

In view of (8.123), we finally obtain

$$\begin{cases} \ddot{\theta} = -\dot{\varphi}^2\sin\theta\cos\theta - \dfrac{g}{R}\sin\theta, \\ \ddot{\varphi} = -\dfrac{2\dot{\varphi}\dot{\theta}\cos\theta}{\sin\theta}, \end{cases} \qquad (8.126)$$

or

$$\begin{cases} \ddot{\theta} + \dot{\varphi}^2\sin\theta\cos\theta + \dfrac{g}{R}\sin\theta = 0, \\ \ddot{\varphi} + 2\dot{\theta}\dot{\varphi}\cot\theta = 0. \end{cases} \qquad (8.127)$$

Integration of this system of second order differential equations can be performed by means of two adequate prime integrals, namely:

i) $p_\varphi = l_z = const.$ (coordinate φ is cyclic);

ii) $E = H = \frac{1}{2}mR^2\dot{\theta}^2 + \dfrac{l_z^2}{2mR^2\sin^2\theta} - mgR\cos\theta$ (neither L nor H explicitly depend on time).

Problem 9

Determine a complete integral of the Hamilton-Jacobi equation for a particle of mass m and electric charge e, moving in the Coulombian field created in vacuum by two fixed centers of charges e_1 and e_2, situated at the distance 2σ from each other.

Solution

The Hamilton-Jacobi equation writes

$$\dfrac{\partial S}{\partial t} + H\left(q_1, ..., q_n, \dfrac{\partial S}{\partial q_1}, ..., \dfrac{\partial S}{\partial q_n}, t\right) = 0, \qquad (8.128)$$

where H is Hamilton's function in which the generalized momenta p_i were replaced by the derivatives $\frac{\partial S}{\partial q_i}$, while $S(q,t)$ is the action integral, as function of generalized coordinates and time.

Being a first-order, partial differential equation, the *general integral* of the Hamilton-Jacobi equation contains an arbitrary additive function. In Mechanics, not the general integral, but a *complete integral* plays an essential role. This is a solution of Hamilton-Jacobi equation, containing as many independent arbitrary constants, as there are independent variables in equation. In our case, the independent variables are the generalized coordinates $q_1, ..., q_n$ and the time t. Therefore, in case of a physical system with n degrees of freedom, a complete integral of the Hamilton-Jacobi equation should contain $n+1$ arbitrary constants. But, since the function S interferes in Hamilton-Jacobi equation only through its derivatives, one of these constants is purely additive and disappears upon differentiation. So, a complete integral of the Hamilton-Jacobi equation writes

$$S(q,t) = f(t, q_1, q_2, ..., q_n, \alpha_1, \alpha_2, ..., \alpha_n) + a, \qquad (8.129)$$

where $\alpha_1, \alpha_2, ..., \alpha_n$ and a are arbitrary constants.

In general, solution to the problem of motion of a mechanical system, by means of the Hamilton-Jacobi formalism, can be found by completing the following steps:

- Determine the Hamiltonian of the system and write the Hamilton-Jacobi equation;
- Search for a complete integral of the Hamilton-Jacobi equation under the form (8.129);
- Take derivatives of the complete integral S with respect to the arbitrary constants $\alpha_1, \alpha_2, ..., \alpha_n$, and equate these derivatives to some other constants $\beta_1, \beta_2, ..., \beta_n$:

$$\frac{\partial S}{\partial \alpha_i} = \beta_i \qquad (i = \overline{1, n}); \qquad (8.130)$$

- Solving the system (8.130), one obtains the generalized coordinates q_i as functions of time and the $2n$ arbitrary constants $\alpha_1, \alpha_2, ..., \alpha_n$ and $\beta_1, \beta_2, ..., \beta_n$, which can be determined by means of initial conditions;
- The generalized momenta as functions of time, $p_i = p_i(t)$ can be determined by means of relations $p_i = \frac{\partial S}{\partial q_i}$.

One of the most frequently used procedures of solving the Hamilton-Jacobi equation is the method of separation of variables. The cyclic variables can always be separated, but also exist circumstances when separation is possible even for non-cyclic coordinates.

A special case of cyclic variables is offered by the conservative systems, characterized by the fact that the Hamiltonian does not explicitly depend on time. The action S then writes

$$S = -Et + W(q), \qquad (8.131)$$

where the *reduced action* $W(q)$ satisfies the *abbreviated* (or *restricted*) Hamilton-Jacobi equation

$$H\left(q_1, ..., q_n, \frac{\partial W}{\partial q_1}, ..., \frac{\partial W}{\partial q_n}\right) = E. \qquad (8.132)$$

The method of separation of variables in equation Hamilton-Jacobi embed all cases of integration based on the existence of cyclic variables. Taking into account that in some cases the separation of variables is possible, even if coordinates are not cyclic, we can conclude that the Hamilton-Jacobi method is one of the most efficient method of solving mechanical problems.

As in all cases of equations of mathematical physics, it is very important to choose a convenient coordinate system, suitable to the problem symmetry. Since our problem presents an elliptic symmetry, it is advisable to work in elliptic coordinates. As we have shown [see (8.85)], in elliptic coordinates λ, μ, φ the Hamiltonian writes

$$H = \frac{1}{2m\sigma^2(\lambda^2 - \mu^2)}\left[(\lambda^2 - 1)p_\lambda^2 + (1 - \mu^2)p_\mu^2\right.$$

$$\left.+ \left(\frac{1}{\lambda^2 - 1} + \frac{1}{1 - \mu^2}\right)p_\varphi^2\right] + V(\lambda, \mu, \varphi). \qquad (8.133)$$

Using Fig.VIII.6, we can write the potential energy V as

$$V = \frac{e_1 e}{4\pi\varepsilon_0 r_1} + \frac{e_2 e}{4\pi\varepsilon_0 r_2} = \frac{e}{4\pi\varepsilon_0}\left(\frac{e_1}{r_1} + \frac{e_2}{r_2}\right). \qquad (8.134)$$

This relation can be put in the following form:

$$V = \frac{a(\lambda) + b(\mu)}{\lambda^2 - \mu^2}, \qquad (8.135)$$

where

$$\begin{cases} a(\lambda) = \left(\dfrac{ee_1}{4\pi\varepsilon_0} + \dfrac{ee_2}{4\pi\varepsilon_0}\right)\dfrac{\lambda}{\sigma}, \\ b(\mu) = \left(\dfrac{ee_1}{4\pi\varepsilon_0} - \dfrac{ee_2}{4\pi\varepsilon_0}\right)\dfrac{\mu}{\sigma}. \end{cases} \qquad (8.136)$$

Indeed, using the relations between cylindrical and elliptic coordinates [see(8.76)]

$$\begin{cases} \rho = \sigma\sqrt{(\lambda^2 - 1)(1 - \mu^2)}; \\ \varphi = \varphi; \\ z = \sigma\lambda\mu, \end{cases} \qquad (8.137)$$

where σ is the *transformation parameter*, and denoting by $\vec{r}_1$ and $\vec{r}_2$ the radius vectors of the points $A_1(0,0,\sigma)$ and $A_2(0,0,-\sigma)$ (see Fig.VIII.6), we have

$$\begin{cases} r_1 = \sqrt{(z-\sigma)^2 + \rho^2}, \\ r_2 = \sqrt{(z+\sigma)^2 + \rho^2}, \end{cases} \quad (8.138)$$

or, in view of (8.137),

$$\begin{cases} \lambda = \dfrac{r_2 + r_1}{2\sigma}, \\ \mu = \dfrac{r_2 - r_1}{2\sigma}. \end{cases} \quad (8.139)$$

Denoting

$$\begin{cases} \alpha_1 = \dfrac{ee_1}{4\pi\varepsilon_0}, \\ \alpha_2 = \dfrac{ee_2}{4\pi\varepsilon_0}, \end{cases} \quad (8.140)$$

relation (8.134) writes

$$V = \frac{\alpha_1}{r_1} + \frac{\alpha_2}{r_2}. \quad (8.141)$$

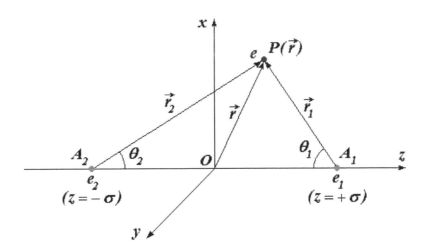

Fig.VIII.6

As one can see,

$$\frac{\alpha_1}{r_1} + \frac{\alpha_2}{r_2} \equiv \frac{a(\lambda) + b(\mu)}{\lambda^2 - \mu^2}. \quad (8.142)$$

Indeed, on the one side

$$\frac{\alpha_1}{r_1} + \frac{\alpha_2}{r_2} = \frac{\alpha_1 r_2 + \alpha_2 r_1}{r_1 r_2},$$

and on the other, in view of (8.139),

$$\frac{a(\lambda) + b(\mu)}{\lambda^2 - \mu^2} = \frac{a+b}{(\lambda+\mu)(\lambda-\mu)} = \frac{\sigma^2(a+b)}{r_1 r_2}.$$

As a final step, one must show that

$$\alpha_1 r_2 + \alpha_2 r_1 = \sigma^2(a+b). \tag{8.143}$$

Since $a = a(\lambda) = \frac{\alpha_1 + \alpha_2}{\sigma}\lambda$ and $b = b(\mu) = \frac{\alpha_1 - \alpha_2}{\sigma}\mu$ relation (8.143) follows immediately.

Since our system is conservative, we are now able to write the abbreviated Hamilton-Jacobi equation (8.132) in elliptic coordinates. Using (8.133) and (8.134), we have

$$\frac{1}{2m\sigma^2(\lambda^2 - \mu^2)} \left[(\lambda^2 - 1)\left(\frac{\partial W}{\partial \lambda}\right)^2 + (1 - \mu^2)\left(\frac{\partial W}{\partial \mu}\right)^2 \right.$$

$$\left. + \left(\frac{1}{\lambda^2 - 1} + \frac{1}{1 - \mu^2}\right)\left(\frac{\partial W}{\partial \varphi}\right)^2 \right] + \frac{a(\lambda) + b(\mu)}{\lambda^2 - \mu^2} = E, \tag{8.144}$$

where the expressions for the generalized momenta

$$p_\lambda = \frac{\partial W}{\partial \lambda}, \qquad p_\mu = \frac{\partial W}{\partial \mu}, \qquad p_\varphi = \frac{\partial W}{\partial \varphi} \tag{8.145}$$

have been considered. Recalling that [see(8.131)]

$$S(q,t) = W(q) - Et, \tag{8.146}$$

our main purpose is to determine the reduced action $W(q)$ by solving the abbreviated Hamilton-Jacobi equation. This can be done by method of reparation of variables. Since φ is also a cyclic coordinate, we search for a solution of the form

$$W(q) = W(\lambda, \mu, \varphi) = p_\varphi \varphi + W_1(\lambda) + W_2(\mu). \tag{8.147}$$

Here p_φ is nothing else but the constant value of the generalized momentum $p_\varphi = \frac{\partial S}{\partial \varphi}$, associated with cyclic coordinate φ, while $W_1(\lambda)$ and $W_2(\mu)$ are solutions of the equations

$$(\lambda^2 - 1)\left(\frac{dW_1}{d\lambda}\right)^2 + \frac{p_\varphi^2}{\lambda^2 - 1} + 2m\sigma^2 a(\lambda) - 2m\sigma^2(\lambda^2 - 1)E = \beta, \tag{8.148}$$

and

$$(1-\mu^2)\left(\frac{dW_2}{d\mu}\right)^2 + \frac{p_\varphi^2}{1-\mu^2} + 2m\sigma^2 b(\mu) + 2m\sigma^2(1-\mu^2)E = -\beta, \quad (8.149)$$

respectively. Here β is an arbitrary constant. The last two equations have been obtained by introducing (8.147) into (8.144), then multiplying by $2m\sigma^2(\lambda^2 - \mu^2)$. After some arrangements of terms, this yields

$$\left[(\lambda^2 - 1)\left(\frac{dW_1}{d\lambda}\right)^2 + \frac{p_\varphi^2}{\lambda^2 - 1} + 2m\sigma^2 a(\lambda) - 2m\sigma^2\lambda^2 E + 2m\sigma^2 E\right]$$

$$+ \left[(1-\mu^2)\left(\frac{dW_2}{d\mu}\right)^2 + \frac{p_\varphi^2}{1-\mu^2} + 2m\sigma^2 b(\mu) + 2m\sigma^2\mu^2 E - 2m\sigma^2 E\right] = 0. \quad (8.150)$$

Since the first square bracket depends only on λ, and the second only on μ, one can write $[..\lambda..] = -[..\mu..] = \beta$, and equations (8.148) and (8.149) follow immediately.

Integrating (8.148) and (8.149), one can easily obtain $W_1(\lambda)$ and $W_2(\mu)$. Introducing these results into (8.147), and then into (8.146), we are left with the following complete integral of the Hamilton-Jacobi equation:

$$S = -Et + p_\varphi\varphi + \int \sqrt{2m\sigma^2 E + \frac{\beta - 2m\sigma^2 a(\lambda)}{\lambda^2 - 1} - \frac{p_\varphi^2}{(\lambda^2 - 1)^2}}\, d\lambda$$

$$+ \int \sqrt{2m\sigma^2 E - \frac{\beta + 2m\sigma^2 b(\mu)}{1 - \mu^2} - \frac{p_\varphi^2}{(1 - \mu^2)^2}}\, d\mu, \quad (8.151)$$

or, by means of (8.136) and (8.140),

$$S = -Et + p_\varphi\varphi + \int \sqrt{2m\sigma^2 E + \frac{\beta - 2m\sigma(\alpha_1 + \alpha_2)\lambda}{\lambda^2 - 1} - \frac{p_\varphi^2}{(\lambda^2 - 1)^2}}\, d\lambda$$

$$+ \int \sqrt{2m\sigma^2 E - \frac{\beta + 2m\sigma(\alpha_1 - \alpha_2)\mu}{1 - \mu^2} - \frac{p_\varphi^2}{(1 - \mu^2)^2}}\, d\mu. \quad (8.152)$$

The complete integral (8.152) depends on the arbitrary constants p_φ, β, and E. Taking derivatives with respect to these quantities, and equating the results to some new constants, one can find the general solution of the equations of motion.

The reader is invited to prove that the constant β is connected to the conservation of the quantity

$$\beta = \sigma^2 p_\rho^2 - M^2 + 2m\sigma(\alpha_1 \cos\theta_1 + \alpha_2 \cos\theta_2),$$

where $\vec{M}$ is the total angular momentum of the particle, and θ_1, θ_2 are the angles shown in Fig.VIII.6.

Problem 10

Show that the transformation

$$\begin{cases} Q = \ln\left(\dfrac{\sin p}{q}\right); \\ P = q \cot p \end{cases} \qquad (8.153)$$

is canonical and find all possible types of generating function.

Solution

To show that the transformation $(q, p) \to (Q, P)$ is canonical, it is sufficient to prove that the absolute value of the Jacobian J of the transformation equals 1. Indeed,

$$J = \begin{vmatrix} \dfrac{\partial Q}{\partial q} & \dfrac{\partial Q}{\partial p} \\ \dfrac{\partial P}{\partial q} & \dfrac{\partial P}{\partial p} \end{vmatrix} = \begin{vmatrix} \dfrac{q}{\sin p}\left(-\dfrac{\sin p}{q^2}\right) & \dfrac{q}{\sin p}\dfrac{\cos p}{q} \\ \cot p & q\left(-\dfrac{1}{\sin^2 p}\right) \end{vmatrix}$$

$$= \begin{vmatrix} -\dfrac{1}{q} & \cot p \\ \cot p & -\dfrac{q}{\sin^2 p} \end{vmatrix} = \dfrac{1}{\sin^2 p} - \cot^2 p = 1.$$

As can be observed,

$$J = \dfrac{\partial Q}{\partial q}\dfrac{\partial P}{\partial p} - \dfrac{\partial Q}{\partial p}\dfrac{\partial P}{\partial q} = \{Q, P\} = 1,$$

where $\{Q, P\}$ is the Poisson bracket of quantities Q and P.

Since our problem concerns one generalized coordinate and one generalized momentum, the maximum number of generating functions is four:

$$F_1(q, Q), \quad F_2(q, P), \quad F_3(p, Q), \quad F_4(p, P).$$

The canonicity condition for the transformation $(q, p) \to (Q, P)$ then furnishes relation

$$p\, dq - P\, dQ = dF_1(q, Q), \qquad (8.154)$$

or, in view of (8.153)

$$(p + \cot p)dq - q\cot^2 p\, dp = dF_1. \tag{8.155}$$

In order to be a total differential, function F_1 must satisfy the relation

$$\frac{\partial}{\partial q}(-q\cot^2 p) = \frac{\partial}{\partial p}(p + \cot p),$$

which proves to be true. By virtue of this property, we can write

$$dF_1 = \frac{\partial F_1}{\partial q}dq + \frac{\partial F_1}{\partial p}dp. \tag{8.156}$$

Comparing (8.156) and (8.155), one obtains

$$\begin{cases} \dfrac{\partial F_1}{\partial q} = p + \cot p, \\ \dfrac{\partial F_1}{\partial p} = -q\cot^2 p. \end{cases} \tag{8.157}$$

Equation $(8.157)_1$ then yields

$$F_1 = \int (p + \cot p)dq + f(Q) = pq + q\cot p + f(Q), \tag{8.158}$$

where $f(Q)$ is an arbitrary function of Q. According to $(8.157)_2$ we then have

$$F_1 = -\int q\cot^2 p\, dp = -q(-p - \cot p) + g(q), \tag{8.159}$$

with $g(q)$ another arbitrary function of q. We therefore found

$$\begin{cases} F_1(q,Q) = qp + q\cot p + f(Q), \\ F_1(q,Q) = qp + q\cot p + g(q), \end{cases}$$

which means that

$$F_1(q,Q) = qp + q\cot p = q\arcsin\left(qe^Q\right) + \frac{\sqrt{1 - q^2 e^{2Q}}}{e^Q}, \tag{8.160}$$

is the generating function of type $F_1(q,Q)$ of the canonical transformation (8.153).

The other three types of generating functions are obtained by means of the appropriate Legendre transformations. Let us start with

$$dF_1 = (p + \cot p)dq - q\cot^2 p\, dp = \left[\arcsin(qe^Q) + \frac{\sqrt{1-q^2e^{2Q}}}{qe^Q}\right]dq$$

$$-\frac{1-q^2e^{2Q}}{qe^{2Q}}\left(\frac{e^Q}{\sqrt{1-q^2e^{2Q}}}dq + \frac{qe^Q}{\sqrt{1-q^2e^{2Q}}}dQ\right),$$

or

$$dF_1 = \arcsin(qe^Q)dq - \frac{\sqrt{1-q^2e^{2Q}}}{e^Q}dQ. \tag{8.161}$$

In order to determine $F_2(q, P)$ let us perform Legendre transformation

$$dF_1 + d(PQ) = d(F_1 + PQ) \equiv dF_2$$

$$= \arcsin(qe^Q)dq - \left(\frac{\sqrt{1-q^2e^{2Q}}}{e^Q} - P\right)dQ + QdP. \tag{8.162}$$

To be independent of Q, the coefficient of dQ in (8.162) must be null, that is

$$P = \frac{\sqrt{1-q^2e^{2Q}}}{e^Q}, \tag{8.163}$$

which yields

$$Q = -\ln\sqrt{q^2+P^2}, \tag{8.164}$$

$$e^Q = (q^2+P^2)^{-1/2}, \tag{8.165}$$

$$e^{2Q} = (q^2+P^2)^{-1}, \tag{8.166}$$

and (8.162) becomes

$$dF_2(q,P) = \arcsin\left[q(q^2+P^2)^{-1/2}\right]dq - \ln\sqrt{q^2+P^2}\, dP. \tag{8.167}$$

One can easily be proved that dF_2 is an exact differential. Indeed,

$$\frac{\partial}{\partial q}\left(-\ln\sqrt{q^2+P^2}\right) = \frac{\partial}{\partial P}\left\{\arcsin\left[q(q^2+P^2)^{-1/2}\right]\right\}.$$

The third generating function $F_3(p, Q)$ can be determined by means of Legendre transformation

$$dF_1 - d(pq) = \arcsin(qe^Q)dq - \frac{\sqrt{1-q^2e^{2Q}}}{e^Q}dQ - q\,dp - p\,dq$$

$$= [\arcsin(qe^Q) - p]\, dq - \frac{\sqrt{1 - q^2 e^{2Q}}}{e^Q} dQ - q\, dp.$$

To depend only on coordinates (p, Q), one must have $p = \arcsin(qe^Q)$, that is $\sin p = qe^Q$. Written in coordinates (p, Q), the generating function $F_3(p, Q) = F_1 - qp$ is given by

$$F_3 = e^{-Q} \sin p(p + \cot p), \tag{8.168}$$

with exact differential

$$dF_3 = -e^{-Q} \sin p\, dp - \frac{\cos p}{e^Q} dQ, \tag{8.169}$$

which can be easily verified, since

$$\frac{\partial}{\partial p}\left(-\frac{\cos p}{e^Q}\right) = \frac{\partial}{\partial Q}\left(-e^{-Q} \sin p\right).$$

To determine the fourth (and final!) type of generating function $F_4 = F_4(p, P)$ of the canonical transformation (8.153) one performs the Legendre transformation

$$dF_1 + d(PQ - pq) = \arcsin\left(qe^Q\right) dq - \frac{\sqrt{1 - q^2 e^{2Q}}}{e^Q} dQ$$

$$+ P\, dQ + Q\, dP - p\, dq - q\, dp$$

$$= \left[\arcsin\left(qe^Q\right) - p\right] dq - \left(\frac{\sqrt{1 - q^2 e^{2Q}}}{e^Q} - P\right) dQ + Q\, dP - q\, dp. \tag{8.170}$$

In order that the generating function $F_4 = F_1 + PQ - pq = F_2 - pq = F_3 + PQ$ depends only on coordinates p and P, the parentheses in (8.170) must be zero, that is

$$\begin{cases} \arcsin\left(qe^Q\right) = p, \\ \dfrac{\sqrt{1 - q^2 e^{2Q}}}{e^Q} = P, \end{cases}$$

which coincide with $(8.153)_1$ and (8.163).

The generating function $F_4(p, P)$ therefore is:

$$F_4 = F_1 + PQ - pq = q \cot p + PQ = P + P \ln\left(\frac{\sin p}{\frac{P}{\cot p}}\right)$$

$$= P + P \ln\left(\frac{\cos p}{P}\right), \tag{8.171}$$

while its differential is

$$dF_4 = -P \tan p\, dp + \ln\left(\frac{\cos p}{P}\right) dP. \tag{8.172}$$

This is an exact differential, because

$$\frac{\partial}{\partial p}\left[\ln\left(\frac{\cos p}{P}\right)\right] = \frac{\partial}{\partial P}(-P\tan p).$$

Summarizing, the four generating functions of the canonical transformation (8.153) are

$$\begin{cases} F_1 = F_1(q, Q) = q \arcsin\left(qe^Q\right) + \dfrac{\sqrt{1 - q^2 e^{2Q}}}{e^Q}, \\ F_2 = F_2(q, P) = q \arcsin\left(\dfrac{q}{\sqrt{q^2 + P^2}}\right) + P - \dfrac{P}{2}\ln(q^2 + P^2), \\ F_3 = F_3(p, Q) = e^{-Q}(p + \cot p)\sin p, \\ F_4 = F_4(p, P) = P + P \ln\left(\dfrac{\cos p}{P}\right), \end{cases}$$

while their exact differentials write

$$\begin{cases} dF_1 = \arcsin(qe^Q)dq - \dfrac{\sqrt{1 - q^2 e^{2Q}}}{e^Q}dQ, \\ dF_2 = \arcsin\left[q(q^2 + P^2)^{-1/2}\right]dq - \ln\sqrt{q^2 + P^2}\, dP, \\ dF_3 = -e^{-Q}\sin p\, dp - \dfrac{\cos p}{e^Q}dQ, \\ dF_4 = -P\tan p\, dp + \ln\left(\dfrac{\cos p}{P}\right) dP. \end{cases}$$

CHAPTER IX

PROBLEMS OF CONTINUOUS SYSTEMS

A. Problems of Classical Electrodynamics

Problem 1

An electrized particle of charge q and mass m moves with velocity $\vec{v}$ in the variable electromagnetic field $(\vec{E}, \vec{B})$, described by the electromagnetic potentials $V(x, y, z, t)$ and $\vec{A}(x, y, z, t)$. Neglecting quantum and relativistic effects, show that the equation of motion of the particle can be written in a Lagrangian form, and determine the Lagrangian function L.

Solution

Since

$$\vec{E} = -\operatorname{grad} V - \frac{\partial \vec{A}}{\partial t}; \qquad (9.1)$$

$$\vec{B} = \operatorname{curl} \vec{A}, \qquad (9.2)$$

the equation of motion of the particle

$$m\ddot{\vec{r}} = q(\vec{E} + \vec{v} \times \vec{B}) \qquad (9.3)$$

also writes

$$m\ddot{\vec{r}} = q\left(-\operatorname{grad} V - \frac{\partial \vec{A}}{\partial t} + \vec{v} \times \operatorname{curl} \vec{A}\right). \qquad (9.4)$$

The last term of (9.4) can be transformed as follows:

$$\vec{v} \times \operatorname{curl}\vec{A} = \varepsilon_{ijk} v_j (\nabla \times \vec{A})_k \vec{u}_i = \varepsilon_{ijk} v_j \varepsilon_{klm}(\partial_l A_m)\vec{u}_i$$

$$= \varepsilon_{kij}\varepsilon_{klm} v_j(\partial_l A_m)\vec{u}_i = (\delta_{il}\delta_{jm} - \delta_{im}\delta_{jl}) v_j (\partial_l A_m)\vec{u}_i$$

$$= \partial_i(v_j A_j)\vec{u}_i - (v_j \partial_j)(A_i \vec{u}_i) = \partial_i(\vec{v}\cdot\vec{A})\vec{u}_i - (\vec{v}\cdot\nabla)\vec{A}$$

$$= \nabla(\vec{v}\cdot\vec{A}) - (\vec{v}\cdot\nabla)\vec{A},$$

where we made allowance for the fact that, within the Lagrangian approach, the generalized coordinates [in our case x_i ($i = \overline{1,3}$)], and generalized velocities [here v_i ($i = \overline{1,3}$)], are independent quantities. Thus v_i, together with $\vec{u}_i = const.$ have been introduced under the derivative sign $\partial_i = \partial/\partial x_i$. Equation (9.4) then becomes

$$m\ddot{\vec{r}} = q\left[-\text{grad}\,V - \frac{\partial\vec{A}}{\partial t} + \nabla(\vec{v}\cdot\vec{A}) - (\vec{v}\cdot\nabla)\vec{A}\right]$$

$$= q\left[-\nabla(V - \vec{v}\cdot\vec{A}) - \left(\frac{\partial\vec{A}}{\partial t} + (\vec{v}\cdot\nabla)\vec{A}\right)\right]$$

$$= -q\left[\nabla(V - \vec{v}\cdot\vec{A}) + \frac{d\vec{A}}{dt}\right]. \tag{9.5}$$

But $\ddot{\vec{r}} = \frac{d\dot{\vec{r}}}{dt} = \frac{d\vec{v}}{dt}$, and (9.5) yields

$$\frac{d}{dt}(m\vec{v} + q\vec{A}) - \nabla\left\{q\left[-V + \vec{v}\cdot\vec{A}\right]\right\} = 0. \tag{9.6}$$

Identifying (9.6) with Lagrange equations of the 2nd kind

$$\frac{d}{dt}\left(\frac{\partial L}{\partial\vec{v}}\right) - \frac{\partial L}{\partial\vec{r}} = 0, \tag{9.7}$$

we obtain

$$\frac{\partial L}{\partial\vec{v}} = m\vec{v} + q\vec{A}; \tag{9.8}$$

$$\frac{\partial L}{\partial\vec{r}} = \nabla\left\{q\left[-V + \vec{v}\cdot\vec{A}\right]\right\} = \frac{\partial}{\partial\vec{r}}\left\{q\left[-V + \vec{v}\cdot\vec{A}\right]\right\}, \tag{9.9}$$

where L is the Lagrangian function, while $\frac{\partial L}{\partial\vec{r}}$ and $\frac{\partial L}{\partial\vec{v}}$ stand for a formal writing of $\frac{\partial L}{\partial\vec{r}} = \left(\frac{\partial L}{\partial x}, \frac{\partial L}{\partial y}, \frac{\partial L}{\partial z}\right)$, and $\frac{\partial L}{\partial\vec{v}} = \left(\frac{\partial L}{\partial v_x}, \frac{\partial L}{\partial v_y}, \frac{\partial L}{\partial v_z}\right)$, respectively.

According to Eq.(9.9), it then follows that

$$L(\vec{r},\vec{v},t) = -qV + q(\vec{v}\cdot\vec{A}) + f(\vec{v}),$$

where $f(\vec{v})$ is an arbitrary function of class C^0 at least. It can be determined by means of (9.8):

$$\frac{df}{d\vec{v}} = m\vec{v} \Rightarrow f(\vec{v}) = \frac{1}{2}m\vec{v}^2,$$

and the Lagrangian becomes

$$L(\vec{r}, \vec{v}, t) = -qV + q(\vec{v} \cdot \vec{A}) + \frac{1}{2}m\vec{v}^2. \tag{9.10}$$

We conclude that the equation of motion (9.3) can be written in the Lagrangian form (9.7), with L given by (9.10). It is important to mention that, in our case,

$$L(\vec{r}, \vec{v}, t) = T(\vec{r}, \vec{v}, t) - U(\vec{r}, \vec{v}, t), \tag{9.11}$$

where $T(\vec{r}, \vec{v}, t)$ is the kinetic energy, and

$$U(\vec{r}, \vec{v}, t) = qV - q(\vec{v} \cdot \vec{A}) \tag{9.12}$$

the *generalized (velocity dependent) potential*.

As well-known, the electromagnetic potentials are defined only up to a gauge transformation

$$V' = V - \frac{\partial \phi}{\partial t}, \qquad \vec{A}' = \vec{A} + \nabla \phi, \tag{9.13}$$

where $\phi(\vec{r}, t)$ is a function of class C^1 in R^3. Let us investigate the transformation $L(V, \vec{A}) \to L(V', \vec{A}')$. We have:

$$L' = -qV' + q(\vec{v} \cdot \vec{A}') + \frac{1}{2}mv^2 = -q\left(V - \frac{\partial \phi}{\partial t}\right) + q\vec{v} \cdot (\vec{A} + \nabla \phi) + \frac{1}{2}mv^2$$

$$= L + q\frac{\partial \phi}{\partial t} + q\vec{v} \cdot \nabla \phi = L + q\left[\frac{\partial \phi}{\partial t} + (\vec{v} \cdot \nabla)\phi\right] = L + \frac{d}{dt}(q\phi). \tag{9.14}$$

The last relation shows that, if two Lagrangians L and L' differ by a total derivative with respect to time of a scalar function $F = F(\vec{r}, t)$, they are equivalent.

Problem 2

Using the Lagrangian technique, determine the equation of motion of a particle of charge q and mass m, moving in the electromagnetic field $(\vec{E}, \vec{B})$.

Solution

According to Lagrangian formalism, the differential equations of motion of the particle write

$$\frac{d}{dt}\left(\frac{\partial T}{\partial \dot{q}_k}\right) - \frac{\partial T}{\partial q_k} = Q_k \quad (k = \overline{1, n}), \tag{9.15}$$

where T is the kinetic energy, Q_k are the generalized forces, and n is the number of degrees of freedom of the system. If the applied force $\vec{F}$ can be derived from *a generalized potential energy function* (or, in short, *a velocity dependent potential*)

$$U(q, \dot{q}, t) = C_j \dot{q}_j + U_0 = U_1 + U_0 \quad (j = \overline{1,n}), \tag{9.16}$$

where $C_j = C_j(q, t)$, then we can add the quantity

$$\frac{\partial U}{\partial q_k} - \frac{d}{dt}\left(\frac{\partial U}{\partial \dot{q}_k}\right)$$

to both sides of (9.15), and obtain

$$\frac{d}{dt}\left[\frac{\partial(T-U)}{\partial \dot{q}_k}\right] - \frac{\partial(T-U)}{\partial q_k} = Q_k - \frac{d}{dt}\left(\frac{\partial U}{\partial \dot{q}_k}\right) + \frac{\partial U}{\partial q_k}. \tag{9.17}$$

If the generalized forces Q_k are given by

$$Q_k = \frac{d}{dt}\left(\frac{\partial U}{\partial \dot{q}_k}\right) - \frac{\partial U}{\partial q_k}, \tag{9.18}$$

and denote

$$L(q, \dot{q}, t) = T(q, \dot{q}, t) - U(q, \dot{q}, t), \tag{9.19}$$

then we are left with the well-known Lagrange equations for natural systems

$$\frac{d}{dt}\left(\frac{\partial L}{\partial \dot{q}_k}\right) - \frac{\partial L}{\partial q_k} = 0. \tag{9.20}$$

Here $L(q, \dot{q}, t) \equiv L(q_1, q_2, ..., q_n, \dot{q}_1, \dot{q}_2, ..., \dot{q}_n, t)$, etc.

To establish the differential equation of motion of the particle of charge q and mass m in the electromagnetic field $(\vec{E}, \vec{B})$, one must first construct the Lagrangian function. Since no constraints act on the particle, as generalized coordinates and generalized velocities one can choose the Cartesian coordinates $q_i \equiv x_i$ $(i = \overline{1,3})$ and the velocity components $\dot{q}_i \equiv \dot{x}_i$ $(i = \overline{1,3})$, respectively.

If only the electric field is present, the Lagrangian writes $L = T - U = T - qV$, where $V(\vec{r}, t)$ is the electric potential. If the magnetic field is also present, we have to consider a new term in the Lagrangian, expressed in terms of the vector potential $\vec{A}(\vec{r}, t)$. To this end, we take advantage of the fact that the Lagrangian is a scalar (invariant) function, so that $\vec{A}$ can only appear as one of the dot products: $\vec{A} \cdot \vec{A}$, $\vec{A} \cdot \vec{r}$, and $\vec{A} \cdot \ddot{\vec{r}}$. Since B^2 does not appear in the equation of motion, the first possibility is excluded. On the other hand, the usual equations of motion met in physics are second-order differential equations, which means that the third possibility is also unacceptable. Therefore, the Lagrangian writes

$$L = \frac{1}{2}mv^2 - qV + q\vec{v}\cdot\vec{A}. \tag{9.21}$$

We then successively have:

$$\frac{\partial L}{\partial \dot{q}_i} \equiv \frac{\partial L}{\partial v_i} = mv_i + qA_i;$$

$$\frac{d}{dt}\left(\frac{\partial L}{\partial \dot{q}_i}\right) \equiv \frac{d}{dt}\left(\frac{\partial L}{\partial \dot{x}_i}\right) = m\ddot{x}_i + q\left(\frac{\partial A_i}{\partial t} + \dot{x}_k\frac{\partial A_i}{\partial x_k}\right);$$

$$\frac{\partial L}{\partial q_i} \equiv \frac{\partial L}{\partial x_i} = -q\frac{\partial V}{\partial x_i} + qv_k\frac{\partial A_k}{\partial x_i} \quad (i,k=1,2,3),$$

and Lagrange's equations (9.20) become

$$m\ddot{x}_i = q\left(-\frac{\partial V}{\partial x_i} - \frac{\partial A_i}{\partial t}\right) + qv_k\left(\frac{\partial A_k}{\partial x_i} - \frac{\partial A_i}{\partial x_k}\right),$$

or

$$m\ddot{x}_i = qE_i + qv_k T_{ik}, \tag{9.22}$$

where by T_{ik} has been denoted the antisymmetric tensor

$$T_{ik} = \frac{\partial A_k}{\partial x_i} - \frac{\partial A_i}{\partial x_k}. \tag{9.23}$$

Let us show that B_j are the components of the axial vector (pseudovector) associated with the antisymmetric tensor (9.23). Indeed,

$$B_j = \frac{1}{2}\varepsilon_{jik}T_{ik} = \frac{1}{2}\varepsilon_{jik}\frac{\partial A_k}{\partial x_i} - \frac{1}{2}\varepsilon_{jik}\frac{\partial A_i}{\partial x_k}$$

$$= \frac{1}{2}\varepsilon_{jik}\frac{\partial A_k}{\partial x_i} - \frac{1}{2}\varepsilon_{jki}\frac{\partial A_k}{\partial x_i} = 2\cdot\frac{1}{2}\varepsilon_{jik}\frac{\partial A_k}{\partial x_i} = (\text{curl}\vec{A})_j.$$

This result explains why we denoted $\text{curl}\vec{A} = \vec{B}$ right from the beginning.

The equation of motion (9.22) then finally writes

$$m\ddot{x}_i = qE_i + qv_k(\varepsilon_{ikl}B_l) = qE_i + q(\vec{v}\times\vec{B})_i,$$

or, in vectorial form
$$m\ddot{\vec{r}} = q(\vec{E} + \vec{v} \times \vec{B}). \qquad (9.24)$$

Problem 3

Solve Problem 2, using the Hamiltonian formalism. In addition, investigate the following two particular cases:

(1) The particle is an electron of charge $-e$, the electromagnetic field is static and uniform, with $\vec{E} \| \vec{B}$, and the initial velocity $\vec{v}_0$ of the particle is orthogonal to the field.

(2) The particle is an electron of charge $-e$, the electromagnetic field is static and uniform, but this time $\vec{E}$ and $\vec{B}$ have different directions, while the initial velocity $\vec{v}_0$ is orthogonal to the plane determined by $\vec{E}$ and $\vec{B}$.

Solution

Let us write the Lagrangian (9.21) in the form
$$L = \frac{1}{2} m v_k v_k - q\phi + q v_k A_k \quad (k = 1, 2, 3), \qquad (9.25)$$

where Einstein's summation convention has been used. According to their definition, the generalized momenta p_i ($i = 1, 2, 3$) are
$$p_i \equiv \frac{\partial L}{\partial \dot{q}_i} = \frac{\partial L}{\partial v_i} = m v_k \delta_{ik} + q A_k \delta_{ik} = m v_i + q A_i \quad (i = 1, 2, 3). \qquad (9.26)$$

The Hamilton therefore is
$$H(q, p, t) \equiv \sum_{k=1}^{n} p_k \dot{q}_k - L(q, \dot{q}, t) = \frac{1}{2} m v_k v_k + q\phi.$$

Since
$$v_k = \frac{1}{m}(p_k - q A_k) \quad (k = 1, 2, 3),$$

we still have
$$H = \frac{1}{2m}(p_k - q A_k)(p_k - q A_k) + q\phi. \qquad (9.27)$$

Hamilton's canonical equations
$$\begin{cases} \dot{q}_i \equiv \dot{x}_i = \dfrac{\partial H}{\partial p_i}; \\ \dot{p}_i = -\dfrac{\partial H}{\partial q_i} \equiv -\dfrac{\partial H}{\partial x_i}, \end{cases} \quad (i = 1, 2, 3) \qquad (9.28)$$

then yield

$$\dot{x}_i = \frac{\partial H}{\partial p_i} = \frac{1}{m}(p_k - qA_k)\delta_{ik} = \frac{1}{m}(p_i - qA_i),$$

$$\dot{p}_i = -\frac{\partial H}{\partial x_i} = \frac{q}{m}(p_k - qA_k)\frac{\partial A_k}{\partial x_i} - q\frac{\partial \phi}{\partial x_i},$$

or, in a synthetic form

$$\begin{cases} \dot{x}_i = \frac{1}{m}(p_i - qA_i); \\ \dot{p}_i = \frac{q}{m}(p_k - qA_k)\frac{\partial A_k}{\partial x_i} - q\frac{\partial \phi}{\partial x_i}. \end{cases} \quad (9.29)$$

Let as now take the time derivative of $(9.29.)_1$. In view of $(9.29)_2$, we then have:

$$m\ddot{x}_i = \dot{p}_i - q\dot{A}_i$$

$$= \frac{q}{m}(p_k - qA_k)\frac{\partial A_k}{\partial x_i} - q\frac{\partial \phi}{\partial x_i} - q\left(\frac{\partial A_i}{\partial t} + \frac{\partial A_i}{\partial x_k}\dot{x}_k\right),$$

or

$$m\ddot{x}_i = -q\frac{\partial \phi}{\partial x_i} - q\frac{\partial A_i}{\partial t} + q\dot{x}_k\left(\frac{\partial A_k}{\partial x_i} - \frac{\partial A_i}{\partial x_k}\right), \quad (9.30)$$

where the functional dependence $\vec{A} = \vec{A}(\vec{r}, t)$ has been taken into account.

Using the relations between field and potentials

$$\vec{E} = -\nabla\phi - \frac{\partial \vec{A}}{\partial t}, \qquad \vec{B} = \nabla \times \vec{A},$$

we still have

$$m\ddot{x}_i = qE_i + qv_k T_{ik},$$

which is precisely equation (9.22), with T_{ik} given by (9.23). From now on, calculations are identical to those developed in Problem 2.

Let us now consider the particular cases (1) and (2):

Case 1

The equation of motion of the electron in the static and uniform electromagnetic field $\vec{E}, \vec{B}$ is [see (9.24)]

$$m\ddot{\vec{r}} = -e(\vec{E} + \vec{v} \times \vec{B}), \quad (9.31)$$

where m is the electron mass, and $-e$ its electric charge. Without loss of generality, let us suppose that $\vec{E}, \vec{B}$ are oriented along the z-axis, and $\vec{v}_0$ along the x-axis. Since

$$\vec{v} \times \vec{B} = \begin{vmatrix} \vec{i} & \vec{j} & \vec{k} \\ \dot{x} & \dot{y} & \dot{z} \\ 0 & 0 & B \end{vmatrix} = \dot{y} B \vec{i} - \dot{x} B \vec{j},$$

the components of vector equation (9.31) along the coordinate axes are

$$\begin{cases} Ox: & m\ddot{x} = -eB\dot{y}, \\ Oy: & m\ddot{y} = eB\dot{x}, \\ Oz: & m\ddot{z} = -eE. \end{cases} \quad (9.32)$$

Recalling that $E = const.$, the last equation of (9.32) yields

$$z(t) = -\frac{eE}{2m} t^2 + \frac{C_1}{m} t + \frac{C_2}{m},$$

where C_1 and C_2 are two arbitrary constants of integration. To find them, one takes $z(0) = 0$, and $\dot{z}(0) = 0$ at the initial moment of time $t = 0$. Then we obtain $C_1 = 0$, $C_2 = 0$, so that

$$z(t) = -\frac{eE}{2m} t^2. \quad (9.33)$$

To integrate the system of coupled ordinary differential equations

$$\begin{cases} m\ddot{x} = -eB\dot{y}, \\ m\ddot{y} = eB\dot{x}, \end{cases}$$

we shall use the complex variable $\xi = x + iy$. Multiplying the last equation by $i = \sqrt{-1}$ and adding the result to the first, we have

$$\ddot{\xi} - i \frac{eB}{m} \dot{\xi} = 0 \quad (9.34)$$

with the solution

$$\xi = ae^{i\omega t} + b, \quad (9.35)$$

where $\omega = \frac{eB}{m}$, while $a = a_x + ia_y$, and $b = b_x + ib_y$ are two complex constants of integration. Returning to the old variables, we can write

$$\begin{cases} x(t) = a_x \cos \omega t - a_y \sin \omega t + b_x, \\ y(t) = a_x \sin \omega t + a_y \cos \omega t + b_y. \end{cases} \quad (9.36)$$

The four arbitrary, real constants of integration a_x, a_y, b_x, b_y can be univocally determined by means of the following four initial conditions:

$$\begin{cases} x(0) = 0, & y(0) = 0, \\ \dot{x}(0) = 0, & \dot{y}(0) = 0. \end{cases} \tag{9.37}$$

Some simple calculations then yield

$$x(t) = \frac{v_0}{\omega} \sin \omega t, \tag{9.38}$$

and

$$y(t) = \frac{v_0}{\omega}(1 - \cos \omega t). \tag{9.39}$$

Using the last two relations, one easily obtains

$$x^2(t) + \left[y(t) - \frac{v_0}{\omega}\right]^2 = \frac{v_0^2}{\omega^2}. \tag{9.40}$$

This means that the projection of the trajectory of electron on xy-plane, at any time, is a *circle* of constant radius $R = v_0/\omega$ and centre at the point $C(x_0, y_0) = C\left(0, \frac{v_0}{\omega}\right)$.

On the other hand,

$$\dot{x}^2(t) + \dot{y}^2(t) = v_0^2,$$

which shows that the motion of rotation about the field is uniform. The angular velocity is

$$\omega = \frac{v_0}{R} = \frac{eB}{m}.$$

The spatial (three-dimensional) motion of the electron is obtained by composing the uniform rotation given by (9.38) and (9.39), with the accelerated motion given by (9.33). The trajectory is, therefore, a helix of variable pitch, wrapped around the circular cylinder of radius R.

Case 2

The motion of the electron in the static, uniform electromagnetic field $\vec{E}, \vec{B}$ is governed by equation (9.31) (we rewrite it, for convenience)

$$m\ddot{\vec{r}} = -e(\vec{E} + \vec{v} \times \vec{B}). \tag{9.41}$$

Suppose, this time, that the z-axis is oriented along $\vec{B}$, the yz-plane is determined by $\vec{E}$ and $\vec{B}$, while the initial conditions demand $\vec{v}_0 = (v_0, 0, 0)$. Since

$$\vec{v} \times \vec{B} = \begin{vmatrix} \vec{i} & \vec{j} & \vec{k} \\ \dot{x} & \dot{y} & \dot{z} \\ 0 & 0 & B \end{vmatrix} = \dot{y} B \vec{i} - \dot{x} B \vec{j},$$

the components of vector equation (9.31) along the coordinate axes are

$$\begin{cases} Ox: & m\ddot{x} = -eB\dot{y}, \\ Oy: & m\ddot{y} = -eE_y + eB\dot{x}, \\ Oz: & m\ddot{z} = -eE_z. \end{cases} \quad (9.42)$$

Using the initial conditions $\vec{r}_0 = (0,0,0)$ and $\vec{v}_0 = (v_0, 0, 0)$, the last equation (9.42) yields

$$z(t) = -\frac{eE_z}{2m} t^2. \quad (9.43)$$

Introducing the complex variable $\xi = x + iy$, the first two equations (9.42) can be compressed to give one equation

$$\ddot{\xi} - i\frac{eB}{m}\dot{\xi} = -i\frac{e}{m} E_y. \quad (9.44)$$

This is a second order linear, non-homogeneous, ordinary differential equation with constant coefficients. Its solution is obtained by adding a particular solution of the non-homogeneous equation to the general solution of the homogeneous equation. Solution of the homogeneous equation has been already obtained [see (9.35)]:

$$\xi_0(t) = a e^{i\omega t} + b,$$

where a and b are two complex constants. Since the non-homogeneous equation contains only derivatives of the unknown variable, we cannot use the term that causes inhomogeneity. So, we shall appeal to the variation of constants method. Following this procedure, since the fundamental set of solutions of the homogeneous equation is

$$\begin{cases} \xi_1(t) = e^{i\omega t}, \\ \xi_2(t) = 1, \end{cases}$$

the general solution to the problem is given by

$$\xi(t) = e^{i\omega t} \int a'(t)\, dt + \int b'(t)\, dt,$$

where $a'(t)$ and $b'(t)$ are the solutions of the first-order, differential equations system

$$\begin{cases} \xi_1(t)a'(t) + \xi_2(t)b'(t) = 0, \\ \xi_1'(t)a'(t) + \xi_2'(t)b'(t) = -i\dfrac{e}{m}E_y, \end{cases}$$

or

$$\begin{cases} e^{i\omega t}a'(t) + b'(t) = 0, \\ i\omega\, e^{i\omega t}a'(t) = -i\dfrac{e}{m}E_y. \end{cases}$$

The second equation yields

$$a(t) = -\frac{ieE_y}{m\omega^2}e^{-i\omega t} + c,$$

while the first equations gives

$$b(t) = \frac{eE_y t}{m\omega} + d,$$

where c and d are two arbitrary, complex constants of integration. The general solution of equation (9.44) therefore is

$$\xi(t) = ce^{i\omega t} - \frac{ieE_y}{m\omega^2} + \frac{eE_y}{m\omega}t + d = ce^{i\omega t} + \frac{E_y}{B}t + g, \qquad (9.45)$$

where g is a new complex constant of integration, whose value can be easily found. Separating the real and imaginary parts in (9.45), we find

$$\begin{cases} x(t) = c_x \cos\omega t - c_y \sin\omega t + \dfrac{E_y}{B}t + g_x, \\ y(t) = c_x \sin\omega t + c_y \cos\omega t + g_y. \end{cases} \qquad (9.46)$$

The initial conditions

$$\begin{cases} x(0) = 0; & y(0) = 0; \\ \dot{x}(0) = v_0; & \dot{y}(0) = 0, \end{cases} \qquad (9.47)$$

lead to

$$\begin{cases} c_x + g_x = 0; & c_y + g_y = 0; \\ -c_y\omega + \dfrac{E_y}{B} = v_0; & c_x\omega = 0, \end{cases}$$

so that

$$\begin{cases} c_x = 0; & g_x = 0; \\ c_y = \dfrac{E_y}{\omega B} - \dfrac{v_0}{\omega}; & g_y = \dfrac{v_0}{\omega} - \dfrac{E_y}{\omega B}. \end{cases}$$

Denoting $\alpha = \frac{v_0}{\omega} - \frac{E_y}{\omega B}$, the relations (9.46) write

$$\begin{cases} x(t) = \alpha \sin \omega t + \dfrac{E_y}{B} t, \\ y(t) = \alpha(1 - \cos \omega t). \end{cases} \qquad (9.48)$$

Introducing a new notation $\alpha = \alpha_0 \frac{E_y}{\omega B}$, with $\alpha_0 = \frac{B}{E_y}\left(v_0 - \frac{E_y}{B}\right)$ in (9.48) and recalling (9.43), we finally have the parametric equations of the trajectory

$$\begin{cases} x(t) = \dfrac{E_y}{\omega B}(\omega t + \alpha_0 \sin \omega t), \\ y(t) = \alpha_0 \dfrac{E_y}{\omega B}(1 - \cos \omega t), \\ z(t) = -\dfrac{eE_z}{2m} t^2. \end{cases} \qquad (9.49)$$

Fig.IX.1 shows an example for the trajectory described by (9.49), for the specified values of the physical quantities. All quantities are expressed in SI unit system. The graphic representations has been performed by means of software *Mathematica*, according to the following command lines:

v₀ = 10^9;
B = 10^-3;
E_y = 10^-4;
E_z = 5*10^-7;
α₀ = v₀*B/E_y -1;
ω = 10^8;
e = -1.6*10^-19;
m = 9.1*10^-31;

x[t_]:= (E_y/ω*B)*(ω*t + α₀*Sin[ω*t]);
y[t_]:= α₀*(E_y/ω*B)*(1- Cos[ω*t]);
z[t_]:= -(e*E_z/2*m)*t^2;

ParametricPlot3D[{x[t], y[t], z[t]}, {t, 0, 0.02}]

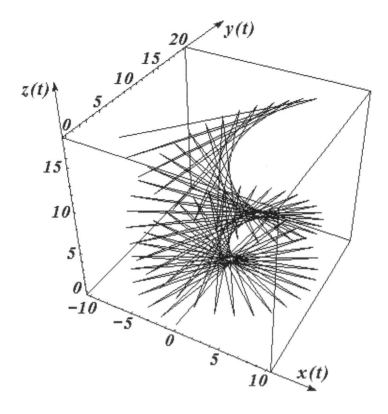

Fig.IX.1

Problem 4

A magnetic dipole of moment μ is placed in the magnetic field $\vec{B}$. Using Hamiltonian approach, determine the interaction energy between the dipole and the field, $W_m = -\vec{\mu} \cdot \vec{B}$.

Solution

Let $\vec{B}$ be the homogeneous and constant magnetic induction field in which is placed the point charge q, and let $\vec{\mu}$ be the magnetic moment produced by a uniform circular motion of q (see Fig.IX.2). Such a magnetic field possesses the vector potential

$$\vec{A} = \frac{1}{2}\vec{B} \times \vec{r}, \tag{9.50}$$

where $\vec{r}$ is the position vector (radius vector) of some point $P(\vec{r})$ of the field with respect to an arbitrary chosen reference frame $Oxyz$ (see Fig.IX.3).

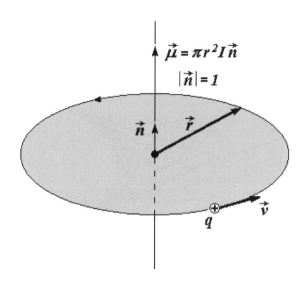

Fig.IX.2

Indeed,

$$\nabla \times \vec{A} = \nabla \times \left(\frac{1}{2}\vec{B} \times \vec{r}\right)$$

$$= \frac{1}{2}[\vec{B}\nabla \cdot \vec{r} - \vec{r}\nabla \cdot \vec{B} + (\vec{r} \cdot \nabla)\vec{B} - (\vec{B} \cdot \nabla)\vec{r}]$$

$$= \frac{1}{2}(3\vec{B} - \vec{B}) = \vec{B},$$

where we took into account that the field $\vec{B}$ is source-free ($\nabla \cdot \vec{B} = 0$) and constant [$(\vec{r} \cdot \nabla)\vec{B} = 0$], while

$$(\vec{B} \cdot \nabla)\vec{r} = B_i \frac{\partial}{\partial x_i}(x_k \vec{u}_k) = B_i \vec{u}_k \delta_{ik} = B_i \vec{u}_i = \vec{B}.$$

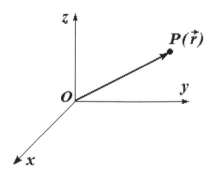

Fig.IX.3

According to (9.27) (see Problem 3), the Hamiltonian of our physical system is

$$H = \frac{1}{2m}(p_k - qA_k)(p_k - qA_k) + q\phi,$$

or, if we use vector notation

$$H = \frac{1}{2m}|\vec{p}|^2 + \frac{q^2}{2m}|\vec{A}|^2 - \frac{q}{m}\vec{p}\cdot\vec{A} + q\phi.$$

By means of formula (9.50), we still have

$$H = \frac{1}{2m}|\vec{p}|^2 + q\phi - \frac{q}{2}\vec{v}\cdot(\vec{B}\times\vec{r}) + \frac{q^2}{8m}|\vec{B}\times\vec{r}|^2. \tag{9.51}$$

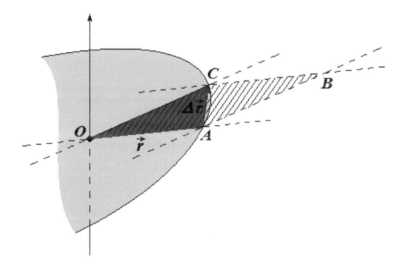

Fig.IX.4

Usually, the last term is much smaller than the rest of the terms, so that it can be neglected. Then we are left with

$$H = H_0 + H_{int}, \tag{9.52}$$

where

$$H_0 = \frac{1}{2m}|\vec{p}|^2 + q\phi \tag{9.53}$$

is the dipole energy when the external field $\vec{B}$ is absent, and

$$H_{int} = -\frac{q}{2}\vec{v}\cdot(\vec{B}\times\vec{r}) \tag{9.54}$$

is the energy of interaction between dipole and the external field.

All we have to do now is to show that the two formulas (9.54) and $W_m = -\vec{\mu} \cdot \vec{B}$ express the same thing. To this end, we shall use the geometric interpretation of the cross product. As far as one knows, the modulus of a cross product of two vectors equals the area of a parallelogram with the vectors for sides (see Fig.IX.4). In the limit $\Delta t \to 0$, the area ΔS swept by the radius vector $\vec{r}$ of the particle is given by

$$\lim_{\Delta t \to 0} \Delta S = \lim_{\Delta t \to 0} \frac{1}{2} |\vec{r} \times \Delta \vec{r}|, \qquad (9.55)$$

where $\Delta \vec{r}$ is the variation of $\vec{r}$ during the time interval Δt. Dividing (9.55) by Δt, one obtains the areolar velocity of the particle

$$\frac{d\vec{S}}{dt} = \lim_{\Delta t \to 0} \frac{\Delta \vec{S}}{\Delta t} = \frac{1}{2} \vec{r} \times \vec{v},$$

and, therefore

$$H_{int} = -\frac{q}{2} \vec{v} \cdot (\vec{B} \times \vec{r}) = -\frac{q}{2} \vec{B} \cdot (\vec{r} \times \vec{v}) = -q \vec{B} \cdot \frac{d\vec{S}}{dt}. \qquad (9.56)$$

Let us now write equation (9.56) for a period T of one complete uniform revolution of the particle, that is

$$H_{int} = -\vec{B} \cdot \frac{q}{T} \Delta \vec{S} = -\vec{B} \cdot I \Delta \vec{S} = -\vec{\mu} \cdot \vec{B}. \qquad (9.57)$$

where $\vec{\mu} = I \Delta \vec{S} = \pi r^2 I \vec{n}$ is the dipole moment of the magnetic sheet, produced by the circular current of radius r. Here $\vec{n}$ is the unit vector of the external normal to the circuit surface.

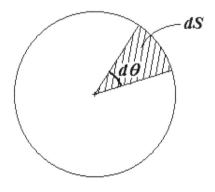

Fig.IX.5

Observation. Equality $\frac{dS}{dt} = \frac{\Delta S}{T}$ which we have previously used is sustained by the uniform feature of the circular motion of the particle. Indeed, in case of uniform, circular motion we have (see Fig.IX.5)

$$dS = \frac{d\theta}{2\pi}\Delta S = \frac{\omega\, dt}{2\pi}\Delta S = \Delta S \frac{dt}{T}.$$

Problem 5

A simple pendulum of length l, made out of a conductor material, moves in such a way that its inferior end slides without friction on a conductor support of circular form. The pendulum arm is a rigid of mass m, moving in the external static and homogeneous magnetic field $|\vec{B}|$, orthogonal at any point to the pendulum rod. One end of the metallic support and the point of support of pendulum are connected by an ideal condenser of capacity C. The electric circuit closes through the pendulum arm, as shown in Fig.IX.6. Neglecting the electric resistance and inductance of the circuit, determine the period T of the pendulum.

Solution

During the infinitesimal time interval dt, the pendulum arm of length l describes on the circular support the distance $ds = l\, d\alpha$ and sweeps the elementary surface of area $dS = l\, ds/2 = l^2 d\alpha/2$ (see Fig.IX.7). Since, at any point, the field $\vec{B}$ is orthogonal to this surface, the elementary magnetic flux through dS is $d\Phi = \vec{B} \cdot d\vec{S} = B\, dS = Bl^2 d\alpha/2$. Since $d\Phi$ is variable, it induces in the pendulum conductor arm the electromotive tension (voltage)

$$e = -\frac{d\Phi}{dt} = -\frac{Bl^2}{2}\frac{d\alpha}{dt} = -\frac{Bl^2}{2}\dot{\alpha}. \qquad (9.58)$$

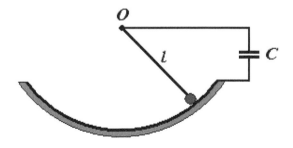

Fig.IX.6

This tension generates in the circuit shown in Fig.IX.7 an electric current, so that on the condenser plates shall appear the time variable electric charge

$$q = C|e| = \frac{CBl^2}{2}\dot{\alpha}. \tag{9.59}$$

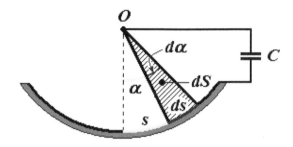

Fig.IX.7

Choosing as generalized coordinate the angle α between the pendulum arm and vertical, the Lagrangian writes

$$L = T - V. \tag{9.60}$$

Here

$$T = \frac{1}{2}I\omega^2 = \frac{1}{2}ml^2\dot{\alpha}^2, \tag{9.61}$$

where $I = ml^2$ is the moment of inertia of the pendulum rod, with respect to point axis of rotation, and

$$V = V_g + V_{em} \tag{9.62}$$

is the total potential energy of the system, where V_g and V_{em} stand for contributions of the gravitational and electromagnetic fields, respectively. If the horizontal plane containing O is chosen as the reference level for the gravitational potential energy (see §3, Chap.III), then

$$V_g = -mgl\cos\alpha, \tag{9.63}$$

while the electromagnetic potential energy is [see (9.58) and (9.59)]

$$V_{em} = \frac{1}{2}qe = -\frac{1}{8}CB^2l^4\dot{\alpha}^2. \tag{9.64}$$

In view of (9.60)-(9.64), the Lagrangian of the system is

$$L = \frac{4ml^2 + CB^2l^4}{8}\dot{\alpha}^2 + mgl\cos\alpha. \tag{9.65}$$

The canonical momentum p_α conjugate to the generalized coordinate α is

$$p_\alpha = \frac{\partial L}{\partial \dot\alpha} = \left(ml^2 + \frac{CB^2 l^4}{4}\right)\dot\alpha, \qquad (9.66)$$

and the Hamiltonian writes

$$H = p_\alpha \dot\alpha - L = \left(\frac{ml^2}{2} + \frac{CB^2 l^4}{8}\right)\dot\alpha^2 - mgl\cos\alpha,$$

or, in terms of α and p_α,

$$H = \frac{p_\alpha^2}{2\left(ml^2 + \frac{CB^2 l^4}{4}\right)} - mgl\cos\alpha. \qquad (9.67)$$

The system of Hamilton's canonical equations then is

$$\begin{cases} \dot\alpha = \dfrac{\partial H}{\partial p_\alpha} = \dfrac{p_\alpha}{ml^2 + \frac{CB^2 l^4}{4}}, \\[2mm] \dot p_\alpha = -\dfrac{\partial H}{\partial \alpha} = -mgl\sin\alpha. \end{cases} \qquad (9.68)$$

The differential equation describing the motion of the pendulum is obtained taking the time derivative of the first equation (9.68), then substituting $\dot p_\alpha$ given by the second equation (9.68). The result is:

$$\ddot\alpha + \frac{mgl\sin\alpha}{ml^2 + \frac{CB^2 l^4}{4}} = 0,$$

or

$$\left(ml^2 + \frac{CB^2 l^4}{4}\right)\ddot\alpha + mgl\sin\alpha = 0. \qquad (9.69)$$

If the angle α is small, one can approximate $\sin\alpha \simeq \alpha$, and the differential equation (9.69) becomes

$$\left(ml^2 + \frac{CB^2 l^4}{4}\right)\ddot\alpha + mgl\alpha = 0. \qquad (9.70)$$

Denoting

$$\omega^2 = \frac{mgl}{ml^2 + \frac{CB^2 l^4}{4}} = \frac{1}{\frac{l}{g}\left(1 + \frac{CB^2 l^2}{4m}\right)}, \qquad (9.71)$$

equation (9.70) writes

$$\ddot\alpha + \omega^2 \alpha = 0. \qquad (9.72)$$

This is the well-known second order, linear, homogeneous, differential equation with constant coefficients known as the *harmonic oscillator equation*. The oscillation period T is

$$T = \frac{2\pi}{\omega} = 2\pi\sqrt{\frac{l}{g}\left(1 + \frac{CB^2l^2}{4m}\right)}, \qquad (9.73)$$

where the term $CB^2l^2/4m$ represents the "contribution" of the electromagnetic effects, in addition to the gravitational influence. Indeed, as we have seen in Chap.III, the period of gravitational pendulum, for small oscillations, is given by

$$T_0 = 2\pi\sqrt{\frac{l}{g}}.$$

B. Problems of Fluid Mechanics

Euler-Lagrange Equations for Continuous Systems

The analytical formalism can be successfully applied in the study of continuous systems (*i.e.* systems with an infinite number of degrees of freedom): fields, fluids, magnetofluids, etc. Let us first deduce the fundamental equations of analytical technique, applied to continuous systems.

Consider the functional

$$J(\varphi) = \int_{D_n} \mathcal{L}\left[x_1, ..., x_n \; ; \; \varphi(x_1, ..., x_n), \frac{\partial\varphi}{\partial x_1}, ..., \frac{\partial\varphi}{\partial x_n}\right] dx_1...dx_n, \qquad (9.74)$$

defined on the bounded domain D_n of a n-dimensional space $\mathbb{R}^n$, where $\mathcal{L}$ is a continuous and differentiable function, admitting as many partial derivatives as necessary, and φ a function of class C^2 in D_n. Assuming that the values of φ on the closed hypersurface S_{n-1} which bounds the domain D_n are given, let us determine the function φ which makes $J(\varphi)$ an extremum.

Suppose that $\varphi(x_1, ..., x_n)$ performs the stationary value of $J(\varphi)$. In this case, for any infinitesimal variation $\varphi \to \varphi + \delta\varphi$, where $\delta\varphi = \epsilon\eta(x_1, ..., x_n)$, with

$$\eta(x_1, ..., x_n)|_{S_{n-1}} = 0, \qquad (9.75)$$

the first variation $\delta J(\varphi)$ of integral (9.74) must be zero. Since

$$\mathcal{L}\left(x, \varphi + \epsilon\eta, \frac{\partial\varphi}{\partial x} + \epsilon\frac{\partial\eta}{\partial x}\right) = \mathcal{L}\left(x, \varphi, \frac{\partial\varphi}{\partial x}\right) + \epsilon\eta\frac{\partial\mathcal{L}}{\partial\varphi} + \epsilon\sum_{i=1}^{n}\frac{\partial\eta}{\partial x_i}\frac{\partial\mathcal{L}}{\partial\varphi_{,i}} + \dots,$$

where $\varphi_{,i} = \partial\varphi/\partial x_i$, the first variation of $J(\varphi)$ is

$$\delta J(\varphi) = \epsilon \int_{D_n} \left(\eta\frac{\partial\mathcal{L}}{\partial\varphi} + \sum_{i=1}^{n}\frac{\partial\eta}{\partial x_i}\frac{\partial\mathcal{L}}{\partial\varphi_{,i}}\right) d\Omega, \qquad (9.76)$$

with $d\Omega = dx_1 dx_2 \dots dx_n$. Integrating by parts the second term in (9.76), we have:

$$\int_{D_n} \sum_{i=1}^{n} \frac{\partial\eta}{\partial x_i}\frac{\partial\mathcal{L}}{\partial\varphi_{,i}} d\Omega$$

$$= \int_{D_n} \sum_{i=1}^{n} \frac{\partial}{\partial x_i}\left(\eta\frac{\partial\mathcal{L}}{\partial\varphi_{,i}}\right) d\Omega - \int_{D_n} \eta\sum_{i=1}^{n} \frac{\partial}{\partial x_i}\left(\frac{\partial\mathcal{L}}{\partial\varphi_{,i}}\right) d\Omega. \qquad (9.77)$$

But

$$d\Omega = dx_1 \dots dx_k \dots dx_n = dx_k dS_k \quad (no\ summation), \qquad (9.78)$$

where $dS_k = dx_1 \dots dx_{k-1} dx_{k+1} \dots dx_n$ is the element of hypersurface orthogonal to dx_k. Using Green-Gauss theorem and the boundary condition (9.75), we see that the first integral on the r.h.s. of (9.77) vanishes:

$$\int_{D_n} \sum_{i=1}^{n} \frac{\partial}{\partial x_i}\left(\eta\frac{\partial\mathcal{L}}{\partial\varphi_{,i}}\right) d\Omega = \int_{S_{n-1}} \sum_{i=1}^{n} \eta\frac{\partial\mathcal{L}}{\partial\varphi_{,i}} dS_i = 0.$$

Thus, the first variation of $J(\varphi)$ is

$$\delta J(\varphi) = \epsilon \int_{D_n} \eta\left[\frac{\partial\mathcal{L}}{\partial\varphi} - \sum_{i=1}^{n} \frac{\partial}{\partial x_i}\left(\frac{\partial\mathcal{L}}{\partial\varphi_{,i}}\right)\right] d\Omega.$$

The necessary and sufficient condition for a stationary value of $J(\varphi)$ requires that, for any function η (except for condition (9.75)), we have

$$\frac{\partial\mathcal{L}}{\partial\varphi} - \sum_{i=1}^{n} \frac{\partial}{\partial x_i}\left(\frac{\partial\mathcal{L}}{\partial\varphi_{,i}}\right) = 0. \qquad (9.79)$$

Assuming that $\mathcal{L}$ is a function of h variables $\varphi^{(1)}(x), ..., \varphi^{(h)}(x)$, let us consider the functional

$$J[\varphi^{(s)}] = \int_{D_n} \mathcal{L}[x, \varphi^{(s)}(x), \varphi^{(s)}_{,x}] dx_1...dx_n \quad (s = \overline{1,h}). \tag{9.80}$$

Following a similar procedure as for a single variable φ, the stationary condition of functional $J[\varphi^{(s)}]$ yields the following system of second-order, partial differential equations:

$$\frac{\partial \mathcal{L}}{\partial \varphi^{(s)}} - \sum_{i=1}^{n} \frac{\partial}{\partial x_i}\left(\frac{\partial \mathcal{L}}{\partial \varphi^{(s)}_{,i}}\right) = 0 \quad (s = \overline{1,h}), \tag{9.81}$$

called the *Euler-Lagrange equations* of our variational problem.

In order to use these equations in CDM (Continuous Deformable Media) mechanics, we choose

$$x_1 = x, \quad x_2 = y, \quad x_3 = z, \quad x_4 = t. \tag{9.82}$$

With this choice, the functional (9.80) becomes

$$J[\varphi^{(s)}] = \int_{t_1}^{t_2}\int_V \mathcal{L}[x,y,z,t; \varphi^{(s)}(x,y,z,t); \varphi^{(s)}_{,x}, \varphi^{(s)}_{,y}, \varphi^{(s)}_{,z}, \varphi^{(s)}_{,t}]$$

$$\times dx\, dy\, dz\, dt \quad (s = \overline{1,h}), \tag{9.83}$$

and Euler-Lagrange equations read:

$$\frac{\partial \mathcal{L}}{\partial \varphi^{(s)}} - \frac{\partial}{\partial x_i}\left(\frac{\partial \mathcal{L}}{\partial \varphi^{(s)}_{,i}}\right) - \frac{\partial}{\partial t}\left(\frac{\partial \mathcal{L}}{\partial \varphi^{(s)}_{,t}}\right) = 0 \quad (s = \overline{1,h}), \tag{9.84}$$

where the summation convention has been used for index $i = 1, 2, 3$.

Comparing (9.83) with the action integral (2.36), we realize that there can be settled an equivalence between them if we choose

$$L = \int_V \mathcal{L}\, dx\, dy\, dz = \int_V \mathcal{L}\, d\tau. \tag{9.85}$$

Therefore function $\mathcal{L}$ stands for the Lagrangian per unit volume. It is called the *Lagrangian density*.

With this notation, Hamilton's principle (2.35) reads

$$\delta \int_{t_1}^{t_2} \left(\int_V \mathcal{L} d\tau \right) dt = 0, \qquad (9.86)$$

and can be used as a fundamental postulate in the study of holonomic CDM, while Euler-Lagrange equations (9.84) are the equations of motion of these systems.

Comparing equations (9.81) with Lagrange's equations for systems with a finite number of degrees of freedom (2.24), we realize that they are different in certain respects. In case of continuous systems the role of generalized coordinates is played by the functions $\varphi^{(s)}$, called *dependent variables* or *variational parameters*, while $x_1, .., x_n$ play the role of *independent variables*. Our choice (9.82) shows that both the space coordinates x, y, z and the time t are now taken as independent parameters, while $\varphi^{(s)}$ are selected from the physical variables which characterize a given system. In view of these considerations, equations (9.81) can be regarded as an infinite chain of Lagrange's-type differential equations, each of them being obtained by a successive fixation of space variables x, y, z. Since $\mathcal{L}$ is a Lagrangian density, all quantities appearing in it must be represented by their densities, such as: mass density ρ, entropy density s, current density $\vec{j}$, etc.

Equations (9.81) are particularly useful in analytical formalism of CDM, because they can be applied not only in the study of condensed media (solid, fluid), but also in the derivation of fundamental equations governing the *fields*.

Observation. Euler-Lagrange's equations (9.81) do not change their form if instead of $\mathcal{L}$ we choose

$$\mathcal{L}'(x, \varphi^{(s)}, \varphi^{(s)}_{,i}) = \mathcal{L}(x, \varphi^{(s)}, \varphi^{(s)}_{,i}) + \sum_{k=1}^{n} \frac{\partial}{\partial x_k} F_k(x, \varphi^{(s)}), \qquad (9.87)$$

provided the integration domain D_n remains unchanged, and the field variables $\varphi^{(s)}$ take fixed values on the boundary S_{n-1} of D_n. To prove this, we integrate the last relation on D_n and, using the generalized Green-Gauss theorem, we obtain:

$$\int_{D_n} \mathcal{L}' d\Omega = \int_{D_n} \mathcal{L} d\Omega + \int_{D_n} \sum_{k=1}^{n} \frac{\partial F_k}{\partial x_k} d\Omega = \int_{D_n} \mathcal{L} d\Omega + \int_{S_{n-1}} \sum_{k=1}^{n} F_k dS_k.$$

Applying now the operator δ to this relation, we still have:

$$\delta \int_{D_n} \mathcal{L}' d\Omega = \delta \int_{D_n} \mathcal{L} d\Omega + \int_{S_{n-1}} \sum_{i=1}^{n}\sum_{k=1}^{n} \left[\frac{\partial F_k}{\partial x_i}\delta x_i + \frac{\partial F_k}{\partial \varphi^{(s)}}\delta \varphi^{(s)}\right] dS_k.$$

But, by hypothesis, on the boundary S_{n-1} we have $\delta x_i = 0$, $\delta \varphi^{(s)} = 0$, and therefore

$$\delta \int_{D_n} \mathcal{L}' d\Omega = \delta \int_{D_n} \mathcal{L} d\Omega$$

which means that the condition of stationary value of $J[\varphi^{(s)}]$

$$\delta J[\varphi^{(s)}] = \delta \int_{D_n} \mathcal{L} d\Omega = 0$$

does not change at the transformation (9.87). As a result, Euler-Lagrange's equations (9.81) do not change their form. In other words, two Lagrangian densities which differ from one another by a divergence term are *equivalent*.

Problem 6

Find the equation of motion of an ideal, compressible fluid, performing isentropic motion in an external potential field[1] (Euler's equation).

Solution

Let us denote by $\mathcal{V}^*(\vec{r}, t)$, $s(\vec{r}, t)$, and $\varepsilon(\vec{r}, t)$ the potential of the exterior field (*e.g.* the gravitational field), entropy and internal energy, taken per unit mass, respectively. Then the kinetic energy density is $\frac{1}{2}\rho|\vec{v}|^2$, where $\vec{v}(\vec{r}, t)$ is the velocity field, while the potential energy is composed by two terms: $\rho\varepsilon$ and $\rho\mathcal{V}^*$, corresponding to internal and external forces, respectively. Nevertheless, the expression

$$\mathcal{L}_o = \frac{1}{2}\rho\vec{v}^2 - \rho(\varepsilon + \mathcal{V}^*) \tag{9.88}$$

cannot be used as a Lagrangian density, because it contains only some of the physical variables which define the system. In turn, this is due

---

[1] J.W.Herivel, Proc. Camb. Phil. Soc., **51**, 1955, p.344.

to the fact that we did not take into consideration the *constraints* acting on the fluid, which in our case are the equation of continuity

$$\frac{\partial \rho}{\partial t} + \nabla \cdot (\rho \vec{v}) = 0, \qquad (9.89)$$

and the equation of conservation of entropy

$$\frac{ds}{dt} = \frac{\partial s}{\partial t} + (\vec{v} \cdot \nabla)s = 0. \qquad (9.90)$$

A suitable Lagrangian density is constructed by using the method of *Lagrangian multipliers*. In this respect we amplify equations (9.89) and (9.90) by the multipliers $\alpha(\vec{r}, t)$ and $\beta(\vec{r}, t)$, respectively, and add the result to (9.88). This yields

$$\mathcal{L} = \frac{1}{2}\rho \vec{v}^2 - \rho(\varepsilon + \mathcal{V}^*) - \alpha\left[\frac{\partial \rho}{\partial t} + \nabla \cdot (\rho \vec{v})\right] - \beta\rho\left(\frac{\partial s}{\partial t} + \vec{v} \cdot \nabla s\right). \qquad (9.91)$$

It is more convenient to our purpose to use the Lagrangian density in a slightly modified form. Taking advantage of the property (9.87), we shall add to (9.91) the divergence

$$\frac{\partial}{\partial x_j}(\alpha \rho v_j) \quad (j = \overline{1, 4}),$$

where we choose $x_1 = x$, $x_2 = y$, $x_3 = z$, $x_4 = t$, $v_1 = v_x$, $v_2 = v_y$, $v_3 = v_z$, $v_4 = 1$. Since

$$-\alpha\left(\frac{\partial \rho}{\partial t} + \vec{v} \cdot \nabla \rho + \rho \nabla \cdot \vec{v}\right) + \nabla \cdot (\alpha \rho \vec{v}) + \frac{\partial}{\partial t}(\alpha \rho) = \rho\left(\frac{\partial \alpha}{\partial t} + \vec{v} \cdot \nabla \alpha\right),$$

we finally obtain

$$\mathcal{L} = \frac{1}{2}\rho \vec{v}^2 - \rho(\varepsilon + \mathcal{V}^*) + \rho\left(\frac{\partial \alpha}{\partial t} + \vec{v} \cdot \nabla \alpha\right) - \beta\rho\left(\frac{\partial s}{\partial t} + \vec{v} \cdot \nabla s\right). \qquad (9.92)$$

Choosing s, ρ, v_x, v_y, v_z as variational parameters $\varphi^{(i)}$, $i = \overline{1, 5}$ in (9.84), we then have:

i) $\varphi^{(1)} = s$. The corresponding Euler-Lagrange equation is

$$\frac{\partial \mathcal{L}}{\partial s} - \frac{\partial}{\partial x_i}\left(\frac{\partial \mathcal{L}}{\partial s_{,i}}\right) - \frac{\partial}{\partial t}\left(\frac{\partial \mathcal{L}}{\partial s_{,t}}\right) = 0. \qquad (9.93)$$

Using the thermodynamic fundamental equation of equilibrium processes
$$T\,ds = d\varepsilon(\rho,s) + p\,d\left(\frac{1}{\rho}\right),$$
we find
$$\frac{\partial \mathcal{L}}{\partial s} = \frac{\partial \mathcal{L}}{\partial \varepsilon}\frac{\partial \varepsilon}{\partial s} = -\rho T\;; \quad \frac{\partial \mathcal{L}}{\partial s_{,i}} = -\beta\rho v_i\;; \quad \frac{\partial \mathcal{L}}{\partial s_{,t}} = -\beta\rho.$$

Introducing these results in (9.93) and using the equation of continuity, after simplifying by $\rho \neq 0$ we arrive at
$$\frac{\partial \beta}{\partial t} + \vec{v}\cdot\nabla\beta = T. \tag{9.94}$$

ii) $\varphi^{(2)} = \rho$. We have
$$\frac{\partial \mathcal{L}}{\partial \rho} - \frac{\partial}{\partial x_i}\left(\frac{\partial \mathcal{L}}{\partial \rho_{,i}}\right) - \frac{\partial}{\partial t}\left(\frac{\partial \mathcal{L}}{\partial \rho_{,t}}\right) = 0.$$

Performing the derivatives
$$\frac{\partial \mathcal{L}}{\partial \rho} = \frac{1}{2}\vec{v}^2 - (\varepsilon + \mathcal{V}^*) - \frac{p}{\rho} + \frac{\partial \alpha}{\partial t} + \vec{v}\cdot\nabla\alpha\;;$$

$$\frac{\partial \mathcal{L}}{\partial \rho_{,i}} = 0\;; \quad \frac{\partial \mathcal{L}}{\partial \rho_{,t}} = 0,$$

we obtain
$$\frac{1}{2}\vec{v}^2 - (\varepsilon + \mathcal{V}^*) - \frac{p}{\rho} + \frac{\partial \alpha}{\partial t} + \vec{v}\cdot\nabla\alpha = 0, \tag{9.95}$$
which is a *Bernoulli-type equation*.

iii) $\varphi^{(3,4,5)} = v_k$ ($k = 1,2,3$). In this case we have three scalar equations
$$\frac{\partial \mathcal{L}}{\partial v_k} - \frac{\partial}{\partial x_i}\left(\frac{\partial \mathcal{L}}{\partial v_{k,i}}\right) - \frac{\partial}{\partial t}\left(\frac{\partial \mathcal{L}}{\partial v_{k,t}}\right) = 0.$$

Since
$$\frac{\partial \mathcal{L}}{\partial v_k} = \rho v_k + \rho\frac{\partial \alpha}{\partial x_k} - \beta\rho\frac{\partial s}{\partial x_k}\;;$$

$$\frac{\partial \mathcal{L}}{\partial v_{k,i}} = 0\;; \quad \frac{\partial \mathcal{L}}{\partial v_{k,t}} = 0,$$

we arrive at
$$\vec{v} = -\nabla\alpha + \beta\nabla s, \qquad (9.96)$$
which is a *Clebsch transformation*. Therefore, the functions $\alpha(\vec{r},t)$, $\beta(\vec{r},t)$, and s play the role of *Clebsch potentials*.

The last step is now to eliminate the multipliers α and β from the equations (9.94) - (9.96). To do this, we shall first replace $\nabla\alpha = -\vec{v} + \beta\nabla s$ into (9.95):
$$-\frac{1}{2}|\vec{v}|^2 - (\varepsilon + \mathcal{V}^*) - \frac{p}{\rho} + \frac{\partial\alpha}{\partial t} + \beta\vec{v}\cdot\nabla s = 0.$$

Applying to this equation the operator *gradient* we still have:
$$-\vec{v}\times\mathrm{curl}\vec{v} - (\vec{v}\cdot\nabla)\vec{v} - \frac{p}{\rho^2}\nabla\rho - T\nabla s - \nabla\mathcal{V}^* - \frac{1}{\rho}\nabla p + \frac{p}{\rho^2}\nabla\rho$$
$$+ \frac{\partial}{\partial t}(\beta\nabla s - \vec{v}) + \beta\nabla(\vec{v}\cdot\nabla s) + (\vec{v}\cdot\nabla s)\nabla\beta = 0.$$

But, by virtue of (9.90) and (9.94):
$$-\vec{v}\times\mathrm{curl}\vec{v} - T\nabla s + \frac{\partial\beta}{\partial t}\nabla s + \beta\nabla\left(\frac{\partial s}{\partial t}\right) + \beta\nabla(\vec{v}\cdot\nabla s) + (\vec{v}\cdot\nabla s)\nabla\beta$$
$$= -\vec{v}\times\mathrm{curl}(\beta\nabla s) - T\nabla s - (\vec{v}\cdot\nabla\beta)\nabla s + T\nabla s + (\vec{v}\cdot\nabla s)\nabla\beta$$
$$= -\vec{v}\times(\nabla\beta\times\nabla s) - (\vec{v}\cdot\nabla\beta)\nabla s + (\vec{v}\cdot\nabla s)\nabla\beta = 0,$$

and therefore we are left with *Euler's equation*
$$\frac{\partial\vec{v}}{\partial t} + (\vec{v}\cdot\nabla)\vec{v} = -\nabla\mathcal{V}^* - \frac{1}{\rho}\nabla p = \vec{F} - \frac{1}{\rho}\nabla p. \qquad (9.97)$$

Observation. Euler's equation can also be written as
$$\frac{\partial\vec{v}}{\partial t} + \frac{1}{2}\nabla(\vec{v}\cdot\vec{v}) - \vec{v}\times(\nabla\times\vec{v}) = \vec{F} - \frac{1}{\rho}\nabla p.$$

An irrotational $[\nabla\times\vec{v} = 0 \Rightarrow \vec{v} = \nabla\varphi(\vec{r},t)]$, stationary $(\frac{\partial\vec{v}}{\partial t} = 0)$ flow of a non-turbulent, perfect, incompressible ($\rho = const.$), and barotropic $[\rho = \rho(p)]$ fluid is then governed by
$$\frac{1}{2}\rho|\vec{v}|^2 + \rho g z + p = const., \qquad (9.97')$$
called *Bernoulli's equation*.

Problem 7

An ideal incompressible liquid of density ρ is situated in a vertical cylindrical container, rotating about its axis with constant angular velocity ω. At rest, the height of the liquid column is h. Supposing that the fluid and the container rotate together, as a solid body, determine:

  a) The shape of the surface of the rotating liquid;
  b) The pressure at every point inside the liquid;
  c) The pressure exerted at every point on the bottom of the container;
  d) The total force and the total pressure exerted by the fluid on the bottom of the container.

Solution

a) Let us attach a reference frame to the rotating liquid, with z-axis along the ascendant vertical, as shown in Fig.IX.8. The components of velocity of some particle of the liquid, with respect to this frame, are:
$$v_x = -\omega y, \quad v_y = \omega x, \quad v_z = 0,$$
because
$$\vec{v} = \vec{\omega} \times \vec{r} = \begin{vmatrix} \vec{i} & \vec{j} & \vec{k} \\ 0 & 0 & \omega \\ x & y & z \end{vmatrix} = \omega x \vec{j} - \omega y \vec{i} = v_x \vec{i} + v_y \vec{j} + v_z \vec{k}.$$

As we already know, the equation of motion of an ideal, incompressible fluid is Euler's equation (9.97)
$$\frac{\partial \vec{v}}{\partial t} + (\vec{v} \cdot \nabla)\vec{v} = \vec{F} - \frac{1}{\rho}\nabla p, \qquad (9.97)$$

where $\vec{F}$ is the mass-specific force
$$\vec{F} = \frac{\Delta \vec{G}}{\Delta m} = \frac{\vec{g}\Delta m}{\Delta m} = \vec{g} = -g\vec{k}.$$

Because there is no time variation of $\vec{v}$, $\frac{\partial \vec{v}}{\partial t} = 0$, and the projections on axes of (9.97) write

$$\begin{cases} Ox: & (\vec{v} \cdot \nabla)v_x = -\dfrac{1}{\rho}\dfrac{\partial p}{\partial x}, \\ Oy: & (\vec{v} \cdot \nabla)v_y = -\dfrac{1}{\rho}\dfrac{\partial p}{\partial y}, \\ Oz: & (\vec{v} \cdot \nabla)v_z = -g - \dfrac{1}{\rho}\dfrac{\partial p}{\partial z}. \end{cases} \qquad (9.98)$$

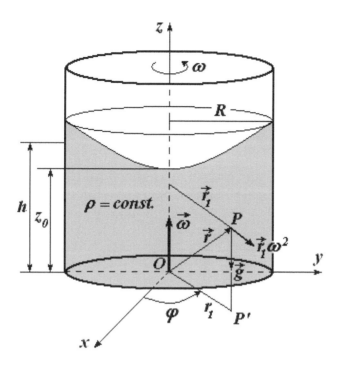

Fig.IX.8

Since

$$(\vec{v} \cdot \nabla)v_x = \left(-\omega y \frac{\partial}{\partial x} + \omega x \frac{\partial}{\partial y}\right)(-\omega y) = -\omega^2 x,$$

$$(\vec{v} \cdot \nabla)v_y = \left(-\omega y \frac{\partial}{\partial x} + \omega x \frac{\partial}{\partial y}\right)(\omega x) = -\omega^2 y,$$

and $v_z = 0$, we are left with the following system

$$\begin{cases} \omega^2 x = \frac{1}{\rho}\frac{\partial p}{\partial x}, \\ \omega^2 y = \frac{1}{\rho}\frac{\partial p}{\partial y}, \\ \frac{1}{\rho}\frac{\partial p}{\partial z} + g = 0. \end{cases} \quad (9.99)$$

Observing that, by integration, equations (9.99) give

$$\begin{cases} \frac{p}{\rho} = \frac{1}{2}\omega^2 x^2 + f_1(y,z), \\ \frac{p}{\rho} = \frac{1}{2}\omega^2 y^2 + f_2(z,x), \\ \frac{p}{\rho} = -gz + f_3(x,y), \end{cases}$$

where f_i ($i = 1, 2, 3$) are arbitrary functions of their arguments, the general solution of the system (9.99) is

$$\frac{p}{\rho} = \frac{1}{2}\omega^2(x^2+y^2) - gz + C, \qquad (9.100)$$

where C is a constant of integration.

Formula (9.100) can also be obtained by working in a non-inertial frame. Consider a frame rotating together with the liquid about z-axis with constant angular velocity ω. Since the fluid acceleration relative to this frame $\vec{a} = \frac{\partial \vec{v}}{\partial t} + (\vec{v}\cdot\nabla)\vec{v}$ equals zero, Euler's equation writes

$$0 = \vec{F}_{tot} - \frac{1}{\rho}\nabla p = \vec{F}_g + \vec{F}_{cf} - \frac{1}{\rho}\nabla p, \qquad (9.101)$$

where $\vec{F}_g = -g\vec{k}$ is the mass-specific gravitational force, and $\vec{F}_{cf} = \omega^2 \vec{r}_1 = \omega^2(x\vec{i} + y\vec{j})$ the mass-specific centrifugal force (N.B. in the non-inertial frame, $\vec{F}_{cf}$ is considered as an applied force). Performing the dot product between (9.101) and $d\vec{r} = dx\,\vec{i} + dy\,\vec{j} + dz\,\vec{k}$, we obtain

$$0 = \left(\vec{F}_g + \vec{F}_{cf} - \frac{1}{\rho}\nabla p\right)\cdot d\vec{r} = -g\,dz + \omega^2(x\,dx + y\,dy) - \frac{1}{\rho}dp$$

$$= d(-gz) + d\left(\frac{1}{2}\omega^2 r_1^2\right) + d\left(-\frac{1}{\rho}p\right),$$

where we took into account that

$$\nabla p \cdot d\vec{r} = \frac{\partial p}{\partial x_i}dx_i = dp,$$

and $r_1^2 = x^2 + y^2$. Integrating, we have

$$\frac{p}{\rho} = \frac{1}{2}\omega^2(x^2+y^2) - gz + C,$$

where C is an arbitrary constant of integration. This is precisely the equation (9.100), as expected.

Since the pressure p of the liquid at the free surface is zero, formula (9.100) yields

$$0 = \frac{\omega^2}{2}(x^2+y^2) - gz + C, \qquad (9.102)$$

or

$$z = \frac{\omega^2}{2g}(x^2+y^2) + C', \qquad (9.102')$$

with $C' = C/g$. We therefore conclude that the shape of the surface of the liquid, rotating about the symmetry z-axis of the cylinder, is a paraboloid of revolution of equation (9.102).

b) The constant of integration C appearing in equation (9.102) can be determined by equalizing the volumes of liquid at rest, $V = \pi R^2 h$, and in its motion. The last one writes

$$V = \int_{(V)} r_1 dr_1 d\varphi dz = \int_0^R r_1 dr_1 \int_0^{2\pi} d\varphi \int_0^{\frac{\omega^2}{2g}(x^2+y^2)+\frac{C}{g}} dz$$

$$= \int_0^R r_1 dr_1 \int_0^{2\pi} d\varphi \int_0^{\frac{\omega^2}{2g}r_1^2+\frac{C}{g}} dz = 2\pi \int_0^R r_1 dr_1 \left(\frac{\omega^2}{2g}r_1^2 + \frac{C}{g}\right)$$

$$= \frac{\pi\omega^2 R^4}{4g} + \frac{\pi C R^2}{g}.$$

Equalizing the two formulas for the fluid volume, we find

$$C = gh - \frac{\omega^2 R^2}{4}.$$

Introducing this result into (9.100), we obtain the pressure at any point inside the rotating liquid

$$p = \rho g(h-z) + \frac{\rho \omega^2}{2}\left(x^2 + y^2 - \frac{R^2}{2}\right). \tag{9.103}$$

c) The pressure of liquid at any point of the bottom of the container is immediately obtained by making $z = 0$ in (9.103):

$$p = p(x,y) = \rho gh + \frac{\rho \omega^2}{2}\left(x^2 + y^2 - \frac{R^2}{2}\right). \tag{9.104}$$

d) The total force exerted by the fluid on the bottom of the container is

$$F\Big|_{z=0,\, S=\pi R^2} = \int_{(S)} dF = \int_{(S)} p(x,y)\, dS = \int_{(S)} p(x,y)\, dxdy$$

$$= \int_{(S)} p(r_1) r_1 dr_1 d\varphi = \int_0^R p(r_1) r_1 dr_1 \int_0^{2\pi} d\varphi$$

$$= 2\pi \int_0^R \left[\rho g h + \frac{\rho\omega^2}{2}\left(r_1^2 - \frac{R^2}{2}\right)\right] r_1 dr_1$$

$$= 2\pi\rho \left(\int_0^R g h r_1 dr_1 + \int_0^R \frac{\omega^2}{2} r_1^3 dr_1 - \int_0^R \frac{\omega^2 R^2}{4} r_1 dr_1\right)$$

$$= \pi R^2 \rho g h = Mg, \tag{9.105}$$

where $M = \rho V = \pi R^2 \rho h$ is the total mass of the liquid. This result shows that the total force equals the weight of the liquid enclosed in the container. The total pressure exerted on bottom of the container is

$$P = p\big|_{z=0, S=\pi R^2} = \frac{F\big|_{z=0, S=\pi R^2}}{S} = \rho g h. \tag{9.106}$$

Problem 8

A layer of viscous liquid of thickness h is upperly limited by a free surface, and by a fixed plane on its lower side. The angle of inclination of the plane with respect to horizontal is α, as shown in Fig.IX.9. Find:
1) Velocity of the liquid under the action of its own weight;
2) Mass flow through a cross section of the liquid layer.

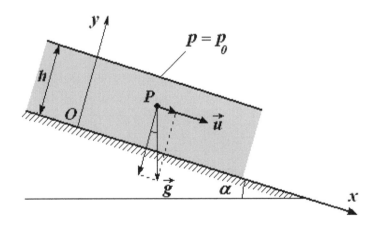

Fig.IX.9

Solution

1) The liquid under investigation is homogeneous, viscous, and incompressible, its motion is plane-parallel and unidimensional, while

the flow is stationary. The main equations describing such a fluid are: the equation of continuity

$$\frac{\partial \rho}{\partial t} + \text{div}(\rho \vec{v}) = 0, \qquad (9.107)$$

and Stokes-Navier equation

$$\rho \left[\frac{\partial \vec{v}}{\partial t} + (\vec{v} \cdot \nabla) \vec{v} \right] = \rho \vec{F} - \nabla p + (\lambda + \mu) \nabla \theta + \mu \Delta \vec{v}, \qquad (9.108)$$

where $\vec{v} = \vec{v}(x, y, z, t)$ is the velocity field, $\vec{F}$ is the mass-specific force, λ and μ are dynamic coefficients of viscosity, and $\theta = \text{div}\vec{v}$. In case of the homogeneous and incompressible fluids (i.e. liquids) $\rho = \rho(\vec{r}, t) = const.$ and (9.107) yields

$$\theta = \text{div}\vec{v} = 0,$$

while Stokes-Navier equation becomes

$$\frac{\partial \vec{v}}{\partial t} + (\vec{v} \cdot \nabla) \vec{v} = \vec{F} - \frac{1}{\rho} \nabla p + \nu \Delta \vec{v}, \qquad (9.108')$$

where $\nu = \mu/\rho$ is the kinematic coefficient of viscosity of the fluid.

Since the fluid flow is stationary and performs in x-direction (see Fig.IX.9), the only non-zero component of the velocity is

$$\vec{v}(x, y, z, t) \equiv \vec{i} v_x = \vec{i} u(x, y).$$

Then equation $\text{div}\vec{v} = 0$ gives

$$\text{div}\vec{v} = \frac{\partial v_x(x, y)}{\partial x} = \frac{\partial u}{\partial x} = 0,$$

that is $u = u(y)$ and, consequently,

$$(\vec{v} \cdot \nabla) \vec{v} = v_x \frac{\partial \vec{v}}{\partial x} = v_x \frac{\partial}{\partial x} \left(\vec{i} u(y) \right) = 0.$$

Taking into account that the motion is stationary ($\frac{\partial \vec{v}}{\partial t} = 0$), equation Stokes-Navier (9.108) becomes

$$\rho \vec{F} - \nabla p + \mu \Delta \vec{v} = 0. \qquad (9.109)$$

Projecting this vector equation on the two axes of the xOy frame attached to the liquid (see Fig.IX.9), we have:

$$\rho g \sin \alpha - \frac{\partial p}{\partial x} + \mu \frac{d^2 u}{dy^2} = 0, \qquad (9.110)$$

and

$$-\rho g \cos \alpha - \frac{\partial p}{\partial y} = 0. \qquad (9.111)$$

This way, our problems turns into a problem of mathematical physics equations. Namely, we have to solve the system (9.110)-(9.111) within the domain $D = \{x, y| -\infty < x < +\infty,\ 0 < y < h\}$, with the following boundary conditions:
(i) $u(y) = 0$, for $y = 0$;
(ii) $\frac{du}{dy} = 0$, for $y = h$;
(iii) $p(x, y) = p_0 = const.$, for $y = h$.

Integrating (9.111), we obtain

$$p = p(x, y) = -\rho g y \cos \alpha + p_1(x).$$

The arbitrary function $p_1(x)$ can be determined by using the boundary condition (iii), which gives

$$p_0 = -\rho g h \cos \alpha + p_1(x),$$

so that

$$p = p(x, y) = p_0 + \rho g (h - y) \cos \alpha. \qquad (9.112)$$

Introducing this result into (9.110), we have

$$\frac{d^2 u}{dy^2} = -\frac{\rho g \sin \alpha}{\mu},$$

and, by integration,

$$u = -\frac{\rho g \sin \alpha}{2\mu} y^2 + C_1 y + C_2.$$

The integration constants C_1 and C_2 are determined by means of conditions (i) and (ii), which lead to

$$C_2 = 0; \quad C_1 = \frac{\rho g \sin \alpha}{\mu} h,$$

so that
$$u = u(y) = \frac{\rho g}{2\mu}(2h - y)y \sin \alpha. \tag{9.113}$$

2) To calculate the mass flow through a cross section of the liquid layer (*i.e.* the mass of liquid going through a unit surface orthogonal to x-axis), we use the definition of this quantity for a stationary flow of a homogeneous and incompressible fluid, which is

$$Q_m = \frac{q_m}{L_z h} = \frac{1}{L_z h} \int_0^h dq_m,$$

where

$$dq_m = \frac{dm}{\Delta t} = \frac{\rho dV}{\Delta t} = \frac{\rho u(y) \Delta t\, L_z dy}{\Delta t} = \rho u(y) L_z dy$$

is the elementary mass crossing the elementary surface of any width L_z (along z-axis) and thickness dy (along y-axis), in unit time. In view of (9.113), we have:

$$Q_m = \frac{1}{L_z h} \int_0^h \rho u(y) L_z dy = \frac{1}{h} \int_0^h \frac{\rho^2 g}{2\mu}(2h-y)y \sin \alpha\, dy$$

$$= \frac{\rho^2 g h^2 \sin \alpha}{3\mu}. \tag{9.114}$$

Problem 9

Investigate the motion of a viscous, heavy liquid, flowing under the action of its own weight, in a rectangular pipe of sides $2a$ (width) and h (height). The angle of inclination of the pipe with respect to horizontal is α (Fig.IX.10). At the surface of liquid the velocity varies according to the law $u(y,h) = u_m \left(1 - \frac{y^2}{a^2}\right)$, where u_m is the maximum velocity at surface, in the middle of the pipe.

Solution

To solve the problem, we appeal to the equation of continuity (9.107), and Stokes-Navier equation (9.108). The components of the mass-specific force $\vec{F}$ and velocity $\vec{v}$ are

$$F_x = g \sin \alpha, \quad F_y = 0, \quad F_z = -g \cos \alpha,$$

and
$$v_x = u, \quad v_y = v_z = 0,$$

respectively. Since the fluid is homogeneous and incompressible (liquid), $\rho = \rho(\vec{r}, t) = const.$, and the equation of continuity (9.107) yields $\operatorname{div} \vec{v} = \frac{\partial u}{\partial x} = 0 \Rightarrow u = u(y, z)$, $-\infty < x < +\infty$. On the other hand, taking into account that the motion is stationary $\left(\frac{\partial \vec{v}}{\partial t} = 0\right)$ and $(\vec{v} \cdot \nabla) \vec{v} = v_x \frac{\partial}{\partial x} \vec{v} = u \frac{\partial}{\partial x} [\vec{i}\, u(y, z)] = 0$, the components of vector equation (9.108) along the coordinate axes are

$$\begin{cases} 0 = \rho g \sin \alpha - \dfrac{\partial p}{\partial x} + \mu \left(\dfrac{\partial^2 u}{\partial y^2} + \dfrac{\partial^2 u}{\partial z^2} \right), \\ 0 = -\dfrac{\partial p}{\partial y}, \\ 0 = -\rho g \cos \alpha - \dfrac{\partial p}{\partial z}. \end{cases} \quad (9.115)$$

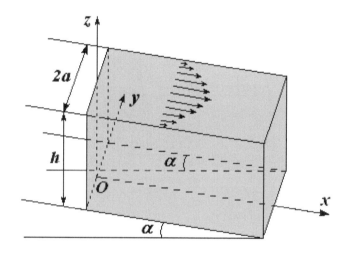

Fig.IX.10

Equation $(9.115)_2$ yields $p = p(x, z)$, while $(9.115)_3$ gives $p = -\rho g z \cos \alpha + p_1$. Since $p = p_0$ for $z = h$, we have $p_1(x) = p_0 + \rho g h \cos \alpha = const.$ In this case $\frac{\partial p}{\partial x} = 0$ and equation $(9.115)_1$ becomes

$$\frac{\partial^2 u}{\partial y^2} + \frac{\partial^2 u}{\partial z^2} = k, \quad -a < y < a, \ 0 < z < h, \quad (9.116)$$

where we denoted $k = -\frac{\rho g \sin \alpha}{\mu}$.

Equation (9.116) is a Poisson-type equation, with the boundary conditions:

(i) $u(y,0) = 0$, $\quad -a \leq y \leq a$;
(ii) $u(-a,z) = 0$, $u(a,z) = 0$, $\quad 0 \leq z \leq h$;
(iii) $u(y,h) = u_m \left(1 - \frac{y^2}{a^2}\right)$, $\quad -a \leq y \leq a$.

Since the motion of the liquid is symmetrical with respect to xz-plane, it is sufficient to be studied only in the domain $0 < y < a$ of the yz-plane. To this end, it is convenient to introduce new, dimensionless coordinates, defined as

$$u = u_m U, \quad y = aY, \quad z = aZ, \quad \frac{h}{a} = l, \quad k = \frac{u_m}{a^2} K.$$

This way, equation (9.116) writes

$$\frac{\partial^2 U}{\partial Y^2} + \frac{\partial^2 U}{\partial Z^2} = K, \quad 0 < Y < 1, \; 0 < Z < l. \tag{9.117}$$

This is a Poisson-type equation for $U = U(Y,Z)$, with the boundary conditions

(i') $U(Y,0) = 0$, $\quad 0 \leq Y \leq 1$;
(ii') $U(1,Z) = 0$, $\quad 0 \leq Z \leq l$;
(iii') $U(Y,l) = 1 - Y^2$, $\quad 0 \leq Y \leq 1$.

The boundary conditions are of the first kind, so that we have to find the solution of the Dirichlet problem for Poisson's equation.

We look for a solution of (9.117) of the form

$$U(Y,Z) = \sum_{n=1}^{\infty} A_n \cos\left(\frac{(2n-1)\pi Y}{2}\right) \zeta_n(Z). \tag{9.118}$$

Since

$$\frac{\partial^2 U}{\partial Y^2} = -\sum_{n=1}^{\infty} A_n \left(\frac{(2n-1)\pi}{2}\right)^2 \cos\left(\frac{(2n-1)\pi Y}{2}\right) \zeta_n(Z),$$

$$\frac{\partial^2 U}{\partial Z^2} = \sum_{n=1}^{\infty} A_n \cos\left(\frac{(2n-1)\pi Y}{2}\right) \frac{d^2 \zeta_n(Z)}{dZ^2},$$

equation (9.117) gives

$$\sum_{n=1}^{\infty} A_n \cos\left(\frac{(2n-1)\pi Y}{2}\right) \left[\frac{d^2 \zeta_n(Z)}{dZ^2} - \left(\frac{(2n-1)\pi}{2}\right)^2 \zeta_n(Z)\right] = K. \tag{9.119}$$

Let us now prove that the r.h.s. of (9.119) can be written as

$$K = K \sum_{n=1}^{\infty} \frac{4(-1)^{n-1}}{(2n-1)\pi} \cos\left(\frac{(2n-1)\pi Y}{2}\right), \qquad (9.120)$$

which is possible because

$$\frac{4}{\pi} \sum_{n=1}^{\infty} (-1)^{n-1} \frac{\cos\left((2n-1)\xi\right)}{2n-1} = \begin{cases} 1, & \text{for } 0 \leq \xi < \frac{\pi}{2}, \\ 0, & \text{for } \xi = \frac{\pi}{2}. \end{cases} \qquad (9.121)$$

To prove (9.121), we shall use the following uniformly, absolutely convergent (for $|z| < 1$), series

$$\frac{1}{1+iz} = 1 - iz - z^2 + iz^3 + ...,$$

$$\frac{1}{1-iz} = 1 + iz - z^2 - iz^3 + ...,$$

$$\frac{1}{1+z^2} = 1 - z^2 + z^4 - z^6 + ... = \sum_{n=1}^{\infty} (-1)^{n-1} z^{2n-2},$$

where $z = \xi + i\eta$ is a complex number. Integrating these relations, we have:

$$\int \frac{1}{1+iz} dz \left[= -i\ln(1+iz) \right] = \int (1 - iz - z^2 + iz^3 + z^4 + ...) \, dz$$

$$= z - i\frac{z^2}{2} - \frac{z^3}{3} + i\frac{z^4}{4} + \frac{z^5}{5} + ...,$$

$$\int \frac{1}{1-iz} dz \left[= i\ln(1-iz) \right] = \int (1 + iz - z^2 - iz^3 + z^4 + ...) \, dz$$

$$= z + i\frac{z^2}{2} - \frac{z^3}{3} - i\frac{z^4}{4} + \frac{z^5}{5} + ...,$$

$$\int \frac{1}{1+z^2} dz \left[= \arctan z \right] = \int (1 - z^2 + z^4 - z^6 + ...) \, dz$$

$$= z - \frac{z^3}{3} + \frac{z^5}{5} - \frac{z^7}{7} + ... = \sum_{n=1}^{\infty} (-1)^{n-1} \frac{z^{2n-1}}{2n-1}.$$

Adding the first two relations, we obtain

$$\int \frac{1}{1+iz} dz + \int \frac{1}{1-iz} dz = -i\ln(1+iz) + i\ln(1-iz) = -i\ln\frac{1+iz}{1-iz}$$

$$= \left(z - i\frac{z^2}{2} - \frac{z^3}{3} + i\frac{z^4}{4} + \frac{z^5}{5} + ..\right) + \left(z + i\frac{z^2}{2} - \frac{z^3}{3} - i\frac{z^4}{4} + \frac{z^5}{5} + ..\right)$$

$$= 2\left(z - \frac{z^3}{3} + \frac{z^5}{5} - ...\right),$$

so that

$$\frac{-i}{2}\ln\frac{1+iz}{1-iz} = z - \frac{z^3}{3} + \frac{z^5}{5} - ... = \sum_{n=1}^{\infty}(-1)^{n-1}\frac{z^{2n-1}}{2n-1}(=\arctan z).$$
(9.122)

Choosing $z = e^{i\xi}$, we can write:

$$\frac{1+iz}{1-iz} = \frac{1+i(\cos\xi + i\sin\xi)}{1-i(\cos\xi + i\sin\xi)}$$

$$= \frac{(1-\sin\xi + i\cos\xi)(1+\sin\xi + i\cos\xi)}{(1+\sin\xi - i\cos\xi)(1+\sin\xi + i\cos\xi)}$$

$$= \frac{i\cos\xi}{1+\sin\xi} = \frac{\cos\xi}{1+\sin\xi}e^{i\frac{\pi}{2}}.$$

Denoting

$$f(z) = \frac{1+ie^{i\xi}}{1-ie^{i\xi}} = \frac{\cos\xi}{1+\sin\xi}e^{i\frac{\pi}{2}},$$

and observing that

$$\ln f(z) = \ln|f(z)| + i\arg f(z),$$

where

$$\arg f(z) = \arctan\frac{\operatorname{Im}[f(z)]}{\operatorname{Re}[f(z)]} = \frac{\pi}{2},$$

the relation (9.122) becomes

$$\frac{-i}{2}\ln f(z) = \frac{-i}{2}\ln\left(\frac{\cos\xi}{1+\sin\xi}e^{i\frac{\pi}{2}}\right)$$

$$= -\frac{i}{2}\ln\left(\frac{\cos\xi}{1+\sin\xi}\right) - \frac{i}{2}\ln e^{i\frac{\pi}{2}} = -\frac{i}{4}\ln\left(\frac{\cos\xi}{1+\sin\xi}\right)^2 + \frac{\pi}{4}.$$

Then we can write

$$-\frac{i}{4}\ln\left(\frac{\cos\xi}{1+\sin\xi}\right)^2 \pm \frac{\pi}{4} = \sum_{n=1}^{\infty}(-1)^{n-1}\frac{e^{i\xi(2n-1)}}{2n-1}$$

$$= \sum_{n=1}^{\infty} (-1)^{n-1} \left[\frac{\cos(2n-1)\xi}{2n-1} + i \frac{\sin(2n-1)\xi}{2n-1} \right],$$

where the "plus" sign corresponds to the case $\xi \in \left[0, \frac{\pi}{2}\right) \cup \left(\frac{3\pi}{2}, 2\pi\right]$, and the "minus" sign to the interval $\xi \in \left(\frac{\pi}{2}, \frac{3\pi}{2}\right)$. Identifying the real and imaginary parts in the last relation, one obtains:

$$\sum_{n=1}^{\infty} (-1)^{n-1} \frac{\cos(2n-1)\xi}{2n-1} = \begin{cases} \frac{\pi}{4}, & \text{for } 0 \leq \xi < \frac{\pi}{2}, \\ 0, & \text{for } \xi = \pm \frac{\pi}{2}, \end{cases}$$

which is relation (9.121), and

$$\sum_{n=1}^{\infty} (-1)^n \frac{\sin(2n-1)\xi}{2n-1} = \frac{1}{4} \ln \left(\frac{\cos \xi}{1 + \sin \xi} \right)^2.$$

Formula (9.121) then yields (9.120) by setting $\xi = \frac{\pi}{2} Y$. By means of (9.120), equation (9.119) then writes

$$\sum_{n=1}^{\infty} A_n \cos \left(\frac{(2n-1)\pi Y}{2} \right) \left[\frac{d^2 \zeta_n(Z)}{dZ^2} - \left(\frac{(2n-1)\pi}{2} \right)^2 \zeta_n(Z) \right]$$

$$= K \sum_{n=1}^{\infty} \frac{4(-1)^{n-1}}{(2n-1)\pi} \cos \left(\frac{(2n-1)\pi Y}{2} \right).$$

Equalizing coefficients of $\cos \left(\frac{(2n-1)\pi Y}{2} \right)$ of both sides, we are left with the following ordinary differential equations:

$$\zeta_n''(Z) - \frac{(2n-1)^2 \pi^2}{4} \zeta_n(Z) = \frac{(-1)^{n-1} 4K}{(2n-1)\pi A_n} \quad (n \in \mathbb{N}^*), \quad (9.123)$$

with boundary conditions $\zeta_n(0) = 0$, and $\sum_{n=1}^{\infty} A_n \cos \left(\frac{(2n-1)\pi Y}{2} \right) \zeta_n(l)$ $= 1 - Y^2$.

For any value of $n \in \mathbb{N}^*$, the general solution of the corresponding non-homogeneous equation (9.123) writes as a sum of the general solution $\zeta_n^{(o)}$ of the homogeneous equation

$$\zeta_n''(Z) - \frac{(2n-1)^2 \pi^2}{4} \zeta_n(Z) = 0 \quad (n \in \mathbb{N}^*) \quad (9.124)$$

and a particular solution $\zeta_n^{(p)}$ of the non-homogeneous equation, that is
$$\zeta_n(Z) = \zeta_n^{(o)} + \zeta_n^{(p)}, \quad n = 1, 2, 3, ...$$

The homogeneous equations (9.124) have solutions of the form
$$\zeta_n^{(o)} = B_n \sinh \frac{(2n-1)\pi Z}{2} + C_n \cosh \frac{(2n-1)\pi Z}{2},$$

while a particular solution of the corresponding non-homogeneous equation must be chosen as a constant (the term producing inhomogeneity is a constant!), for each $n \in \mathbb{N}^*$:
$$\zeta_n^{(p)} = P_n = (const.)_n. \tag{9.125}$$

The constants P_n must verify the inhomogeneous equations (9.123)
$$-\frac{(2n-1)^2 \pi^2}{4} P_n = \frac{(-1)^{n-1} 4K}{(2n-1)\pi A_n},$$

which gives
$$P_n = \frac{(-1)^n 16K}{(2n-1)^3 \pi^3 A_n}.$$

Therefore, the general solutions of (9.123) are
$$\zeta_n(Z) = B_n \sinh \frac{(2n-1)\pi Z}{2} + C_n \cosh \frac{(2n-1)\pi Z}{2} + \frac{(-1)^n 16K}{(2n-1)^3 \pi^3 A_n}. \tag{9.126}$$

The arbitrary integration constants B_n and C_n are determined from the boundary conditions:
$$\zeta_n(0) = 0,$$

and
$$\sum_{n=1}^{\infty} A_n \cos\left(\frac{(2n-1)\pi Y}{2}\right) \zeta_n(l) = 1 - Y^2.$$

The first boundary condition yields
$$\zeta_n(0) = C_n + \frac{(-1)^n 16K}{(2n-1)^3 \pi^3 A_n} = 0,$$

that is
$$C_n = -\frac{(-1)^n 16K}{(2n-1)^3 \pi^3 A_n} = \frac{(-1)^{n-1} 16K}{(2n-1)^3 \pi^3 A_n}, \tag{9.127}$$

and $\zeta_n(Z)$ becomes

$$\zeta_n(Z) = B_n \sinh \frac{(2n-1)\pi Z}{2} + \frac{(-1)^{n-1}16K}{(2n-1)^3\pi^3 A_n} \cosh \frac{(2n-1)\pi Z}{2}$$

$$+ \frac{(-1)^n 16K}{(2n-1)^3\pi^3 A_n}$$

$$= B_n \sinh \frac{(2n-1)\pi Z}{2} + \frac{(-1)^n 16K}{(2n-1)^3\pi^3 A_n}\left[1 - \cosh \frac{(2n-1)\pi Z}{2}\right]. \tag{9.128}$$

The second boundary condition then gives

$$\sum_{n=1}^{\infty} A_n \cos\left(\frac{(2n-1)\pi Y}{2}\right) \zeta_n(l) = \sum_{n=1}^{\infty} A_n \cos\left(\frac{(2n-1)\pi Y}{2}\right)$$

$$\times \left\{ B_n \sinh \frac{(2n-1)\pi l}{2} + \frac{(-1)^n 16K}{(2n-1)^3\pi^3 A_n}\left[1 - \cosh \frac{(2n-1)\pi l}{2}\right]\right\}$$

$$= \sum_{n=1}^{\infty} \left\{ A_n B_n \sinh \frac{(2n-1)\pi l}{2} + \frac{(-1)^n 16K}{(2n-1)^3\pi^3}\left[1 - \cosh \frac{(2n-1)\pi l}{2}\right]\right\}$$

$$\times \cos\left(\frac{(2n-1)\pi Y}{2}\right) = 1 - Y^2. \tag{9.129}$$

Next, we shall express $1 - Y^2$ as a series of the same type as that on the l.h.s. To this end, we first observe that, since the series [see (9.121)]

$$\sum_{n=1}^{\infty} (-1)^{n-1} \frac{\cos(2n-1)\xi}{2n-1} = \frac{\pi}{4}$$

is absolutely and uniformly convergent, it can be integrated term by term. Therefore:

$$\int \left[\sum_{n=1}^{\infty} (-1)^{n-1} \frac{\cos(2n-1)\xi}{2n-1}\right] d\xi$$

$$= \sum_{n=1}^{\infty} \left\{\int \left[(-1)^{n-1} \frac{\cos(2n-1)\xi}{2n-1}\right] d\xi\right\}$$

$$= \sum_{n=1}^{\infty} (-1)^{n-1} \frac{\sin(2n-1)\xi}{(2n-1)^2} = \frac{\pi}{4}\xi + K_1.$$

To determine the integration constant K_1 we impose condition for the last relation to be valid for any ξ, including $\xi = 0$. This gives $K_1 = 0$, and the last relation becomes

$$\sum_{n=1}^{\infty} (-1)^{n-1} \frac{\sin(2n-1)\xi}{(2n-1)^2} = \frac{\pi}{4}\xi.$$

A new integration yields

$$\int \left[\sum_{n=1}^{\infty} (-1)^{n-1} \frac{\sin(2n-1)\xi}{(2n-1)^2} \right] d\xi$$

$$= \sum_{n=1}^{\infty} \left\{ \int \left[(-1)^{n-1} \frac{\sin(2n-1)\xi}{(2n-1)^2} \right] d\xi \right\}$$

$$= -\sum_{n=1}^{\infty} (-1)^{n-1} \frac{\cos(2n-1)\xi}{(2n-1)^3} = \frac{\pi}{8}\xi^2 + K_2,$$

or

$$\sum_{n=1}^{\infty} (-1)^{n-1} \frac{\cos(2n-1)\xi}{(2n-1)^3} = -\frac{\pi}{8}\xi^2 + K_3,$$

where $K_3 = -K_2$ is a new constant of integration. Using the already known procedure, we set $\xi = 0$ and obtain

$$K_3 = \sum_{n=1}^{\infty} \frac{(-1)^{n-1}}{(2n-1)^3} = \frac{\pi^3}{32},$$

so that

$$\sum_{n=1}^{\infty} (-1)^{n-1} \frac{\cos(2n-1)\xi}{(2n-1)^3} = \frac{\pi}{8}\left(\frac{\pi^2}{4} - \xi^2\right), \quad 0 \le \xi \le \frac{\pi}{2}.$$

Taking now ξ of the form $\xi = \frac{\pi}{2}Y$, we still have

$$1 - Y^2 = \sum_{n=1}^{\infty} \frac{(-1)^{n-1} 32}{(2n-1)^3 \pi^3} \cos\left(\frac{(2n-1)\pi Y}{2}\right).$$

With these results, (9.129) becomes

$$\sum_{n=1}^{\infty} \left\{ A_n B_n \sinh \frac{(2n-1)\pi l}{2} + \frac{(-1)^n 16K}{(2n-1)^3 \pi^3} \left[1 - \cosh \frac{(2n-1)\pi l}{2} \right] \right\}$$

355

$$\times \cos\left(\frac{(2n-1)\pi Y}{2}\right) = \sum_{n=1}^{\infty} \frac{(-1)^{n-1}32}{(2n-1)^3\pi^3} \cos\left(\frac{(2n-1)\pi Y}{2}\right).$$

Identifying the coefficients of $\cos\left(\frac{(2n-1)\pi Y}{2}\right)$ of both sides, we still have

$$A_n B_n \sinh\frac{(2n-1)\pi l}{2} + \frac{(-1)^n 16K}{(2n-1)^3\pi^3}\left[1 - \cosh\frac{(2n-1)\pi l}{2}\right]$$

$$= \frac{(-1)^{n-1}32}{(2n-1)^3\pi^3},$$

which gives

$$B_n = \left\{\frac{(-1)^{n-1}32}{(2n-1)^3\pi^3} + \frac{(-1)^{n-1}16K}{(2n-1)^3\pi^3}\left[1 - \cosh\frac{(2n-1)\pi l}{2}\right]\right\}$$

$$\bigg/ A_n \sinh\frac{(2n-1)\pi l}{2}$$

$$= \frac{(-1)^{n-1}16}{(2n-1)^3\pi^3}\left[2 + K\left(1 - \cosh\frac{(2n-1)\pi l}{2}\right)\right] \bigg/ A_n \sinh\frac{(2n-1)\pi l}{2}.$$

Introducing B_n into (9.128), we have

$$\zeta_n(Z) = \left\{\frac{(-1)^{n-1}16}{(2n-1)^3\pi^3}\left[2 + K\left(1 - \cosh\frac{(2n-1)\pi l}{2}\right)\right]\right.$$

$$\bigg/ A_n \sinh\frac{(2n-1)\pi l}{2}\bigg\} \sinh\frac{(2n-1)\pi Z}{2}$$

$$+ \frac{(-1)^n 16K}{(2n-1)^3\pi^3 A_n}\left[1 - \cosh\frac{(2n-1)\pi Z}{2}\right]$$

$$= \frac{1}{A_n}\frac{\sinh\frac{(2n-1)\pi Z}{2}}{\sinh\frac{(2n-1)\pi l}{2}}$$

$$\times \frac{(-1)^{n-1}16}{(2n-1)^3\pi^3}\left[2 + K\left(1 - \cosh\frac{(2n-1)\pi l}{2}\right)\right]$$

$$+ \frac{1}{A_n}\frac{(-1)^n 16K}{(2n-1)^3\pi^3}\left[1 - \cosh\frac{(2n-1)\pi Z}{2}\right].$$

To simplify this formula, let us denote

$$\beta(Z) = 1 - \cosh \frac{(2n-1)\pi Z}{2}$$

and

$$\gamma(Z) = \sinh \frac{(2n-1)\pi Z}{2}.$$

We then have:

$$\zeta_n(Z) = \frac{1}{A_n} \frac{(-1)^{n-1} 16}{(2n-1)^3 \pi^3} \left\{ [2 + K\beta(l)] \frac{\gamma(Z)}{\gamma(l)} - K\beta(Z) \right\}. \quad (9.130)$$

In view of (9.130), the solution (9.118) writes

$$U(Y, Z) = \sum_{n=1}^{\infty} \frac{(-1)^{n-1} 16}{(2n-1)^3 \pi^3} \left\{ [2 + K\beta(l)] \frac{\gamma(Z)}{\gamma(l)} - K\beta(Z) \right\}$$

$$\times \cos \left(\frac{(2n-1)\pi Y}{2} \right).$$

A new notation

$$\delta(Y) = \cos \left(\frac{(2n-1)\pi Y}{2} \right)$$

finally yields

$$U(Y, Z) = \sum_{n=1}^{\infty} \frac{(-1)^{n-1} 16\, \delta(Y)}{(2n-1)^3 \pi^3} \left\{ [2 + K\beta(l)] \frac{\gamma(Z)}{\gamma(l)} - K\beta(Z) \right\}.$$
(9.131)

Our final purpose is to determine $u = u(y, z)$. Let us then go back to the dimensional quantities $u = u_m U$, $y = aY$, $z = aZ$, $\frac{h}{a} = l$, and $k = \frac{u_m}{a^2} K$. Since

$$U(Y, Z) = \sum_{n=1}^{\infty} \frac{(-1)^{n-1} 16}{(2n-1)^3 \pi^3} \cos \left(\frac{(2n-1)\pi Y}{2} \right)$$

$$\times \left\{ \left[2 + K \left(1 - \cosh \frac{(2n-1)\pi l}{2} \right) \right] \frac{\sinh \frac{(2n-1)\pi Z}{2}}{\sinh \frac{(2n-1)\pi l}{2}} \right.$$

$$\left. - K \left(1 - \cosh \frac{(2n-1)\pi Z}{2} \right) \right\},$$

we have

$$u(y,z) = u_m \sum_{n=1}^{\infty} \frac{(-1)^{n-1} 16}{(2n-1)^3 \pi^3} \cos\left(\frac{(2n-1)\pi y}{2a}\right)$$

$$\times \left\{ \left[2 + \frac{ka^2}{u_m}\left(1 - \cosh\frac{(2n-1)\pi h}{2a}\right)\right] \frac{\sinh\frac{(2n-1)\pi z}{2a}}{\sinh\frac{(2n-1)\pi h}{2a}} \right.$$

$$\left. - \frac{ka^2}{u_m}\left(1 - \cosh\frac{(2n-1)\pi z}{2a}\right) \right\},$$

or, still,

$$u(y,z) = u_m \sum_{n=1}^{\infty} \frac{(-1)^{n-1} 16}{(2n-1)^3 \pi^3} \cos\left(\frac{(2n-1)\pi y}{2a}\right)$$

$$\times \left\{ \left[2 - \frac{\rho g a^2 \sin\alpha}{\mu u_m}\left(1 - \cosh\frac{(2n-1)\pi h}{2a}\right)\right] \frac{\sinh\frac{(2n-1)\pi z}{2a}}{\sinh\frac{(2n-1)\pi h}{2a}} \right.$$

$$\left. + \frac{\rho g a^2 \sin\alpha}{\mu u_m}\left(1 - \cosh\frac{(2n-1)\pi z}{2a}\right) \right\}. \quad (9.132)$$

Taking $a = h = 1\,m$, $\mu = \rho \sin\alpha$, $u_m = 1\,m \cdot s^{-1}$ and $g \approx 10\,m \cdot s^{-2}$, we finally obtain

$$u(y,z) = \sum_{n=1}^{\infty} \frac{(-1)^{n-1} 32}{(2n-1)^3 \pi^3} \cos\left(\frac{(2n-1)\pi y}{2}\right)$$

$$\times \left\{ \left[1 - 5\left(1 - \cosh\frac{(2n-1)\pi}{2}\right)\right] \frac{\sinh\frac{(2n-1)\pi z}{2}}{\sinh\frac{(2n-1)\pi}{2}} \right.$$

$$\left. + 5\left(1 - \cosh\frac{(2n-1)\pi z}{2}\right) \right\}. \quad (9.133)$$

Fig.IX.11 shows the curves of equal velocity, obtained by means of software *Mathematica*, if the following command lines are used:

```
Remove["Global`*"];
Unprotect[In, Out];
Clear[In, Out];

k  = (-1)^(n-1)*32/((2*n-1)^3*Pi^3);
e1 = Cos[((2*n-1)*Pi*y)/2];
e2 = 1-Cosh[((2*n-1)*Pi)/2];
e3 = Sinh[((2*n-1)*Pi*z)/2];
e4 = Sinh[((2*n-1)*Pi)/2];
e5 = 1-Cosh[((2*n-1)*Pi*z)/2];

NSolve[∑_{n=1}^{∞} (k *e1*((1-5*e2)*(e3/e4)+5*e5)) == 0.4, z];
G1 = Plot[Evaluate[ z /. %], {y, 0, 1}, Frame->True, PlotRange->All];
NSolve[∑_{n=1}^{∞} (k *e1*((1-5*e2)*(e3/e4)+5*e5)) == 0.6, z];
G2 = Plot[Evaluate[ z /. %], {y, 0, 1}, Frame->True, PlotRange->All];
NSolve[∑_{n=1}^{∞} (k *e1*((1-5*e2)*(e3/e4)+5*e5)) == 0.8, z];
G3 = Plot[Evaluate[ z /. %], {y, 0, 1}, Frame->True, PlotRange->All];
NSolve[∑_{n=1}^{∞} (k *e1*((1-5*e2)*(e3/e4)+5*e5)) == 1., z];
G4 = Plot[Evaluate[ z /. %], {y, 0, 1}, Frame->True, PlotRange->All];
NSolve[∑_{n=1}^{∞} (k *e1*((1-5*e2)*(e3/e4)+5*e5)) == 1.2, z];
G5 = Plot[Evaluate[ z /. %], {y, 0, 1}, Frame->True, PlotRange->All];
NSolve[∑_{n=1}^{∞} (k *e1*((1-5*e2)*(e3/e4)+5*e5)) == 1.4, z];
G6 = Plot[Evaluate[ z /. %], {y, 0, 1}, Frame->True, PlotRange->All];
NSolve[∑_{n=1}^{∞} (k *e1*((1-5*e2)*(e3/e4)+5*e5)) == 1.6, z];
G7 = Plot[Evaluate[ z /. %], {y, 0, 1}, Frame->True, PlotRange->All];
Show[G1, G2, G3, G4, G5, G6, G7]
```

As one can see, the flow velocity of the viscous fluid, under the action of its own gravity, is maximum at the depth of about 0.35 m.

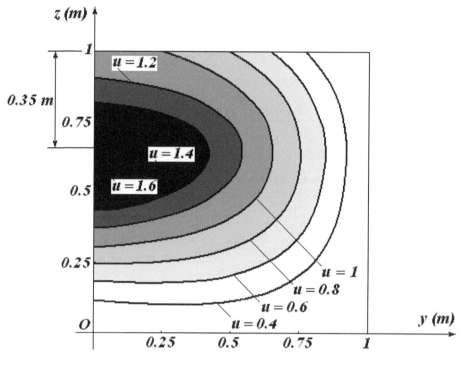

Fig.IX.11

C. Problems of Magnetofluid Dynamics and Quantum Mechanics

Hamilton's Canonical Equations for Continuous Systems

The purpose of this chapter is to write Hamilton's canonical equations for continuous systems (fields, fluids, plasma) and show how this formalism can be applied in Magnetohydrodynamics. Recalling definition of the Hamiltonian for discrete systems

$$H(q,p,t) = \sum_{i=1}^{n} p_i \dot{q}_i - L(q,\dot{q},t),$$

where q_i ($i = \overline{1,n}$) are the generalized coordinates, $\dot{q}_i$ ($i = \overline{1,n}$) the generalized velocities, and p_i ($i = \overline{1,n}$) the generalized momenta, let us define the *Hamiltonian density*, that is the Hamiltonian per unit volume, as

$$\mathcal{H} = \sum_{s=1}^{h} \pi_{(s)} \varphi^{(s)}_{,t} - \mathcal{L}, \qquad (9.134)$$

where $\mathcal{L}$ is the Lagrangian density, $\varphi^{(s)}_{,t}$ are the partial derivatives with respect to time of the variational parameters $\varphi^{(s)}$, and $\pi_{(s)}$ stand for the *momentum densities* associated with $\varphi^{(s)}$

$$\pi_{(s)} = \frac{\partial \mathcal{L}}{\partial \varphi^{(s)}_{,t}}. \tag{9.135}$$

If we choose $x_1 = x$, $x_2 = y$, $x_3 = z$, $x_4 = t$, the functional dependence of the Lagrangian density $\mathcal{L}$ is

$$\mathcal{L} = \mathcal{L}\big[x_i, t, \varphi^{(s)}(x_i, t), \varphi^{(s)}_{,i}, \varphi^{(s)}_{,t}\big] \quad (i = 1, 2, 3;\ s = \overline{1, h}), \tag{9.136}$$

where $\varphi^{(s)}_{,i} = \partial \varphi^{(s)} / \partial x_i$. In this case, we have

$$\mathcal{H} = \mathcal{H}\big[x_i, t, \varphi^{(s)}(x_i, t), \varphi^{(s)}_{,i}, \pi_{(s)}\big] \quad (i = 1, 2, 3;\ s = \overline{1, h}). \tag{9.137}$$

The Hamiltonian of a continuous system then writes

$$H = \int_V \mathcal{H}\big[x_i, t, \varphi^{(s)}, \varphi^{(s)}_{,i}, \pi_{(s)}\big]\, d\tau. \tag{9.138}$$

Recalling that $\varphi^{(s)}$ are continuous and derivable functions of the independent variables x, y, z, t, let us perform an arbitrary variation δH of H for some fixed values of x, y, z ($\delta x_i = 0$). In view of (9.137), and using the summation convention for index $i = 1, 2, 3$, we have

$$\delta H = \int_V \delta \mathcal{H}\, d\tau = \int_V \bigg[\frac{\partial \mathcal{H}}{\partial t}\delta t + \sum_{s=1}^{h} \frac{\partial \mathcal{H}}{\partial \varphi^{(s)}}\delta \varphi^{(s)}$$
$$+ \sum_{s=1}^{h} \frac{\partial \mathcal{H}}{\partial \varphi^{(s)}_{,i}}\delta \varphi^{(s)}_{,i} + \sum_{s=1}^{h} \frac{\partial \mathcal{H}}{\partial \pi_{(s)}}\delta \pi_{(s)}\bigg]\, d\tau. \tag{9.139}$$

The variation δH can be written in an alternative form. Using (9.134) and (9.136), we find:

$$\delta H = \int_V \delta\bigg(\sum_{s=1}^{h} \pi_{(s)} \varphi^{(s)}_{,t} - \mathcal{L}\bigg)\, d\tau = \int_V \bigg[\sum_{s=1}^{h}\Big(\pi_{(s)}\delta \varphi^{(s)}_{,t} + \varphi^{(s)}_{,t}\delta \pi_{(s)}\Big)$$
$$- \frac{\partial \mathcal{L}}{\partial t}\delta t - \sum_{s=1}^{h}\bigg(\frac{\partial \mathcal{L}}{\partial \varphi^{(s)}}\delta \varphi^{(s)} + \frac{\partial \mathcal{L}}{\partial \varphi^{(s)}_{,t}}\delta \varphi^{(s)}_{,t} + \frac{\partial \mathcal{L}}{\partial \varphi^{(s)}_{,i}}\delta \varphi^{(s)}_{,i}\bigg)\bigg]\, d\tau. \tag{9.140}$$

But
$$\int_V \sum_{s=1}^h \frac{\partial \mathcal{L}}{\partial \varphi_{,i}^{(s)}} \delta\varphi_{,i}^{(s)} \, d\tau = \int_V \frac{\partial}{\partial x_i} \Big(\sum_{s=1}^h \frac{\partial \mathcal{L}}{\partial \varphi_{,i}^{(s)}} \delta\varphi^{(s)}\Big) d\tau$$
$$- \int_V \sum_{s=1}^h \frac{\partial}{\partial x_i} \Big(\frac{\partial \mathcal{L}}{\partial \varphi_{,i}^{(s)}}\Big) \delta\varphi^{(s)} \, d\tau. \qquad (9.141)$$

The first integral on the r.h.s can be transformed into a surface integral over the boundary S of the domain of volume V. Since $\varphi^{(s)}$ have fixed values on S, we obtain

$$\int_V \frac{\partial}{\partial x_i} \Big(\sum_{s=1}^h \frac{\partial \mathcal{L}}{\partial \varphi_{,i}^{(s)}} \delta\varphi^{(s)}\Big) \delta\tau = \oint_S \sum_{s=1}^h \frac{\partial \mathcal{L}}{\partial \varphi_{,i}^{(s)}} \delta\varphi^{(s)} dS_i = 0. \qquad (9.142)$$

Introducing these results into (9.140) and performing some reduction of terms, by means of Euler-Lagrange's equations (9.81) we arrive at

$$\delta H = \int_V \Big\{ \sum_{s=1}^h \varphi_{,t}^{(s)} \delta\pi_{(s)} - \frac{\partial \mathcal{L}}{\partial t}\delta t - \sum_{s=1}^h \Big[\frac{\partial \mathcal{L}}{\partial \varphi^{(s)}} - \frac{\partial}{\partial x_i}\Big(\frac{\partial \mathcal{L}}{\partial \varphi_{,i}^{(s)}}\Big)\Big]\delta\varphi^{(s)} \Big\} d\tau$$
$$= \int_V \Big[\sum_{s=1}^h \big(\varphi_{,t}^{(s)} \delta\pi_{(s)} - \pi_{(s),t}\delta\varphi^{(s)}\big) - \frac{\partial \mathcal{L}}{\partial t}\delta t\Big] d\tau. \qquad (9.143)$$

A similar integration by parts can be done in (9.139), which yields

$$\delta H = \int_V \Big\{ \sum_{s=1}^h \frac{\partial \mathcal{H}}{\partial \pi_{(s)}}\delta\pi_{(s)} + \frac{\partial \mathcal{H}}{\partial t}\delta t$$
$$+ \sum_{s=1}^h \Big[\frac{\partial \mathcal{H}}{\partial \varphi^{(s)}} - \frac{\partial}{\partial x_i}\Big(\frac{\partial \mathcal{H}}{\partial \varphi_{,i}^{(s)}}\Big)\Big]\delta\varphi^{(s)} \Big\} d\tau. \qquad (9.144)$$

Equalizing the coefficients of the same arbitrary variations $\delta\varphi^{(s)}$, $\delta\pi_{(s)}$, and δt in (9.143) and (9.144), we obtain the following system of equations

$$\varphi_{,t}^{(s)} = \frac{\partial \mathcal{H}}{\partial \pi_{(s)}}; \quad \pi_{(s),t} = -\frac{\partial \mathcal{H}}{\partial \varphi^{(s)}} + \frac{\partial}{\partial x_i}\Big(\frac{\partial \mathcal{H}}{\partial \varphi_{,i}^{(s)}}\Big) \quad (i=1,2,3;\ s=\overline{1,h}),$$
$$(9.145)$$

as well as the identity
$$\frac{\partial \mathcal{H}}{\partial t} = -\frac{\partial \mathcal{L}}{\partial t}. \tag{9.146}$$

The system of $2h$ partial differential equations (9.145) is analogous to the Hamilton's system of canonical equations (2.40), while the identity (9.146) is also valid for physical systems with a finite number of degrees of freedom.

Equations (9.145) can be written in a symmetric form by making allowance for the notion of *functional derivative*. To do this, we shall calculate the partial derivative of $L = \int_V \mathcal{L} \, d\tau$ with respect to $\varphi^{(s)}(x_j)$ ($j = \overline{1,4}$) for $x_1 = x$, $x_2 = y$, $x_3 = z$ fixed. Let us consider the family of functions $\varphi^{(s)}(x_j, \varepsilon)$, where ε is a parameter chosen in such a way that

$$\varphi^{(s)}(x_j, \varepsilon)|_S = \varphi^{(s)}(x_j, 0) \equiv \varphi^{(s)}(x_j), \tag{9.147}$$

where S is the closed surface bounding the domain of volume V. Consider the derivative

$$\frac{dL}{d\varepsilon} = \int_V \sum_{s=1}^{h} \left[\frac{\partial \mathcal{L}}{\partial \varphi^{(s)}} \frac{\partial \varphi^{(s)}}{\partial \varepsilon} + \frac{\partial \mathcal{L}}{\partial \varphi^{(s)}_{,i}} \frac{\partial \varphi^{(s)}_{,i}}{\partial \varepsilon} \right] d\tau.$$

But

$$\int_V \sum_{s=1}^{h} \frac{\partial \mathcal{L}}{\partial \varphi^{(s)}_{,i}} \frac{\partial \varphi^{(s)}_{,i}}{\partial \varepsilon} d\tau = \int_V \sum_{s=1}^{h} \frac{\partial \mathcal{L}}{\partial \varphi^{(s)}_{,i}} \frac{\partial}{\partial x_i} \left(\frac{\partial \varphi^{(s)}}{\partial \varepsilon} \right) d\tau$$

$$= \int_V \sum_{s=1}^{h} \frac{\partial}{\partial x_i} \left(\frac{\partial \mathcal{L}}{\partial \varphi^{(s)}_{,i}} \frac{\partial \varphi^{(s)}}{\partial \varepsilon} \right) d\tau - \int_V \sum_{s=1}^{h} \frac{\partial}{\partial x_i} \left(\frac{\partial \mathcal{L}}{\partial \varphi^{(s)}_{,i}} \right) \frac{\partial \varphi^{(s)}}{\partial \varepsilon} d\tau$$

$$= \oint_S \sum_{s=1}^{h} \frac{\partial \mathcal{L}}{\partial \varphi^{(s)}_{,i}} \frac{\partial \varphi^{(s)}}{\partial \varepsilon} dS_i - \int_V \sum_{s=1}^{h} \frac{\partial}{\partial x_i} \left(\frac{\partial \mathcal{L}}{\partial \varphi^{(s)}_{,i}} \right) \frac{\partial \varphi^{(s)}}{\partial \varepsilon} d\tau.$$

According to condition (9.147), the surface integral vanishes, and hence

$$\frac{dL}{d\varepsilon} = \int_V \sum_{s=1}^{h} \left[\frac{\partial \mathcal{L}}{\partial \varphi^{(s)}} - \frac{\partial}{\partial x_i} \left(\frac{\partial \mathcal{L}}{\partial \varphi^{(s)}_{,i}} \right) \right] \frac{\partial \varphi^{(s)}}{\partial \varepsilon} d\tau. \tag{9.148}$$

The choice of $\varphi^{(s)}(x_j, \varepsilon)$ tells us that they are equal to $\varphi^{(s)}(x_j)$ everywhere, except for a vicinity Q of the fixed point x, y, z. The same property holds valid for $\partial \varphi^{(s)}/\partial \varepsilon$. At the limit $Q \to 0$, we can write

$$\lim_{Q \to 0} \frac{dL/d\varepsilon}{\partial \varphi^{(s)}/\partial \varepsilon} = \frac{\delta L}{\delta \varphi^{(s)}} = \frac{\partial L}{\partial \varphi^{(s)}} - \frac{\partial}{\partial x_i}\left(\frac{\partial \mathcal{L}}{\partial \varphi^{(s)}_{,i}}\right) \quad (i = \overline{1,4}). \quad (9.149)$$

This expression is called the *functional derivative* or *variational derivative* of the Lagrangian L with respect to the field variables $\varphi^{(s)}$. Using this definition, Euler-Lagrange's equations (9.145) become

$$\frac{\delta L}{\delta \varphi^{(s)}} = 0. \quad (9.150)$$

Observing that $\mathcal{H}$ does not depend on $\varphi^{(s)}_{,t}$ and $\partial \pi_{(s)}/\partial x_i$, the functional derivatives of the Hamiltonian H with respect to $\pi_{(s)}$ and $\varphi^{(s)}$ are

$$\frac{\delta H}{\delta \pi_{(s)}} = \frac{\partial \mathcal{H}}{\partial \pi_{(s)}} \; ; \quad \frac{\delta H}{\delta \varphi^{(s)}} = \frac{\partial \mathcal{H}}{\partial \varphi^{(s)}} - \frac{\partial}{\partial x_j}\left(\frac{\partial \mathcal{H}}{\partial \varphi^{(s)}_{,j}}\right) \quad (j = 1,2,3),$$
$$(9.151)$$

which allows us to write equations (9.145) in a symmetric form

$$\varphi^{(s)}_{,t} = \frac{\delta H}{\delta \pi_{(s)}} \; ; \quad \pi_{(s),t} = -\frac{\delta H}{\delta \varphi^{(s)}}. \quad (9.152)$$

Problem 10

Using the Hamiltonian approach, find the equation of motion of an ideal magnetofluid, performing isentropic motion in an external electromagnetic field. It is assumed that the gravitational field is also taken into account.

Solution

The first step is to write the Lagrangian density of our problem. This should be composed by three terms: one term $\mathcal{L}^f_o$ corresponding to the fluid, one term $\mathcal{L}^{em}_o$ corresponding to the electromagnetic field, and one term $\mathcal{L}_{int}$ expressing the interaction between the two fields:

$$\mathcal{L} = \mathcal{L}^f_o + \mathcal{L}^{em}_o + \mathcal{L}_{int}. \quad (9.153)$$

In this formulation, the usual electromagnetic potentials $\vec{A}$, ϕ [see (9.1) and (9.2)] can be chosen as variational parameters.

But this choice is not unique. As an alternative approach to the variational formalism of our problem, we shall use a different representation of the electromagnetic field, which makes it possible to simplify the Lagrangian density and, consequently, the solution of this application.

Since our model implies the existence of conduction, convection, and displacement current densities, the usual Maxwell's source equations

$$\nabla \times \vec{B} = \mu_0 \vec{j}; \quad \nabla \cdot \vec{E} = \frac{\rho^e}{\varepsilon_0} \qquad (9.154)$$

should be written as

$$\frac{1}{\mu_o} \nabla \times \vec{B} = \vec{j} + \rho^e \vec{v} + \epsilon_o \frac{\partial \vec{E}}{\partial t} \; ; \quad \epsilon_o \, div \, \vec{E} = \rho^e. \qquad (9.155)$$

These equations can be expressed in a symmetric form, similar to that of the source-free Maxwell's equations

$$\nabla \times \vec{E} = -\frac{\partial \vec{B}}{\partial t}; \quad \nabla \cdot \vec{B} = 0 \qquad (9.156)$$

by using the Lagrangian density

$$\mathcal{L}' = \frac{1}{2}\epsilon_o |\vec{E}|^2 - \frac{1}{2\mu_o}|\vec{B}|^2 + \vec{P} \cdot \left(\vec{E} + \vec{v} \times \vec{B} - \frac{1}{\lambda}\vec{j}\right)$$

$$- \vec{M} \cdot \left(curl \, \vec{E} + \frac{\partial \vec{B}}{\partial t}\right) - \psi \, div \, \vec{B}, \qquad (9.157)$$

where the source-free Maxwell's equations (9.156), and Ohm's law

$$\vec{E} + \vec{v} \times \vec{B} = \frac{1}{\lambda}\vec{j} \qquad (9.158)$$

have been used as equations of constraint, while $\vec{P}(\vec{r},t)$, $\vec{M}(\vec{r},t)$, and $\psi(\vec{r},t)$ are Lagrangian multipliers. If we choose as variational parameters E_i, B_i ($i = 1,2,3$), the Euler - Lagrange equations (9.81) yield:

$$\vec{E} = \frac{1}{\epsilon_o}(\nabla \times \vec{M} - \vec{P}), \quad \vec{B} = \mu_o\left(\nabla\psi + \vec{P} \times \vec{v} + \frac{\partial \vec{M}}{\partial t}\right). \qquad (9.159)$$

These relations define the electromagnetic field $\vec{E}$, $\vec{B}$ in terms of the *generalized antipotentials* $\vec{M}$, ψ. As one can see, the appearance of the terms $\vec{P}$ and $\vec{P} \times \vec{v}$ generalizes the usual antipotentials, defined in the case $\vec{j} = 0$, $\rho^e = 0$. Taking the *divergence* of $(9.159)_1$ and the *curl* of $(9.159)_2$, we find the source Maxwell's equation in the symmetric form

$$\nabla \times \left(\frac{1}{\mu_o}\vec{B} - \vec{P} \times \vec{v}\right) = \frac{\partial}{\partial t}(\epsilon_o \vec{E} + \vec{P}),$$

$$\nabla \cdot (\epsilon_o \vec{E} + \vec{P}) = 0. \qquad (9.160)$$

These equations have been first obtained by *Calkin*[1], who called vector field $\vec{P}$ the *'polarization'*. Comparing (9.160) with (9.157), we can write

$$\vec{j} = \vec{v}\nabla \cdot \vec{P} + \nabla \times (\vec{P} \times \vec{v}) + \frac{\partial \vec{P}}{\partial t}, \quad \rho^e = -\nabla \cdot \vec{P}, \qquad (9.161)$$

which satisfy identically the equation of continuity

$$\frac{\partial \rho^e}{\partial t} + \nabla \cdot (\vec{j} + \rho^e \vec{v}) = 0. \qquad (9.162)$$

Like the usual electromagnetic potentials, the antipotentials $\vec{M}$, ψ can be related by a Lorentz-type condition. Introducing (9.159) into Maxwell's source-free equations (9.156), we have

$$\Delta \vec{M} - \epsilon_o \mu_o \frac{\partial^2 \vec{M}}{\partial t^2} = \nabla\left(\nabla \cdot \vec{M} + \epsilon_o \mu_o \frac{\partial \psi}{\partial t}\right)$$

$$+ \epsilon_o \mu_o \frac{\partial}{\partial t}(\vec{P} \times \vec{v}) - \nabla \times \vec{P}.$$

In order that $\vec{M}$ satisfies the homogeneous D'Alembert's wave equation, the following two conditions must be fulfilled:

$$\mathrm{div}\,\vec{M} + \epsilon_o \mu_o \frac{\partial \psi}{\partial t} = 0; \qquad (9.163)$$

$$\epsilon_o \mu_o \frac{\partial}{\partial t}(\vec{P} \times \vec{v}) = \mathrm{curl}\,\vec{P}. \qquad (9.164)$$

Relation (9.163) is the *Lorentz condition* for our antipotentials, and we shall use it as a *constraint* in the Lagrangian density.

The advantage of this representation is that the Lagrangian density of the electromagnetic field

$$\mathcal{L}_o = \frac{1}{2}\epsilon_o|\vec{E}|^2 - \frac{1}{2\mu_o}|\vec{B}|^2, \qquad (9.165)$$

---

[1] M.G.Calkin, *An Action Principle for Magnetohydrodynamics*, Can. J. Phys., 41, 1963, p. 2241.

in which $\vec{E}$, $\vec{B}$ are given in terms of $\vec{M}$, ψ, *includes* the interaction between field and particles. This is possible because the interaction between the electromagnetic field and the point sources has been replaced by an interaction between the electromagnetic and the 'polarization' fields. Using Herivel's Lagrangian density (9.92), we postulate the Lagrangian density for our model to be

$$\mathcal{L} = \frac{1}{2\mu_o}|\vec{B}|^2 - \frac{1}{2}\epsilon_o|\vec{E}|^2 - \frac{1}{2\epsilon_o}\left(\nabla \cdot \vec{M} + \epsilon_o\mu_o\frac{\partial \psi}{\partial t}\right)^2$$

$$+ \frac{1}{2}\rho|\vec{v}|^2 - \rho(\varepsilon + \mathcal{V}^*) + \rho\left(\frac{\partial \alpha}{\partial t} + \vec{v}\cdot\nabla\alpha\right) - \beta\rho\left(\frac{\partial s}{\partial t} + \vec{v}\cdot\nabla s\right). \quad (9.166)$$

Before going further, we need to define the explicit relation between velocity and the 'polarization' fields. Using Euler-Lagrange equations (9.84) with $\varphi^{(s)} \equiv v_i$ ($i = 1, 2, 3$) as variational parameters, we have:

$$\frac{\partial \mathcal{L}}{\partial v_{i,k}} = 0 \; ; \quad \frac{\partial \mathcal{L}}{\partial v_{i,t}} = 0 \; ; \quad \frac{\partial \mathcal{L}}{\partial v_i} = \frac{\partial \mathcal{L}}{\partial B_k}\frac{\partial B_k}{\partial v_i} + \rho v_i + \rho\alpha_{,i} - \beta\rho s_{,i} = 0.$$

Since

$$\frac{\partial \mathcal{L}}{\partial B_k} = \frac{1}{\mu_o}B_k \; ;$$

$$\frac{\partial B_k}{\partial v_i} = \frac{\partial}{\partial v_i}[\mu_o(\psi_{,k} + \epsilon_{kjm}P_j v_m + M_{k,t})] = \mu_o\epsilon_{kjm}\delta_{im}P_j = \mu_o\epsilon_{kji}P_j,$$

we finally find

$$v_j = -\alpha_{,j} + \beta s_{,j} - \frac{1}{\rho}\epsilon_{jkm}B_k P_m, \quad (9.167)$$

or, equivalently

$$\vec{v} = -\nabla\alpha + \beta\nabla s - \frac{1}{\rho}\vec{B}\times\vec{P}, \quad (9.168)$$

which is a *generalized Clebsch transformation*[1].

According to (9.135), the *momentum densities* π_{M_j}, π_{P_j}, π_ψ, π_α, π_s, associated with the field variables M_j, P_j, ψ, α, s, are:

$$\pi_{M_j} = \frac{\partial \mathcal{L}}{\partial M_{j,t}} = B_j \; ; \quad \pi_{P_j} = \frac{\partial \mathcal{L}}{\partial P_{j,t}} = 0 \; ;$$

---

[1] Merches, I., *Variational principle in magnetohydrodynamics*, Phys. Fluids 12 (10), 2225, (1969).

$$\pi_\psi = \frac{\partial \mathcal{L}}{\partial \psi_{,t}} = -\mu_o(M_{i,i} + \epsilon_o\mu_o\psi_{,t}) = 0 ; \qquad (9.169)$$

$$\pi_\alpha = \frac{\partial \mathcal{L}}{\partial \alpha_{,t}} = \rho ; \quad \pi_s = \frac{\partial \mathcal{L}}{\partial s_{,t}} = -\beta\rho.$$

Using (9.134), we now can write the Hamiltonian density $\mathcal{H}$:

$$\mathcal{H} = \sum_{s=1}^{6} \pi_{(s)}\varphi^{(s)}_{,t} - \mathcal{L} = B_j M_{j,t} - \mu_o(M_{j,j} + \epsilon_o\mu_o\psi_{,t})\psi_{,t}+$$

$$\frac{1}{2\epsilon_o}(M_{j,j} + \epsilon_o\mu_o\psi_{,t})^2 + \rho\alpha_{,t} - \beta\rho s_{,t} - \frac{1}{2\mu_o}B_j B_j + \frac{1}{2}\epsilon_o E_j E_j - \frac{1}{2}\rho v_j v_j$$

$$+\rho(\varepsilon + \mathcal{V}^*) - \rho(\alpha_{,t} + v_j\alpha_{,j}) + \beta\rho(s_{,t} + v_j s_{,j}).$$

The form of the Hamiltonian density can be simplified if we observe that, by virtue of (9.167), we can write

$$B_j M_{j,t} = B_j\left[\frac{1}{\mu_o}B_j - (\vec{P}\times\vec{v})_j - \psi_{,j}\right]$$

$$= \frac{1}{\mu_o}B_j B_j + v_j(\rho v_j + \rho\alpha_{,j} - \beta\rho s_{,j}) - B_j\psi_{,j}.$$

Introducing this result into $\mathcal{H}$, we obtain the Hamiltonian density in terms of the field variables, and their partial derivatives with respect to coordinates and time:

$$\mathcal{H} = \frac{1}{2\mu_o}B_j B_j + \frac{1}{2}\epsilon_o E_j E_j - B_j\psi_{,j}$$

$$+\frac{1}{2\epsilon_o}[M_{j,j}M_{k,k} - \epsilon_o^2\mu_o^2(\psi_{,t})^2] + \frac{1}{2}\rho v_j v_j + \rho(\varepsilon + \mathcal{V}^*). \qquad (9.170)$$

In order to apply the Hamiltonian technique, it is necessary to express $\mathcal{H}$ in terms of the field variables M_j, P_j, ψ, α, s, and their conjugate momentum densities π_{M_j}, π_{P_j}, π_ψ, π_α, π_s. Using $(9.159)_1$, and (9.169), we get the Hamiltonian density in the final form:

$$\mathcal{H} = \frac{1}{2\mu_o}\pi_{M_j}\pi_{M_j} + \frac{1}{2\epsilon_o}(\epsilon_{jkm}M_{m,k} - P_j)(\epsilon_{jli}M_{i,l} - P_j) - \pi_{M_j}\psi_{,j}$$

$$-\frac{1}{2\epsilon_o\mu_o}\pi_\psi(M_{j,j} - \epsilon_o\mu_o\psi_{,t}) + \frac{1}{2}\pi_\alpha\left(-\alpha_{,j} - \frac{1}{\pi_\alpha}\pi_s s_{,j} - \frac{1}{\pi_\alpha}\epsilon_{jkm}\pi_{M_k}P_m\right)$$

$$\times \left(-\alpha_{,j} - \frac{1}{\pi_\alpha}\pi_s s_{,j} - \frac{1}{\pi_\alpha}\epsilon_{jli}\pi_{M_l}P_i \right) + \pi_\alpha \varepsilon + \pi_\alpha \mathcal{V}^*, \qquad (9.171)$$

where $\varepsilon = \varepsilon(\pi_\alpha, s)$. Utilizing the calculation following formula (9.96) and applying canonical equations (9.145), we obtain the following system:

$$\pi_{M_j,t} = -\epsilon_{jkm}E_{m,k} \;;$$

$$M_{j,t} = \frac{1}{\mu_o}\pi_{M_j} - \psi_{,j} - \epsilon_{jkm}P_k v_m \;;$$

$$\pi_{P_j,t} = 0 = E_j + \epsilon_{jkm}v_k \pi_{M_m} \;;$$

$$P_{j,t} = 0 \;;$$

$$\pi_{\psi,t} = -\pi_{M_j,t} \;;$$

$$\psi_{,t} = -\frac{1}{2\epsilon_o\mu_o}(M_{j,j} - \epsilon_o\mu_o\psi_{,t}) \;;$$

$$\pi_{\alpha,t} = -(\pi_\alpha v_j)_{,j} \;;$$

$$\alpha_{,t} = -\frac{1}{2}v_j v_j - v_j \alpha_{,j} + \varepsilon + \mathcal{V}^* + \frac{1}{\pi_\alpha}p \;;$$

$$\pi_{s,t} = -T\pi_\alpha - (\pi_s v_j)_{,t} \;;$$

$$s_{,t} = -v_j s_{,j}.$$

Rearranging these equations and using the vector notation, we have:

$$\frac{\partial \vec{B}}{\partial t} = -\nabla \times \vec{E} \;; \quad \nabla \cdot \vec{B} = 0,$$

$$\vec{E} + \vec{v} \times \vec{B} = 0,$$

$$\vec{B} = \mu_o\left(\nabla \psi + \vec{P} \times \vec{v} + \frac{\partial \vec{M}}{\partial t} \right),$$

$$\nabla \cdot \vec{M} + \epsilon_o\mu_o\frac{\partial \psi}{\partial t} = 0,$$

$$\frac{\partial s}{\partial t} + \vec{v} \cdot \nabla s = 0,$$

$$\frac{\partial \rho}{\partial t} + \nabla \cdot (\rho\vec{v}) = 0,$$

$$\frac{\partial \vec{P}}{\partial t} = 0,$$

$$\frac{\partial \beta}{\partial t} + \vec{v} \cdot \nabla \beta = T,$$

$$\frac{1}{2}|\vec{v}|^2 - \mathcal{V}^* - \varepsilon - \frac{p}{\rho} + \frac{\partial \alpha}{\partial t} + \vec{v} \cdot \nabla \alpha = 0.$$

We therefore found the source-free Maxwell's equations (9.156), Ohm's law for infinite conductivity, the field $\vec{B}$ in terms of the antipotentials $\vec{M}$, ψ, the Lorentz condition (9.163), the equation of conservation of entropy (9.90), the equation of continuity (9.89), as well as the equations (9.94) and (9.95), which we also obtained in the case of uncharged fluids.

Recalling that our final purpose is to obtain the equation of motion, the next step consists in eliminating the Lagrangian multipliers from the equations

$$\vec{v} + \nabla \alpha - \beta \nabla s + \frac{1}{\rho}\vec{B} \times \vec{P} = 0,$$

$$\frac{1}{2}|\vec{v}|^2 - \epsilon - \mathcal{V}^* - \frac{p}{\rho} + \frac{\partial \alpha}{\partial t} + \vec{v} \cdot \nabla \alpha = 0, \qquad (9.172)$$

$$\frac{\partial \beta}{\partial t} + \vec{v} \cdot \nabla \beta = T.$$

Extracting $\nabla \alpha$ from $(9.172)_1$ and introducing this expression into $(9.172)_2$, then taking the *gradient* of the result, we have

$$\nabla(\beta \vec{v} \cdot \nabla s) + \frac{\partial}{\partial t}(\nabla \alpha) - \frac{1}{2}\nabla |\vec{v}|^2 = \nabla\left(\varepsilon + \mathcal{V}^* + \frac{p}{\rho} + \frac{1}{\rho}\vec{v} \cdot \vec{B} \times \vec{P}\right),$$

or, by using again $(9.172)_1$

$$\frac{\partial}{\partial t}(\beta \nabla s) - \frac{\partial \vec{v}}{\partial t} - \frac{\partial}{\partial t}\left(\frac{1}{\rho}\vec{B} \times \vec{P}\right) - \frac{1}{2}\nabla|\vec{v}|^2 + \nabla(\beta \vec{v} \cdot \nabla s)$$

$$= \nabla\left(\varepsilon + \mathcal{V}^* + \frac{p}{\rho} + \frac{1}{\rho}\vec{v} \cdot \vec{B} \times \vec{P}\right).$$

Using the vector formula

$$\nabla(\vec{A} \cdot \vec{B}) = \vec{A} \times (\nabla \times \vec{B}) + \vec{B} \times (\nabla \times \vec{A})$$

$$+ (\vec{A} \cdot \nabla)\vec{B} + (\vec{B} \cdot \nabla)\vec{A},$$

and the relation (9.163), we still have

$$\frac{\partial \vec{v}}{\partial t} + (\vec{v}\cdot\nabla)\vec{v} = -\nabla\mathcal{V}^* - \frac{1}{\rho}\nabla p + \frac{\partial}{\partial t}\left(\frac{1}{\rho}\vec{B}\times\vec{P}\right)$$

$$+\vec{v}\times\left[\nabla\times\left(\frac{1}{\rho}\vec{B}\times\vec{P}\right)\right] + \nabla\left(\frac{1}{\rho}\vec{v}\cdot\vec{B}\times\vec{P}\right). \qquad (9.173)$$

If $\vec{a}$, $\vec{b}$, $\vec{c}$ are any three vector fields, it is not difficult to prove the following vector identity:

$$\vec{a}\times[\nabla\times(\vec{b}\times\vec{c})] + \vec{b}\times[\nabla\times(\vec{c}\times\vec{a})] + \vec{c}\times[\nabla\times(\vec{a}\times\vec{b})]$$
$$= (\vec{b}\times\vec{c})\nabla\cdot\vec{a} + (\vec{c}\times\vec{a})\nabla\cdot\vec{b} + (\vec{a}\times\vec{b})\nabla\cdot\vec{c} + \nabla(\vec{c}\cdot\vec{a}\times\vec{b}). \qquad (9.174)$$

Multiplying this relation by $\frac{1}{\rho}$ and performing some simple calculations, we arrive at

$$\nabla\left(\frac{1}{\rho}\vec{c}\cdot\vec{a}\times\vec{b}\right) - (\vec{c}\cdot\vec{a}\times\vec{b})\nabla\frac{1}{\rho} + \left[\nabla\times\left(\frac{1}{\rho}\vec{a}\times\vec{b}\right)\right]\times\vec{c}$$

$$-\left[\nabla\left(\frac{1}{\rho}\right)\times(\vec{a}\times\vec{b})\right]\times\vec{c} + (\vec{a}\times\vec{b})\frac{1}{\rho}\nabla\cdot\vec{c}$$

$$= \frac{1}{\rho}[\nabla\times(\vec{a}\times\vec{c}) + \vec{c}\,\nabla\cdot\vec{a}]\times\vec{b} + \frac{1}{\rho}\vec{a}\times[\nabla\times(\vec{b}\times\vec{c}) + \vec{c}\,\nabla\cdot\vec{b}].$$

If in this expression we put $\vec{a} = \vec{P}$, $\vec{b} = \vec{B}$, $\vec{c} = \vec{v}$ and replace the term $\frac{1}{\rho}\nabla\cdot\vec{v}$ by means of the equation of continuity, we still have

$$\nabla\left(\frac{1}{\rho}\vec{v}\cdot\vec{P}\times\vec{B}\right) + \vec{v}\times\left[\nabla\times\left(\frac{1}{\rho}\vec{B}\times\vec{P}\right)\right] + \frac{\partial}{\partial t}\left(\frac{1}{\rho}\vec{P}\times\vec{B}\right)$$

$$+\vec{v}\times\left[(\vec{P}\times\vec{B})\times\nabla\left(\frac{1}{\rho}\right)\right] - \left[\nabla\left(\frac{1}{\rho}\right)\times(\vec{P}\times\vec{B})\right]\times\vec{v}$$

$$= \frac{1}{\rho}\frac{\partial}{\partial t}(\vec{P}\times\vec{B}) + \frac{1}{\rho}[\nabla\times(\vec{P}\times\vec{v}) + \vec{v}\,\nabla\cdot\vec{P}]\times\vec{B} + \frac{1}{\rho}\vec{P}\times[\nabla\times(\vec{B}\times\vec{v}) + \vec{v}\,\nabla\cdot\vec{B}],$$

or, after some reduction of terms

$$\nabla\left(\frac{1}{\rho}\vec{v}\cdot\vec{P}\times\vec{B}\right) + \vec{v}\times\left[\nabla\times\left(\frac{1}{\rho}\vec{B}\times\vec{P}\right)\right] + \frac{\partial}{\partial t}\left(\frac{1}{\rho}\vec{P}\times\vec{B}\right)$$

$$= \frac{1}{\rho}\left[\frac{\partial\vec{P}}{\partial t} + \nabla\times(\vec{P}\times\vec{v}) + \vec{v}\,\nabla\cdot\vec{P}\right]\times\vec{B}$$

$$+\frac{1}{\rho}\vec{P}\times\left[\frac{\partial\vec{B}}{\partial t}+\nabla\times(\vec{B}\times\vec{v})+\vec{v}\,\nabla\cdot\vec{B}\right].$$

In view of Maxwell's equation (9.156)$_2$, the induction equation for infinite conductivity (9.156)$_1$, and the relation (9.161)$_1$, the r.h.s. of the last equation reduces to $\frac{1}{\rho}\vec{j}\times\vec{B}$. Introducing this result into (9.173), we finally arrive at

$$\rho\left[\frac{\partial\vec{v}}{\partial t}+(\vec{v}\cdot\nabla)\,\vec{v}\right]=\vec{j}\times\vec{B}-\nabla p+\rho\vec{F}',$$

which is the expected equation of motion. Here $\vec{F}'=-\mathrm{grad}\,\mathcal{V}^*$ is the external (*e.g.* gravitational) mass-specific force (see Problem 6).

Problem 11

Using the analytical formalism, derive time-dependent Schrödinger equation for a microparticle of mass m, described by the wave function $\psi(\vec{r},t)$, using the following two procedures:

a) The Lagrangian approach for continuous systems, by means of the Lagrangian density

$$\mathcal{L}=\frac{\hbar^2}{2m}\nabla\psi\cdot\nabla\psi^*+V\psi\psi^*+\frac{\hbar}{2i}(\psi^*\psi_{,t}-\psi\psi^*_{,t}); \quad (9.175)$$

b) The Hamiltonian method for continuous systems.

Solution

a) In this case, Schrödinger equation for the microparticle of mass m, described by the wave function ψ, follows from the Euler-Lagrange equation

$$\frac{\partial\mathcal{L}}{\partial\psi^*}-\frac{\partial}{\partial x_i}\left(\frac{\partial\mathcal{L}}{\partial\psi^*_{,i}}\right)-\frac{\partial}{\partial t}\left(\frac{\partial\mathcal{L}}{\partial\psi^*_{,t}}\right)=0 \quad (i=1,2,3),$$

or, using the formal vector formulation for spacial derivatives

$$\frac{\partial\mathcal{L}}{\partial\psi^*}-\nabla\left(\frac{\partial\mathcal{L}}{\partial(\nabla\psi^*)}\right)-\frac{\partial}{\partial t}\left(\frac{\partial\mathcal{L}}{\partial\psi^*_{,t}}\right)=0. \quad (9.176)$$

Using (9.175), we have:

$$\frac{\partial\mathcal{L}}{\partial\psi^*}=V\psi+\frac{\hbar}{2i}\psi_{,t};$$

$$\frac{\partial \mathcal{L}}{\partial (\nabla \psi^*)} = \frac{\hbar^2}{2m} \nabla \psi;$$

$$\nabla \left(\frac{\partial \mathcal{L}}{\partial (\nabla \psi^*)} \right) = \frac{\hbar^2}{2m} \nabla \cdot \nabla \psi = \frac{\hbar^2}{2m} \Delta \psi;$$

$$\frac{\partial \mathcal{L}}{\partial \psi^*_{,t}} = -\frac{\hbar}{2i} \psi;$$

$$\frac{\partial}{\partial t} \left(\frac{\partial \mathcal{L}}{\partial \psi^*_{,t}} \right) = -\frac{\hbar}{2i} \psi_{,t}.$$

Introducing these results into (9.176), we obtain

$$V\psi + \frac{\hbar}{2i} \frac{\partial \psi}{\partial t} - \frac{\hbar^2}{2m} \Delta \psi + \frac{\hbar}{2i} \frac{\partial \psi}{\partial t} = 0,$$

or

$$i\hbar \frac{\partial \psi}{\partial t} = -\frac{\hbar^2}{2m} \Delta \psi + V\psi, \qquad (9.177)$$

which is the equation we were looking for. In operatorial form, equation (9.177) can be written as

$$\hat{E}\psi = \hat{H}\psi, \qquad (9.178)$$

where

$$\hat{E} = i\hbar \frac{\partial}{\partial t} \qquad (9.179)$$

is the *energy operator*, and

$$\hat{H} = -\frac{\hbar^2}{2m} \Delta + V \qquad (9.180)$$

is the (Hermitian) Hamiltonian operator for our physical system: a particle of mass m moving in the field with potential energy V.

As one can see, Schrödinger equation in its general form

$$i\hbar \frac{\partial}{\partial t} \psi(\vec{r}, t) = \hat{H} \psi(\vec{r}, t)$$

is closely related to the Hamilton-Jacobi equation

$$\frac{\partial}{\partial t} S(q_i, t) = H \left(q_i, \frac{\partial S}{\partial q_i}, t \right),$$

where S is the action, and H the Hamilton function. Here the generalized coordinates q_i ($i = 1, 2, 3$), used in the Hamilton-Jacobi formalism, can be set to the position in Cartesian coordinates $\vec{r} = \vec{r}(x, y, z)$. Substituting $\psi = \sqrt{\rho(\vec{r}, t)} \exp iS(\vec{r}, t)/\hbar$, where ρ is the probability density, into Schrödinger equation, and then taking the limit $\hbar \to 0$, we get the Hamilton-Jacobi equation. We therefore can conclude that the motion of the particle of mass m, described by Schrödinger equation, in the limit $\hbar \to 0$ is governed by Hamilton-Jacobi equation.

Using a similar procedure, we find Schrödinger equation for the wave function ψ^*, which is

$$-i\hbar \frac{\partial \psi^*}{\partial t} = -\frac{\hbar^2}{2m} \Delta \psi^* + V \psi^*, \qquad (9.181)$$

or

$$\hat{E}^* \psi^* = \hat{H}^* \psi^* = \hat{H} \psi^*, \qquad (9.182)$$

where $\hat{E}^* = -i\hbar \frac{\partial}{\partial t}$, and $\hat{H}^* = \hat{H}$.

b) The Hamiltonian density for a continuous system is defined by [see (9.134)]:

$$\mathcal{H} = \sum_{s=1}^{h} \pi_{(s)} \varphi_{,t}^{(s)} - \mathcal{L}, \qquad (9.183)$$

where

$$\pi_{(s)} = \frac{\partial \mathcal{L}}{\partial \varphi_{,t}^{(s)}}$$

are the generalized momentum densities. In this case, Hamilton's canonical equations [see (9.152)] write

$$\varphi_{,t}^{(s)} = \frac{\delta H}{\delta \pi_{(s)}} \; ; \quad \pi_{(s),t} = -\frac{\delta H}{\delta \varphi^{(s)}}, \qquad (9.184)$$

where

$$\frac{\delta}{\delta \varphi^{(s)}} = \frac{\partial}{\partial \varphi^{(s)}} - \frac{\partial}{\partial x_j}\left(\frac{\partial}{\partial \varphi_{,j}^{(s)}}\right) \quad (j = 1, 2, 3), \qquad (9.185)$$

is the functional (or variational) derivative with respect to the field variables $\varphi^{(s)}$.

By choosing ψ^* as field variable, we have:

$$\pi_{\psi^*} = \frac{\partial \mathcal{L}}{\partial \psi_{,t}^*} = -\frac{\hbar}{2i} \psi, \qquad (9.186)$$

and the Hamiltonian density writes

$$\mathcal{H} = \pi_{\psi^*}\psi^*_{,t} - \mathcal{L} = -\frac{\hbar^2}{2m}\nabla\psi\cdot\nabla\psi^* - V\psi\psi^* - \frac{\hbar}{2i}\psi^*\psi_{,t}$$

$$= -\frac{\hbar^2}{2m}\nabla\psi\cdot\nabla\psi^* + \frac{2i}{\hbar}\psi^*\pi_{\psi^*} - \frac{\hbar}{2i}\psi^*\psi_{,t}, \qquad (9.187)$$

where $\mathcal{L}$ is given by (9.175). Using (9.187), we then have:

$$\frac{\partial\mathcal{H}}{\partial\psi^*} = \frac{2i}{\hbar}V\pi_{\psi^*} - \frac{\hbar}{2i}\psi_{,t},$$

$$\frac{\partial\mathcal{H}}{\partial(\nabla\psi^*)} = -\frac{\hbar^2}{2m}\nabla\psi,$$

and therefore

$$\nabla\left(\frac{\partial\mathcal{H}}{\partial(\nabla\psi^*)}\right) = -\frac{\hbar^2}{2m}\nabla\cdot\nabla\psi = -\frac{\hbar^2}{2m}\Delta\psi.$$

The functional derivative of $\mathcal{H}$ with respect to the field variable ψ^* is

$$\frac{\delta\mathcal{H}}{\delta\psi^*} = \frac{\partial\mathcal{H}}{\partial\psi^*} - \nabla\left(\frac{\partial\mathcal{H}}{\partial(\nabla\psi^*)}\right)$$

$$= \frac{2i}{\hbar}V\pi_{\psi^*} - \frac{\hbar}{2i}\psi_{,t} + \frac{\hbar^2}{2m}\Delta\psi,$$

so that

$$(\pi_{\psi^*})_{,t} = -\frac{\delta\mathcal{H}}{\delta\psi^*} = -\frac{2i}{\hbar}V\pi_{\psi^*} + \frac{\hbar}{2i}\psi_{,t} - \frac{\hbar^2}{2m}\Delta\psi,$$

or, in view of (9.186),

$$-\frac{\hbar}{2i}\psi_{,t} = -\frac{2i}{\hbar}V\left(-\frac{\hbar}{2i}\psi\right) + \frac{\hbar}{2i}\psi_{,t} - \frac{\hbar^2}{2m}\Delta\psi,$$

that is

$$i\hbar\frac{\partial\psi}{\partial t} = -\frac{\hbar^2}{2m}\Delta\psi + V\psi,$$

which is precisely Schrödinger equation.

A similar calculation can be performed if the wave function ψ is taken as field variable. Indeed, we have:

$$\pi_\psi = \frac{\partial \mathcal{L}}{\partial \psi_{,t}} = \frac{\hbar}{2i}\psi^*,$$

so that the Hamiltonian density writes

$$\mathcal{H} = \pi_\psi \psi_{,t} - \mathcal{L} = -\frac{\hbar^2}{2m}\nabla\psi\cdot\nabla\psi^* - V\psi\psi^* + \frac{\hbar}{2i}\psi\psi^*_{,t}$$

$$= -\frac{\hbar^2}{2m}\nabla\psi\cdot\nabla\psi^* - \frac{2i}{\hbar}V\psi\pi_\psi + \frac{\hbar}{2i}\psi\psi^*_{,t}.$$

Thus,

$$\frac{\partial \mathcal{H}}{\partial \psi} = -\frac{2i}{\hbar}V\pi_\psi + \frac{\hbar}{2i}\psi^*_{,t};$$

$$\frac{\partial \mathcal{H}}{\partial(\nabla\psi)} = -\frac{\hbar^2}{2m}\nabla\psi^*;$$

$$\nabla\left(\frac{\partial \mathcal{H}}{\partial(\nabla\psi)}\right) = -\frac{\hbar^2}{2m}\Delta\psi^*.$$

The functional derivative of $\mathcal{H}$ with respect to ψ then is

$$\frac{\delta \mathcal{H}}{\delta \psi} = \frac{\partial \mathcal{H}}{\partial \psi} - \nabla\left(\frac{\partial \mathcal{H}}{\partial(\nabla\psi)}\right)$$

$$= -\frac{2i}{\hbar}V\pi_\psi + \frac{\hbar}{2i}\psi^*_{,t} + \frac{\hbar^2}{2m}\Delta\psi^*,$$

so that

$$(\pi_\psi)_{,t} = -\frac{\delta \mathcal{H}}{\delta \psi} = \frac{2i}{\hbar}V\pi_\psi - \frac{\hbar}{2i}\psi^*_{,t} - \frac{\hbar^2}{2m}\Delta\psi^*,$$

or, in terms of ψ^* and its derivatives,

$$-i\hbar\frac{\partial \psi^*}{\partial t} = -\frac{\hbar^2}{2m}\Delta\psi^* + V\psi^*,$$

which is Schrödinger equation for ψ. In operatorial form, this equation writes

$$\hat{E}^*\psi^* = \hat{H}^*\psi^* = \hat{H}\psi^*,$$

with

$$\hat{E}^* = \left(i\hbar\frac{\partial}{\partial t}\right)^* = -i\hbar\frac{\partial}{\partial t}$$

and

$$\hat{H}^* = \hat{H} = -\frac{\hbar^2}{2m}\Delta + V = \frac{1}{2m}\left(\frac{\hbar}{i}\nabla\right)^2 + V = \frac{\hat{\vec{p}}^{\,2}}{2m} + V,$$

where $\hat{\vec{p}} = \frac{\hbar}{i}\nabla$ is the momentum operator.

Appendix A

GREEK ALPHABET

No	Capital letter	Small letter	Name
1	A	α	alfa
2	B	β	beta
3	Γ	γ	gamma
4	Δ	δ	delta
5	E	ϵ (*variant* : ε)	epsilon
6	Z	ζ	dzeta
7	H	η	eta
8	Θ	θ (*variant* : ϑ)	theta
9	I	ι	iota
10	K	κ	kappa
11	Λ	λ	lambda
12	M	μ	miu
13	N	ν	niu
14	Ξ	ξ	csi
15	O	o	omicron
16	Π	π (*variant* : ϖ)	pi
17	P	ρ (*variant* : ϱ)	rho
18	Σ	σ (*variant* : ς)	sigma
19	T	τ	tau
20	Υ	υ	upsilon
21	Φ	ϕ (*variant* : φ)	fi
22	X	χ	hi
23	Ψ	ψ	psi
24	Ω	ω	omega

Appendix B

ELEMENTS OF TENSOR CALCULUS

B.1. Scalar and vector quantities

The quantities usually met in physics can be classified in terms of their tensor nature: *scalar* = 0th rank (order), *vector* = 1st rank, 2nd rank, etc. tensor quantities[1]. Note the difference between mathematical definition of a *tensor* (scalar, vector, etc.), and *tensor physical quantities*. It is also essential to mention the nature of the space on which a tensor is defined. Our present discussion is carried on the Euclidean three-dimensional space E_3.

Any physical quantity which is completely determined by knowing a (real or complex) number, associated with it and expressed in physical units, is called *scalar physical quantity*. The fundamental property of scalar quantities is that they do not depend on any coordinate system. In this category fall: mass, mechanical work, temperature, entropy, electric charge, etc. These quantities are usually called, simply, *scalars*.

Any quantity which, in addition to its numerical value (*magnitude*, or *modulus*), is characterized by a spatial orientation (direction and sense), is called *vector quantity* or, simply, *vector*. For example, the mechanical force acting on a rigid body is determined by its magnitude, direction and sense. Quantities like: linear velocity, acceleration, momentum, electric field intensity, etc., are also vectors.

The physical reality displays quantities whose determination requires more than three associated real numbers. These quantities are called *tensor quantities* and can be found, for example, in the study of rigid body and continuous deformable media mechanics. Here are some examples: the inertia tensor $I_{\alpha\beta}$, $(\alpha, \beta = 1, 2, 3)$, Cauchy's stress tensor T_{ik} $(i, k = 1, 2, 3)$, the momentum density flow tensor $\Pi_{ik} = \rho v_i v_k + p\delta_{ik}$ $(i, k = 1, 2, 3)$, etc.

---

[1] In various branches of physics we meet even tensors of higher rank. For instance, the elasticity tensor (also called *stiffness* tensor) C_{ijkl} $(i, j, k, l = 1, 2, 3)$ is a fourth rank tensor.

In the Euclidean three-dimensional space E_3, a *vector* is a mathematical entity represented by an oriented segment of straight line, the algebraic components on the axes being the *vector components*. A vector is usually denoted by an arrow over a letter (*e.g.* $\vec{F}$ - the force vector) or a group of two letters (*e.g.* $\overrightarrow{AB}$), etc. In the last example, A is the *origin* (or *starting point*, or *tail*) of the vector, B its *terminal point* (also called *head* or *tip*), the *sense* goes from A to B, while the vector *modulus* is determined by the length of the segment $\|AB\|$. Usually, the modulus of a vector $\vec{A}$ is denoted by $|\vec{A}|$, or, simpler, by A.

A vector with magnitude one (in a certain unit system) is called *unit vector*. In geometry and physics, the *versor* of an axis Δ or of a vector $\vec{A}$ is a unit vector indicating its direction and sense.

In terms of its origin, a vector can be:

1) *free*, if its origin (point of application) is arbitrary (in the laboratory reference frame). Examples: the gravitational acceleration, the magnetic induction of the Earth, etc.

2) *bound*, if its tail is fixed. For example, a force applied to a certain point of an elastic body, or the electric field intensity attached to an electric point-charge, are bound vectors.

3) *sliding*, if its origin can be moved anywhere on its direction. A classical example in this respect is the force acting on a rigid body. The angular velocity $\vec{\omega}$ describing rotation of a rigid body about a fixes axis is also an example of sliding vector.

There are various methods used to describe a vector quantity: descriptive, geometric, numerical, etc. For instance, a free vector is fully characterized by its three projections on axes of a coordinate frame in E_3 or, equivalently, its magnitude and the two angles formed by the vector with anyone of the coordinate axes (the third angle can be determined by means of the other two angles, on condition that the property of acute or obtuse angle is known). To define a sliding vector one must know its magnitude, the two angles between the vector direction and any two axes, as well as two coordinates of the intersection point of the vector direction with any of coordinates planes. In case of a bound vector, one must know six quantities: three coordinates of the origin, and three of the head, or, equivalently, three coordinates of the origin and three projections of the vector on axes.

B.2. Elements of vector algebra

Position of an arbitrary point P in space can be univoquely defined by a vector with its point of application at the origin of a conveniently chosen coordinate system, and head at point P. It is usually called *position vector* or *radius vector*, being denoted by $\vec{r}$. An *axis* is a straight line on which a sense has been chosen, denoted by an arrow. The opposite sense is called *negative*.

Consider an axis (Δ) and an arbitrary vector $\vec{a}$, with origin at A and head at B. The *projection* of $\vec{a}$ on Δ is the vector defined by the orthogonal projections A' and B' of A and B on axis Δ (see Fig.B.1).

Let a_Δ be the algebraic magnitude of vector projection $\overrightarrow{A'B'}$. By definition, the sign of the magnitude is " $+$ " if the axis and the projection have the same sense, and " $-$ " if the senses are opposite. As observed, we can write $a_\Delta = a \cos \alpha$, which mean that projection a_Δ is positive, negative, or zero, in terms of $\alpha < \frac{\pi}{2}$, $> \frac{\pi}{2}$, and $\frac{\pi}{2}$, respectively. We shall use notation

$$a_\Delta = pr_\Delta \vec{a}. \qquad (B.1)$$

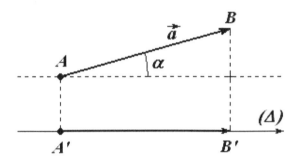

Fig.B.1

If $\vec{a}$ equals the unit vector $\vec{i}$, then (B.1) writes

$$pr_\Delta \vec{a} = \cos \alpha. \qquad (B.2)$$

Let $Oxyz$ be a right-handed orthogonal frame, and $\vec{a}$ an arbitrary vector with origin at P_1 and its terminal point at P_2 (see Fig.B.2). The magnitudes of the vector projections $\vec{a}_x, \vec{a}_y, \vec{a}_z$ on axes are called *components* of the vector $\vec{a}$. According to Fig.B.2, we have

$$a_x = x_2 - x_1, \quad a_y = y_2 - y_1, \quad a_z = z_2 - z_1, \qquad (B.3)$$

where x_1, y_1, z_1 and x_2, y_2, z_2 are the coordinates of the terminal points of $\vec{a}$.

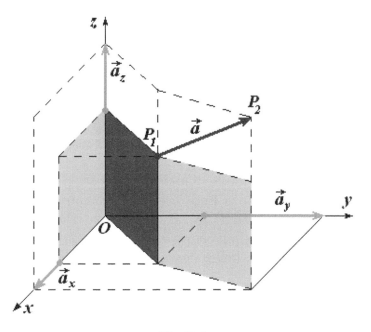

Fig.B.2

Two (or more) vectors with common magnitudes and orientations, but parallel directions, are called *equipolent* (or *equivalent*). The equipolent vectors are determined by the same relations (B.3). The vector projections $\vec{a}_x, \vec{a}_y, \vec{a}_z$ are *the vector components* of $\vec{a}$.

A relation of equality between two vectors can be established only for vectors of the same type. For instance, two sliding vectors $\vec{a}$ and $\vec{b}$ are equal if they have the same direction, sense, and magnitude. This equality is expressed by the vector relation $\vec{a} = \vec{b}$, or, equivalently, by the scalar relations

$$a_x = b_x, \quad a_y = b_y, \quad a_z = b_z,$$

obtained by projecting the vector relation $\vec{a} = \vec{b}$ on coordinate axes. The equality relation has the following properties:
1. Reflexivity: $\vec{a} = \vec{a}$
2. Symmetry: if $\vec{a} = \vec{b}$, then $\vec{b} = \vec{a}$, and reciprocally.
3. Transitivity: if $\vec{a} = \vec{b}$ and $\vec{b} = \vec{c}$, then $\vec{a} = \vec{c}$.

Vectors with the same direction are called *collinear*. If two collinear vectors have opposite senses, they are called *opposite*.

B.2.1. Vector operations

The most common operations with free vectors are:
- addition (subtraction);
- multiplication by a scalar;

- scalar (dot) product;
- vector (cross) product;
- mixed product.

By means of operations of scalar and vector products can also be defined some other vector operations: dot product of a vector product, double vector product, etc.

Vector addition

Consider two free vectors $\vec{a}$ and $\vec{b}$. To draw their sum, we take two vectors equipolent to $\vec{a}$ and $\vec{b}$, with their common origin at an arbitrary point O (see Fig.B.3). By definition, the diagonal of the parallelogram constructed on the two vectors is called *resultant vector* or *vector sum* of $\vec{a}$ and $\vec{b}$, being denoted by $\vec{c} = \vec{a} + \vec{b}$.

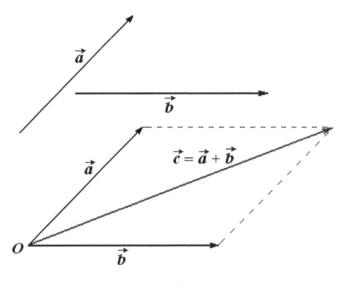

Fig.B.3

The resultant of two vectors can also be drawn by using the *polygon rule*. At the terminal point of one vector one constructs a vector equipollent to the second one: the resultant is then a vector with origin at the origin of the first vector, and head at the head of the second vector. The parallelogram rule can be generalized for n vectors $\vec{a}_1, \vec{a}_2, ..., \vec{a}_n$. In this case, the vector sum $\vec{a} = \vec{a}_1 + \vec{a}_2 + ... + \vec{a}_n = \sum_{i=1}^{n} \vec{a}_i$ is traced by applying $n-1$ times the parallelogram rule, as shown in Fig.B.4. As observed, the vectors $\vec{a}_1, \vec{a}_2, ..., \vec{a}_n$ are placed on sides of a polygonal contour. If the contour is closed, then the resultant is zero.

The following properties of the vector sum are obvious:
- commutativity: $\vec{a} + \vec{b} = \vec{b} + \vec{a}$;

- associativity: $(\vec{a} + \vec{b}) + \vec{c} = \vec{a} + (\vec{b} + \vec{c})$.

The above considerations show that a vector can be fully defined by means of its three projections on axes of a Cartesian coordinate system $Oxyz$. According to Fig.B.2, we have

$$\vec{a} = \vec{a}_x + \vec{a}_y + \vec{a}_z. \qquad (B.4)$$

If $\vec{i}$, $\vec{j}$, $\vec{k}$ are the versors of the axes Ox, Oy, and Oz, respectively, then we can write

$$\vec{a}_x = a_x \vec{i}, \quad \vec{a}_y = a_y \vec{j}, \quad \vec{a}_z = a_z \vec{k},$$

and $\vec{a}$ becomes

$$\vec{a} = a_x \vec{i} + a_y \vec{j} + a_z \vec{k}, \qquad (B.5)$$

which is the *analytical expression* of vector $\vec{a}$. Using notations

$$a_x = a_1, \quad a_y = a_2, \quad a_z = a_3,$$

$$\vec{i} = \vec{u}_1, \quad \vec{j} = \vec{u}_2, \quad \vec{k} = \vec{u}_3,$$

we can also write (B.5) as

$$\vec{a} = a_1 \vec{u}_1 + a_2 \vec{u}_2 + a_3 \vec{u}_3 = \sum_{i=1}^{3} a_i \vec{u}_i. \qquad (B.6)$$

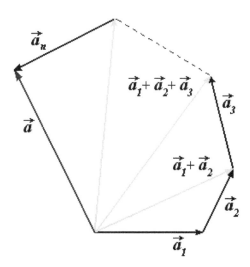

Fig.B.4

Here and hereafter we shall use Einstein's summation convention: when an index is repeated in a term that implies a sum over all possible values for that index, then the summation is implicitly understood and the summation symbol is omitted. This way, relation (B.6) becomes

$$\vec{a} = a_i \vec{u}_i \quad (i = \overline{1,3}). \tag{B.7}$$

The position vector of any point P relative to a Cartesian frame $Oxyz$ can be written according to (B.7). Position of the point P is uniquely determined either by coordinates x, y, z, or by its position vector $\vec{r}$, with origin at O and head at P. In fact, to any point corresponds a position vector $\vec{r}$ with respect to coordinate frame $Oxyz$. Observing that x, y, z are the components of $\vec{r}$ on axes, and making allowance for (B.6), the analytical expression of $\vec{r}$ writes

$$\vec{r} = x\vec{i} + y\vec{j} + z\vec{k} = x_i \vec{u}_i \quad (i = \overline{1,3}). \tag{B.8}$$

Let us consider three vectors $\vec{a}$, $\vec{b}$, $\vec{c}$ with the components

$$\vec{a} = \vec{a}(a_x, a_y, a_z), \quad \vec{b} = \vec{b}(b_x, b_y, b_z), \quad \vec{c} = \vec{c}(c_x, c_y, c_z). \tag{B.9}$$

Using (B.7), we realize that the vector relation $\vec{c} = \vec{a} + \vec{b}$ is equivalent to the following three scalar relations:

$$c_i = a_i + b_i \quad (i = \overline{1,3}). \tag{B.10}$$

Vector subtraction

If $\vec{a}$ and $\vec{b}$ are two free vectors, then relation

$$\vec{d} = \vec{a} - \vec{b}$$

define the operation of *subtraction*. In fact, this operation is a consequence of the vector addition, since the difference $\vec{a} - \vec{b}$ can be written as $\vec{a} + (-\vec{b})$, where $-\vec{b}$ is the opposite of $\vec{b}$. The resultant is obtained by means of the parallelogram rule: the vector difference $\vec{d}$ has its origin at the head of the subtrahend vector $\vec{b}$, and its terminal point at the head of the minuend vector $\vec{a}$. Projecting on axes the vector relation $\vec{d} = \vec{a} - \vec{b}$, we obtain the following three equivalent scalar relations:

$$d_i = a_i - b_i \quad (i = \overline{1,3}). \tag{B.11}$$

Multiplication of a vector by a scalar

Let $\vec{a}$ be a free vector, and λ a real scalar. Their product, denoted by $\lambda\vec{a}$, is a vector collinear to $\vec{a}$, of magnitude $|\lambda\vec{a}| = |\lambda||\vec{a}|$, having the same sense as $\vec{a}$ if $\lambda > 0$, and opposite sense if $\lambda < 0$. If $\lambda = 0$, or $\vec{a} = 0$ (null vector), we have $\lambda\vec{a} = 0$.

The fact that two vectors $\vec{a}$ and $\vec{b}$ are collinear can be expressed as

$$\vec{b} = \lambda\vec{a}, \qquad (B.12)_1$$

or

$$\vec{a} = \mu\vec{b}. \qquad (B.12)_2$$

Multiplication of a vector by a scalar has the following properties:
i) $(\lambda + \mu)\vec{a} = \lambda\vec{a} + \mu\vec{a}$;
ii) $\lambda(\vec{a} + \vec{b}) = \lambda\vec{a} + \lambda\vec{b}$;
iii) $\lambda(\mu\vec{a}) = \mu(\lambda\vec{a}) = \lambda\mu\vec{a}$;
iv) $1 \cdot \vec{a} = \vec{a} \cdot 1 = \vec{a}$.

Consider n vectors $\vec{a}_1, \vec{a}_2, ..., \vec{a}_n$. If there exist n scalars $\lambda_1, \lambda_2, ..., \lambda_n$, not all zero, such that

$$\lambda_1\vec{a}_1 + \lambda_2\vec{a}_2 + ... + \lambda_n\vec{a}_n = 0, \qquad (B.13)$$

then vectors $\vec{a}_1, \vec{a}_2, ..., \vec{a}_n$ are said to be *linearly dependent*. If relation (B.13) holds only for $\lambda_1 = \lambda_2 = ... = \lambda_n = 0$, then $\vec{a}_1, \vec{a}_2, ..., \vec{a}_n$ are *linearly independent*. As an example, consider the linear combination of two non-collinear vectors $\vec{a}_1$ and $\vec{a}_2$:

$$\vec{a}_3 = \lambda_1\vec{a}_1 + \lambda_2\vec{a}_2, \qquad (B.14)$$

where λ_1 and λ_2 are not simultaneously zero. In this case $\vec{a}_3$ is a vector situated in the plane determined by directions of $\vec{a}_1$ and $\vec{a}_2$, which means that $\vec{a}_1$, $\vec{a}_2$, and $\vec{a}_3$ are coplanar. The coplanarity condition (B.14) can also be written as

$$\lambda_1\vec{a}_1 + \lambda_2\vec{a}_2 + \lambda_3\vec{a}_3 = 0, \qquad (B.15)$$

which means that the three vectors are *linearly dependent*. Reciprocally, if three vectors are linearly dependent, then they are coplanar. The vector equation (B.13) shows that vectors $\vec{a}_1, \vec{a}_2, ..., \vec{a}_n$ are linearly independent, ($\lambda_1 = \lambda_2 = ... = \lambda_n = 0$), if the rank of the matrix coefficients equals the number of unknowns $\lambda_1, \lambda_2, ..., \lambda_n$. Therefore, in the Euclidean space E_3 exist at most three linearly independent vectors, which form a basis of this space.

Consider the system of versors $\vec{i}, \vec{j}, \vec{k}$ of axes of the Cartesian system $Oxyz$. These vectors are linearly independent, meaning that they form a basis in E_3, and, consequently, any vector $\vec{a}$ can be written in terms of its components a_x, a_y, a_z and the basis elements $\vec{i}, \vec{j}, \vec{k}$ according to (B.5):
$$\vec{a} = a_x \vec{i} + a_y \vec{j} + a_z \vec{k}.$$

Scalar (dot) product

The *scalar* (or *dot*, or *inner*) *product* of two vectors $\vec{a}$ and $\vec{b}$, denoted $\vec{a} \cdot \vec{b}$, is defined as

$$\vec{a} \cdot \vec{b} = |\vec{a}||\vec{b}| \cos\left(\widehat{\vec{a}, \vec{b}}\right) = a_x b_x + a_y b_y + a_z b_z = a_i b_i \quad (i = \overline{1,3}). \quad (B.16)$$

According to definition (B.16), the dot product is a *scalar*.

Properties:
i) the dot product is commutative: $\vec{a} \cdot \vec{b} = \vec{b} \cdot \vec{a}$;
ii) the dot product is distributive with respect to vector addition:
$$\vec{a} \cdot (\vec{b} + \vec{c}) = \vec{a} \cdot \vec{b} + \vec{a} \cdot \vec{c};$$

iii) if $\vec{a} = \vec{b}$, then $\vec{a} \cdot \vec{a} = |\vec{a}|^2 = a^2$;
iv) for any scalar λ, we have:
$$\vec{a} \cdot (\lambda \vec{b}) = \vec{b} \cdot (\lambda \vec{a}) = \lambda (\vec{a} \cdot \vec{b});$$

v) if $\vec{a} \cdot \vec{b} = 0$, then either the vectors are orthogonal, or at list one of them is null.
vi) if $\vec{u}_1, \vec{u}_2, \vec{u}_3$ are the versors of an orthogonal reference frame, we have:
$$\vec{u}_1 \cdot \vec{u}_2 = \vec{u}_2 \cdot \vec{u}_3 = \vec{u}_3 \cdot \vec{u}_1 = 0,$$

and
$$\vec{u}_1^2 = \vec{u}_2^2 = \vec{u}_3^2 = 1.$$

The last two relations can be written in a condensed form
$$\vec{u}_i \cdot \vec{u}_k = \delta_{ik}, \quad (B.17)$$

where
$$\delta_{ik} = \begin{cases} 1, & i = k, \\ 0, & i \neq k \end{cases} \quad (B.18)$$

is the *Kronecker symbol*.

Using (B.5) and (B.18), one can express the analytic form of the dot product of vectors $\vec{a}(a_x, a_y, a_z)$ and $\vec{b}(b_x, b_y, b_z)$ as

$$\vec{a} \cdot \vec{b} = (a_i \vec{u}_i) \cdot (b_j \vec{u}_j) = a_i b_j \delta_{ij} = a_i b_i \quad (i = \overline{1,3}). \qquad (B.19)$$

In particular, if $\vec{a} = \vec{b}$,

$$\vec{a} \cdot \vec{a} = |\vec{a}|^2 = a^2 = a_i a_i = a_1^2 + a_2^2 + a_3^2. \qquad (B.20)$$

The magnitude of projection of the vector $\vec{a}$ on axis Ox_i ($i = 1, 2, 3$) of a Cartesian frame $Oxyz$ is given by the scalar product

$$(\vec{a})_i = \vec{a} \cdot \vec{u}_i = a_k \vec{u}_k \cdot \vec{u}_i = a_k \delta_{ik} = a_i, \qquad (B.21)$$

where $\vec{u}_i$ is the versor of the x_i-axis.

In physics, the dot product is used, for example, to express the elementary work done by a force $\vec{F}$ which, acting on a body, displaces its tail by quantity $d\vec{r}$: $dL = \vec{F} \cdot d\vec{r}$, or the infinitesimal flow of the magnetic induction $\vec{B}$ through surface $d\vec{S}$: $d\Phi = \vec{B} \cdot d\vec{S}$.

Vector (cross) product

Let $\vec{a}$ and $\vec{b}$ be two free vectors. Their *vector* or *cross product* is a vector denoted $\vec{c} = \vec{a} \times \vec{b}$, with the following properties:
i) its direction is normal to the plane determined by $\vec{a}$ and $\vec{b}$;
ii) its sense is given by the *right-hand screw rule* (or *right-hand rule*);
iii) its magnitude equals the area of the parallelogram that the vectors span (vectors $\vec{a}$ and $\vec{b}$ have the same origin at O - see Fig.B.5)

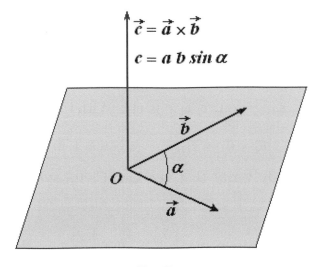

Fig.B.5

By definition, the cross product of vectors $\vec{a}$ and $\vec{b}$ is given by

$$\vec{c} = \vec{a} \times \vec{b} = |\vec{a}||\vec{b}| \sin\left(\widehat{\vec{a},\vec{b}}\right) \vec{u}_c = \varepsilon_{ijk} a_j b_k \vec{u}_i, \qquad (B.22)$$

where $\vec{u}_c$ is the versor of vector $\vec{c}$, and ε_{ijk} ($i,j,k = 1,2,3$) is the completely antisymmetric unit third rank pseudotensor (the Levi-Civita symbol).

The cross product of two vectors has the following properties:
i) it is *anticommutative*: $\vec{a} \times \vec{b} = -\vec{b} \times \vec{a}$;
ii) if λ is a scalar, then: $(\lambda \vec{a}) \times \vec{b} = \vec{a} \times (\lambda \vec{b})$;
iii) it is *distributive* with respect to vector addition:

$$\vec{a} \times (\vec{b} + \vec{c}) = \vec{a} \times \vec{b} + \vec{a} \times \vec{c};$$

iv) if one of the vectors is null, or the two vectors are collinear, the cross product is also null. In the last case we have:

$$\vec{a} \times \vec{b} = \vec{a} \times (\lambda \vec{a}) = \lambda(\vec{a} \times \vec{a}) = 0;$$

v) the versors $\vec{u}_1$, $\vec{u}_2$, and $\vec{u}_3$ of an orthogonal reference frame obey the following rules:

$$\begin{cases} \vec{u}_1 \times \vec{u}_1 = 0, & \vec{u}_2 \times \vec{u}_2 = 0, & \vec{u}_3 \times \vec{u}_3 = 0; \\ \vec{u}_1 \times \vec{u}_2 = \vec{u}_3, & \vec{u}_2 \times \vec{u}_3 = \vec{u}_1, & \vec{u}_3 \times \vec{u}_1 = \vec{u}_2. \end{cases} \qquad (B.23)$$

These relations can be written in a condensed form by means of the *Levi-Civita symbol*, defined as

$$\varepsilon_{ijk} = \begin{cases} +1, & \text{if } i,j \text{ and } k \text{ are an even permutation of 1,2, and 3;} \\ -1, & \text{if } i,j \text{ and } k \text{ are an odd permutation of 1,2, and 3;} \\ 0, & \text{if any of the indices are equal,} \end{cases}$$

e.g. $\varepsilon_{123} = +1$, $\varepsilon_{312} = -1$, $\varepsilon_{233} = 0$, etc. With this notation, relations (B.23) write

$$\vec{u}_i \times \vec{u}_j = \varepsilon_{ijk} \vec{u}_k \quad (i,j,k = \overline{1,3}). \qquad (B.24)$$

One can easily verify that the Levi-Civita symbol satisfies the following relation:

$$\varepsilon_{ijk} \varepsilon_{pqs} = \begin{vmatrix} \delta_{ip} & \delta_{iq} & \delta_{is} \\ \delta_{jp} & \delta_{jq} & \delta_{js} \\ \delta_{kp} & \delta_{kq} & \delta_{ks} \end{vmatrix}. \qquad (B.25)$$

Setting $i = p$ and summing over this index, we have:

$$\varepsilon_{ijk}\varepsilon_{iqs} = \delta_{jq}\delta_{ks} - \delta_{js}\delta_{kq},$$

where all indices go from 1 to 3. Making $j = q$ in the last relation and summing again, we obtain

$$\varepsilon_{ijk}\varepsilon_{ijs} = \delta_{jj}\delta_{ks} - \delta_{js}\delta_{kj} = 3\delta_{ks} - \delta_{ks} = 2\delta_{ks}.$$

By virtue of this property, let us multiply (B.24) by ε_{ijs} and perform summation over indices i and j. The result is

$$\varepsilon_{ijs}\vec{u}_i \times \vec{u}_j = \varepsilon_{ijs}\varepsilon_{ijk}\vec{u}_k = 2\delta_{sk}\vec{u}_k = 2\vec{u}_s,$$

which yields

$$\vec{u}_s = \frac{1}{2}\varepsilon_{sij}\vec{u}_i \times \vec{u}_j \quad (i,j,s = \overline{1,3}).$$

The analytic form of the cross product, in view of (B.24), therefore is:

$$\vec{a} \times \vec{b} = (a_i\vec{u}_i) \times (b_j\vec{u}_j) = \varepsilon_{ijk}a_ib_j\vec{u}_k \quad (i,j,s = \overline{1,3}). \qquad (B.26)$$

Another modality to express the cross product of vectors $\vec{a}$ and $\vec{b}$ is to write (B.26) as follows:

$$\vec{a} \times \vec{b} = \varepsilon_{ijk}a_ib_j\vec{u}_k$$

$$= (a_yb_z - a_zb_y)\vec{i} + (a_zb_x - a_xb_z)\vec{j} + (a_xb_y - a_yb_x)\vec{k}$$

$$= \begin{vmatrix} \vec{i} & \vec{j} & \vec{k} \\ a_x & a_y & a_z \\ b_x & b_y & b_z \end{vmatrix}, \qquad (B.27)$$

which means that the cross product can be expressed as a third order symbolic determinant.

Projecting the cross product $\vec{a} \times \vec{b}$ onto direction $\vec{u}_p$, i.e. performing the dot product $\vec{a} \times \vec{b} \cdot \vec{u}_p$, we have

$$\left(\vec{a} \times \vec{b}\right)_p = \vec{a} \times \vec{b} \cdot \vec{u}_p = \varepsilon_{ijk}a_ib_j\vec{u}_k \cdot \vec{u}_p$$

$$= \varepsilon_{ijk}a_ib_j\delta_{kp} = \varepsilon_{ijp}a_ib_j = \varepsilon_{pij}a_ib_j.$$

If the vectors $\vec{a}$ and $\vec{b}$ are collinear, the determinant (B.27) is null, because the second and third rows of the determinant have proportional elements.

In physics, the cross product is used to express, for example, the following quantities:

- Moment of a force $\vec{F}$ with respect to a pole O: $\vec{M} = \vec{r} \times \vec{F}$, where $\vec{r}$ is "connecting" the pole O with the point of application of $\vec{F}$;

- Vortex vector[1] $\vec{\Omega} = \frac{1}{2}\nabla \times \vec{v} = \frac{1}{2}\operatorname{curl}\vec{v}$ in fluid mechanics, where $\vec{v}$ is the velocity field of the fluid;

- Magnetic induction $\vec{B}$ of a linear current, $\vec{B} = \frac{\mu}{4\pi}\oint_{(\Gamma)} \frac{I d\vec{l} \times \vec{r}}{r^3}$.

where μ is the magnetic permeability of the medium in which the electric circuit is displaced, $\vec{dl}$ is an arc/line element whose sense is given by the electric current of intensity I of the linear electric circuit Γ, and $\vec{r}$ is the vector "connecting" the origin of $\vec{dl}$ with the point where the induction $\vec{B}$ is determined;

- Energy flux density of an electromagnetic field, *i.e. Poynting's vector* $\vec{S} = \vec{E} \times \vec{H}$, where $\vec{E}$ and $\vec{H}$ are the electric and magnetic components of the electromagnetic field, respectively.

Mixed product of three vectors

Let $\vec{a}$, $\vec{b}$ and $\vec{c}$ be three free vectors, with their origin at O. The quantity

$$\vec{a} \cdot (\vec{b} \times \vec{c}) = a_i \varepsilon_{ijk} b_j c_k$$

$$= (a_i \vec{u}_i) \cdot (b_j \vec{u}_j \times c_k \vec{u}_k)$$

$$= a_i b_j c_k \vec{u}_i \cdot (\vec{u}_j \times \vec{u}_k)$$

$$= \begin{vmatrix} a_x & a_y & a_z \\ b_x & b_y & b_z \\ c_x & c_y & c_z \end{vmatrix}, \qquad (B.28)$$

where[2]

$$\vec{u}_i \cdot (\vec{u}_j \times \vec{u}_k) = \varepsilon_{ijk},$$

is called *the mixed product* of the vectors $\vec{a}, \vec{b}$, and $\vec{c}$.

---

[1] In fact, $\vec{\Omega}$ is not a polar vector, but a pseudovector.

[2] This way, we can give a different definition to the Levi-Civita symbol, as being the mixed product of versors of an orthogonal frame on the Euclidean space E_3: $\varepsilon_{ijk} = \vec{u}_i \cdot (\vec{u}_j \times \vec{u}_k)$, as the Kronecker symbol δ_{ij} is given by the dot product $\delta_{ij} = \vec{u}_i \cdot \vec{u}_j$.

According to (B.28), the properties of the mixed product are closely connected to the properties of determinants: the mixed product is null if two rows of the determinant are proportional, and changes its sign if two rows are inverted. Also, a cyclic permutation of vectors (rows in the determinant) leaves the mixed product unchanged:

$$\vec{a} \cdot (\vec{b} \times \vec{c}) = \vec{b} \cdot (\vec{c} \times \vec{a}) = \vec{c} \cdot (\vec{a} \times \vec{b}).$$

In physics, the mixed product of three vectors is used - for example - to express the moment of a force $\vec{F}$ about an axis Δ:

$$M_\Delta = \vec{u}_\Delta \cdot (\vec{r} \times \vec{F}) = \vec{u}_\Delta \cdot \vec{M}_O,$$

where $\vec{u}_\Delta$ is the versor of axis Δ and $\vec{M}_O$ the moment of $\vec{F}$ about an arbitrary point (pole) O on the axis. It can be shown that the choice of the point O is arbitrary.

Mixed product of four vectors
(Dot product of cross products)

The mixed product of four vectors can be obtained from the mixed product of three vectors, by replacing the vector under dot product with a cross product, such as: $(\vec{a} \times \vec{b}) \cdot (\vec{c} \times \vec{d})$. Using analytical method, we have:

$$(\vec{a} \times \vec{b}) \cdot (\vec{c} \times \vec{d}) = (\vec{a} \times \vec{b})_i (\vec{c} \times \vec{d})_i = \varepsilon_{ijk} a_j b_k \varepsilon_{ipq} c_p d_q$$

$$= (\delta_{jp}\delta_{kq} - \delta_{jq}\delta_{kp}) a_j b_k c_p d_q = a_j c_j b_k d_k - a_j d_j b_k c_k$$

$$= (\vec{a} \cdot \vec{c})(\vec{b} \cdot \vec{d}) - (\vec{a} \cdot \vec{d})(\vec{b} \cdot \vec{c}). \qquad (B.29)$$

Double cross product

Consider three free vectors $\vec{a}, \vec{b}$ and $\vec{c}$, with their common origin at some point O. The double cross product of these vectors is defined as

$$\vec{d} = \vec{a} \times (\vec{b} \times \vec{c}). \qquad (B.30)$$

According to our previous considerations one observes that, from the geometric point of view, the vector $\vec{d}$ is orthogonal to both vectors $\vec{a}$ and $(\vec{b} \times \vec{c})$, therefore is coplanar with $\vec{b}$ and $\vec{c}$. Using analytical approach, we have:

$$\vec{a} \times (\vec{b} \times \vec{c}) = \varepsilon_{ijk} a_j (\vec{b} \times \vec{c})_k \vec{u}_i = \varepsilon_{ijk} \varepsilon_{klm} a_j b_l c_m \vec{u}_i$$

$$= (\delta_{il}\delta_{jm} - \delta_{im}\delta_{jl}) a_j b_l c_m \vec{u}_i = a_j c_j (b_i \vec{u}_i) - a_j b_j (c_i \vec{u}_i)$$

$$= (\vec{a} \cdot \vec{c})\vec{b} - (\vec{a} \cdot \vec{b})\vec{c}. \qquad (B.31)$$

In physics, the double cross product is used, for example, in order to express the macroscopic force of interaction between two linear electric circuits:

$$\vec{F}_{12} = -\vec{F}_{21} = \frac{\mu}{4\pi} \oint_{\Gamma_1} \oint_{\Gamma_2} \frac{\vec{dl_2} \times (\vec{dl_1} \times \vec{r}_{12})}{r_{12}^3},$$

where μ is the magnetic permeability of the medium where the electric circuits are placed, I_1 and I_2 are the current intensities, $\vec{dl_1}$ and $\vec{dl_2}$ are two vector arc elements of circuits Γ_1 and Γ_2, respectively, while $\vec{r}_{12} = -\vec{r}_{21}$ is connecting the tails of $\vec{dl_1}$ and $\vec{dl_2}$.

Lagrange's identity

Squaring the cross product of two free vectors $\vec{a}$ and $\vec{b}$, we have:

$$(\vec{a} \times \vec{b})^2 = (\vec{a} \times \vec{b}) \cdot (\vec{a} \times \vec{b}) = (\vec{a} \times \vec{b})_i (\vec{a} \times \vec{b})_i$$

$$= \varepsilon_{ijk} a_j b_k \varepsilon_{imn} a_m b_n = (\delta_{jm}\delta_{kn} - \delta_{jn}\delta_{km}) a_j a_m b_k b_n$$

$$= a_j a_j b_k b_k - a_j a_k b_k b_j = (a_j a_j)(b_k b_k) - (a_j b_j)(a_k b_k)$$

$$= \vec{a}^2 \vec{b}^2 - (\vec{a} \cdot \vec{b})^2, \qquad (B.32)$$

called *Lagrange's identity*. This is a particular case of Lagrange's identity written for any two sets $\{a_1, a_2, ..., a_n\}$ and $\{b_1, b_2, ..., b_n\}$ of real or complex numbers:

$$\left(\sum_{k=1}^{n} a_k^2\right)\left(\sum_{k=1}^{n} b_k^2\right) - \left(\sum_{k=1}^{n} a_k b_k\right)^2$$

$$= \sum_{i=1}^{n-1} \sum_{j=i+1}^{n} (a_i b_j - a_j b_i)^2 = \frac{1}{2} \sum_{i=1}^{n} \sum_{j=1, j \neq i}^{n} (a_i b_j - a_j b_i)^2.$$

B.3. Tensors and pseudotensors

Since teaching experience of the authors showed that some students meet difficulties in understanding the notion of *pseudotensor*, in this paragraph we shall focus our attention on this subject. For a physicist, the concepts of *pseudoscalar* and *pseudovector*, as particular cases of a pseudotensor (a pseudoscalar is a pseudotensor of rank zero, while a pseudovector is a pseudotensor of rank one), are of significant importance. If we agree to call *scalar of the first type* a true scalar, then a pseudoscalar can be named *scalar of the second type*. In the same way, a vector can be *polar (vector of the first type)* or *axial (vector of the second type)*.

From a strictly mathematical viewpoint, these definitions are rather simple, but in physics all these concepts have to be connected to our intuition. Let us begin our investigation with the notion of *tensor*.

There are various possibilities to define the notion of tensor, but in our case the most appropriate is to start from the *general coordinate transformation* concept. Without going too deep in the subject, we shall underline the most important definitions and their physical significance.

Position in space of an arbitrary point (particle) P can be determined by the system of three Cartesian coordinates x_1, x_2, x_3 with respect to the three-orthogonal reference frame $Oxyz$ (see Fig.B.6). Since both position of O and direction of coordinate axes are arbitrary (due to homogeneity and isotropy of vacuum space), it is necessary to define the law of transformation of the coordinates of point P, when passing from the initial coordinate system $S(Ox_1x_2x_3)$ to another system $S'(Ox'_1x'_2x'_3)$. Obviously, the "new" coordinates x'_i ($i = \overline{1,3}$) are functions of the "old" coordinates x_i ($i = \overline{1,3}$):

$$x'_i = f_i(x_1, x_2, x_3) \quad (i = \overline{1,3}).$$

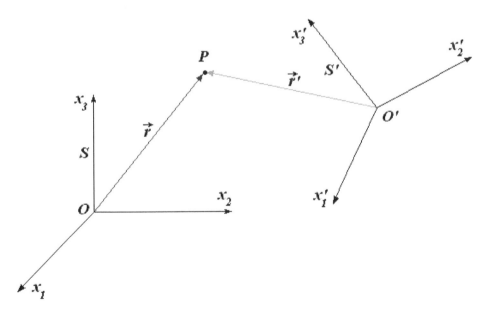

Fig.B.6

The relations $x'_i = const.$ ($i = \overline{1,3}$) represent equations of three planes with respect to S only if the above transformation is linear, that is

$$x'_i = R_{ij}x_j + X_i \quad (i,j = \overline{1,3}),$$

where X_i are coordinates of the origin O' of the frame S', with respect to the frame S. The nine quantities R_{ij} can be considered as a matrix, called *matrix of coordinate transformation*. In order to determine the same distance between any two points, no matter which coordinate system is considered, the quantities R_{ij} have to obey the *orthogonality condition*

$$R_{ij}R_{ik} = \delta_{jk} \quad (i,j,k = \overline{1,3}).$$

This condition diminishes up to three the number of independent parameters R_{ij}. The linear transformation $x'_i = R_{ij}x_j + X_i$, where the quantities R_{ij} obey the orthogonality condition, is called *non-homogeneous coordinate transformation*. If $X_i = 0$, the transformation is called *homogeneous*.

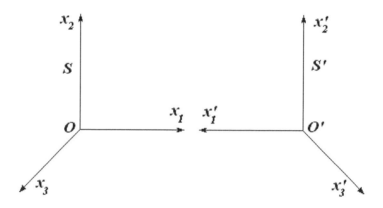

Fig.B.7

It can be shown that

$$\det \hat{R} \equiv (\det R_{ij}) = \pm 1.$$

Transformations characterized by $\det \hat{R} = +1$ are called *proper transformations*, while those with $\det \hat{R} = -1$ are named *improper transformations*. For example, translations $x'_i = X_i$ ($i = \overline{1,3}$) and/or rotations $x'_i = R_{ij}x_j$ ($i,j = \overline{1,3}$) are proper transformations, while inversions (reflections in a mirror of the coordinate axes) are improper transformations. Fig B.7 shows a graphic representation of an improper transformation, whose matrix is

$$\hat{R} = \begin{pmatrix} -1 & 0 & 0 \\ 0 & 1 & 0 \\ 0 & 0 & 1 \end{pmatrix}.$$

It can be proved that the set of non-homogeneous orthogonal transformations form a group structure with respect to the operation of composition of these transformations, called *non-homogeneous orthogonal coordinate transformation group*. The properties of this group, as well as of its main subgroups, can be found in the recommended literature.

For a better understanding of the concepts of pseudoscalar and pseudovector, as particular cases of a pseudotensor, it is necessary to define the notion of *frame orientation*. A frame S is called *right-handed* if rotation of a right screw (or drill) oriented along Ox_3 by an angle of $90°$ produces superposition of axis Ox_1 over axis Ox_2 (see Fig.B.8). In other words, if $\vec{u}_1 \times \vec{u}_2 = \vec{u}_3$, where $\vec{u}_1, \vec{u}_2$ and $\vec{u}_3$ are versors of the axes Ox_1, Ox_2 and Ox_3, respectively, then $Oxyz$ is a *right-handed reference frame*. This way, connection between the cross product operation and the frame orientation becomes more obvious. A left-handed frame can be obtained by an odd number of inversions of coordinate axes.

Within the framework of orthogonal coordinate transformations, the above considerations can be expressed in a concise form as follows. Consider a right-handed frame S. Then any frame S' obtained by an orthogonal coordinate transformation with

→ $\det \hat{R} = +1$ is a right-handed frame;
→ $\det \hat{R} = -1$ is a left-handed frame.

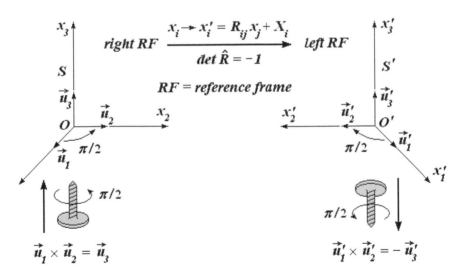

Fig.B.8

According to these observations, one can define the main algebraic quantities as follows:

Definition. A *scalar* is a quantity characterized by a number which remains unchanged under any coordinate transformation. A scalar physical quantity is defined in the same way, but the unit system has to be mentioned.

Definition. A scalar with the same algebraic form in any coordinate system is called *invariant*.

Definition. A vector $\vec{v}$ is a quantity characterized by an ordered system of three numbers $\{v_1, v_2, v_3\}$, called *vector components*, which under an orthogonal coordinate transformation change according to the law

$$v'_i = R_{ij} v_j \quad (i, j = \overline{1, 3}),$$

where v'_i are the vector components in S', and R_{ij} are the elements of the transformation matrix.

Definition. A n*th rank tensor* in the Euclidean space E_m is a quantity characterized by an ordered set of m^n numbers $t_{i_1 i_2 \ldots i_n}$ ($i_1, i_2, \ldots, i_n = \overline{1, m}$), called *tensor components*, which under an orthogonal coordinate transformation change according to the law

$$t'_{i_1 i_2 \ldots i_n} = R_{i_1 j_1} R_{i_2 j_2} \ldots R_{i_n j_n} \, t_{j_1 j_2 \ldots j_n}$$

$$(i_1, i_2, \ldots, i_n, j_1, j_2, \ldots, j_n = \overline{1, m}).$$

Definition. A n*th rank pseudotensor* in the Euclidean space E_m is a quantity characterized by an ordered set of m^n numbers $t^*_{i_1 i_2 \ldots i_n}$ ($i_1, i_2, \ldots, i_n = \overline{1, m}$), called *pseudotensor components*, which under an orthogonal coordinate transformation change according to the law

$$t^{*'}_{i_1 i_2 \ldots i_n} = (\det \hat{R}) R_{i_1 j_1} R_{i_2 j_2} \ldots R_{i_n j_n} \, t^*_{j_1 j_2 \ldots j_n}$$

$$(i_1, i_2, \ldots, i_n, j_1, j_2, \ldots, j_n = \overline{1, m}).$$

Consequently, under a proper orthogonal transformation ($\det \hat{R} = +1$) pseudotensors transform in the same way as ordinary tensors, while under an improper transformation ($\det \hat{R} = -1$) appears a change of sign. A pseudovector is also called an *axial vector*, while an ordinary vector is called *a polar vector*. For example, under an inversion of axes (improper orthogonal transformation) $x_i \to x'_i = -x_i$ the components of a vector transform according to $v'_i = -v_i$ ($i = \overline{1, 3}$), while a pseudovector obeys the rule $v^{*'}_i = v^*_i$ ($i = \overline{1, 3}$). Keeping in mind the connection between frame orientation and the proper/improper character of coordinate transformation, we can say that tensors change their sign when orientation changes, while pseudotensors don't.

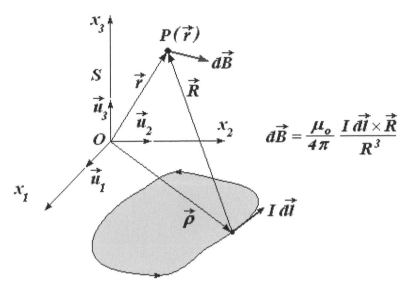

Fig.B.9

To illustrate these theoretical considerations, let us take into account a physical application, namely the formula expressing the axial vector of magnetic induction produced in vacuum by a continuous electric current I, situated in the plane $x_1 O x_2$ (see Fig.B.9). As well-known, the appropriate formula is the Biot-Savart law

$$d\vec{B} = \frac{\mu_o}{4\pi} \frac{I\,d\vec{l} \times \vec{R}}{R^3} = \frac{\mu_o}{4\pi} \frac{I\,d\vec{l} \times (\vec{r} - \vec{\rho})}{|\vec{r} - \vec{\rho}|^3},$$

where vector $\vec{R}$ has its tip at the point where the field is determined, and its tail at the origin of the current element $I\,d\vec{l}$. Since the electric current lies in $x_1 O x_2$ plane, we have

$$I\,d\vec{l} \times \vec{R} = \begin{vmatrix} \vec{u}_1 & \vec{u}_2 & \vec{u}_3 \\ I dl_1 & I dl_2 & 0 \\ X_1 & X_2 & X_3 \end{vmatrix}$$

$$= I X_3 dl_2 \vec{u}_1 + I X_2 dl_1 \vec{u}_3 - I X_1 dl_2 \vec{u}_3 - I X_3 dl_1 \vec{u}_2,$$

leading to the following components of the elementary magnetic field $d\vec{B}$ at the point $P(\vec{r})$

$$\begin{cases} dB_1 = +\dfrac{\mu_o I}{4\pi} \dfrac{X_3 dl_2}{R^3}; \\ dB_2 = -\dfrac{\mu_o I}{4\pi} \dfrac{X_3 dl_1}{R^3}; \\ dB_3 = +\dfrac{\mu_o I}{4\pi} \dfrac{(X_2 dl_1 - X_1 dl_2)}{R^3}. \end{cases}$$

Let us now perform the improper orthogonal transformation

$$x_i \to x'_i = -x_i \quad (i = \overline{1,3}).$$

As a result, the scalar quantities remain unchanged ($|\vec{R}'| = |\vec{R}|$, $I' = I$, $\mu'_o = \mu_o$), but the vector quantities change their sign ($\vec{dl'} = -\vec{dl}$, $\vec{R}' = -\vec{R}$), so that

$$d\vec{B}' = \frac{\mu'_o}{4\pi} \frac{I' \vec{dl'} \times \vec{R}'}{R'^3} = \frac{\mu_o}{4\pi} \frac{I}{R^3} [(-\vec{dl}) \times (-\vec{R})] = \frac{\mu_o}{4\pi} \frac{I}{R^3} (\vec{dl} \times \vec{R}) = d\vec{B},$$

that is

$$\begin{cases} dB'_1 = dB_1, \\ dB'_2 = dB_2, \\ dB'_3 = dB_3, \end{cases}$$

which is in complete agreement with the above considerations: under an improper orthogonal transformation (or, equivalently, under the change of the frame orientation) the pseudotensor components remain unchanged.

Let us now get back to the special connection between the cross product operation and the pseudo/non-pseudo character of tensors. As we have seen, a mirror-image of a reference frame means the change of orientation of axes. But this change determines the change of tensors sign and maintains the sign of pseudotensors. Since the effect of inversion can algebraically be expressed by means of the cross product, the connection between the operation of cross product and pseudo/non-pseudo character of tensors becomes obvious.

As we have seen, a right-handed reference frame is characterized by $\vec{u}_1 \times \vec{u}_2 = \vec{u}_3$. Performing one (or an odd number of) inversion(s) of coordinate axes (e.g. $x'_3 = -x_3$), the frame orientation changes, becoming a left-handed frame: $\vec{u}_1 \times \vec{u}_2 = -\vec{u}_3$. Due to the non-commutative property of the cross product, nothing changes if the inversion affects any other coordinate axis. Indeed, if we perform inversion $x'_2 = -x_2$, then (see Fig.B.10)

$$\vec{u}_3 = \vec{u}_2 \times \vec{u}_1 = -\vec{u}_1 \times \vec{u}_2,$$

or

$$\vec{u}_1 \times \vec{u}_2 = -\vec{u}_3,$$

which is precisely the same relation as in case of inversion of Ox_1.

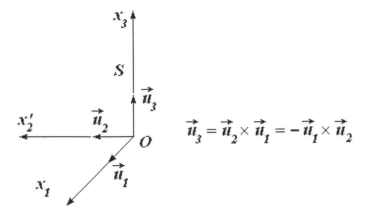

Fig.B.10

If, on the contrary, the number of inversions is even, then neither the frame orientation, nor the pseudo/non-pseudo character of tensors is affected. Algebraically, this property can also be expressed by means of the cross product. Indeed, if two inversions (say, of axes Ox_1 and Ox_3) take place, then according to Fig.B.11 we can write $\vec{u}_1 \times \vec{u}_2 = \vec{u}_3$, meaning that the new frame is also right-handed, as it was before inversions.

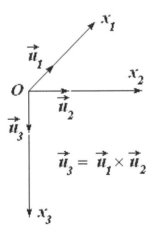

Fig.B.11

The above discussion suggests a simple and useful method of identification of the character of tensors (pseudo, or non-pseudo). To this end, we must observe that the change of the frame orientation (followed by the change of sign of tensor components, on the one hand, and by maintenance of the sign of pseudotensor components, on the other) depends on the number (even or odd) of inversions of coordinate axes. This way, if we conventionally associate minus sign to pseudotensor quantities, and plus sign to tensor quantities, and observe that, as a result of this association, the cross product has the character of "pseudo-", due to its anticommutativity (associated with (-1) sign), then the problem is solved: the character of tensor quantities (of any kind) is given by the "resultant" sign associated with expressions defining these quantities $[(+1)$ or $(-1)]$. Here are a few examples.

1. As we know, the position vector of the tail of a force $\vec{F}$, with respect to a point (pol) is a polar vector (as well as the force $\vec{F}$). In view of the above defined convention, to both vectors one associates the sign $(+1)$. Since the operation of cross product has a character of pseudovector, it is associated with sign (-1). Consequently, the moment $\vec{M}$ of the force $\vec{F}$ relative to a pol is a pseudovector. Indeed,

$$\underbrace{\vec{M}}_{(-1)} = \underbrace{\vec{r}}_{(+1)} \underbrace{\times}_{(-1)} \underbrace{\vec{F}}_{(+1)}.$$

2. The operator *nabla* $\left(\nabla = \vec{i}\frac{\partial}{\partial x} + \vec{j}\frac{\partial}{\partial y} + \vec{k}\frac{\partial}{\partial z}\right)$ is a polar vector, and so is the linear velocity of a moving particle of fluid, therefore the quantity called *vorticity*

$$\underbrace{\vec{\Omega}}_{(-1)} = \tfrac{1}{2} \underbrace{\nabla}_{(+1)} \underbrace{\times}_{(-1)} \underbrace{\vec{v}}_{(+1)}$$

is a pseudovector (or *axial vector*). Applying the same procedure, one can show that the curl of any polar vector is a pseudovector.

3. The force of interaction between two circuits Γ_1 and Γ_2 traveled by the electric currents I_1 and I_2 is a polar vector. Indeed, since the current elements $I_1 d\vec{l}_1$ and $I_2 d\vec{l}_2$ are polar vectors, and so is the position vector of $I_2 d\vec{l}_2$ with respect to the origin of $I_1 d\vec{l}_1$, $\vec{r}_{12} = -\vec{r}_{21}$, we have

$$\underbrace{\vec{F}_{12}}_{(+1)} = \underbrace{-\vec{F}_{21}}_{(+1)}$$

$$= \frac{\mu}{4\pi} I_1 I_2 \oint\limits_{(\Gamma_1)} \oint\limits_{(\Gamma_2)} \frac{1}{r_{12}^3} \underbrace{d\vec{l}_2}_{(+1)} \underbrace{\times}_{(-1)} \underbrace{(d\vec{l}_1}_{(+1)} \underbrace{\times}_{(-1)} \underbrace{\vec{r}_{12})}_{(+1)}.$$

4. The moment of a force $\vec{F}$ with respect to an axis Δ is:

$$M_\Delta = \vec{u}_\Delta \cdot (\vec{r} \times \vec{F}) = \vec{u}_\Delta \cdot \vec{M}_O,$$

where $\vec{u}_\Delta$ is the versor of Δ, while $\vec{M}_O$ is the moment of $\vec{F}$ with respect to an arbitrary point (pol) of the axis. This quantity is a pseudoscalar. Indeed,

$$\underbrace{M_\Delta}_{(-1)} = \underbrace{\vec{u}_\Delta}_{(+1)} \cdot (\underbrace{\vec{r}}_{(+1)} \underbrace{\times}_{(-1)} \underbrace{\vec{F}}_{(+1)}).$$

5. If $\vec{a}, \vec{b}$ and $\vec{d}$ are axial vectors, and $\vec{c}$ a polar vector, then $(\vec{a} \times \vec{b}) \cdot (\vec{c} \times \vec{d})$ is a pseudoscalar. Indeed,

$$\underbrace{p}_{(-1)} = (\underbrace{\vec{a}}_{(-1)} \underbrace{\times}_{(-1)} \underbrace{\vec{b}}_{(-1)}) \cdot (\underbrace{\vec{c}}_{(-1)} \underbrace{\times}_{(-1)} \underbrace{\vec{d}}_{(-1)}).$$

Next we shall give some examples of physical quantities, of 0th and 1st rank tensor, together with their units in SI.

i) Examples of scalar physical quantities: mass, $m[kg]$; temperature, $T[K]$; pressure, $p[N \cdot m^{-2}]$; mechanical work, $L[J]$; concentration of a chemical substance in a solution, $c[mol \cdot m^{-3}]$; power, $P[W]$; density of a substance, $\rho[kg \cdot m^{-3}]$; magnetic permeability of vacuum, $\mu_0[H \cdot m^{-1}]$; electric potential in a point, $V[V]$; electric flux, $\Phi_e[V \cdot m]$, etc.

ii) Examples of pseudoscalar physical quantities: moment of a force with respect to an axis, $M_\Delta[N \cdot m]$; magnetic flux, $\Phi_m[T \cdot m^2]$, etc.

iii) Examples of vector physical quantities: linear velocity, $\vec{v}[m \cdot s^{-1}]$; Newtonian force, $\vec{F}[N]$; linear acceleration, $\vec{a}[m \cdot s^{-2}]$; electric field intensity, $\vec{E}[V \cdot m^{-1}]$; electromagnetic flux density (Poynting vector), $\vec{Y}[J \cdot m^{-2} \cdot s^{-1}]$; electric field induction, $\vec{D}[C \cdot m^{-2} = N \cdot m^{-1} \cdot V^{-1}]$; electromagnetic momentum density, $\vec{g}[kg \cdot m^{-2} \cdot s^{-1}]$, etc.

iv) Examples of pseudovector physical quantities: magnetic induction, $\vec{B}[T]$; angular velocity, $\vec{\omega}[rad \cdot s^{-1}]$; moment of a force relative to a pole, $\vec{M}[n \cdot m]$; fluid vorticity, $\Omega[s^{-1}]$, etc.

B.4. Elements of vector analysis

B.4.1. Vector functions

Consider a vector $\vec{a}$ depending on a real, variable parameter t. If to each value of t corresponds a vector $\vec{a}$, we say that $\vec{a}$ is a vector function of t: $\vec{a} = \vec{a}(t)$. If $\vec{r} = \vec{r}(t)$ is the position vector of some point P with respect to the Cartesian coordinate system $Oxyz$, then the curve described by P is given by the parametric equations $x = x(t)$, $y = y(t)$, $z = z(t)$.

Let P be a material point (particle) whose position is defined by vector $\vec{a}(t)$ with respect to origin O of the frame $Oxyz$. A variation of the parameter t produces a variation of $\vec{a}(t)$, so that its head describes a curve C. If radius vector is proportional to the velocity of the moving particle, then the curve $\vec{a} = \vec{a}(t)$ is called *hodograph* or *velocity diagram* (see Fig.B.12).

B.4.2. Derivative and differential of a vector function

Let $\vec{a} = \vec{a}(t)$ be a continuous vector function of parameter t. Then its *derivative with respect to t* is defined as

$$\frac{d\vec{a}}{dt} = \lim_{\Delta t \to 0} \frac{\Delta \vec{a}}{\Delta t},$$

where

$$\Delta \vec{a} = \vec{a}(t + \Delta t) - \vec{a}(t)$$

is the variation $\Delta \vec{a}$ of the function $\vec{a}(t)$, corresponding to a variation Δt of the parameter t (which determines displacement of the particle from position P_1 to position P_2), while $\frac{\Delta \vec{a}}{\Delta t}$ is a vector collinear to $\Delta \vec{a}$.

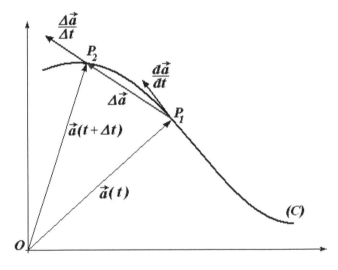

Fig.B.12

Quantity $\frac{\Delta \vec{a}}{\Delta t}$ is a vector parallel to tangent to the curve C at point P, its sense being given by the increasing variation of parameter t. The magnitude (modulus) of derivative $\frac{d\vec{a}}{dt}$ is

$$\left| \frac{d\vec{a}}{dt} \right| = \sqrt{\left(\frac{da_x}{dt}\right)^2 + \left(\frac{da_y}{dt}\right)^2 + \left(\frac{da_z}{dt}\right)^2}.$$

Differential of a vector function $\vec{a} = \vec{a}(t)$ can be defined by relation $d\vec{a} = \frac{d\vec{a}}{dt} dt$, where dt is an elementary, arbitrary variation of parameter

t. If vector $\vec{a}$ is a function of several variables $x_1, x_2, ..., x_n$, the *total differential* of $\vec{a}(x_1, x_2, ..., x_n)$ is defined as

$$d\vec{a} = \frac{\partial \vec{a}}{\partial x_i} dx_i \quad (i = \overline{1, n}),$$

where $\frac{\partial}{\partial x_i}$ is the operator of partial derivative with respect to x_i. Keeping in mind the properties of vector operations and of the differentiation of scalar functions, we have:

$$\frac{d}{dt}(\vec{a} \pm \vec{b}) = \frac{d\vec{a}}{dt} \pm \frac{d\vec{b}}{dt};$$

$$\frac{d}{dt}(\lambda \vec{a}) = \lambda \frac{d\vec{a}}{dt} \quad (\lambda = const.);$$

$$\frac{d}{dt}(\vec{a} \cdot \vec{b}) = \vec{a} \cdot \frac{d\vec{b}}{dt} + \vec{b} \cdot \frac{d\vec{a}}{dt};$$

$$\frac{d}{dt}(\vec{a} \times \vec{b}) = \vec{a} \times \frac{d\vec{b}}{dt} + \frac{d\vec{a}}{dt} \times \vec{b};$$

$$\frac{d}{dt}[\vec{a} \cdot (\vec{b} \times \vec{c})] = \frac{d\vec{a}}{dt} \cdot (\vec{b} \times \vec{c}) + \vec{a} \cdot \frac{d}{dt}(\vec{b} \times \vec{c});$$

$$\frac{d}{dt}[\vec{a} \times (\vec{b} \times \vec{c})] = \frac{d\vec{a}}{dt} \times (\vec{b} \times \vec{c}) + \vec{a} \times \frac{d}{dt}(\vec{b} \times \vec{c}).$$

B.4.3. Integral of a vector function

The (definite) integral of a continuous vector function $\vec{a}(t)$ can be defined in much the same way as of scalar functions, except that the integral is a vector. In Euclidean space E_3 the following three types of integrals are defined:

(a) Curvilinear (line) integral: $\int_{P_1}^{P_2} \vec{a} \cdot d\vec{r}$, along a curve C, between points P_1 and P_2, where vector $\vec{a}$ has its origin on the curve, and $d\vec{r}$ is a vector element of C. This integral is called *circulation* of vector $\vec{a}$ along curve C, between points P_1 and P_2. If C is a closed curve, circulation is denoted by $\oint_{(C)} \vec{a} \cdot d\vec{r}$.

(b) Double integral: $\int_S \vec{a} \cdot d\vec{S}$, where $\vec{a}$ is a vector with its origin at S, and $d\vec{S}$ an oriented surface element. The integral is called *flux of vector* $\vec{a}$ through surface S. If S is a closed surface, the integral is denoted $\oint_S \vec{a} \cdot d\vec{S}$.

(c) Triple integral: $\int_D \vec{a}\, d\tau$, where D is a domain of volume V, $d\tau$ a volume element, and $\vec{a}$ has its origin at an arbitrary point of D.

B.4.4. Scalar and vector fields

If a physical quantity has a well-determined value in any point of a spatial domain D, then the multitude of these values define *the field* of the physical quantity. For example, a heated body produces a temperature field. In each point of the body, the temperature has a certain value. Or, as another example, in each point of a flowing liquid can be determined the velocity of some fluid particle at that point. To define a field, one must establish a correspondence between each point of the domain D and the value of physical quantity at that point. This correspondence defines a *point function*. Since each point is determined by its coordinates x, y, z (or, equivalently, by its position vector $\vec{r}$) relative to the origin of a coordinate reference system $Oxyz$, the fields of various physical quantities are determined by knowledge of the specified (scalar, vector, or tensor) functions depending on the variables x, y, z.

If to each point $P \in D$ of Euclidean space E_3 one can associate a scalar quantity $\varphi(P)$, then we say that in domain D has been defined a *scalar field* $\varphi(x, y, z)$. Such a scalar function is, for example, temperature: at each point of some domain D the temperature has a certain value $T(P)$. Correspondence between the points of the considered domain and the values of temperature at those points defines a scalar function $T(P) = T(x, y, z)$. Another example is offered by pressure p of the air: at each point of the surrounding atmosphere one can settle a scalar function $p(P) = p(x, y, z)$.

If to each point $P \in D$ one can associate a vector quantity $\vec{a}(P) = \vec{a}(x.y.z)$, we say that in D has been defined a *vector field*. Such a field, exists for example, in the region surrounding an electrized body: at each point of this domain one can define a vector field called *electric field intensity* $\vec{E}$. Velocities associated to each molecule of a fluid in motion is another example of vector field.

A scalar φ (or vector $\vec{A}$) field is called *non-stationary* if φ (respectively, $\vec{A}$) depend explicitly on time. If not, the field is called *stationary*. Functions φ and $\vec{A}$, together with their partial derivatives, are supposed to be continuous.

B.5. First-order vector differential operators [1]

B.5.1. Gradient

Let $\varphi(x, y, z)$ be a function of class C^1 (the first derivatives exist and are continuous), defined on a certain domain $D \subset E_3$. The above investigation shows that $\varphi(x, y, z)$ defines in D a scalar field. The differential of φ is

$$d\varphi = \frac{\partial \varphi}{\partial x}dx + \frac{\partial \varphi}{\partial y}dy + \frac{\partial \varphi}{\partial z}dz = \frac{\partial \varphi}{\partial x_i}dx_i \quad (i = \overline{1,3}). \quad (B.33)$$

In view of (B.16), this relation can be written as

$$d\varphi = \left(\frac{\partial \varphi}{\partial x}\vec{i} + \frac{\partial \varphi}{\partial y}\vec{j} + \frac{\partial \varphi}{\partial z}\vec{k}\right) \cdot \left(dx\,\vec{i} + dy\,\vec{j} + dz\,\vec{k}\right)$$

$$= \left(\frac{\partial \varphi}{\partial x_1}\vec{u}_1 + \frac{\partial \varphi}{\partial x_2}\vec{u}_2 + \frac{\partial \varphi}{\partial x_3}\vec{u}_3\right) \cdot \left(dx_1\vec{u}_1 + dx_2\vec{u}_2 + dx_3\vec{u}_3\right),$$

therefore it can be considered as the dot product of vectors

$$d\vec{r} = dx\,\vec{i} + dy\,\vec{j} + dz\,\vec{k}$$

and

$$\vec{A} = \frac{\partial \varphi}{\partial x}\vec{i} + \frac{\partial \varphi}{\partial y}\vec{j} + \frac{\partial \varphi}{\partial z}\vec{k} = \vec{u}_i \frac{\partial \varphi}{\partial x_i} \quad (i = \overline{1,3}). \quad (B.34)$$

The vector field $\vec{A}$, whose Cartesian components are $\frac{\partial \varphi}{\partial x_i}$ ($i = \overline{1,3}$), is called *gradient* of scalar field φ and usually is denoted by gradφ. With this notation, relation (B.33) writes

$$d\varphi = \text{grad}\varphi \cdot d\vec{r}. \quad (B.35)$$

Vector gradφ determines variation of function φ about a point, while its component on x_i-axis, which is $\frac{\partial \varphi}{\partial x_i}$, represents the rate of variation of φ on x_i-direction. The direction of gradφ at some point is direction of the most rapid variation of scalar function φ, starting from that point, while the sense of gradφ is given by the increase of

---

[1] In vector calculus, "grad", "div", and "curl" are usually called *operators*. If fact, the only true operator is *nabla*, and $\nabla \varphi = \text{grad}\varphi$, $\nabla \cdot \vec{A} = \text{div}\vec{A}$, and $\nabla \times \vec{B} = \text{curl}\vec{B}$ are expressions obtained as a result of different ways of application of *nabla* upon scalar or vector quantities. Note that the Laplacian Δ is, also, a true operator.

φ. Relation (B.33) shows that the vector gradφ can be obtained by applying the operator

$$\nabla = \vec{i}\frac{\partial}{\partial x} + \vec{j}\frac{\partial}{\partial y} + \vec{k}\frac{\partial}{\partial z} = \vec{u}_i\frac{\partial}{\partial x_i} \quad (i = \overline{1,3}) \tag{B.36}$$

to the scalar function $\varphi(x_i)$. This operator, denoted by symbol ∇, is called *nabla operator* or *Hamilton's operator*. By means of *nabla*, the gradient writes

$$\text{grad}\varphi = \vec{u}_i\frac{\partial \varphi}{\partial x_i} = \nabla\varphi \quad (i = \overline{1,3}). \tag{B.37}$$

The vector field $\vec{A}$ defined by (B.37) is called *potential field*, and φ is the *potential* of this field. For example, using relation between the electrostatic field intensity $\vec{E}$ and the electrostatic potential V, $\vec{E} = -\text{grad}V$, we can determine one quantity, if the other is known. (The sign "minus" is introduced by convention, to show that the field lines of $\vec{E}$ go out of "plus" point charges and enter the "minus" point charges).

Definition (B.36) shows that ∇ is a linear, vector differential operator. Projecting $\nabla\varphi(x_i)$ on x_i-direction, we have:

$$(\text{grad}\varphi)_i = (\nabla\varphi)\cdot \vec{u}_i = \left(\vec{u}_k\frac{\partial \varphi}{\partial x_k}\right)\cdot\vec{u}_i = \frac{\partial \varphi}{\partial x_k}\vec{u}_k\cdot\vec{u}_i = \frac{\partial \varphi}{\partial x_k}\delta_{ik} = \frac{\partial \varphi}{\partial x_i},$$

or, by means of the commonly used notations in vector analysis

$$\text{grad}_i\varphi = \nabla_i\varphi = \frac{\partial \varphi}{\partial x_i} = \partial_i\varphi = \varphi_{,i}. \tag{B.38}$$

Equipotential surfaces

Suppose that function $\varphi(P)$ is uniform and continuous in D, together with its partial derivatives up to the second order, and consider all points where this function has the same value. These points form a surface in D. Indeed, if x, y, z are the coordinates of P, relation $\varphi(P) = K(const.)$ can also be written as

$$\varphi(x,y,z) = K, \tag{B.39}$$

which is the implicit form of a surface in $D \subset E_3$. In view of (B.35) and (B.39), we then have

$$d\varphi = 0 = \text{grad}\varphi \cdot d\vec{r}. \tag{B.40}$$

This relation is valid at any point of surface (B.39). Since equality (B.40) has to be true for any vector of the dot product, and because $d\vec{r}$ lies in the plane tangent to surface (B.39), it follows that at any point of this surface the vector $\nabla\varphi$ is oriented along the normal to surface. Giving values to the constant K in (B.39) one obtains a family of surfaces satisfying condition (B.40), called *equipotential surfaces* or *level surfaces*. Equation (B.40) is called *equation of equipotential surfaces*.

Field lines

Consider a stationary vector field $\vec{A}$ and an arbitrary curve given by its parametric equations $x_i = x_i(s)$ ($i = \overline{1,3}$). If $\vec{A}$ is tangent to the curve at any of its points, then the curve is called *field line* or *line of force* of the field $\vec{A}$. The *differential equations of field lines* are obtained from the obvious relation $\vec{A} \times d\vec{s} = 0$. Projecting this vector relation on the axes of a Cartesian orthogonal frame, we have

$$\frac{dx}{A_x} = \frac{dy}{A_y} = \frac{dz}{A_z}. \qquad (B.41)$$

According to (B.40), the vector $\vec{A} = \text{grad}\varphi$ is orthogonal to the arc element $d\vec{r}$ which, in turn, is tangent to equipotential surfaces. Since the field lines of $\vec{A} = \text{grad}\varphi$ and $d\vec{r}$ are collinear, it follows that they are also orthogonal to the equipotential surfaces $\varphi = const$.

Directional derivative

Let (Δ) be an axis whose direction and sense are given by the versor $\vec{u}_\Delta = \frac{d\vec{r}}{ds}$, where $ds = |d\vec{r}|$ is the magnitude of the line element of the axis (Δ). The *directional derivative* of a scalar function $\varphi = \varphi(\vec{r})$ along the given axis (Δ) is the projection of $\text{grad}\varphi$ on the axis:

$$\text{grad}\varphi \cdot \vec{u}_\Delta = \frac{d\varphi}{d\vec{r}} \cdot \frac{d\vec{r}}{ds} = \frac{d\varphi}{ds}.$$

Here, a formal writing for the gradient, $\text{grad}\varphi = \frac{d\varphi}{d\vec{r}}$, has been used.

B.5.2. Divergence of a vector field

Since ∇ is a vector operator, one can perform the dot product $\nabla \cdot \vec{A}$, where $\vec{A} = \vec{A}(A_x, A_y, A_z)$ is an arbitrary vector field. This operation is called *divergence of $\vec{A}$* and is denoted by $\text{div}\vec{A}$. Note that, even if the scalar product is commutative, $\vec{A} \cdot \nabla \neq \nabla \cdot \vec{A}$. Indeed, the divergence of $\vec{A}$ is a well-defined quantity (scalar or pseudoscalar, depending on the tensor nature of $\vec{A}$)

$$\nabla \cdot \vec{A} = \left(\vec{u}_i \frac{\partial}{\partial x_i}\right) \cdot (A_k \vec{u}_k) = \delta_{ik} \frac{\partial A_k}{\partial x_i} = \frac{\partial A_i}{\partial x_i}.$$

$$= \frac{\partial A_x}{\partial x} + \frac{\partial A_y}{\partial y} + \frac{\partial A_z}{\partial z}, \qquad (B.42)$$

while the quantity

$$\vec{A} \cdot \nabla = \left(A_x \vec{i} + A_y \vec{j} + A_z \vec{k} \right) \cdot \left(\vec{i} \frac{\partial}{\partial x} + \vec{j} \frac{\partial}{\partial y} + \vec{k} \frac{\partial}{\partial z} \right)$$

$$= A_x \frac{\partial}{\partial x} + A_y \frac{\partial}{\partial y} + A_z \frac{\partial}{\partial z}$$

is, in its turn, an operator. Using notations (B.38), we can also write

$$\text{div}\vec{A} = \nabla_i A_i = \frac{\partial A_i}{\partial x_i} = \partial_i A_i = A_{i,i} \quad (i = \overline{1,3}). \qquad (B.43)$$

A vector field $\vec{A}$ with zero divergence, $\text{div}\vec{A} = 0$, is called *divergence-free* (or *source-free* or *solenoidal*). The lines of such a field are closed curves. If $\text{div}\vec{A} \neq 0$, we have a *source-field*, with the following two possibilities:

$$\text{div}\vec{A} \neq 0 \quad \begin{matrix} \nearrow & \text{div}\vec{A} > 0, \\ \searrow & \text{div}\vec{A} < 0. \end{matrix}$$

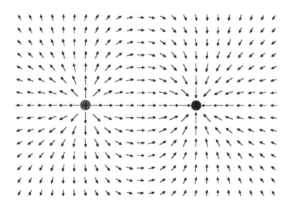

Fig.B.13a

In the first case ($\text{div}\vec{A} > 0$) the sources of the field are positive (*springs*, or *wellheads*), while in the second sources are negative (or *sinks*). For a source-field, there always exists a field line which does not close, but goes to infinity. It is conventionally accepted that the lines of a source field emerge from positive sources, and enter the negative sources. Obviously, the field lines of a solenoidal field are closed curves. Such a field is, for example, the magnetic field $\vec{B}$ of a solenoid (Fig.B.13a), similar to the field of a permanent magnet, shown in (Fig.B.13b).

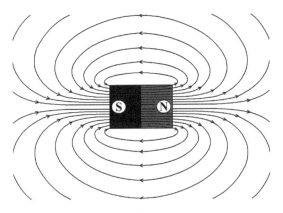

Fig.B.13b

No matter how small is a magnet, it always has two poles, north and south. P.A.M.Dirac predicted the existence of the magnetic monopole, but its real existence has not been detected so far. The almost perfect analogy between electrostatic end magnetostatic fields, for the simplest case of two electric/magnetic charges, is graphically sketched in Fig.B.13a and Fig.B.13b.

The main property of the electrostatic field is expressed by Gauss's law $\text{div}\vec{E} = \frac{1}{\varepsilon}\rho \neq 0$, where $\rho = \rho(\vec{r}) = \lim_{\Delta\tau \to 0} \frac{\Delta q}{\Delta\tau} = \frac{dq}{d\tau}$ is the electric charge density, while Gauss's law for the magnetostatic field writes $\text{div}\vec{B} = 0$. In those regions of space where no electric charges are present ($\rho = 0$), the two equations have the same form: $\text{div}\vec{E} = 0$, and $\text{div}\vec{B} = 0$. In other words, in such regions of space the two fields behave identically, which allows one to speak not about electric and magnetic field separately, but about an *electromagnetic field*. The mathematical "equivalence" of the two fields allowed their unification under the same theory. This was done by J.C. Maxwell, who wrote his celebrated system of equations[1].

---

[1] Theory developed by J.C.Maxwell was the first successful classical unified field theory. Later, in the Standard Model, the weak and the electromagnetic interactions have been combined into a unified electroweak theory. In its turn, this theory was unified with theory of strong interactions (quantum chromodynamics), leading to a new theory capable to give a unitary description of electromagnetic and nuclear (weak and strong) interactions. Gravity has yet to be successfully included in a theory of everything. Simply trying to combine the graviton with the strong and electroweak interactions runs into fun-

B.5.3. Curl of a vector field

Consider a vector field $\vec{A}(A_x, A_y, A_z)$ and let us perform the cross product $\nabla \times \vec{A}$. The result of this operation is a vector field $\vec{B}$, $\vec{B} = \nabla \times \vec{A}$, called *curl of the vector field* $\vec{A}$. If $\vec{A}$ is a polar vector, then $\vec{B}$ is an axial vector. The analytical expression of $\vec{B} = \nabla \times \vec{A}$ is obtained by means of (B.27):

$$\vec{B} = \nabla \times \vec{A} = \varepsilon_{ijk} \partial_j A_k \vec{u}_i$$

$$= \left(\frac{\partial A_z}{\partial y} - \frac{\partial A_y}{\partial z}\right)\vec{i} + \left(\frac{\partial A_x}{\partial z} - \frac{\partial A_z}{\partial x}\right)\vec{j} + \left(\frac{\partial A_y}{\partial x} - \frac{\partial A_x}{\partial y}\right)\vec{k}$$

$$= \begin{vmatrix} \vec{i} & \vec{j} & \vec{k} \\ \frac{\partial}{\partial x} & \frac{\partial}{\partial y} & \frac{\partial}{\partial z} \\ A_x & A_y & A_z \end{vmatrix}. \qquad (B.44)$$

Performing dot product of (B.44) and versor $\vec{u}_s$, we obtain the projection of $\vec{B} = \mathrm{curl}\vec{A}$ on the direction defined by $\vec{u}_s$:

$$(\mathrm{curl}\vec{A})_s = (\varepsilon_{ijk}\partial_j A_k \vec{u}_i) \cdot \vec{u}_s = \varepsilon_{ijk}\partial_j A_k \delta_{si} = \varepsilon_{sjk}\partial_j A_k. \qquad (B.45)$$

For example, the component along x-axis of $\mathrm{curl}\vec{A}$ is:

$$(\mathrm{curl}\vec{A})_1 = (\mathrm{curl}\vec{A})_x = \varepsilon_{1jk}\partial_j A_k = \partial_2 A_3 - \partial_3 A_2 = \frac{\partial A_z}{\partial y} - \frac{\partial A_y}{\partial z}.$$

A vector field $\vec{A}$ having the property $\mathrm{curl}\vec{A} = 0$ is called *irrotational* or *curl-free*. Such a field is, for example, the electrostatic field intensity $\vec{E}$.

---

damental difficulties since the resulting theory is not renormalizable. Theoretical physicists have not yet formulated a widely accepted, consistent theory that combines general relativity and quantum mechanics. The incompatibility of the two theories remains an outstanding problem in the field of physics. Some theoretical physicists currently believe that a quantum theory of general relativity may require frameworks other than field theory itself, such as string theory or loop quantum gravity. A Grand Unified Theory (GUT) is a model in particle physics in which at high energy, the three gauge interactions of the Standard Model which define the electromagnetic, weak, and strong interactions, together with gravitational interactions, are merged into one single theory. At the present there is no accepted unified field theory, and thus it remains an open line of research.

B.6. Vector identities

Here are the most frequently used identities of vector analysis, together with their justification.

1. $\operatorname{grad}(\varphi\psi) = \varphi\operatorname{grad}\psi + \psi\operatorname{grad}\varphi.$ (B.46)

   Indeed,

$$\operatorname{grad}(\varphi\psi) = \vec{u}_i\frac{\partial(\varphi\psi)}{\partial x_i} = \vec{u}_i\frac{\partial\psi}{\partial x_i}\varphi + \vec{u}_i\frac{\partial\varphi}{\partial x_i}\psi = \varphi\operatorname{grad}\psi + \psi\operatorname{grad}\varphi.$$

2. $\operatorname{div}(\varphi\vec{A}) = \varphi\operatorname{div}\vec{A} + \vec{A}\cdot\operatorname{grad}\varphi.$ (B.47)

   Indeed,

$$\operatorname{div}(\varphi\vec{A}) = \frac{\partial(\varphi A_i)}{\partial x_i} = \varphi\frac{\partial A_i}{\partial x_i} + A_i\frac{\partial\varphi}{\partial x_i} = \varphi\operatorname{div}\vec{A} + \vec{A}\cdot\operatorname{grad}\varphi.$$

3. $\operatorname{curl}(\varphi\vec{A}) = \varphi\operatorname{curl}\vec{A} + (\operatorname{grad}\varphi)\times\vec{A}.$ (B.48)

   Indeed,

$$\operatorname{curl}(\varphi\vec{A}) = \varepsilon_{ijk}\partial_j(\varphi A_k)\vec{u}_i = \varphi(\varepsilon_{ijk}\partial_j A_k\vec{u}_i) + \varepsilon_{ijk}(\partial_j\varphi)A_k\vec{u}_i$$

$$= \varphi\operatorname{curl}\vec{A} + (\operatorname{grad}\varphi)\times\vec{A}.$$

4. $\operatorname{div}(\vec{A}\times\vec{B}) = \vec{B}\cdot\operatorname{curl}\vec{A} - \vec{A}\cdot\operatorname{curl}\vec{B}.$ (B.49)

   Indeed,

$$\operatorname{div}(\vec{A}\times\vec{B}) = \partial_i(\vec{A}\times\vec{B})_i = \partial_i(\varepsilon_{ijk}A_j B_k)$$

$$= \varepsilon_{ijk}A_j\partial_i B_k + \varepsilon_{ijk}B_k\partial_i A_j = -A_j\varepsilon_{jik}\partial_i B_k + B_k\varepsilon_{kij}\partial_i A_j$$

$$= -A_j(\operatorname{curl}\vec{B})_j + B_k(\operatorname{curl}\vec{A})_k = -\vec{A}\cdot\operatorname{curl}\vec{B} + \vec{B}\cdot\operatorname{curl}\vec{A}.$$

5. $\operatorname{grad}(\vec{A}\cdot\vec{B}) = \vec{A}\times(\operatorname{curl}\vec{B}) + \vec{B}\times(\operatorname{curl}\vec{A})$
   $\qquad\qquad + (\vec{A}\cdot\nabla)\vec{B} + (\vec{B}\cdot\nabla)\vec{A}.$ (B.50)

   We first observe that a component of the gradient reads

$$\partial_j(\vec{A}\cdot\vec{B}) = \partial_j(A_k B_k) = A_k\partial_j B_k + B_k\partial_j A_k.$$

Next, let us multiply relation $(\operatorname{curl}\vec{B})_s = \epsilon_{slm}\partial_l B_m$ by ϵ_{sjk} and perform summation over indices l and m:

$$\epsilon_{sjk}(\operatorname{curl}\vec{B})_s = (\delta_{lj}\delta_{mk} - \delta_{lk}\delta_{mj})\partial_l B_m = \partial_j B_k - \partial_k B_j.$$

Using this result, we still have:

$$\partial_j(\vec{A}\cdot\vec{B}) = \epsilon_{jks}[A_k(\operatorname{curl}\vec{B})_s + B_k(\operatorname{curl}\vec{A})_s] + (A_k\partial_k)B_j + (B_k\partial_k)A_j,$$

or, in a vector form

$$\mathrm{grad}(\vec{A}\cdot\vec{B}) = \vec{A}\times\mathrm{curl}\,\vec{B} + \vec{B}\times\mathrm{curl}\,\vec{A} + (\vec{A}\cdot\nabla)\vec{B} + (\vec{B}\cdot\nabla)\vec{A}.$$

6. $\mathrm{curl}(\vec{A}\times\vec{B}) = \vec{A}\,\mathrm{div}\vec{B} - \vec{B}\,\mathrm{div}\vec{A} + (\vec{B}\cdot\nabla)\vec{A} - (\vec{A}\cdot\nabla)\vec{B}.$ (B.51)
Indeed,

$$\mathrm{curl}(\vec{A}\times\vec{B}) = \vec{u}_i\varepsilon_{ijk}\partial_j(\vec{A}\times\vec{B})_k = \vec{u}_i\varepsilon_{ijk}\varepsilon_{klm}\partial_j(A_l B_m)$$

$$= \vec{u}_i\varepsilon_{ijk}\varepsilon_{klm}A_l\partial_j B_m + \vec{u}_i\varepsilon_{ijk}\varepsilon_{klm}B_m\partial_j A_l$$

$$= \vec{u}_i(\delta_{il}\delta_{jm} - \delta_{im}\delta_{jl})A_l\partial_j B_m + \vec{u}_i(\delta_{il}\delta_{jm} - \delta_{im}\delta_{jl})B_m\partial_j A_l$$

$$= \vec{u}_i A_i\partial_j B_j - \vec{u}_i A_j\partial_j B_i + \vec{u}_i B_j\partial_j A_i - \vec{u}_i B_i\partial_j A_j$$

$$= (A_i\vec{u}_i)(\partial_j B_j) - (B_i\vec{u}_i)(\partial_j A_j) + (B_j\partial_j)(A_i\vec{u}_i) - (A_j\partial_j)(B_i\vec{u}_i)$$

$$= \vec{A}\nabla\cdot\vec{B} - \vec{B}\nabla\cdot\vec{A} + (\vec{B}\cdot\nabla)\vec{A} - (\vec{A}\cdot\nabla)\vec{B}.$$

7. $\mathrm{div}(\mathrm{curl}\vec{A}) = 0.$ (B.52)
Indeed,

$$\mathrm{div}(\mathrm{curl}\vec{A}) = \partial_i(\mathrm{curl}\vec{A})_i = \varepsilon_{ijk}\partial_i(\partial_j A_k) = \frac{1}{2}\bigl(\varepsilon_{ijk}\partial_i\partial_j A_k + \varepsilon_{ijk}\partial_i\partial_j A_k\bigr)$$

$$= \frac{1}{2}\bigl(\varepsilon_{ijk}\partial_i\partial_j A_k - \varepsilon_{ijk}\partial_j\partial_i A_k\bigr) = \frac{1}{2}\bigl(\varepsilon_{ijk}\partial_i\partial_j A_k - \varepsilon_{ijk}\partial_i\partial_j A_k\bigr) = 0,$$

where we have supposed that $\vec{A}(x,y,z)$ satisfy the conditions demanded by Schwartz' theorem, which "assure" the equality of mixed second-order derivatives of A_k, that is

$$\frac{\partial^2 A_k}{\partial x_i \partial x_j} = \frac{\partial^2 A_k}{\partial x_j \partial x_i} \iff \partial_i\partial_j A_k = \partial_j\partial_i A_k.$$

Therefore, since the pseudotensor ε_{ijk} is antisymmetric, and $\partial_i\partial_j A_k$ symmetric in the same pair of indices (ij), we have

$$\varepsilon_{ijk}\partial_i\partial_j A_k = \varepsilon_{ijk}\frac{\partial^2 A_k}{\partial x_i \partial x_j} = 0.$$

8. $\mathrm{curl}(\mathrm{grad}\varphi) = 0.$ (B.53)
Indeed,

$$\mathrm{curl}(\mathrm{grad}\varphi) = \vec{u}_i\varepsilon_{ijk}\partial_j(\mathrm{grad}\varphi)_k = \vec{u}_i\varepsilon_{ijk}\partial_j\partial_k\varphi \equiv \vec{u}_i\varepsilon_{ijk}\frac{\partial^2\varphi}{\partial x_j \partial x_k} = 0,$$

where the pair of summation indices in the product $\varepsilon_{ijk}\frac{\partial^2 \varphi}{\partial x_j \partial x_k}$ is (jk).

9. $\operatorname{div}(\operatorname{grad}\varphi) = \nabla \cdot (\nabla\varphi) = (\nabla \cdot \nabla)\varphi = \nabla^2 \varphi = \Delta\varphi,$ (B.54)
where the operator (denoted as Δ)

$$\nabla^2 = \Delta = \frac{\partial^2}{\partial x^2} + \frac{\partial^2}{\partial y^2} + \frac{\partial^2}{\partial z^2} = \frac{\partial^2}{\partial x_i \partial x_i} \equiv \partial_i \partial_i$$

is called the *Laplace operator* or *Laplacian* (in Cartesian coordinates). Equation

$$\Delta\varphi = 0$$

is called *Laplace's equation*. Solutions of Laplace's equation are called *harmonic functions*. This operator can be generalized in four dimensions, by choosing three spacial coordinates and an additional fourth coordinate for the time (temporal coordinate). It is called *D'Alembert's operator* or *Dalembertian*, being denoted by $\square$. Taking $x_1 = x$, $x_2 = y$, $x_3 = z$, and $x_4 = ivt$, where v is a constant velocity, we have

$$\square = \sum_{\mu=1}^{4} \frac{\partial^2}{\partial x_\mu \partial x_\mu} = \frac{\partial^2}{\partial x_1^2} + \frac{\partial^2}{\partial x_2^2} + \frac{\partial^2}{\partial x_3^2} + \frac{\partial^2}{\partial x_4^2} = \Delta - \frac{1}{v^2}\frac{\partial^2}{\partial t^2}.$$

Choosing $v = c$, where c is velocity of light in vacuum, we meet the Minkowski space-time.

10. $\operatorname{curl}(\operatorname{curl}\vec{A}) = \operatorname{grad}(\operatorname{div}\vec{A}) - \Delta\vec{A}.$ (B.55)
Indeed,

$$\operatorname{curl}(\operatorname{curl}\vec{A}) = \vec{u}_i \varepsilon_{ijk} \partial_j (\operatorname{curl}\vec{A})_k = \vec{u}_i \varepsilon_{ijk} \varepsilon_{klm} \partial_j \partial_l A_m$$
$$= \vec{u}_i (\delta_{il}\delta_{jm} - \delta_{im}\delta_{jl})\partial_j \partial_l A_m = \vec{u}_i \partial_m \partial_i A_m - \vec{u}_i \partial_j \partial_j A_i$$
$$= \vec{u}_i \partial_i \partial_m A_m - \partial_j \partial_j A_i \vec{u}_i$$
$$= \vec{u}_i \partial_i (\operatorname{div}\vec{A}) - \partial_j \partial_j \vec{A} = \nabla(\nabla \cdot \vec{A}) - \Delta\vec{A}.$$

To conclude, here are some final remarks. *Nabla* (or *del*) is a vector differential operator. When applied to a scalar field, nabla denotes the gradient. When applied to a vector field, this can be done by means of dot product (divergence), or cross product (curl). The dot product of nabla with itself is an operator called Laplacian. In its turn, the Laplacian can be applied to both scalar and vector[1] fields. The following Table synthesizes these final remarks.

---

[1] As and example, here is the equation satisfied by the vector potential $\vec{A}$ in an isotropic and homogeneous medium : $\Delta \vec{A} = -\mu \vec{j}$, where $\vec{j}$ is the current density vector.

	∇ (nabla operator)	φ (scalar field)	$\vec{A}$ (vector field)
∇	–	$\nabla\varphi$ gradient (vector)	–
$\nabla\cdot$	$\nabla\cdot\nabla=\nabla^2=\Delta$ laplacian (scalar)	–	$\nabla\cdot\vec{A}$ divergence (scalar)
$\nabla\times$	–	–	$\nabla\times\vec{A}$ curl (vector)

Table B.1. Various ways of acting of the nabla operator.

B.7. Orthogonal curvilinear coordinates. First-order differential operators in orthogonal curvilinear coordinates

Let $\vec{r}$ be the radius-vector of a point $P(\vec{r})$ in tridimensional Euclidean space E_3 and x_1, x_2, x_3 its Cartesian coordinates with respect to a three-orthogonal Cartesian frame $Ox_1x_2x_3$. Consider, also, three independent real parameters q_1, q_2, q_3, so that coordinates x_i ($i = \overline{1,3}$) are functions of class C^1 of these parameters:

$$x_i = x_i(q_1, q_2, q_3) \quad (i = \overline{1,3}). \tag{B.56}$$

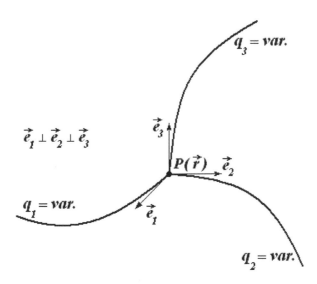

Fig.B.14

Suppose that between the set of coordinates x_i on the one hand, and of the parameters q_j, on the other, exists a one-to-one correspondence (this is an alternate name for a *bijection*), that is to a set of coordinates x corresponds a single set of parameters q, and vice-versa. Under these circumstances, transformation (B.56) is locally reversible if the functional determinant (Jacobian) is different from zero:

$$J = \frac{\partial(x_1, x_2, x_3)}{\partial(q_1, q_2, q_3)} \neq 0. \qquad (B.57)$$

If two out of the three parameters q_j ($j = \overline{1,3}$), say q_2 and q_3, have fixed values, then one obtains the line (curve) of coordinate q_1. This means that through any point in space pass three lines of coordinates (see Fig.B.14). Parameters q_1, q_2 and q_3 are called *curvilinear coordinates*, or *general coordinates* of the point P.

If $\vec{e}_i$ ($i = \overline{1,3}$) are vectors tangent to the curve of coordinates q_i ($i = \overline{1,3}$), then according to (B.57) are linearly independent, therefore they form a basis in E_3. Indeed,

$$\vec{e}_1 \cdot (\vec{e}_2 \times \vec{e}_3) = \frac{\partial \vec{r}}{\partial q_1} \cdot \left(\frac{\partial \vec{r}}{\partial q_2} \times \frac{\partial \vec{r}}{\partial q_3} \right)$$

$$= \begin{vmatrix} \frac{\partial x_1}{\partial q_1} & \frac{\partial x_2}{\partial q_1} & \frac{\partial x_3}{\partial q_1} \\ \frac{\partial x_1}{\partial q_2} & \frac{\partial x_2}{\partial q_2} & \frac{\partial x_3}{\partial q_2} \\ \frac{\partial x_1}{\partial q_3} & \frac{\partial x_2}{\partial q_3} & \frac{\partial x_3}{\partial q_3} \end{vmatrix} = \frac{\partial(x_1, x_2, x_3)}{\partial(q_1, q_2, q_3)} = J \neq 0.$$

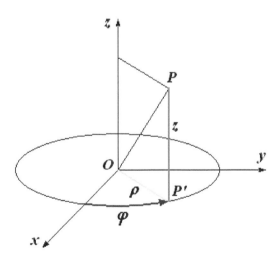

Fig.B.15

If at any point of the domain $D^{(q)}$, defined by the set of all possible values of parameters (q), the vectors $\vec{e}_i$ form a right-handed orthogonal trieder, then the three curvilinear coordinates define an *orthogonal curvilinear coordinate system*. If $\vec{e}_i$ are linearly independent, but are not orthogonal to each other, then q_i ($i = \overline{1,3}$) define a *general curvilinear coordinate system*.

Two of the most commonly used orthogonal curvilinear coordinate systems are *cylindrical* and *spherical coordinates*. The cylindrical coordinates of a point P are ρ, φ and z (see Fig.B.15). The range of variation of these parameters is:

$$D^{(cyl)} = \left\{ (\rho, \varphi, z) \,\middle|\, \rho \geq 0,\ 0 \leq \varphi < 2\pi,\ -\infty < z < +\infty \right\}.$$

In this case, relations (B.56) write

$$\begin{cases} x_1 \equiv x = \rho \cos \varphi, \\ x_2 \equiv y = \rho \sin \varphi, \\ x_3 \equiv z = z, \end{cases} \qquad (B.58)$$

while the inverse relations $q_j = q_j(x_1, x_2, x_3)$ ($j = \overline{1,3}$) are

$$\begin{cases} q_1 \equiv \rho = \sqrt{x^2 + y^2}, \\ q_2 \equiv \varphi = \arctan \dfrac{y}{x}, \\ q_3 \equiv z = z. \end{cases}$$

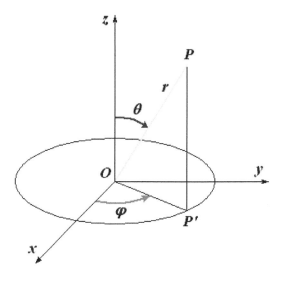

Fig.B.16

The *spherical coordinates* which define position of some point P are r, θ, φ (see Fig.B.16). The range of variation of these parameters is:

$$D^{(spher)} = \left\{ (r, \theta, \varphi) \,\middle|\, r \geq 0,\ 0 \leq \theta \leq \pi,\ 0 \leq \varphi < 2\pi \right\}.$$

Relations (B.56) then write:

$$\begin{cases} x_1 \equiv x = r\sin\theta\cos\varphi, \\ x_2 \equiv y = r\sin\theta\sin\varphi, \\ x_3 \equiv z = r\cos\theta, \end{cases} \quad (B.59)$$

and the inverse relations $q_j = q_j(x_1, x_2, x_3)$ $(j = \overline{1,3})$ are

$$\begin{cases} q_1 \equiv r = \sqrt{x^2 + y^2 + z^2}, \\ q_2 \equiv \theta = \arccos\dfrac{z}{\sqrt{x^2 + y^2 + z^2}}, \\ q_2 \equiv \varphi = \arctan\dfrac{y}{x}. \end{cases}$$

In order to define a new coordinate system, parameters (q_1, q_2, q_3) have to vary between certain limits, so that the set of points of space and the set of curvilinear coordinates $q_j = q_j(x_1, x_2, x_3)$ be in univocal correspondence. Let $\hat{e}_1$, $\hat{e}_2$ and $\hat{e}_3$ be the versors of the three vectors $\vec{e}_i$ $(i = \overline{1,3})$. In the new coordinate system, the radius vector $\vec{r}$ of some point $P(\vec{r})$ is given by

$$\vec{r} = \vec{r}(q_1, q_2, q_3). \quad (B.60)$$

Since $\dfrac{\partial \vec{r}}{\partial q_i}$ $(i = \overline{1,3})$ are vectors tangent to the coordinate lines q_i $(i = \overline{1,3})$, we can write

$$\frac{\partial \vec{r}}{\partial q_i} = \vec{e}_i = h_i \hat{e}_i \quad (i = \overline{1,3}), \quad (B.61)$$

where we denoted

$$h_i = \left|\frac{\partial \vec{r}}{\partial q_i}\right| = \sqrt{\left(\frac{\partial x_1}{\partial q_i}\right)^2 + \left(\frac{\partial x_2}{\partial q_i}\right)^2 + \left(\frac{\partial x_3}{\partial q_i}\right)^2} \quad (i = \overline{1,3}). \quad (B.62)$$

These quantities are called *Lamé coefficients*. They stand for the moduli of the vectors $\vec{e}_i$, tangent to the coordinate lines q_i.

Let us also consider the versors $\vec{n}_1$, $\vec{n}_2$, and $\vec{n}_3$ oriented along normals to the surfaces $q_1 = q_1^o = const.$, $q_2 = q_2^o = const.$, and $q_3 = q_3^o = const.$, respectively. Since vectors grad q_i are also normal to surfaces $q_i = q_i^o = const.$, denoting by g_i the moduli of vectors gradq_i, we can write

$$\operatorname{grad} q_i = g_i \vec{n}_i \quad \text{(no summation;} \quad i = \overline{1,3}). \quad (B.63)$$

Since
$$\text{grad } q_i = \frac{\partial q_i}{\partial x_1}\vec{e}_1 + \frac{\partial q_i}{\partial x_2}\vec{e}_2 + \frac{\partial q_i}{\partial x_3}\vec{e}_3 \equiv \frac{\partial q_i}{\partial x}\vec{i} + \frac{\partial q_i}{\partial y}\vec{j} + \frac{\partial q_i}{\partial z}\vec{k},$$

we have:
$$g_i^2 = \left(\frac{\partial q_i}{\partial x_1}\right)^2 + \left(\frac{\partial q_i}{\partial x_2}\right)^2 + \left(\frac{\partial q_i}{\partial x_3}\right)^2$$
$$= \left(\frac{\partial q_i}{\partial x}\right)^2 + \left(\frac{\partial q_i}{\partial y}\right)^2 + \left(\frac{\partial q_i}{\partial z}\right)^2. \qquad (B.64)$$

The quantities g_i ($i = \overline{1,3}$) are called *differential parameters of the first order*.

Definition. We say that vectors $\vec{a}_1, \vec{a}_2$ and $\vec{a}_3$, on the one side, and vectors $\vec{b}_1, \vec{b}_2$ and $\vec{b}_3$, on the other, form a system of reciprocal vectors with respect to each other, if the following relations are fulfilled:

$$\vec{a}_i \cdot \vec{b}_i = 1 \quad \text{(no summation;} \quad i = \overline{1,3}),$$

$$\vec{a}_i \cdot \vec{b}_j = 0 \quad (i = \overline{1,3};\ i \neq j).$$

Let us show that $\text{grad} q_i$ ($i = \overline{1,3}$) and $\frac{\partial \vec{r}}{\partial q_i}$ ($i = \overline{1,3}$) form a system of reciprocal vectors. To this end, we must prove that

$$\text{grad} q_i \cdot \frac{\partial \vec{r}}{\partial q_i} = 1 \quad \text{(no summation;} \quad i = \overline{1,3}),$$

and
$$\text{grad} q_i \cdot \frac{\partial \vec{r}}{\partial q_j} = 0 \quad (i = \overline{1,3};\ i \neq j).$$

Taking the dot product between vector relation

$$d\vec{r} = \frac{\partial \vec{r}}{\partial q_1} dq_1 + \frac{\partial \vec{r}}{\partial q_2} dq_2 + \frac{\partial \vec{r}}{\partial q_3} dq_3$$

and $\text{grad} q_i$, we have
$$dq_i = \text{grad} q_i \cdot d\vec{r}$$
$$= \left(\text{grad} q_i \cdot \frac{\partial \vec{r}}{\partial q_1}\right) dq_1 + \left(\text{grad} q_i \cdot \frac{\partial \vec{r}}{\partial q_2}\right) dq_2 + \left(\text{grad} q_i \cdot \frac{\partial \vec{r}}{\partial q_3}\right) dq_3$$

and, since dq_1, dq_2 and dq_3 are independent, it follows that only the coefficient of dq_i ($i = \overline{1,3}$) equals 1, the other two coefficients being zero, which completes the proof.

In case of a Cartesian system of coordinates, versors $\hat{e}_1 \equiv \vec{i}$, $\hat{e}_2 \equiv \vec{j}$, and $\hat{e}_3 \equiv \vec{k}$ are orthogonal two-by-two. If the origin O of the frame "travels" through the whole space, each of them form a system of equipolent vectors. Unlike Cartesian coordinates, in case of curvilinear orthogonal coordinates the versors $\hat{e}_1, \hat{e}_2$, and $\hat{e}_3$ form a right-handed orthogonal system at any point P in space, but the versors associated to some other point $P' \neq P$ are not equipolent to those with origin at P. In addition, vectors $\vec{e}_1$, $\vec{e}_2$ and $\vec{e}_3$ do not keep their magnitude when the point P "travels" throughout the space. Finally, in case of any other curvilinear coordinates both vector systems $\vec{e}_1, \vec{e}_2, \vec{e}_3$, and $\vec{n}_1, \vec{n}_2, \vec{n}_3$ have *any* orientation (but remain, nevertheless, linearly independent).

Coordinate lines of the *orthogonal curvilinear coordinates* are orthogonal two-by-two at any point P of space. It then follows that, in case of orthogonal curvilinear coordinates, the versors $\vec{n}_1, \vec{n}_2$ and $\vec{n}_3$ coincide with $\hat{e}_1, \hat{e}_2$, and $\hat{e}_3$, respectively, being orthogonal two-by-two. Therefore, we have:

$$\vec{n}_i = \hat{e}_i; \quad \hat{e}_i \cdot \hat{e}_j = 0; \quad \vec{n}_i \cdot \vec{n}_j = 0 \quad (i \neq j).$$

Following these relations, in order to be orthogonal, it is necessary and sufficient for a system of curvilinear coordinates to satisfy

$$\frac{\partial \vec{r}}{\partial q_i} \cdot \frac{\partial \vec{r}}{\partial q_j} = \frac{\partial x}{\partial q_i} \cdot \frac{\partial x}{\partial q_j} + \frac{\partial y}{\partial q_i} \cdot \frac{\partial y}{\partial q_j} + \frac{\partial z}{\partial q_i} \cdot \frac{\partial z}{\partial q_j} = 0 \quad (i \neq j), \quad (B.65)$$

or,

$$\mathrm{grad}q_i \cdot \mathrm{grad}q_j = \frac{\partial x}{\partial q_i} \cdot \frac{\partial x}{\partial q_j} + \frac{\partial y}{\partial q_i} \cdot \frac{\partial y}{\partial q_j} + \frac{\partial z}{\partial q_i} \cdot \frac{\partial z}{\partial q_j} = 0 \quad (i \neq j). \quad (B.65')$$

If the curvilinear coordinates are orthogonal, one can find a relationship between Lamé's coefficients h_i, and the first-order differential parameters g_i. Using relations $\mathrm{grad}\, q_i = g_i \hat{e}_i$ (no summation), and $\frac{\partial \vec{r}}{\partial q_i} = h_i \hat{e}_i$ (no summation), and reciprocal relations as well, we obtain:

$$\mathrm{grad}\, q_i \cdot \frac{\partial \vec{r}}{\partial q_i} = g_i \hat{e}_i \cdot h_i \hat{e}_i = g_i h_i = 1,$$

so that

$$g_i = \frac{1}{h_i} \quad (i = 1, 2, 3), \quad (B.66)$$

and, in particular,

$$grad\, q_i = \frac{\hat{e}_i}{h_i} \quad (\text{no summation};\ i = \overline{1,3}).$$

Line (arc) element in orthogonal curvilinear coordinates

In a system of orthogonal curvilinear coordinates we can write

$$d\vec{r} = \frac{\partial \vec{r}}{\partial q_1}dq_1 + \frac{\partial \vec{r}}{\partial q_2}dq_2 + \frac{\partial \vec{r}}{\partial q_3}dq_3$$

$$= h_1 dq_1 \hat{e}_1 + h_2 dq_2 \hat{e}_2 + h_3 dq_3 \hat{e}_3. \quad (B.67)$$

The components $|d\vec{r}_1| \equiv |d\vec{s}_1| = |h_1 dq_1 \hat{e}_1| = ds_1 = h_1 dq_1$, $ds_2 = h_2 dq_2$, and $ds_3 = h_3 dq_3$ of the vector $d\vec{r}$ along directions defined by $\hat{e}_1, \hat{e}_2$ and $\hat{e}_3$ are called *curvilinear components* of $d\vec{r}$. Squaring (B.67), we obtain

$$d\vec{r}^2 = d\vec{r} \cdot d\vec{r} \equiv ds^2 = h_1^2(dq_1)^2 + h_2^2(dq_2)^2 + h_3^2(dq_3)^2,$$

or, by convention,

$$ds^2 = h_1^2 dq_1^2 + h_2^2 dq_2^2 + h_3^2 dq_3^2. \quad (B.68)$$

This is the squared line element (called *metric*) in orthogonal curvilinear coordinates.

Surface element in orthogonal curvilinear coordinates

The surface element in curvilinear coordinates can be obtained by using the definition of the surface element in Cartesian coordinates. As we know, from geometric point of view, the cross product of two vectors $\vec{a}$ and $\vec{b}$ is a (pseudo)vector whose direction is given by the right-hand screw rule, and magnitude equal to the area of the parallelogram that the vectors span (see Fig.B.17).

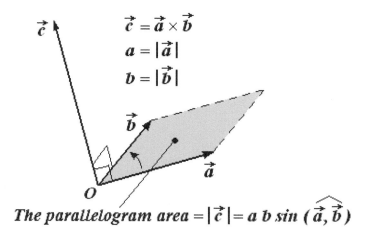

Fig.B.17

In view of these remarks, a vector surface element in orthogonal curvilinear coordinates is given by the cross product of two line elements, also considered in curvilinear coordinates. For a better understanding, let us first consider a Cartesian coordinate system. In such a system, there exists three vector surface elements: $d\vec{S}_x = dydz\vec{i}$, $d\vec{S}_y = dzdx\vec{j}$, and $d\vec{S}_z = dxdy\vec{k}$, given by cross product of three line elements: $d\vec{r}_x = dx\,\vec{i}$, $d\vec{r}_y = dy\,\vec{j}$, and $d\vec{r}_z = dz\,\vec{k}$. More precisely, the vector surface element $d\vec{S}_x = d\vec{r}_y \times d\vec{r}_z = dy\,dz(\vec{j} \times \vec{k}) = dy\,dz\,\vec{i}$ is an elementary vector whose magnitude equals the area of the rectangle determined by $d\vec{r}_y$ and $d\vec{r}_z$, its direction being orthogonal to the area $dy\,dz$, and sense given by the right-hand screw rule.

The other two vector surface elements are obtained in a similar manner (see Fig.B.18):

$$d\vec{S}_y = d\vec{r}_z \times d\vec{r}_x = dz\,dx(\vec{k} \times \vec{i}) = dzdx\vec{j},$$

and

$$d\vec{S}_z = d\vec{r}_x \times d\vec{r}_y = dx\,dy(\vec{i} \times \vec{j}) = dxdy\vec{k}.$$

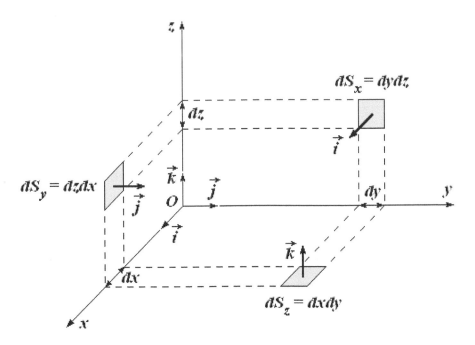

Fig.B.18

The vector surface element therefore is

$$d\vec{S} = d\vec{S}_x + d\vec{S}_y + d\vec{S}_z = dy\,dz\,\vec{i} + dz\,dx\,\vec{j} + dx\,dy\,\vec{k}.$$

By analogy with Cartesian coordinates, the surface element constructed on line elements $d\vec{r}_i = h_i dq_i \hat{e}_i$ and $d\vec{r}_j = h_j dq_j \hat{e}_j$ (no summation) is

$$d\widetilde{S}_{ij} = d\vec{r}_i \times d\vec{r}_j = h_i h_j dq_i dq_j (\hat{e}_i \times \hat{e}_j),$$

where we have used symbol $\sim$ over S to display the tensor character of cross product of the two line elements. The vector associated with the antisymmetric (pseudo)tensor $d\widetilde{S}_{ij}$ is [see (B.91)]:

$$d\vec{S}_i = \frac{1}{2}\varepsilon_{ijk} d\widetilde{S}_{jk} = \frac{1}{2}\varepsilon_{ijk} d\vec{r}_j \times d\vec{r}_k \equiv \frac{1}{2}\varepsilon_{ijk} d\vec{s}_j \times d\vec{s}_k.$$

For example, taking $i = 1$, we have:

$$d\vec{S}_1 \equiv d\vec{S}_{q_1} = \frac{1}{2}\varepsilon_{1jk} d\widetilde{S}_{jk}$$

$$= \frac{1}{2}\left(\varepsilon_{123} d\vec{r}_2 \times d\vec{r}_3 + \varepsilon_{132} d\vec{r}_3 \times d\vec{r}_2\right)$$

$$= \frac{1}{2}\left(h_2 h_3 dq_2 dq_3\, \hat{e}_2 \times \hat{e}_3 - h_2 h_3 dq_2 dq_3\, \hat{e}_3 \times \hat{e}_2\right)$$

$$= h_2 h_3 dq_2 dq_3\, \hat{e}_2 \times \hat{e}_3 = h_2 h_3 dq_2 dq_3\, \hat{e}_1. \qquad (B.69)$$

Proceeding in the same manner, the other two vector surface elements $d\vec{S}_2 = h_3 h_1 dq_3 dq_1 \hat{e}_2$ and $d\vec{S}_3 = h_1 h_2 dq_1 dq_2 \hat{e}_3$ are obtained. The vector surface element in orthogonal curvilinear coordinates then writes

$$d\vec{S} = h_2 h_3 dq_2 dq_3 \hat{e}_1 + h_3 h_1 dq_3 dq_1 \hat{e}_2 + h_1 h_2 dq_1 dq_2 \hat{e}_3. \qquad (B.70)$$

In the above relation $h_i h_j dq_i dq_k$ is the vector surface element normal to the "plane" $q_i P q_j$ $(i, j = \overline{1,3})$, its orientation being given by versor $\hat{e}_k$ $(k = \overline{1,3})$.

Volume element in orthogonal curvilinear coordinates

The volume element in orthogonal curvilinear coordinates can be obtained by generalizing the volume element written in Cartesian coordinates,

$$dV = d\vec{r}_1 \cdot (d\vec{r}_2 \times d\vec{r}_3) = dx dy dz \left[\vec{i} \cdot \left(\vec{j} \times \vec{k}\right)\right] = dx dy dz,$$

taking into account the geometric interpretation of the mixed product of three vectors.

Consider two infinitely near points $M(q_1, q_2, q_3)$ and $N(q_1 + dq_1, q_2 + dq_2, q_3 + dq_3)$. Tracing the three coordinate surfaces passing through M, and the three passing through N, one obtains an elementary curvilinear parallelepiped, with edges $\frac{\partial \vec{r}}{\partial q_1} dq_1$, $\frac{\partial \vec{r}}{\partial q_2} dq_2$, and $\frac{\partial \vec{r}}{\partial q_3} dq_3$. The volume of this elementary parallelepiped is given by the mixed product

$$dV = d\vec{s}_1 \cdot (d\vec{s}_2 \times d\vec{s}_3) = h_1 h_2 h_3 dq_1 dq_2 dq_3 \left[\hat{e}_1 \cdot (\hat{e}_2 \times \hat{e}_3)\right]$$

$$= h_1 h_2 h_3 dq_1 dq_2 dq_3. \qquad (B.71)$$

Examples of orthogonal curvilinear coordinates

1. Cylindrical coordinates

Let us apply the previous general considerations to cylindrical coordinates, and calculate Lamé's coefficients, the arc element, the metric, the surface and volume elements. Since

$$\begin{cases} x = \rho\cos\varphi, \ y = \rho\sin\varphi, \ z = z, \\ q_1 = \rho, \ q_2 = \varphi, \ q_3 = z, \end{cases}$$

the radius vector $\vec{r}$ of some point P of the space writes

$$\vec{r} = x\vec{i} + y\vec{j} + z\vec{k} = \rho\cos\varphi\,\vec{i} + \rho\sin\varphi\,\vec{j} + z\vec{k},$$

so that

$$\frac{\partial \vec{r}}{\partial \rho} = \cos\varphi\,\vec{i} + \sin\varphi\,\vec{j},$$

$$\frac{\partial \vec{r}}{\partial \varphi} = -\rho\sin\varphi\,\vec{i} + \rho\cos\varphi\,\vec{j},$$

$$\frac{\partial \vec{r}}{\partial z} = \vec{k}.$$

One first observes that

$$\frac{\partial \vec{r}}{\partial \rho}\cdot\frac{\partial \vec{r}}{\partial \varphi} = 0; \quad \frac{\partial \vec{r}}{\partial z}\cdot\frac{\partial \vec{r}}{\partial \varphi} = 0; \quad \frac{\partial \vec{r}}{\partial \rho}\cdot\frac{\partial \vec{r}}{\partial z} = 0,$$

which, in agreement with (B.65), prove that our frame is an orthogonal coordinate system. Next, one can calculate Lamé's coefficients

$$h_1 = 1, \quad h_2 = \rho, \quad h_3 = 1,$$

the arc element

$$d\vec{r} = d\rho\,\hat{e}_1 + \rho d\varphi\,\hat{e}_2 + dz\,\hat{e}_3 = d\rho\,\vec{u}_\rho + \rho d\varphi\,\vec{u}_\varphi + dz\vec{k},$$

the metric

$$ds^2 = d\rho^2 + \rho^2 d\varphi^2 + dz^2,$$

the vector surface element

$$d\vec{S} = \rho d\varphi dz\,\hat{e}_1 + d\rho dz\,\hat{e}_2 + \rho d\rho d\varphi\,\hat{e}_3$$

$$= \rho d\varphi dz\,\vec{u}_\rho + d\rho dz\,\vec{u}_\varphi + \rho d\rho d\varphi\,\vec{u}_z,$$

and the volume element
$$dV = \rho d\rho d\varphi dz.$$

Spherical coordinates

Following the same steps as shown above, we have:

$$x = r\sin\theta\cos\varphi, \quad y = r\sin\theta\sin\varphi, \quad z = r\cos\theta,$$

$$q_1 = r, \quad q_2 = \theta, \quad q_3 = \varphi,$$

$$\vec{r} = r\sin\theta\cos\varphi\vec{i} + r\sin\theta\sin\varphi\vec{j} + r\cos\theta\vec{k},$$

$$\frac{\partial\vec{r}}{\partial r} = \sin\theta\cos\varphi\vec{i} + \sin\theta\sin\varphi\vec{j} + \cos\theta\vec{k},$$

$$\frac{\partial\vec{r}}{\partial\theta} = r\cos\theta\cos\varphi\vec{i} + r\cos\theta\sin\varphi\vec{j} - r\sin\theta\vec{k},$$

$$\frac{\partial\vec{r}}{\partial\varphi} = -r\sin\theta\sin\varphi\vec{i} + r\sin\theta\cos\varphi\vec{j},$$

One observes that

$$\frac{\partial\vec{r}}{\partial r}\cdot\frac{\partial\vec{r}}{\partial\theta} = 0, \quad \frac{\partial\vec{r}}{\partial\varphi}\cdot\frac{\partial\vec{r}}{\partial\theta} = 0, \quad \frac{\partial\vec{r}}{\partial r}\cdot\frac{\partial\vec{r}}{\partial\varphi} = 0,$$

which, in agreement with (B.65), show that this is also a system of orthogonal curvilinear coordinates. Next, let us calculate: Lamé's coefficients

$$h_1 = 1, \quad h_2 = r, \quad h_3 = r\sin\theta,$$

the arc element

$$d\vec{r} \equiv d\vec{s} = dr\hat{e}_1 + rd\theta\hat{e}_2 + r\sin\theta d\varphi\hat{e}_3$$

$$= dr\vec{u}_r + rd\theta\vec{u}_\theta + r\sin\theta d\varphi\vec{u}_\varphi,$$

the metric

$$ds^2 = dr^2 + r^2 d\theta^2 + r^2\sin^2\theta d\varphi^2,$$

the vector surface element

$$d\vec{S} = r^2\sin\theta d\theta d\varphi\hat{e}_1 + r\sin\theta drd\varphi\hat{e}_2 + rdrd\theta\hat{e}_3$$

$$= r^2\sin\theta d\theta d\varphi\vec{u}_r + r\sin\theta drd\varphi\vec{u}_\theta + rdrd\theta\vec{u}_\varphi,$$

and the volume element

$$dV = r^2 \sin\theta dr d\theta d\varphi.$$

Gradient, curl, divergence and Laplacian in orthogonal curvilinear coordinates (OCC)

Gradient in OCC

As well known, finding solution to a physical problem is considerably facilitated if one takes into account its symmetry. Since usually we have to do with applications involving spherical, cylindrical, polar plane, etc. symmetry, it is necessary to know the expressions of the differential operators (gradient, curl, divergence, Laplacian) in these coordinates.

Let $\varphi = \varphi(x,y,z) \equiv \varphi(\vec{r})$ be a scalar field of class C^1 on a domain $D \subset E_3$. Then, if q_1, q_2, q_3 are the system of OCC, we have:

$$\operatorname{grad}\varphi(q_1,q_2,q_3) = \frac{\partial \varphi}{\partial q_1}\operatorname{grad}q_1 + \frac{\partial \varphi}{\partial q_2}\operatorname{grad}q_2 + \frac{\partial \varphi}{\partial q_3}\operatorname{grad}q_3.$$

Since

$$\operatorname{grad}q_i = \frac{\hat{e}_i}{h_i} \quad \text{(no summation)},$$

we can write

$$\operatorname{grad}\varphi(q_1,q_2,q_3) = \frac{1}{h_1}\frac{\partial \varphi}{\partial q_1}\hat{e}_1 + \frac{1}{h_2}\frac{\partial \varphi}{\partial q_2}\hat{e}_2 + \frac{1}{h_3}\frac{\partial \varphi}{\partial q_3}\hat{e}_3. \qquad (B.72)$$

For example, the gradient in cylindrical coordinates writes

$$\operatorname{grad}\phi(\rho,\varphi,z) = \frac{\partial \phi}{\partial \rho}\vec{u}_\rho + \frac{1}{\rho}\frac{\partial \phi}{\partial \varphi}\vec{u}_\varphi + \frac{\partial \phi}{\partial z}\vec{k}, \qquad (B.73)$$

while in spherical coordinates is

$$\operatorname{grad}\phi(r,\theta,\varphi) = \frac{\partial \phi}{\partial r}\vec{u}_r + \frac{1}{r}\frac{\partial \phi}{\partial \theta}\vec{u}_\theta + \frac{1}{r\sin\theta}\frac{\partial \phi}{\partial \varphi}\vec{u}_\varphi. \qquad (B.74)$$

Curl in OCC

Consider the vector field $\vec{V} = V_i\hat{e}_i = V_1\hat{e}_1 + V_2\hat{e}_2 + V_3\hat{e}_3$ of class C^1 on the domain $D \subset E_3$, and let us calculate curl $\vec{V}$. We have:

$$\operatorname{curl}\vec{V} = \operatorname{curl}(V_1\hat{e}_1 + V_2\hat{e}_2 + V_3\hat{e}_3) = V_1\operatorname{curl}\hat{e}_1 + (\operatorname{grad}V_1) \times \hat{e}_1$$

$$+V_2\operatorname{curl}\hat{e}_2 + (\operatorname{grad}V_2) \times \hat{e}_2 + V_3\operatorname{curl}\hat{e}_3 + (\operatorname{grad}V_3) \times \hat{e}_3$$
$$= V_i\operatorname{curl}\hat{e}_i + (\operatorname{grad}V_i) \times \hat{e}_i. \qquad (B.75)$$

To express $\operatorname{curl}\hat{e}_i$ we use the already known relation

$$\operatorname{grad}q_i = \frac{1}{h_i}\hat{e}_i \quad (\text{no summation;} \quad i = \overline{1,3}).$$

According to vector identity (B.53)

$$\operatorname{curl}(\operatorname{grad}q_i) = 0 \quad \forall q_i,$$

and therefore

$$\operatorname{curl}(\operatorname{grad}q_i) = \operatorname{curl}\left(\frac{1}{h_i}\hat{e}_i\right)$$
$$= \frac{1}{h_i}\operatorname{curl}\hat{e}_i + \operatorname{grad}\left(\frac{1}{h_i}\right) \times \hat{e}_i = 0 \quad (\text{no summation}),$$

so that

$$\frac{1}{h_i}\operatorname{curl}\hat{e}_i = -\operatorname{grad}\left(\frac{1}{h_i}\right) \times \hat{e}_i \quad (\text{no summation}). \qquad (B.76)$$

But

$$\operatorname{grad}\left(\frac{1}{h_i}\right) = -\frac{1}{h_i^2}\operatorname{grad}h_i \quad (\text{no summation}),$$

and (B.76) becomes

$$\frac{1}{h_i}\operatorname{curl}\hat{e}_i = \frac{1}{h_i^2}\operatorname{grad}h_i \times \hat{e}_i \quad (\text{no summation;} \quad i = 1,2,3).$$

Introducing this relation into (B.75), we can write

$$\operatorname{curl}\vec{V} = V_i\operatorname{curl}\hat{e}_i + (\operatorname{grad}V_i) \times \hat{e}_i$$
$$= \frac{V_i}{h_i}\operatorname{grad}h_i \times \hat{e}_i + (\operatorname{grad}V_i) \times \hat{e}_i$$
$$= \left(\frac{V_i}{h_i}\operatorname{grad}h_i + \operatorname{grad}V_i\right) \times \hat{e}_i = \frac{1}{h_i}\operatorname{grad}(V_ih_i) \times \hat{e}_i$$
$$= \frac{1}{h_1}\operatorname{grad}(V_1h_1) \times \hat{e}_1 + \frac{1}{h_2}\operatorname{grad}(V_2h_2) \times \hat{e}_2 + \frac{1}{h_3}\operatorname{grad}(V_3h_3) \times \hat{e}_3.$$

Consequently,
$$\mathrm{curl}\vec{V} = \frac{1}{h_i}\mathrm{grad}(V_i h_i) \times \hat{e}_i \quad (i = \overline{1,3}). \tag{B.77}$$

For example:
$$(\mathrm{curl}\vec{V})_1 = \mathrm{curl}\vec{V} \cdot \hat{e}_1$$

$$= \left[\frac{1}{h_1}\mathrm{grad}(V_1 h_1) \times \hat{e}_1 + \frac{1}{h_2}\mathrm{grad}(V_2 h_2) \times \hat{e}_2 + \frac{1}{h_3}\mathrm{grad}(V_3 h_3) \times \hat{e}_3\right] \cdot \hat{e}_1$$

$$= \left[\frac{1}{h_2}\mathrm{grad}(V_2 h_2) \times \hat{e}_2\right] \cdot \hat{e}_1 + \left[\frac{1}{h_3}\mathrm{grad}(V_3 h_3) \times \hat{e}_3\right] \cdot \hat{e}_1$$

$$= \left(\mathrm{grad}V_2 \times \hat{e}_2 + \frac{V_2}{h_2}\mathrm{grad}h_2 \times \hat{e}_2\right) \cdot \hat{e}_1$$

$$+ \left(\mathrm{grad}V_3 \times \hat{e}_3 + \frac{V_3}{h_3}\mathrm{grad}h_3 \times \hat{e}_3\right) \cdot \hat{e}_1$$

$$= \left[\left(\frac{1}{h_1}\frac{\partial V_2}{\partial q_1}\hat{e}_1 + \frac{1}{h_2}\frac{\partial V_2}{\partial q_2}\hat{e}_2 + \frac{1}{h_3}\frac{\partial V_2}{\partial q_3}\hat{e}_3\right) \times \hat{e}_2\right] \cdot \hat{e}_1$$

$$+ \frac{V_2}{h_2}\left[\left(\frac{1}{h_1}\frac{\partial h_2}{\partial q_1}\hat{e}_1 + \frac{1}{h_2}\frac{\partial h_2}{\partial q_2}\hat{e}_2 + \frac{1}{h_3}\frac{\partial h_2}{\partial q_3}\hat{e}_3\right) \times \hat{e}_2\right] \cdot \hat{e}_1$$

$$+ \left[\left(\frac{1}{h_1}\frac{\partial V_3}{\partial q_1}\hat{e}_1 + \frac{1}{h_2}\frac{\partial V_3}{\partial q_2}\hat{e}_2 + \frac{1}{h_3}\frac{\partial V_3}{\partial q_3}\hat{e}_3\right) \times \hat{e}_3\right] \cdot \hat{e}_1$$

$$+ \frac{V_3}{h_3}\left[\left(\frac{1}{h_1}\frac{\partial h_3}{\partial q_1}\hat{e}_1 + \frac{1}{h_2}\frac{\partial h_3}{\partial q_2}\hat{e}_2 + \frac{1}{h_3}\frac{\partial h_3}{\partial q_3}\hat{e}_3\right) \times \hat{e}_3\right] \cdot \hat{e}_1$$

$$= \left(\frac{1}{h_3}\frac{\partial V_2}{\partial q_3} + \frac{V_2}{h_2 h_3}\frac{\partial h_2}{\partial q_3}\right)(\hat{e}_3 \times \hat{e}_2) \cdot \hat{e}_1$$

$$+ \left(\frac{1}{h_2}\frac{\partial V_3}{\partial q_2} + \frac{V_3}{h_2 h_3}\frac{\partial h_3}{\partial q_2}\right)(\hat{e}_2 \times \hat{e}_3) \cdot \hat{e}_1$$

$$= -\frac{1}{h_3}\frac{\partial V_2}{\partial q_3} - \frac{V_2}{h_2 h_3}\frac{\partial h_2}{\partial q_3} + \frac{1}{h_2}\frac{\partial V_3}{\partial q_2} + \frac{V_3}{h_2 h_3}\frac{\partial h_3}{\partial q_2}$$

$$= \frac{1}{h_2 h_3}\left[\frac{\partial}{\partial q_2}(V_3 h_3) - \frac{\partial}{\partial q_3}(V_2 h_2)\right].$$

The other two components are obtained in an analogous way:
$$(\mathrm{curl}\vec{V})_2 = \frac{1}{h_3 h_1}\left[\frac{\partial}{\partial q_3}(V_1 h_1) - \frac{\partial}{\partial q_1}(V_3 h_3)\right],$$

and
$$(\text{curl}\vec{V})_3 = \frac{1}{h_1 h_2}\left[\frac{\partial}{\partial q_1}(V_2 h_2) - \frac{\partial}{\partial q_2}(V_1 h_1)\right].$$

Therefore,
$$\text{curl}\vec{V} = \frac{1}{h_2 h_3}\left[\frac{\partial}{\partial q_2}(V_3 h_3) - \frac{\partial}{\partial q_3}(V_2 h_2)\right]\hat{e}_1$$
$$+ \frac{1}{h_3 h_1}\left[\frac{\partial}{\partial q_3}(V_1 h_1) - \frac{\partial}{\partial q_1}(V_3 h_3)\right]\hat{e}_2$$
$$+ \frac{1}{h_1 h_2}\left[\frac{\partial}{\partial q_1}(V_2 h_2) - \frac{\partial}{\partial q_2}(V_1 h_1)\right]\hat{e}_3.$$

For example, in cylindrical coordinates $\text{curl}\vec{V}$ writes
$$\text{curl}\vec{V} = \left(\frac{1}{\rho}\frac{\partial V_z}{\partial \varphi} - \frac{\partial V_\varphi}{\partial z}\right)\vec{u}_\rho + \left(\frac{\partial V_\rho}{\partial z} - \frac{\partial V_z}{\partial \rho}\right)\vec{u}_\varphi$$
$$+ \frac{1}{\rho}\left[\frac{\partial(\rho V_\varphi)}{\partial \rho} - \frac{\partial V_\rho}{\partial \varphi}\right]\vec{k},$$

while in spherical coordinates writes
$$\text{curl}\vec{V} = \frac{1}{r\sin\theta}\left[\frac{\partial}{\partial \theta}(V_\varphi \sin\theta) - \frac{\partial V_\theta}{\partial \varphi}\right]\vec{u}_r$$
$$+ \left[\frac{1}{r\sin\theta}\frac{\partial V_r}{\partial \varphi} - \frac{1}{r}\frac{\partial}{\partial r}(rV_\varphi)\right]\vec{u}_\theta + \frac{1}{r}\left[\frac{\partial}{\partial r}(rV_\theta) - \frac{\partial V_r}{\partial \theta}\right]\vec{u}_\varphi.$$

Divergence in OCC

Consider the vector field $\vec{V} = V_i \hat{e}_i = V_1 \hat{e}_1 + V_2 \hat{e}_2 + V_3 \hat{e}_3$ of class C^1 on the domain $D \subset E_3$. The divergence of $\vec{V}$ is
$$\text{div}\vec{V} = \text{div}(V_i \hat{e}_i) = \text{div}(V_1 \hat{e}_1 + V_2 \hat{e}_2 + V_3 \hat{e}_3)$$
$$= V_1 \text{div}\hat{e}_1 + (\text{grad}V_1)\cdot\hat{e}_1 + V_2\text{div}\hat{e}_2 + (\text{grad}V_2)\cdot\hat{e}_2 + V_3\text{div}\hat{e}_3 + (\text{grad}V_3)\cdot\hat{e}_3$$
$$= V_i \text{div}\hat{e}_i + (\text{grad}V_i)\cdot\hat{e}_i. \quad (B.78)$$

To calculate $\text{div}\hat{e}_i$ ($i = \overline{1,3}$) we take into account that: $\hat{e}_1 = \hat{e}_2 \times \hat{e}_3$, $\hat{e}_2 = \hat{e}_3 \times \hat{e}_1$, $\hat{e}_3 = \hat{e}_1 \times \hat{e}_2$. For $i=1$, we have:
$$\text{div}\,\hat{e}_1 = \text{div}(\hat{e}_2 \times \hat{e}_3) = \hat{e}_3 \cdot \text{curl}\hat{e}_2 - \hat{e}_2 \cdot \text{curl}\hat{e}_3$$

$$= \hat{e}_3 \cdot \frac{1}{h_2}\mathrm{grad}h_2 \times \hat{e}_2 - \hat{e}_2 \cdot \frac{1}{h_3}\mathrm{grad}h_3 \times \hat{e}_3$$

$$= \hat{e}_2 \times \hat{e}_3 \cdot \left(\frac{\mathrm{grad}h_2}{h_2} + \frac{\mathrm{grad}h_3}{h_3}\right) = \hat{e}_1 \cdot \left(\frac{grad(h_2 h_3)}{h_2 h_3}\right).$$

By circular permutations, we also have:

$$\mathrm{div}\hat{e}_2 = \hat{e}_2 \cdot \left(\frac{\mathrm{grad}(h_3 h_1)}{h_3 h_1}\right),$$

and

$$\mathrm{div}\hat{e}_3 = \hat{e}_3 \cdot \left(\frac{\mathrm{grad}(h_1 h_2)}{h_1 h_2}\right).$$

Introducing these results into (B.78), we obtain:

$$\mathrm{div}\vec{V} = V_1 \hat{e}_1 \cdot \frac{\mathrm{grad}(h_2 h_3)}{h_2 h_3} + \mathrm{grad}V_1 \cdot \hat{e}_1 + V_2 \hat{e}_2 \cdot \frac{\mathrm{grad}(h_3 h_1)}{h_3 h_1}$$

$$+ \mathrm{grad}V_2 \cdot \hat{e}_2 + V_3 \hat{e}_3 \cdot \frac{\mathrm{grad}(h_1 h_2)}{h_1 h_2} + \mathrm{grad}V_3 \cdot \hat{e}_3$$

$$= \hat{e}_1 \cdot \left[V_1 \frac{\mathrm{grad}(h_2 h_3)}{h_2 h_3} + \mathrm{grad}V_1\right] + \hat{e}_2 \cdot \left[V_2 \frac{\mathrm{grad}(h_3 h_1)}{h_3 h_1} + \mathrm{grad}V_2\right]$$

$$+ \hat{e}_3 \cdot \left[V_3 \frac{\mathrm{grad}(h_1 h_2)}{h_1 h_2} + \mathrm{grad}V_3\right]$$

$$= \hat{e}_1 \cdot \frac{\mathrm{grad}(V_1 h_2 h_3)}{h_2 h_3} + \hat{e}_2 \cdot \frac{\mathrm{grad}(V_2 h_3 h_1)}{h_3 h_1} + \hat{e}_3 \cdot \frac{\mathrm{grad}(V_3 h_1 h_2)}{h_1 h_2}$$

$$= \frac{1}{h_1 h_2 h_3}\left[\mathrm{grad}(V_1 h_2 h_3) \cdot h_1 \hat{e}_1 + \mathrm{grad}(V_2 h_3 h_1) \cdot h_2 \hat{e}_2 \right.$$

$$\left. + \mathrm{grad}(V_3 h_1 h_2) \cdot h_3 \hat{e}_3\right]. \quad (B.79)$$

Since

$$\mathrm{grad}\varphi \cdot h_1 \hat{e}_1 = h_1 \hat{e}_1 \cdot \left(\frac{\partial \varphi}{\partial q_1}\mathrm{grad}q_1 + \frac{\partial \varphi}{\partial q_2}\mathrm{grad}q_2 + \frac{\partial \varphi}{\partial q_3}\mathrm{grad}q_3\right)$$

$$= h_1 \hat{e}_1 \cdot \left(\frac{\partial \varphi}{\partial q_1}\frac{\hat{e}_1}{h_1} + \frac{\partial \varphi}{\partial q_2}\frac{\hat{e}_2}{h_2} + \frac{\partial \varphi}{\partial q_3}\frac{\hat{e}_3}{h_3}\right) = \frac{\partial \varphi}{\partial q_1},$$

and, analogously,
$$\text{grad}\varphi \cdot h_2 \hat{e}_2 = \frac{\partial \varphi}{\partial q_2},$$
$$\text{grad}\varphi \cdot h_3 \hat{e}_3 = \frac{\partial \varphi}{\partial q_3},$$

it follows from (B.79) that
$$\text{div}\vec{V} = \frac{1}{h_1 h_2 h_3} \left[\frac{\partial}{\partial q_1}(V_1 h_2 h_3) + \frac{\partial}{\partial q_2}(V_2 h_3 h_1) + \frac{\partial}{\partial q_3}(V_3 h_1 h_2) \right]. \tag{B.80}$$

For example, the divergence in cylindrical coordinates is
$$\text{div}\vec{V} = \frac{1}{\rho}\frac{\partial}{\partial \rho}(\rho V_\rho) + \frac{1}{\rho}\frac{\partial V_\varphi}{\partial \varphi} + \frac{\partial V_z}{\partial z},$$
whereas in spherical coordinates is written as
$$\text{div}\vec{V} = \frac{1}{r^2 \sin\theta}\left[\frac{\partial}{\partial r}(r^2 \sin\theta\, V_r) + \frac{\partial}{\partial \theta}(r \sin\theta\, V_\theta) + \frac{\partial}{\partial \varphi}(r\, V_\varphi)\right].$$

Laplacian in OCC

Following the definition, we can write:
$$\Delta\Phi = \text{div}(\text{grad}\Phi). \tag{B.81}$$
Using (B.72), (B.80), and (B.81), we have:
$$\Delta\Phi = \frac{1}{h_1 h_2 h_3}\left[\frac{\partial}{\partial q_1}\left(\frac{h_2 h_3}{h_1}\frac{\partial \Phi}{\partial q_1}\right) + \frac{\partial}{\partial q_2}\left(\frac{h_3 h_1}{h_2}\frac{\partial \Phi}{\partial q_2}\right)\right.$$
$$\left. + \frac{\partial}{\partial q_3}\left(\frac{h_1 h_2}{h_3}\frac{\partial \Phi}{\partial q_3}\right)\right]. \tag{B.82}$$

For example, in cylindrical coordinates the Laplacian writes
$$\Delta\Phi = \frac{1}{\rho}\left[\frac{\partial}{\partial \rho}\left(\rho\frac{\partial \Phi}{\partial \rho}\right) + \frac{\partial}{\partial \varphi}\left(\frac{1}{\rho}\frac{\partial \Phi}{\partial \varphi}\right) + \frac{\partial}{\partial z}\left(\rho\frac{\partial \Phi}{\partial z}\right)\right]$$
$$= \frac{1}{\rho}\frac{\partial}{\partial \rho}\left(\rho\frac{\partial \Phi}{\partial \rho}\right) + \frac{1}{\rho^2}\frac{\partial^2 \Phi}{\partial \varphi^2} + \frac{\partial^2 \Phi}{\partial z^2},$$
while in spherical coordinates is expressed as
$$\Delta\Phi = \frac{1}{r^2 \sin\theta}\left[\frac{\partial}{\partial r}\left(r^2 \sin\theta\frac{\partial \Phi}{\partial r}\right) + \frac{\partial}{\partial \theta}\left(\sin\theta\frac{\partial \Phi}{\partial \theta}\right)\right.$$
$$\left. + \frac{\partial}{\partial \varphi}\left(\frac{1}{\sin\theta}\frac{\partial \Phi}{\partial \varphi}\right)\right]$$
$$= \frac{1}{r^2}\frac{\partial}{\partial r}\left(r^2 \frac{\partial \Phi}{\partial r}\right) + \frac{1}{r^2 \sin\theta}\frac{\partial}{\partial \theta}\left(\sin\theta\frac{\partial \Phi}{\partial \theta}\right) + \frac{1}{r^2 \sin^2\theta}\frac{\partial^2 \Phi}{\partial \varphi^2}.$$

B.8. Some applications of vector analysis

Exercise 1

Let $\vec{r}(x,y,z)$ be the radius vector of the point P, with respect to the origin of the Cartesian frame $Oxyz$. Calculate: $\operatorname{curl}\vec{r}$, $\operatorname{div}\vec{r}$, $\operatorname{grad} r$, $\operatorname{grad}\left(\frac{1}{r}\right)_{r\neq 0}$, $\Delta\left(\frac{1}{r}\right)_{r\neq 0}$, and $\Delta\left(\frac{1}{r}\right)$.

Solution

$$\operatorname{curl}\vec{r} = \vec{u}_i \varepsilon_{ijk}\partial_j x_k = \vec{u}_i \varepsilon_{ijk}\delta_{jk} = 0;$$

$$\operatorname{div}\vec{r} = \frac{\partial x_i}{\partial x_i} = 3;$$

$$\operatorname{grad} r = \vec{u}_i \frac{\partial r}{\partial x_i} = \vec{u}_i \frac{\partial r}{\partial x_i}\sqrt{x_i x_i}$$

$$= \vec{u}_i \frac{1}{2\sqrt{x_k x_k}} 2x_k \frac{\partial x_k}{\partial x_i} = \frac{x_k \delta_{ki}}{r}\vec{u}_i = \frac{x_i \vec{u}_i}{r} = \frac{\vec{r}}{r} = \vec{u}_r;$$

$$\operatorname{grad}\left(\frac{1}{r}\right)_{r\neq 0} = \vec{u}_i \frac{\partial}{\partial x_i}\left(\frac{1}{r}\right) = -\vec{u}_i \frac{1}{r^2}\frac{\partial r}{\partial x_i}$$

$$= -\vec{u}_i \frac{1}{r^2}\frac{\partial}{\partial x_i}\sqrt{x_k x_k} = -\vec{u}_i \frac{1}{r^2}\frac{x_k \delta_{ki}}{r} = -\frac{x_i \vec{u}_i}{r^3} = -\frac{\vec{r}}{r^3};$$

$$\Delta\left(\frac{1}{r}\right)_{r\neq 0} = \operatorname{div}\left[\operatorname{grad}\left(\frac{1}{r}\right)\right] = -\operatorname{div}\left(\frac{\vec{r}}{r^3}\right)$$

$$= -\frac{1}{r^3}\operatorname{div}\vec{r} - \vec{r}\cdot\nabla\left(\frac{1}{r^3}\right) = -\frac{3}{r^3} - \vec{r}\cdot\left(-\frac{1}{r^6}\right)\frac{\partial r^3}{\partial x_i}\vec{u}_i$$

$$= -\frac{3}{r^3} + \frac{\vec{r}\cdot\vec{u}_i}{r^6}\frac{\partial}{\partial x_i}(x_k x_k)^{3/2} = -\frac{3}{r^3} + \frac{\vec{r}\cdot\vec{u}_i}{r^6}\frac{3}{2}(x_k x_k)^{1/2} 2x_k \frac{\partial x_k}{\partial x_i}$$

$$= -\frac{3}{r^3} + \frac{3}{r^5}\vec{r}\cdot\vec{r} = 0.$$

To determine $\Delta\left(\frac{1}{r}\right)$ for any value of r, that is including value $r = 0$, let us appeal to some simple knowledge of electrostatics. As well-known, Poisson's equation for an electric charge of volume density $\rho(\vec{r})$, situated in vacuum, writes

$$\Delta V(\vec{r}) = -\frac{1}{\varepsilon_0}\rho(\vec{r}). \tag{B.83}$$

In case of a point charge distribution Q, whose position is defined by the radius vector $\vec{r}_0$, we can write

$$\rho(\vec{r}) = Q\,\delta(\vec{r} - \vec{r}_0), \tag{B.84}$$

where $\delta(\vec{r}-\vec{r_0}) = \delta(x-x_0)\delta(y-y_0)\delta(z-z_0)$ is the *delta Dirac distribution*. Since in this case

$$V(\vec{r}) = \frac{Q}{4\pi\varepsilon_0}\frac{1}{|\vec{r}-\vec{r_0}|}, \qquad (B.85)$$

by substituting (B.84) and (B.85) into (B.83), we have:

$$\Delta\left(\frac{Q}{4\pi\varepsilon_0}\frac{1}{|\vec{r}-\vec{r_0}|}\right) = -\frac{\rho(\vec{r})}{\varepsilon_0} = -\frac{Q}{\varepsilon_0}\delta(\vec{r}-\vec{r_0}),$$

or

$$\Delta\left(\frac{1}{|\vec{r}-\vec{r_0}|}\right) = -4\pi\,\delta(\vec{r}-\vec{r_0}).$$

If the electric point charge is situated at the origin of the coordinate system ($\vec{r_0} = 0$), the desired equation follows immediately:

$$\Delta\left(\frac{1}{r}\right) = -4\pi\delta(r).$$

Exercise 2

If $\vec{A}$ is a constant vector, calculate: $\mathrm{div}(\vec{A}\times\vec{r})$, $\mathrm{curl}(\vec{A}\times\vec{r})$, and $\mathrm{grad}(\vec{A}\cdot\vec{r})$.

Solution

According to the vector identity (C.49), we have

$$\mathrm{div}(\vec{A}\times\vec{r}) = \vec{r}\cdot\mathrm{curl}\vec{A} - \vec{A}\cdot\mathrm{curl}\vec{r} = \vec{r}\cdot 0 - \vec{A}\cdot 0 = 0.$$

Using (B.51), we can write

$$\mathrm{curl}(\vec{A}\times\vec{r}) = \vec{A}\,\mathrm{div}\,\vec{r} - \vec{r}\,\mathrm{div}\vec{A} + (\vec{r}\cdot\nabla)\vec{A} - (\vec{A}\cdot\nabla)\vec{r}$$

$$= \vec{A}\,\mathrm{div}\vec{r} - (\vec{A}\cdot\nabla)\vec{r} = 3\vec{A} - \vec{A} = 2\vec{A}.$$

Finally, by means of (B.52), we have

$$\mathrm{grad}(\vec{A}\cdot\vec{r}) = \vec{A}\times(\nabla\times\vec{r}) + \vec{r}\times(\nabla\times\vec{A}) + (\vec{A}\cdot\nabla)\vec{r} + (\vec{r}\cdot\nabla)\vec{A} = (\vec{A}\cdot\nabla)\vec{r} = \vec{A}.$$

Exercise 3

If $\vec{a}$ is a constant vector, and $\vec{r}$ the radius vector of some point in the Cartesian frame $Oxyz$, calculate:

(a) divergence and curl of the vector $\vec{V} = \vec{a}\times(\vec{r}\times\vec{a})$;

(b) divergence and curl of the vector $\vec{V} = \vec{r} \times (\vec{a} \times \vec{r})$;
(c) divergence and curl of the vector $\vec{V} = (\vec{a} \times \vec{r}) + (\vec{a} \cdot \vec{r})\vec{a}$;
(d) divergence and curl of the vector $\vec{V} = \frac{\vec{a} \times \vec{r}}{r^2}$;
(e) divergence and curl of the vector $\vec{V} = \operatorname{grad} r$.

Solution

(a) We have:

$$\operatorname{div}\vec{V} = \operatorname{div}\big[\vec{a} \times (\vec{r} \times \vec{a})\big] = (\vec{r} \times \vec{a}) \cdot \operatorname{curl}\vec{a} - \vec{a} \cdot \operatorname{curl}(\vec{r} \times \vec{a})$$

$$= -\vec{a} \cdot \operatorname{curl}(\vec{r} \times \vec{a}) = -\vec{a} \cdot \big[\vec{r}\operatorname{div}\vec{a} - \vec{a}\operatorname{div}\vec{r}$$

$$+ (\vec{a} \cdot \nabla)\vec{r} - (\vec{r} \cdot \nabla)\vec{a}\big] = -\vec{a} \cdot (-3\vec{a} + \vec{a}) = 2a^2,$$

and

$$\operatorname{curl}\vec{V} = \operatorname{curl}\big[\vec{a} \times (\vec{r} \times \vec{a})\big]$$

$$= \vec{a}\operatorname{div}(\vec{r} \times \vec{a}) - (\vec{r} \times \vec{a})\operatorname{div}\vec{a} + \big[(\vec{r} \times \vec{a}) \cdot \nabla\big]\vec{a} - (\vec{a} \cdot \nabla)(\vec{r} \times \vec{a})$$

$$= -(\vec{a} \cdot \nabla)(\vec{r} \times \vec{a}) = -a_l \partial_l(\varepsilon_{ijk} x_j a_k)\vec{u}_i = -\varepsilon_{ijk} a_l a_k \delta_{lj}\vec{u}_i$$

$$= -\varepsilon_{ijk} a_j a_k \vec{u}_i = -\frac{1}{2}\big(\varepsilon_{ijk} a_j a_k + \varepsilon_{ijk} a_j a_k\big)\vec{u}_i$$

$$= -\frac{1}{2}\big(\varepsilon_{ijk} a_j a_k + \varepsilon_{ikj} a_k a_j\big)\vec{u}_i$$

$$= -\frac{1}{2}\big(\varepsilon_{ijk} a_j a_k - \varepsilon_{ijk} a_j a_k\big)\vec{u}_i = 0.$$

(b) In an analogous way, we can write:

$$\operatorname{div}\vec{V} = \operatorname{div}[\vec{r} \times (\vec{a} \times \vec{r})] = (\vec{a} \times \vec{r}) \cdot \operatorname{curl}\vec{r} - \vec{r} \cdot \operatorname{curl}(\vec{a} \times \vec{r})$$

$$= -\vec{r} \cdot (2\vec{a}) = -2\vec{r} \cdot \vec{a},$$

and

$$\operatorname{curl}\vec{V} = \operatorname{curl}\big[\vec{r} \times (\vec{a} \times \vec{r})\big] = \vec{r}\operatorname{div}(\vec{a} \times \vec{r})$$

$$- (\vec{a} \times \vec{r})\operatorname{div}\vec{r} + \big[(\vec{a} \times \vec{r}) \cdot \nabla\big]\vec{r} - (\vec{r} \cdot \nabla)(\vec{a} \times \vec{r})$$

$$= -3(\vec{a} \times \vec{r}) + (\vec{a} \times \vec{r})_i \partial_i(x_k \vec{u}_k) - x_j \partial_j(\vec{u}_i \varepsilon_{ilm} a_l x_m)$$

$$= -3(\vec{a} \times \vec{r}) + (\vec{a} \times \vec{r})_i \vec{u}_k \delta_{ik} - \vec{u}_i \varepsilon_{ilm} x_j a_l \delta_{jm}$$

$$= -3(\vec{a} \times \vec{r}) + (\vec{a} \times \vec{r}) - (\vec{a} \times \vec{r}) = -3(\vec{a} \times \vec{r}).$$

(c) For $\vec{V} = (\vec{a} \times \vec{r}) + (\vec{a} \cdot \vec{r})\vec{a}$, we have:

$$\text{div}\,\vec{V} = \text{div}\,(\vec{a} \times \vec{r}) + \text{div}\left[(\vec{a} \cdot \vec{r})\vec{a}\right]$$

$$= \text{div}\left[(\vec{a} \cdot \vec{r})\vec{a}\right]$$

$$= (\vec{a} \cdot \vec{r})\text{div}\,\vec{a} + \vec{a} \cdot \text{grad}(\vec{a} \cdot \vec{r}) = \vec{a} \cdot \vec{a} = a^2,$$

and

$$\text{curl}\,\vec{V} = \text{curl}\,(\vec{a} \times \vec{r}) + \text{curl}\left[(\vec{a} \cdot \vec{r})\vec{a}\right]$$

$$= 2\vec{a} + (\vec{a} \cdot \vec{r})\text{curl}\,\vec{a} + \text{grad}(\vec{a} \cdot \vec{r}) \times \vec{a} = 2\vec{a}.$$

(d) Let us now calculate divergence and curl of $\vec{V} = \frac{\vec{a} \times \vec{r}}{r^2}$, with $\vec{a} = const$. We have:

$$\text{div}\,\vec{V} = \text{div}\left(\frac{\vec{a} \times \vec{r}}{r^2}\right) = \frac{1}{r^2}\text{div}(\vec{a} \times \vec{r}) + (\vec{a} \times \vec{r}) \cdot \text{grad}\left(\frac{1}{r^2}\right)$$

$$= (\vec{a} \times \vec{r}) \cdot \text{grad}\left(\frac{1}{r^2}\right) = (\vec{a} \times \vec{r}) \cdot \left[\vec{u}_i \frac{\partial}{\partial x_i}\left(\frac{1}{r^2}\right)\right]$$

$$= -(\vec{a} \times \vec{r}) \cdot \vec{u}_i \left(\frac{2}{r^3}\right)\frac{1}{2r}2x_k\frac{\partial x_k}{\partial x_i}$$

$$= -\frac{2}{r^4}\left[(\delta_{ik}x_k\vec{u}_i) \cdot (\vec{a} \times \vec{r})\right] = -\frac{2}{r^4}\left[\vec{r} \cdot (\vec{a} \times \vec{r})\right] = 0,$$

and

$$\text{curl}\,\vec{V} = \text{curl}\left(\frac{\vec{a} \times \vec{r}}{r^2}\right) = \frac{1}{r^2}\text{curl}(\vec{a} \times \vec{r}) + \text{grad}\left(\frac{1}{r^2}\right) \times (\vec{a} \times \vec{r})$$

$$= \frac{2\vec{a}}{r^2} - \frac{2}{r^4}\left[\vec{r} \times (\vec{a} \times \vec{r})\right] = \frac{2\vec{a}}{r^2} - \frac{2}{r^4}\left[r^2\vec{a} - (\vec{a} \cdot \vec{r})\vec{r}\right] = \frac{2}{r^4}\left[(\vec{a} \cdot \vec{r})\vec{r}\right].$$

(e) Finally, for $\vec{V} = \text{grad}\,r$, we have:

$$\text{div}\vec{V} = \text{div}(\text{grad}r) = \text{div}\left(\frac{\vec{r}}{r}\right) = \frac{1}{r}\text{div}\,\vec{r} + \vec{r} \cdot \text{grad}\left(\frac{1}{r}\right)$$

$$= \frac{3}{r} + \vec{r} \cdot \left(-\frac{\vec{r}}{r^3}\right) = \frac{3}{r} - \frac{1}{r} = \frac{2}{r},$$

and

$$\text{curl}\vec{V} = \text{curl}(\text{grad}r) = 0.$$

Exercise 4

Given the vector field $\vec{w}$ with properties $\operatorname{div}\vec{w} = 0$, and $\operatorname{curl}\vec{w} = 0$, show that $\vec{E} = \operatorname{grad}(\vec{r}\cdot\vec{w}) + \operatorname{curl}(\vec{r}\times\vec{w}) + \vec{w} = 0$.

Solution

Using vector identities (B.50) and (B.51), we have:

$$\vec{E} = \operatorname{grad}(\vec{r}\cdot\vec{w}) + \operatorname{curl}(\vec{r}\times\vec{w}) + \vec{w}$$

$$= \vec{r}\times\operatorname{curl}\vec{w} + \vec{w}\times\operatorname{curl}\vec{r} + (\vec{r}\cdot\nabla)\vec{w} + (\vec{w}\cdot\nabla)\vec{r}$$

$$+ \vec{r}\operatorname{div}\vec{w} - \vec{w}\operatorname{div}\vec{r} + (\vec{w}\cdot\nabla)\vec{r} - (\vec{r}\cdot\nabla)\vec{w} + \vec{w}$$

$$= 2(\vec{w}\cdot\nabla)\vec{r} - 2\vec{w} = 2w_i\partial_i(x_k\vec{u}_k) - 2\vec{w} = 2\vec{w} - 2\vec{w} = 0.$$

Exercise 5

Given the vector field $\vec{v}$ with properties $\operatorname{div}\vec{v} = \vec{a}^2$, and $\operatorname{curl}\vec{v} = 2\vec{a}$, where $\vec{a}$ is a constant vector, show that $L = \operatorname{div}(\vec{v}\times\operatorname{curl}\vec{v}) - 4\operatorname{div}\vec{v} = 0$.

Solution

In view of vector identity (B.49), we have:

$$L = \operatorname{div}(\vec{v}\times\operatorname{curl}\vec{v}) - 4\operatorname{div}\vec{v} = \operatorname{curl}\vec{v}\cdot\operatorname{curl}\vec{v} - \vec{v}\cdot\operatorname{curl}(\operatorname{curl}\vec{v}) - 4\operatorname{div}\vec{v}$$

$$= 4\vec{a}\cdot\vec{a} - 2\vec{v}\cdot\operatorname{curl}\vec{a} - 4\vec{a}^2 = 0.$$

B.9. Isomorphism between the set of second rank antisymmetric tensors and the set of vectors, defined on the same space E_3

Consider a second rank covariant tensor A_{ij} ($i,j = \overline{1,3}$), defined on the three-dimensional Euclidean space E_3. In general, a tensor has n^m components. Here n represents the number of dimensions of the space, and $m = p + q$ indicates the tensor rank, where p stands for the number of covariant indices, and q for the number of contravariant indices. In our case, $q = 0$, $m = p = 2$, and $n = 3$, so that we have $3^2 = 9$ components. In matrix form, the tensor A_{ij} writes

$$(A_{ij}) = \begin{pmatrix} A_{11} & A_{12} & A_{13} \\ A_{21} & A_{22} & A_{23} \\ A_{31} & A_{32} & A_{33} \end{pmatrix}. \quad (B.86)$$

If the tensor A_{ij} is antisymmetric

$$A_{ij} = -A_{ji} \quad \forall i,j = \overline{1,3}, \quad (B.87)$$

the elements disposed on the principal diagonal are zero, and we are left with
$$(A_{ij}) = \begin{pmatrix} 0 & A_{12} & A_{13} \\ A_{21} & 0 & A_{23} \\ A_{31} & A_{32} & 0 \end{pmatrix}. \qquad (B.88)$$

In view of (B.87), this relation can be written in eight equivalent ways, as follows:

$$(A_{ij})_1 = \begin{pmatrix} 0 & A_{12} & A_{13} \\ -A_{12} & 0 & A_{23} \\ -A_{13} & -A_{23} & 0 \end{pmatrix};$$

$$(A_{ij})_2 = \begin{pmatrix} 0 & -A_{21} & A_{13} \\ A_{21} & 0 & A_{23} \\ -A_{13} & -A_{23} & 0 \end{pmatrix};$$

$$(A_{ij})_3 = \begin{pmatrix} 0 & A_{12} & -A_{31} \\ -A_{12} & 0 & A_{23} \\ A_{31} & -A_{23} & 0 \end{pmatrix};$$

$$(A_{ij})_4 = \begin{pmatrix} 0 & A_{12} & A_{13} \\ -A_{12} & 0 & -A_{32} \\ -A_{13} & A_{32} & 0 \end{pmatrix};$$

$$(A_{ij})_5 = \begin{pmatrix} 0 & -A_{21} & -A_{31} \\ A_{21} & 0 & A_{23} \\ A_{31} & -A_{23} & 0 \end{pmatrix};$$

$$(A_{ij})_6 = \begin{pmatrix} 0 & -A_{21} & A_{13} \\ A_{21} & 0 & -A_{32} \\ -A_{13} & A_{32} & 0 \end{pmatrix};$$

$$(A_{ij})_7 = \begin{pmatrix} 0 & A_{12} & -A_{31} \\ -A_{12} & 0 & -A_{32} \\ A_{31} & A_{32} & 0 \end{pmatrix};$$

$$(A_{ij})_8 = \begin{pmatrix} 0 & -A_{21} & -A_{31} \\ A_{21} & 0 & -A_{32} \\ A_{31} & A_{32} & 0 \end{pmatrix}.$$

These relations show that a second-rank antisymmetric tensor, defined on a three dimensional space, has only three essentially distinct components. It can be shown that, in general, a p-times covariant and

q-times contravariant tensor, defined on a n-dimensional space, antisymmetric in a indices, has $C_n^a n^{p+q-a}$ essentially distinct components. In our case, $C_3^2 3^{2+0-2} = C_3^2 = \frac{3 \cdot 2}{2} = 3$.

A similar relation can be found for a tensor symmetric in s indices, in which case the number of essentially distinct components is $\overline{C}_n^s n^{p+q-s}$, where

$$\overline{C}_n^s = \frac{n(n+1)(n+2)...(n+s-1)}{s!}$$

is the number of combinations with repetition of n elements taken s times.

Coming now back, out of eight possibilities of choosing A_{ij} we shall consider the third, that is

$$(A_{ij})_3 = \begin{pmatrix} 0 & A_{12} & -A_{31} \\ -A_{12} & 0 & A_{23} \\ A_{31} & -A_{23} & 0 \end{pmatrix}. \qquad (B.89)$$

This choice was made on purpose, because according to (B.89) this is the only possibility of having as distinct components A_{12}, A_{23}, and A_{31}.

Since in a three-dimensional space vectors have three components, one can consider the three essentially distinct components of a second rank antisymmetric tensor as the components of a vector in this space, which are:

$$\begin{matrix} A_{12} \to v_3, & & v_1 = A_{23}, \\ A_{23} \to v_1, & \Rightarrow & v_2 = A_{31}, \\ A_{31} \to v_2, & & v_3 = A_{12}. \end{matrix} \qquad (B.90)$$

This way, the choice of indices agrees with the cyclic permutation of indices 1, 2 and 3. In one relation, correspondences shown by (B.90) can be written as

$$v_i = \frac{1}{2}\varepsilon_{ijk}A_{jk}, \qquad (B.91)$$

where ε_{ijk} $(i,j,k = \overline{1,3})$ is the Levi-Civita symbol. This relation can be easily verified for any group of indices.

Conversely, we can ask whether the reciprocal correspondence is valid, that is if the three distinct components v_i $(i = \overline{1,3})$ of a vector in a three-dimensional space could stand for the three essentially distinct components of a second rank antisymmetric tensor on the same space. The answer is affirmative and the correspondence writes

$$v_1 \to A_{23}, \qquad A_{12} = v_3,$$
$$v_2 \to A_{31}, \quad \to \quad A_{23} = v_1, \qquad (B.92)$$
$$v_3 \to A_{12}, \qquad A_{31} = v_2.$$

These correspondences can also be written in a single relation as

$$A_{ij} = \varepsilon_{ijk} v_k. \qquad (B.93)$$

The two relations (B.91) and (B.93) express the isomorphism between the set of second rank antisymmetric tensors and the set of vectors on E_3.

In fact, for the sake of rigurosity, to a second rank antisymmetric tensor, defined on E_3, one can associate a *pseudovector (axial vector)* and conversely, to an ordinary vector one can associate a second rank *pseudotensor*.

REFERENCES

1. Arnold, V.I.: Mathematical Methods of Classical Mechanics, Springer-Verlag, Berlin Heidelberg (1997)
2. Arthur, W., Fenster, S.: Mechanics Halt, Rimhart Winston Inc. (1969)
3. Baratesch, H.J.: Mathematische Formelen, 16th edn., Kln, Leipzig, Buch-und Zeit-Verlagsgesellschaft mbH (1992)
4. Bartlett, W.H.C.: Elements of Analytical Mechanics, Scholarly Publishing Office, University of Michigan Library (2006)
5. Balan, St.: Collection of Problems in Mechanics, 2nd edn., Did. and Ped. Printing House, Bucharest (1972)
6. Bose, S.K., Chattoraj, D.: Elementary Analytical Mechanics, Alpha Science International Ltd. (2000)
7. Bradbury, T.C.: Theoretical Mechanics, John Wiley and Sons, Inc., New-York (1968)
8. Bradford, N.C.: Mechanics, Wiley, London (1973)
9. Bradeanu, P., Pop, I., Bradeanu, D.: Problems and Exercises in Theoretical Mechanics, Technical Printing House, Bucharest (1978)
10. Bucholtz, N.N., Voronkov, I.M., Minakov, I.A.: Collection of Problems in Mechanics, Technical Printing House, Bucharest (1962)
11. Burlacu, L., David, D.Gh.: Collection of Problems in Analytical Mechanics, Bucharest University Press, Bucharest (1988)
12. Chaichian, M., Merches, I., Tureanu, A.: Mechanics: An Intensive Course, Springer-Verlag, Berlin Heidelberg (2012)
13. Chaichian, M., Perez Rojas, H., Tureanu, A.: From the Cosmos to Quarks: Basic Concepts in Physics, Springer-Verlag, Berlin Heidelberg (2014)
14. Chandrasekhar, S., Fermi, E.: Problems of Gravitational Stability in the Presence of a Magnetic Field, Ap. J. 118: 116 (1953), Bibcode: 1953ApJ118..116C, doi:10.1086/145732, Retrieved March 24 (2012)
15. Collins, G.W.: The Virial Theorem in Stellar Astrophysics, Pachart Press (1978)

16. Cronin, J.A., Greenberg, D.F., Telegdi, V.L.: University of Chicago Graduate Problems in Physics with Solutions, Addison-Wesley, Reading, Mass. (1967)

17. de Lange, O., Pierrus, J.: Solved Problems in Classical Mechanics: Analytical and Numerical Solutions with Comments, Oxford University Press, USA (2010)

18. Fassano, A., Marmi, S.: Analytical Mechanics: An Introduction, Oxford University Press (2006)

19. Forray, M.J.: Variational Calculus in Science and Engineering, McGraw-Hill, New-York (1968)

20. Fox, W.R.: Introduction to Fluid Mechanics, Wiley, New-York (1973)

21. Fox, R.W., McDonald, A.T., Pritchard, P.J.: Introduction to Fluid Mechanics, Wiley, New-York (2009)

22. Gantmacher, F.: Lectures in Analytical Mechanics, Mir, Moskow (1975)

23. Gignoux, C., Silvestre-Brac, B.: Solved Problems in Lagrangian and Hamiltonian Mechanics, Grenoble Sciences, Springer (2009)

24. Goldstein, H.: Classical Mechanics, Addison-Wesley Publishing Co. Mass., 2nd edn. (1980)

25. Grechko, L.G., Sugakov, V.I., Tomasevich, O.F., Fedorchenko, A.M.: Problems in Theoretical Physics, Mir Publishers, Moscow (1977)

26. Greenwood, D.T.: Principles of Dynamics, 2nd edn., Englewood Cliffs (1988), Dover, New-York (1997)

27. Hand, L.N., Finch, J.D.: Analytical Mechanics, Cambridge University Press (1998)

28. Ionescu-Pallas, N.: Introduction to Modern Theoretical Mechanics, Academy Printing House, Bucharest (1969)

29. Irodov, I., Saveliev, I., Zamcha, O.: Recueil de problèmes de physique générale, Éditions Mir, Moscou (1976)

30. Jeffery, D.J.: Classical Mechanics Problems, Portpentagram Publishing (self-published), New Mexico (2001)

31. Johns, O.D.: Analytical Mechanics for Relativity and Quantum Mechanics, Oxford University Press (2005)

32. Kibble, T.W.: Classical Mechanics, McGraw-Hill, London (1973), Addison-Wesley Longman (1986)

33. Kittel, C., Kight, W.D., Ruderman, M.A.: Mechanics, Berkeley Physics Course, vol. I, McGraw-Hill, New-York (1973)

34. Kotkin, L.G., Serbo, V.G.: Collection of Problems in Classical Mechanics, Pergamon Press, Oxford, 1st edn. (1971)

35. Lagrange, J.L.: Analytical Mechanics, Kluwer Academic Publishers (2010)

36. Landau, L.D., Lifshitz, E.M.: Fluid Mechanics, 2nd edn. Pergamon (1987)

37. Landau, L.D., Lifshitz, E.M.: The Classical Theory of Fields, Pergamon (1987)

38. Landau, L.D., Lifshitz, E.M.: Mechanics, 3rd edn., Pergamon Press (1976)

39. Lurie, A.I.: Analytical Mechanics, Springer-Verlag, Berlin Heidelberg (2002)

40. Marion, J.: Classical Dynamics, Academic Press Inc., New-York (1965)

41. Merches, I., Burlacu, L.: Applied Analytical Mechanics, The Voice of Bucovina Press, Iasi (1995)

42. Meshcherskii, I.V.: A Collection of Problems of Mechanics, Pergamon Press, Oxford (1965)

43. Moore, E.L.: Theoretical Mechanics, Wiley (1983)

44. Morin, D.: Introduction to Classical Mechanics with Problems and Solutions, Cambridge University Press (2008)

45. Papastavridis, J.G.: Analytical Mechanics: A Comprehensive Treatise on the Dynamics of Constrained Systems for Engineers, Physicists and Mathematicians, Oxford University Press (2002)

46. Pellegrini, C., Cooper, R.K.: Modern Analytical Mechanics, Springer-Verlag, Berlin Heidelberg (1999)

47. Pollard, H.: Mathematical Introduction to Celestial Mechanics, Englewood Cliffs, N.J.: PrenticeHall, Inc. (1966)

48. Rossberg, K.: A First Course in Analytical Mechanics, Wiley, New-York (1983)

49. Saletan, E.J., Cromer, A.H.: Theoretical Mechanics, John Wiley, New-York (1971)

50. Serrin, J.: Mathematical Principles of Classical Fluid Mechanics, Handbuch der Physik, Springer-Verlag, Berlin, vol. VIII/1, Fluid Dynamics I (1959)

51. Spiegel, M.R.: Theory and Problems of Theoretical Mechanics, McGraw-Hill, New-York (1967)

52. Ter Haar, D.: Elements of Hamiltonian Mechanics, North-Holland Publishing Co., Amsterdam (1964)

53. Tomasevich, O.F.: A Collection of Problems in Theoretical Physics, Kiev University Press, Kiev (1958)

54. Torok, J.S.: Analytical Mechanics: With an Introduction to Dynamical Systems, Wiley-Interscience (1999)

55. Vladimirov, V., Mikhailov, V., Chabounine, M., Karimova, Kh., Sidorov, Y., Vacharine, A.: Recueil de problèmes d'équations de physique mathématique, Éditions Mir, Moscou (1976)

56. Wess, J.: Theoretische Mechanik, Springer-Verlag, Berlin Heildelberg (2007)

57. Ziwet, A., Field, P.: Introduction to Analytical Mechanics, Scholarly Publishing Office, University of Michigan (2005)

For Product Safety Concerns and Information please contact our
EU representative GPSR@taylorandfrancis.com Taylor & Francis
Verlag GmbH, Kaufingerstraße 24, 80331 München, Germany